Coastal Zone Management

Coastal Zone Management
Global Perspectives, Regional Processes, Local Issues

Edited by

Mu. Ramkumar
R. Arthur James
David Menier
K. Kumaraswamy

Elsevier
Radarweg 29, PO Box 211, 1000 AE Amsterdam, Netherlands
The Boulevard, Langford Lane, Kidlington, Oxford OX5 1GB, United Kingdom
50 Hampshire Street, 5th Floor, Cambridge, MA 02139, United States

© 2019 Elsevier Inc. All rights reserved.

No part of this publication may be reproduced or transmitted in any form or by any means, electronic or mechanical, including photocopying, recording, or any information storage and retrieval system, without permission in writing from the publisher. Details on how to seek permission, further information about the Publisher's permissions policies and our arrangements with organizations such as the Copyright Clearance Center and the Copyright Licensing Agency, can be found at our website: www.elsevier.com/permissions.

This book and the individual contributions contained in it are protected under copyright by the Publisher (other than as may be noted herein).

Notices
Knowledge and best practice in this field are constantly changing. As new research and experience broaden our understanding, changes in research methods, professional practices, or medical treatment may become necessary.

Practitioners and researchers must always rely on their own experience and knowledge in evaluating and using any information, methods, compounds, or experiments described herein. In using such information or methods they should be mindful of their own safety and the safety of others, including parties for whom they have a professional responsibility.

To the fullest extent of the law, neither the Publisher nor the authors, contributors, or editors, assume any liability for any injury and/or damage to persons or property as a matter of products liability, negligence or otherwise, or from any use or operation of any methods, products, instructions, or ideas contained in the material herein.

Library of Congress Cataloging-in-Publication Data
A catalog record for this book is available from the Library of Congress

British Library Cataloguing-in-Publication Data
A catalogue record for this book is available from the British Library

ISBN: 978-0-12-814350-6

For information on all Elsevier publications
visit our website at https://www.elsevier.com/books-and-journals

Working together
to grow libraries in
developing countries

www.elsevier.com • www.bookaid.org

Publisher: Candice Janco
Acquisition Editor: Louisa Hutchins
Editorial Project Manager: Michelle W. Fisher
Production Project Manager: Omer Mukthar
Cover Designer: Christian J. Bilbow

Typeset by SPi Global, India

Cover photo credit: Geosciences Ocean Laboratory, UMR CNRS 6538, Author: Camille Traini, Salt Marshes of Billiers and Ambon (Morbihan, France)

Contents

Contributors xvii

Chapter 1: Coastal Zone Management During Changing Climate and Rising Sea Level: Transcendence of Institutional, Geographic, and Subject Field Barriers Is the Key.. 1
Mu. Ramkumar, David Menier, K. Kumaraswamy
   1 Introduction 1
   2 Transcendence of Local Issues to a Global Scale Through Regional
      Processes 2
      2.1 Spatial-Temporal Scale of Cause-Effect 2
      2.2 Transcendence of Causes and Effects 3
      2.3 Need to Have Unified System of Coastal Environmental Management 5
   3 Transcendence of Subject, Institutional, and Geographic Areas
      Are the Key 7
   4 Conclusions 9
   Acknowledgments 9
   References 10

PART 1 Sealevel Cycles and Oceanography 13

Chapter 2: Development of Ideas and New Trends in Modern Sea Level Research: The Pre-Quaternary, Quaternary, Present, and Future 15
Nils-Axel Mörner
   1 Introduction 15
   2 Development of Ideas 16
      2.1 Earth's Rotation 16
      2.2 Discussions in the 18th Century 16
      2.3 Observations in the 19th Century 20
   3 Handling Sea Level Research 22
   4 Quaternary Sea Level Changes 23
      4.1 The Postglacial Period and Definition of *Eustasy* 24
      4.2 The Glacial/Interglacial Cycles 27
      4.3 The Problem With Global Glacial Isostasy 29
      4.4 Summary of Quaternary Sea Level Variables 31
   5 Pre-Quaternary Sea Level Changes—An Alternative View 32

    5.1 The Paris 1980 Illusory Consensus..35
    5.2 The New Synthesis...35
    5.3 Lessons From Field Studies in the 1970s...36
    5.4 Deep-Sea Hiatus..42
    5.5 Rates of Sea Level Changes..43
    5.6 The "State of the Art"...44
    5.7 Concluding Comments on Cretaceous Sea Level Changes..............................47
  6 Additional Notes on Tertiary Sea Level Changes..48
    6.1 The Big Drop at Around 30 Ma..49
    6.2 The Messinian "Salinity Crisis" and Sea Level Changes................................50
    6.3 The Pliocene 3 Ma Highstand...50
  7 The Future as Seen From the Past...51
  8 Conclusions..54
  Acknowledgment...55
  References...55

PART 2 Shoreline and Coastal Changes..63

Chapter 3: Typology and Mechanisms of Coastal Erosion in Siliciclastic Rocks of the Northwest Borneo Coastline (Sarawak, Malaysia): A Field Approach............65
Dominique Dodge-Wan, R. Nagarajan
  1 Introduction...65
  2 Study Area...66
  3 Methodology..68
  4 Results...69
    4.1 General...69
    4.2 Geology..70
    4.3 Typology of Coastal Erosion...75
  5 Discussion..93
  6 Conclusions and Implications...95
  Acknowledgments...96
  References...96
  Further Reading..98

Chapter 4: Monitoring Spatial and Temporal Scales of Shoreline Changes in the Cuddalore Region, India..99
Subbarayan Saravanan, K.S.S. Parthasarathy, S.R. Vishnuprasath
  1 Introduction...99
  2 Study Area...100
  3 Methodology..102
    3.1 Shoreline Delineation and Image Processing Techniques..............................103
    3.2 Shoreline Change Analysis...103
  4 Results...105
  5 Discussion..106

6 Conclusions..107
References...110
Further Reading..112

Chapter 5: Shoreline Evolution Under the Influence of Oceanographic and Monsoon Dynamics: The Case of Terengganu, Malaysia..........................113

Effi Helmy Ariffin, Mouncef Sedrati, Nurul Rabitah Daud, Manoj Joseph Mathew,
Mohd Fadzil Akhir, Nor Aslinda Awang, Rosnan Yaacob, Numair Ahmed Siddiqui,
Mohd Lokman Husain

1 Introduction..113
    1.1 Regional Geological Setting...115
2 Methodology..118
    2.1 Shoreline Evolution...118
    2.2 Sediment Mean Size Measurement..118
    2.3 Physical Setting...119
3 Result..119
    3.1 Magnitude of Coastal Evolution...119
    3.2 Grain Size Distributions..119
    3.3 Main Causes of Erosion Along the Terengganu Coast.......................120
4 Discussions...125
    4.1 Coastal Processes..125
    4.2 Coastal Defense Structures and Erosion Control................................126
5 Conclusions..128
Acknowledgments..128
References..128

Chapter 6: Spatial and Statistical Analyses of Clifftop Retreat in the Gulf of Morbihan and Quiberon Peninsula, France: Implications on Cliff Evolution and Coastal Zone Management..131

Soazig Pian, David Menier

1 Introduction..131
2 Study Area..132
    2.1 The Gulf of Morbihan Cliffs...132
    2.2 The Quiberon Peninsula Cliffs..134
3 Materials and Methods..134
    3.1 Measuring Clifftop Retreat From Analysis of Aerial Photographs....135
    3.2 Construction of the Spatial Database..136
    3.3 Spatial and Statistical Analysis...139
4 Results...141
    4.1 Clifftop Retreat Mapping..141
    4.2 Spatial Associations Inferred From the Nearest-Neighbor Analysis..142
    4.3 Analysis of Statistical Relationships Between Clifftop Retreat Values and the Studied Factors...142
5 Discussion..148

6 Conclusion...151
Acknowledgments...152
References..152

Chapter 7: Erosional Responses of Eastern and Western Coastal Regions of India, Under Global, Regional, and Local Scale Causes..155

K.Ch.V. Naga Kumar, G. Demudu, V.P. Dinesan, Girish Gopinath, P.M. Deepak,
K. Lakshmanadinesh, Kakani Nageswara Rao

1 Introduction..155
2 Study Area..156
3 Methods of Study...159
4 Results...161
   4.1 Coastal Erosion in the AP and Kerala States..161
   4.2 Coastal Erosion in the Krishna Delta...161
   4.3 Coastal Erosion in Visakhapatnam...162
5 Discussion...166
   5.1 Global-Scale Activities and Coastal Erosion—The AP and Kerala Coasts......166
   5.2 Regional Activities and Coastal Erosion—Krishna Delta Coast.....................168
   5.3 Local Activities and Coastal Erosion—Visakhapatnam City Coast................170
6 Conclusion..172
Acknowledgments...173
References..173
Further Reading..178

Chapter 8: Influences of Inherited Structures, and Longshore Hydrodynamics Over the Spatio-Temporal Coastal Dynamics Along the Gâvres-Penthîevre, South Brittany, France...181

Soazig Pian, David Menier, Hervé Régnauld, Mu. Ramkumar, Mouncef Sedrati,
Manoj Joseph Mathew

1 Introduction..181
2 Regional Setting...182
3 Material and Methods...185
   3.1 Waves and Currents..185
   3.2 Analyses of Beach Changes Over a Short Time Scale................................186
   3.3 Analysis of Coastline Variation...189
   3.4 Identification of Sediment Cell Boundaries...190
4 Results...192
   4.1 Sediment Transport..192
   4.2 Beach Profile Evolution...192
   4.3 Temporal Coastline Variations..196
5 Discussion...196
   5.1 Coastal Dynamics as Revealed by Profiling..196
   5.2 Dynamics of Sediment Transport and Control of Beach Morphology........199
6 Conclusions..202

Acknowledgments..202
References..202

Chapter 9: Temporal Trends of Breaker Waves and Beach Morphodynamics Along the Central Tamil Nadu Coast, India..................207
V. Joevivek, N. Chandrasekar

   1 Introduction..207
   2 Study Area..208
   3 Data and Methods...209
      3.1 Field Data..209
      3.2 Breaker Wave Type and Beach Morphodynamic State Model...........211
   4 Results..212
      4.1 Wave Climate..212
      4.2 Foreshore Slope..212
      4.3 Particle Settling Velocity...215
      4.4 Grain Size Distribution..215
      4.5 Trends in Breaker Wave Type and Beach Morphodynamic State......220
   5 Discussion..220
   6 Conclusions...226
   Acknowledgments..226
   References..227
   Further Reading..229

PART 3 Coastal Hydrogeology..231

Chapter 10: Assessing Coastal Aquifer to Seawater Intrusion: Application of the GALDIT Method to the Cuddalore Aquifer, India.........233
Subbarayan Saravanan, Parthasarathy K.S.S., S. Sivaranjani

   1 Introduction..233
   2 Regional Geological Setting of the Study Area..235
      2.1 Hydrogeology of the Study Area...235
   3 Methodology..237
      3.1 Computation of the GALDIT Index...238
   4 Results..239
      4.1 Groundwater Occurrence (G)..239
      4.2 Aquifer Hydraulic Conductivity (A)..239
      4.3 Depth to Water Table (L)..239
      4.4 Distance From the Shore/River (D)...241
      4.5 Impact of Existing Status of Saltwater Intrusion (I)..........................243
      4.6 Thickness of the Aquifer (T)...245
   5 Discussion..246
   6 Conclusions...247
   References..248
   Further Reading..250

Chapter 11: An Assessment of the Administrative-Legal, Physical-Natural, and Socio-Economic Subsystems of the Bay of Saint Brieuc, France: Implications for Effective Coastal Zone Management..251

Nerea Del Estal, Manoj Mathew, David Menier, Erwan Gensac, Hugo Delanoe, Romain LeGall, Thomas Joostens, Angélique Rouille, Béatrice Guillement, Gaëlle Lemaitre

- 1 Introduction..251
- 2 The Subsystems..252
  - 2.1 Administrative and Legal Subsystem..252
  - 2.2 Physical and Natural Subsystem..255
  - 2.3 Socio-Economic Subsystem..263
- 3 Conclusions..275
- References..275

PART 4 Coastal Sediment Geochemistry..277

Chapter 12: Geochemical Characterization of Beach Sediments of Miri, NW Borneo, SE Asia: Implications on Provenance, Weathering Intensity, and Assessment of Coastal Environmental Status..279

R. Nagarajan, A. Anandkumar, S.M. Hussain, M.P. Jonathan, Mu. Ramkumar, S. Eswaramoorthi, A. Saptoro, Han Bing Chua

- 1 Introduction..279
- 2 Regional Setting..283
- 3 Materials and Methods..287
  - 3.1 Calculation of Environmental Indices..289
- 4 Results..290
  - 4.1 Grain Size..290
  - 4.2 Major Oxides..293
  - 4.3 Trace Elements..293
  - 4.4 UCC Normalization..293
  - 4.5 Rare Earth Elements..294
  - 4.6 Systematic Paleontology..294
- 5 Discussion..296
  - 5.1 Weathering and Mobility of Elements..296
  - 5.2 Sorting and Recycling..298
  - 5.3 Provenance..300
  - 5.4 Statistical Analysis..302
  - 5.5 Environmental Significance..313
- 6 Implications on Local and Regional Scale Coastal Zone Management Strategies..318
- 7 Conclusions..319
- Acknowledgments..320
- References..320
- Further Reading..330

Chapter 13: Multimarker Pollution Studies Along the East Coast of Southern India..331

Murugaiah Santhosh Gokul, Hans-Uwe Dahms, Krishnan Muthukumar, Thanamegam Kaviarasan, Santhaseelan Henciya, Sivanandham Vignesh, Thiyagarajamoorthy Dhinesh Kumar, Rathinam Arthur James

1 Introduction..331
2 Materials and Methods...333
   2.1 Study Area..333
   2.2 Sample Collection and Preparation..333
   2.3 Isolation of Pollution Indicators and Epidemiological Survey.................333
   2.4 Trace Metal Analysis..335
   2.5 Antibiotic Resistant Strains..335
3 Result and Discussion...335
   3.1 Pollution Indicators..335
   3.2 Antibiotic-Resistant Strains..338
   3.3 Trace Metal Occurrence..340
   3.4 Correlation Matrix and Factor Analysis...342
4 Conclusion..344
References..344

Chapter 14: Chromium Fractionation in the River Sediments and its Implications on the Coastal Environment: A Case Study in the Cauvery Delta, Southeast Coast of India..347

S. Dhanakumar, R. Mohanraj

1 Introduction..347
2 Regional Setting..349
3 Methods and Materials...350
4 Results..352
5 Discussion...355
6 Conclusion..358
References..358

Chapter 15: Seasonal Variations of Groundwater Geochemistry in Coastal Aquifers, Pondicherry Region, South India...361

S. Chidambaram, S. Pethaperumal, R. Thilagavathi, V. Dhanu Radha, C. Thivya, M.V. Prasanna, K. Tirumalesh, Banaja Rani Panda

1 Introduction..361
2 Study Area..362
   2.1 Methodology..363
   2.2 Results and Discussion..366
   2.3 Spatial Distribution of EC...366
   2.4 Chadda's Hydrogeochemical Process Evaluation..................................371
   2.5 Isotope Studies...371
   2.6 Statistical Analysis...375

    3 Conclusion...377
    References..378

Chapter 16: Anthropogenic Influence of Heavy Metal Pollution on the Southeast Coast of India..381
R. Rajaram, A. Ganeshkumar
    1 Introduction...381
    2 Study Area...383
    3 Materials and Methods..385
        3.1 Collection and Preservation of Samples...385
        3.2 Analysis of Heavy Metals in Water and Sediment Samples....................385
        3.3 Multivariate Statistical Analysis..386
    4 Results..386
        4.1 Variation of Heavy Metal Concentration in Water Samples....................386
        4.2 Variation of Heavy Metal Concentration in Sediment Samples..............387
    5 Source Identification..389
        5.1 Analysis of Interrelationship Between Heavy Metal Concentrations......390
    6 Discussion..392
    7 Conclusion...393
    Acknowledgments...395
    References..395
    Further Reading..399

PART 5 Coastal Zone Management Concepts and Applications..401

Chapter 17: Adaptation Strategies to Address Rising Water Tables in Coastal Environments Under Future Climate and Sea-Level Rise Scenarios......................403
Alex K. Manda, Wendy A. Klein
    1 Introduction...403
    2 Current Options...406
    3 Conclusions...408
    References..408
    Further Reading..409

Chapter 18: Analytic Hierarchy Process to Weigh Groundwater Management Criteria in Coastal Regions..411
Wendy A. Klein, Alex K. Manda
    1 Introduction...411
    2 Methods...412
        2.1 Establishing Criteria for Groundwater Management................................412
        2.2 Survey Instrument..414
        2.3 Analysis..415

3 Results and Discussion...419
4 Shortcomings..426
5 Conclusions..427
Acknowledgments...428
References..428
Further Reading..429

Chapter 19: Interlinking of Rivers as a Strategy to Mitigate Coeval Floods and Droughts: India in Focus With Perspectives on Coastal Zone Management..431

Mu. Ramkumar, David Menier, K. Kumaraswamy, M. Santosh, K. Balasubramani, R.A. James

1 Introduction...431
2 The History of ILR Program...434
3 Regional Setting and Dynamics..435
   3.1 Geology of the Indian Subcontinent..435
   3.2 Geomorphology of India..440
   3.3 Tectono-Geomorphic Evolution of India and the Cenozoic-Recent River Basins...443
   3.4 Climate..444
   3.5 Coastal Hydrodynamics...446
4 Discussion..448
   4.1 Spatio-Temporal Variations of Precipitation.......................................449
   4.2 Political Impediments on Nationalization and ILR..............................450
   4.3 Uniqueness of River Basins and Geological Impediments..................450
   4.4 Self-Regulating River Dynamics and Plausible Impacts of ILR...........452
   4.5 Interlinked Nature of River Basin Dynamics and the Potential Threat by ILR..452
   4.6 Probable Impacts on Deltaic and Coastal Regions by ILR..................456
   4.7 Knowledge of Hazard Mitigation by Ancestors..................................460
5 Recommendations for Sustainable Development.......................................463
Acknowledgments...465
References..465

Chapter 20: Utility of Landsat Data for Assessing Mangrove Degradation in Muthupet Lagoon, South India...471

Subbarayan Saravanan, R. Jegankumar, Ayyakkannu Selvaraj, Jacinth Jennifer J, Parthasarathy K.S.S.

1 Introduction...471
2 Study Area...473
3 Methodology..473
   3.1 Image Preprocessing..473
   3.2 NDVI Map Generation..475

    3.3 NDWI Map Generation..475
    3.4 Land Use and Land Cover Classification..475
    3.5 Land Use/Cover Change Detection..476
  4 Results and Discussion..476
    4.1 Spectral Indices...476
    4.2 Change Detection..477
  5 Conclusions..480
  References..483
  Further Reading...484

Chapter 21: Recent Morphobathymetrical Changes of the Vilaine Estuary (South Brittany, France): Discrimination of Natural and Anthropogenic Forcings and Assessment for Future Trends..485

Evelyne Goubert, David Menier, Camille Traini, Manoj Mathew, Romain Le Gall

  1 Introduction..485
  2 Study Area..487
    2.1 Regional Geological, Geomorphic, Oceanographic, and Climatic Settings.......487
    2.2 Geomorphological and Sedimentological Settings..487
    2.3 The Main Features of the Actual Sedimentary Dynamic..................................489
  3 Data and Methods..491
    3.1 Methodological Approach...491
    3.2 Bathymetric Data..491
    3.3 Meteorological and Hydrologic Data..491
  4 Results..492
    4.1 Morphobathymetrical Evolution and Sedimentary Balance from 1960 to 2013..492
    4.2 The Relationship Between the Spatio-Temporal Sedimentary Balance and the Climatic and Hydrodynamic Forcings..492
  5 Discussion..497
  6 Conclusions..497
  Acknowledgments...498
  References..498

Chapter 22: Impact of Seaweed Farming on Socio-Economic Development of a Fishing Community in Palk Bay, Southeast Coast of India............................501

S. Rameshkumar, R. Rajaram

  1 Introduction..501
  2 Commercial Importance of Seaweed..502
  3 Significance of Seaweed Farming..503
  4 Seaweed Farming in Tamil Nadu...503
  5 Culture Methods..503
    5.1 Long-Line Rope Method..504
    5.2 Net Bag Method..505

- 5.3 Floating Bamboo Raft Method ... 505
- 5.4 Bottom-Culture Method ... 505
- 6 Discussion ... 505
  - 6.1 Role of Government and Nongovernment Sectors for the Development of Seaweed Farming in Tamil Nadu ... 506
  - 6.2 Employment Opportunities and Self-Help Groups (SHGs) Model in Seaweed Cultivation on the Tamil Nadu Coast ... 507
  - 6.3 Environment Impact Assessment (EIA) Study for Invasive Seaweed Farming ... 510
- 7 Conclusions ... 512
- References ... 512
- Further Reading ... 513

Chapter 23: Habitat Risk Assessment Along Coastal Tamil Nadu, India—An Integrated Methodology for Mitigating Coastal Hazards ... 515
A. BalaSundareshwaran, S. Abdul Rahaman, K. Balasubramani, K. Kumaraswamy, Mu. Ramkumar

- 1 Introduction ... 515
- 2 Study Area ... 517
- 3 Data and Methodology ... 518
  - 3.1 Analytical Hierarchy Process (AHP) ... 519
  - 3.2 Fuzzy Linear Membership ... 523
  - 3.3 Coastal Vulnerability Index (CVI) ... 523
- 4 Results ... 524
  - 4.1 Physical Indicators ... 524
  - 4.2 Socio-Economic Indicators ... 527
  - 4.3 Environmental and Climatic Indicators ... 530
  - 4.4 Coastal Vulnerability Index (CVI) ... 532
- 5 Discussion ... 535
- 6 Conclusions ... 537
- Web Sources ... 539
- Acknowledgments ... 539
- References ... 539
- Further Reading ... 542

Index ... *543*

Contributors

S. Abdul Rahaman Department of Geography, Bharathidasan University, Tiruchirappalli, Tamil Nadu, India

Mohd Fadzil Akhir Institute of Oceanography and Environment, Universiti Malaysia Terengganu, Kuala Terengganu, Malaysia

A. Anandkumar Department of Applied Geology, Curtin University, Miri, Malaysia

Effi Helmy Ariffin Géosciences Océan Laboratory UMR CNRS 6538, University of South Brittany, Vannes Cedex, France; School of Marine and Environmental Science; Institute of Oceanography and Environment, Universiti Malaysia Terengganu, Kuala Terengganu, Malaysia

Nor Aslinda Awang National Hydraulic Research Institute of Malaysia, Ministry of Natural Resources and Environment, Seri Kembangan, Malaysia

K. Balasubramani Department of Geography, Central University of Tamil Nadu, Thiruvarur, India

A. BalaSundareshwaran Department of Geography, Bharathidasan University, Tiruchirappalli, Tamil Nadu, India

N. Chandrasekar Centre for GeoTechnology, Manonmaniam Sundaranar University; Francis Xavier Engineering college, Tirunelveli, India

S. Chidambaram Department of Earth Sciences, Annamalai University, Chidambaram, India; Water Research Center, Kuwait Institute for Scientific Research, Kuwait City, Kuwait

Han Bing Chua Department of Chemical Engineering, Curtin University, Miri, Malaysia

Hans-Uwe Dahms Department of Biomedical Science and Environmental Biology, KMU—Kaohsiung Medical University, Kaohsiung, City, Taiwan, ROC

Nurul Rabitah Daud Institute of Oceanography and Environment, Universiti Malaysia Terengganu, Kuala Terengganu; Universiti Teknologi MARA, Shah Alam, Malaysia

P.M. Deepak Geomatics Division, Centre for Water Resources Development and Management (CWRDM), Kozhikode, India

Nerea Del Estal Géosciences Océan Laboratory UMR CNRS 6538, University of South Brittany, Vannes cedex, France; Mott MacDonald Limited, Mott MacDonald House, Croydon, United Kingdom

Hugo Delanoe Géosciences Océan Laboratory UMR CNRS 6538, University of South Brittany, Vannes cedex, France

G. Demudu Department of Geography, Andhra University, Visakhapatnam, India

S. Dhanakumar Department of Environmental Science, PSG College of Arts and Science, Coimbatore, India

V. Dhanu Radha Water Research Center, Kuwait Institute for Scientific Research, Kuwait City, Kuwait

Contributors

V.P. Dinesan   Geomatics Division, Centre for Water Resources Development and Management (CWRDM), Kozhikode, India

Dominique Dodge-Wan   Department of Applied Geology, Curtin University, Miri, Malaysia

S. Eswaramoorthi   Department of Applied Geology, Curtin University, Miri, Malaysia

A. Ganeshkumar   Department of Marine Science, Bharathidasan University, Tiruchirappalli, India

Erwan Gensac   Géosciences Océan Laboratory UMR CNRS 6538, University of South Brittany, Vannes cedex, France

Murugaiah Santhosh Gokul   Department of Marine Science, Bharathidasan University, Tiruchirappalli, India

Girish Gopinath   Geomatics Division, Centre for Water Resources Development and Management (CWRDM), Kozhikode, India

Evelyne Goubert   Géosciences Océan Laboratory UMR CNRS 6538, University of South Brittany, Vannes cedex, France

Béatrice Guillement   Géosciences Océan Laboratory UMR CNRS 6538, University of South Brittany, Vannes cedex, France

Santhaseelan Henciya   Department of Marine Science, Bharathidasan University, Tiruchirappalli, India

Mohd Lokman Husain   School of Marine and Environmental Science; Institute of Oceanography and Environment, Universiti Malaysia Terengganu, Kuala Terengganu, Malaysia

S.M. Hussain   Department of Geology, University of Madras, Guindy Campus, Chennai, India

Jacinth Jennifer J   Department of Civil Engineering, NIT Tiruchirappalli, Tiruchirappalli, India

Rathinam Arthur James   Department of Marine Science, Bharathidasan University, Tiruchirappalli, India

R. Jegankumar   Department of Geography, Bharathidasan University, Tiruchirappalli, India

V. Joevivek   Akshaya college of Engineering and Technology, Coimbatore; Centre for GeoTechnology, Manonmaniam Sundaranar University, Tirunelveli, India

M.P. Jonathan   Centro Interdisciplinario de Investigaciones y Estudios sobre Medio Ambiente y Desarrollo (CIIEMAD), Instituto Politécnico Nacional (IPN), Mexico City, Mexico

Thomas Joostens   Géosciences Océan Laboratory UMR CNRS 6538, University of South Brittany, Vannes cedex, France

Parthasarathy K.S.S.   Department of Applied Mechanics and Hydraulics, NIT Surathkal, Mangalore, India

Thanamegam Kaviarasan   Department of Marine Science, Bharathidasan University, Tiruchirappalli, India

Wendy A. Klein   Coastal Resources Management Program; Department of Geological Sciences; Institute for Coastal Science and Policy, East Carolina University, Greenville, NC, United States

Thiyagarajamoorthy Dhinesh Kumar   Department of Marine Science, Bharathidasan University, Tiruchirappalli, India

K. Kumaraswamy   Department of Geography, Bharathidasan University, Tiruchirappalli, India

K. Lakshmanadinesh   Geomatics Division, Centre for Water Resources Development and Management (CWRDM), Kozhikode, India

Romain Le Gall Géosciences Océan Laboratory UMR CNRS 6538, University of South Brittany, Vannes Cedex, France

Gaëlle Lemaitre Géosciences Océan Laboratory UMR CNRS 6538, University of South Brittany, Vannes cedex, France

Alex K. Manda Department of Geological Sciences; Institute for Coastal Science and Policy; Coastal Resources Management Program, East Carolina University, Greenville, NC, United States

Manoj Joseph Mathew Institute of Oceanography and Environment, Universiti Malaysia Terengganu, Kuala Terengganu; South-East Asia Carbonate Research Laboratory, Universiti Teknologi PETRONAS, Tronoh, Malaysia; Géosciences Océan, Laboratory UMR CNRS 6538, University of South Brittany, Vannes Cedex, France

David Menier Géosciences Océan Laboratory UMR CNRS 6538, University of South Brittany, Vannes Cedex, France

R. Mohanraj Department of Environmental Management, Bharathidasan University, Tiruchirappalli, India

Nils-Axel Mörner Paleogeophysics & Geodynamics, Stockholm, Sweden

Krishnan Muthukumar Department of Marine Science, Bharathidasan University, Tiruchirappalli, India

K.Ch.V. Naga Kumar Geomatics Division, Centre for Water Resources Development and Management (CWRDM), Kozhikode, India

R. Nagarajan Department of Applied Geology, Curtin University, Miri, Malaysia

Kakani Nageswara Rao Department of Geo-Engineering, Andhra University, Visakhapatnam, India

Banaja Rani Panda Department of Earth Sciences, Annamalai University, Chidambaram, India

Parthasarathy K.S.S. Department of Applied Mechanics and Hydraulics, NIT Surathkal, Mangalore, India

S. Pethaperumal State Groundwater Unit and Soil Conservation, Department of Agriculture, Pondicherry, India

Soazig Pian Géosciences Océan Laboratory UMR CNRS 6538, University of South Brittany, Vannes Cedex, France

M.V. Prasanna Department of Applied Geology, Faculty of Engineering and Science, Curtin University, Miri, Malaysia

R. Rajaram Department of Marine Science, Bharathidasan University, Tiruchirappalli, India

S. Rameshkumar Department of Marine Science, Bharathidasan University, Tiruchirapalli, India

Mu. Ramkumar Department of Geology, Periyar University, Salem, India

Hervé Régnauld Géosciences Océan Laboratory UMR CNRS 6538, University of South Brittany, Vannes Cedex; LETG-Rennes Laboratory (Littoral, Environment, Remote Sensing Laboratory), 2 UMR CNRS 6554, University of Rennes, Rennes Cedex, France

Angélique Rouille Géosciences Océan Laboratory UMR CNRS 6538, University of South Brittany, Vannes cedex, France

M. Santosh School of Earth Sciences and Resources, China University of Geosciences Beijing, Beijing, P.R. China; Department of Earth Sciences, University of Adelaide, Adelaide, SA, Australia

A. Saptoro Department of Chemical Engineering, Curtin University, Miri, Malaysia

Contributors

Subbarayan Saravanan   Department of Civil Engineering, NIT Trichy, Tiruchirappalli, India

Mouncef Sedrati   Géosciences Océan Laboratory UMR CNRS 6538 University of South Brittany, Vannes Cedex, France

Ayyakkannu Selvaraj   Department of Civil Engineering, NIT Tiruchirappalli, Tiruchirappalli, India

Numair Ahmed Siddiqui   Department of Geosciences; Institute of Hydrocarbon Recovery, Universiti Teknologi PETRONAS, Tronoh, Malaysia

S. Sivaranjani   Department of Geography, Bharathidasan University, Tiruchirappalli, India

R. Thilagavathi   Department of Earth Sciences, Annamalai University, Chidambaram, India

C. Thivya   Department of Geology, University of Madras, Chennai, India

K. Tirumalesh   SOE, IAD, Bhabha Atomic Research Center, Mumbai, India

Camille Traini   University of Bayreuth, Institute of Geography, Bayreuth, Germany

Sivanandham Vignesh   Department of Marine Science, Bharathidasan University, Tiruchirappalli, India

S.R. Vishnuprasath   Bharathidasan University, Trichirappalli, India

Rosnan Yaacob   School of Marine and Environmental Science, Universiti Malaysia Terengganu, Kuala Terengganu, Malaysia

CHAPTER 1

Coastal Zone Management During Changing Climate and Rising Sea Level: Transcendence of Institutional, Geographic, and Subject Field Barriers Is the Key

Mu. Ramkumar*, David Menier[†], K. Kumaraswamy[‡]

*Department of Geology, Periyar University, Salem, India, [†]Géosciences Océan Laboratory UMR CNRS 6538, University of South Brittany, Vannes Cedex, France, [‡]Department of Geography, Bharathidasan University, Tiruchirappalli, India

1 Introduction

Sprawling between the high-tide water mark toward land, and up to 200m isobaths toward the ocean, and accounting for about 18% of Earth's surface and 25% of primary global productivity, coastal zones are environmentally sensitive ecological niches. These zones constitute a mere 5% of the total landmass, but sustain three-quarters of the world's population (among which almost half is urban!) and yield more than half of the global gross domestic product (Vorosmarty et al., 2009). Costanza et al. (1997) estimated the potential economic value of transitional coastal zones to be more than USD 22,000 ha^{-1} yaer^{-1}. Notwithstanding this productivity and proliferation, about 4.6% of the world population is living under the threat of potential loss of about 9.3% of the global gross domestic product (Hinkel et al., 2014) by coastal flooding due to the rising sea level. These simple statistics exemplify the importance and perils of coastal zones. The stakes are rising as the population growth and economic pressure in the coastal zones continue to increase (Brommer and Bochev-van der Burgh, 2009).

Coastal regions and populations are exposed to pressures and hazards from both land and sea, making the coastal zones the most transformed and imperiled social-ecological systems on Earth, characterized by pervasive, unsustainable practices (Cummins et al., 2014). In addition, these zones are under the influence of ever-changing sea levels, waves, tides, and currents; in addition to their inherent susceptibility to the dynamics of the lithosphere, hydrosphere, biosphere, and atmosphere, making them complex systems. This complexity poses difficulty in managing coastal regions for sustainable development. The stakes are higher because of the dense population of infrastructure, industry, and human settlements along the coasts (Douvere, 2008; Diedrich et al., 2010). In this regard, "sustainability" of management practice, that is, a balanced approach to the management (*sensu* McKenna et al., 2008) of social, economic, and environmental issues concerned with coastal zones becomes tenuous, and demands a diverse approach, suitable for tackling local issues, and operating under the regional processes of litho, hydro, bio, and atmospheric interactions. The approach also needs to be constrained under the global-scale changing climate, including rising sea levels, which are increasing anthropogenic pressure and resultant feedback of coastal zones. In short, it necessitates sustainable management of the region. Cicin-Sain and Belfiore (2005) defined the sustainable management of coastal zones as "a dynamic process for the sustainable management and use of coastal zones, taking into account, at the same time, the fragility of coastal ecosystems and landscapes, the diversity of activities and uses, their interactions, the maritime orientation of certain activities and uses, and their impact on both the marine and land part." This definition provides for integrated coastal management as a continuous and dynamic process by which decisions are made for the sustainable use, development, and protection of coastal zones. The process involves collaboration and co-ordination of different sectors of society, including researchers, governmental and non-governmental bodies, users, and inhabitants (Foucat, 2002).

Recognizing these diverse needs for sustainable management, a wide range of studies are being conducted on documenting, characterizing, and predicting the occurrences and patterns of litho, hydro, bio, and atmospheric processes, and the resultant responses of coastal zones, at various spatial and temporal scales. In this chapter, we demonstrate how coastal zone management strategies transcend the spatial and temporal scales and why integrated management should be implemented for sustenance.

2 Transcendence of Local Issues to a Global Scale Through Regional Processes

2.1 Spatial-Temporal Scale of Cause-Effect

According to Brommer and Bochev-van der Burgh (2009), in light of projected global climate change (IPCC, 2007), it is of paramount importance to understand the long-term (decades to centuries) and large-scale (10–10^2 km^2) evolution of coastal zones for sustainable coastal management and related coastal impact assessments. Within these sustainable visions, shoreline management plans should be developed targeting the determination of future coastal defence policies and strategic long-term planning for shoreline evolution. Forecasts of coastal changes and quantitative risk assessments spanning this time interval are key requirements. However, present coastal research still mainly focuses on forecasting coastal system evolution in response to changes in hydrodynamic processes and sea level on rather small temporal scales, for example, the tidal cycle. The overall trend or direction the system evolves to on longer temporal scales can greatly affect and alter the impact of smaller-scale processes; but still, little is known on how to account for changes in long-term system evolution (Brommer and Bochev-van der Burgh, 2009).

Cowell et al. (2003) presented the spatial and temporal scales of coastal processes into observable landforms (Fig. 1) and interpreted them into exogenic and endogenic processes. As shown in Fig. 1, the processes operate and the products result from a variety of spatial and temporal scales. From these, it can be surmised that the coastal zone management practices, applicable to a geographic location for a time-duration, may not be applicable and or effective for other settings and timeframes. Hence, understanding the spatial and temporal scales of processes that are in operation in a setting, followed by designing and implementing management strategies, could be the key for success. Through a review, Mumby and Harborne (1999) demonstrated a case in this regard. These authors detailed the problems of usage of nonstandard scales in defining habitats, and presented a list of challenges and pitfalls of coastal zone management. They also elaborated on how nonsystematic classification of habitats and ambiguous documentation creates problems on several scales. First, meaningful interpretation of the habitat classification scheme may be difficult on the scale of individual habitat maps. This difficulty applies to managers using the scheme for planning, and field surveyors attempting to adopt it in situ. Second, integrating several habitat maps on, say, a national scale is difficult because there is little or no standardization in terms. Thus, not only it is difficult to decide when two terms are synonymous, but the lack of quantitative detail also obscures actual differences in habitat types, thereby decreasing the probability that habitats will be distinguished correctly. These integration problems may be ameliorated if a national organization coordinates or undertakes the mapping (Mumby et al., 1995), but the problems tend to be exacerbated on international scales.

2.2 Transcendence of Causes and Effects

Though global warming (IPCC, 2007) and the attendant sea level rise (Allen and Komar, 2006; Rao et al., 2008) are known, the variability of their consequences with reference to different geographic, climatic, and other settings is poorly understood. For example, Kallepalli et al. (2017) demonstrated how global warming causes a locally enhanced sea level rise, and aggravates coastal erosion, depending on local-scale land subsidence and lithological characteristics. The phenomenal rise in coastal tourism in recent decades (Hall, 2001) has brought remote and inaccessible regions under the influence of intense anthropogenically influenced changes in physical, chemical, and biological paradigms of coastal ecosystems. The adverse effect of the rise in sea level due to global warming on coastal environments

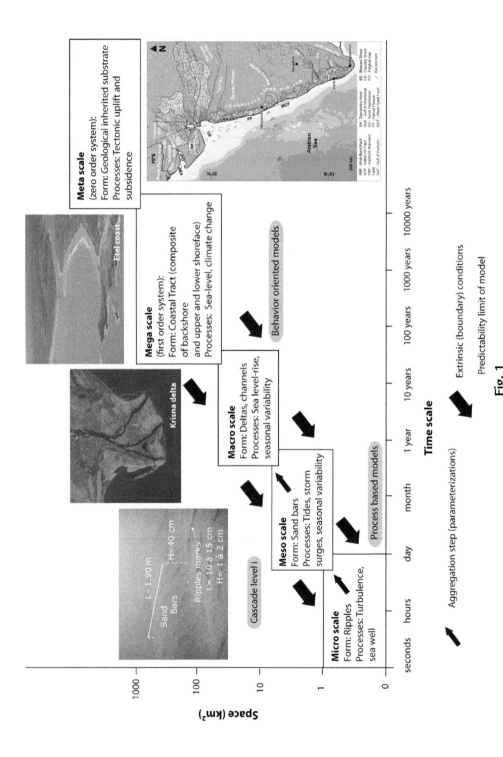

Fig. 1

Coastal processes and their spatial-temporal scales. *Modified from Cowell, P.J., Stive, M.J.F., Niedoroda, A.W., de Vriend, D.J., Swift, D.J.P., Kaminsky, G.M., Capobianco, M., 2003. The coastal-tract (Part 1): a conceptual approach to aggregated modeling of low-order coastal change. J. Coast. Res. 19, 812–827.*

becomes multiplied, also due to the land use changes in inland regions, and resultant variability of sediment influx (Ramkumar et al., 2015a; Rao et al., 2008) into the river systems, estuaries, and coastal ocean.

The benefit of nourishing and protecting coastal vegetation systems such as seagrass beds, mangrove swamps, mudflats, and marsh fields, and monitoring sediment carbon dynamics to offset rising atmospheric CO_2 for mitigation of global warming was demonstrated by Brown et al. (2016). Despite constituting one-third of terrestrial vegetational area (Donato et al., 2011; McLeod et al., 2011), these coastal systems have higher primary productivity (Bouillon et al., 2008; Alongi, 2014), and an ability to redistribute themselves in relation to changing sea levels (McLeod et al., 2011; Sanders et al., 2012; Duarte et al., 2013; Leopold et al., 2013); thus acting as effective carbon sinks (Bouillon et al., 2008; Alongi, 2014). Thus, owing to the dynamic equilibrium of coastal zones and the interlinked nature of the litho-bio-hydro-atmosphere at the coastal zone, the impacts of the consequences are amplified, often detrimentally, due to the intensive anthropogenic intervention into the natural processes along the coast. A few of the consequences include, but are not limited to, the retreat of Arctic ice sheets and the reduction of the polar bear population (Courtland, 2008); a significant upward altitudinal shift of species at an average of 29 m per shift since the year 1905 (Lenoir et al., 2008); a significant rise in epidemics and metabolic disorders due to the deterioration of air quality (Hoyle, 2008); and an increase in incidences of droughts and geohazards that result in the escalation of food prices due to a disparity in demand-supply (Parry et al., 2008). The study of Weng (2000) is a classic work that documented the origin and propagation of agriculture in relation to climate and sea level changes during the Holocene era, and the development and widespread practice of rice cultivation, horticulture, and the dyke-pond system of human-environment interaction. The study also explains the adverse effect of the imprudent use of technology over natural the environment that manifested through increased incidents of geohazards in riparian regions, including coastal tracts.

These observations and reviews demonstrated the connectedness of spatially distal regions, besides the transcendence of local-regional events into regional-global scale repercussions. An effective approach for addressing these repercussions may involve the ability to recognize this connectedness in coastal environmental systems, and then designing appropriate spatial-temporal scale remediation measures and integrating the isolated efforts into a regional-global scale strategy.

2.3 Need to Have Unified System of Coastal Environmental Management

Coppola (2011) stated that the vast majority of environmental management systems are concerned with specific individual issues and local areas, and thus ignore the fact that the natural world is an interconnected and dependent system of which humans are a part, albeit a highly influential one. Considering environmental issues on an individual and separate basis will never be as effective as considering the issues, at least initially, at a systems level, where all of the inputs can be taken into account. Because most environmental issues today involve at least some level of human influence, it would be wise to manage human interactions with the environment at wider scales with holistic policies that are as integrated and wide-reaching as is feasible. For example, estuaries are situated on the border of the land and sea, they are zones of rapidly fluctuating environmental conditions and are markedly influenced by catchment and oceanic processes, as well as in-estuary uses (Traini et al., 2015). These features complicate their management and need monitoring and assessment at a basin-scale rather than the coastal part alone (Ramkumar, 2003, 2004a,b, 2007; Ramkumar et al., 2009, 2015a,b). According to Coppola, many aspects of environmental management are currently conducted in isolation. Managers focus on their own responsibilities, and all too often fail to consider other related issues, and ignore the interrelated nature of the environment as a whole. Nowhere is this truer than in coastal zones. These dynamic and highly inter-connected regions are frequently affected by human land use decisions. While ecosystem-based management has made significant strides in theory and design in recent years, straightforward and effective tools are needed to bring it into wider use. These views are aptly supported by the recent review by Roy et al. (2015), who opined that land-use and land-cover change is a key focus area for the global change community because of its significant impacts on climate change, biogeochemical cycles, biodiversity, and water resources. The land-use and land-cover changes are driven by variations in multi-scale interacting driving factors, such as biophysical conditions of the land, demography, technology, affluence, political structures, economy, and people's attitudes and values. These driving factors vary with geography and time. The land-use and land-cover changes are also heterogeneous both spatially and temporally. Therefore, improved representation of both spatial and temporal dimensions of land-use and land-cover changes is crucial for a better understanding of human influence on the natural environment.

At its core, environmental management is achieved through the design and control of human behavior. Land use planning provides an excellent tool for the management of a variety of influential human activities by controlling and designing the ways in which humans use land and natural resources. In its present state, land use planning falls short of its potential as an environmental and natural resource management tool. This is primarily due to a lack of coordination and the failure of land use planners to consider the environment in their charge holistically (Coppola, 2011). Case studies demonstrating the forward and feedback mechanisms of local cause-regional processes and regional and or global phenomenon and local-regional consequences include, but are not limited to, Weng (2000), Hall (2001), Ramkumar (2000, 2003), Ramkumar et al. (1999, 2015b), Baskaran (2004), Hema Malini and Rao (2004), Lewsey et al. (2004), Ericson et al. (2006), Jayanthi (2009), Lichter et al. (2010), Rossi et al. (2011), Ghosh and Datta (2012), Hinkel et al. (2014), Roy et al. (2015), Brown et al. (2016), Conrad et al. (2017), Kallepalli et al. (2017), and Sreenivasulu et al. (2018).

3 Transcendence of Subject, Institutional, and Geographic Areas Are the Key

It is ironic that a statement made few decades ago (Dickert and Tuttle, 1985) "several theoretical, analytical, and institutional difficulties have impeded the development and application of the assessment of cumulative environmental impacts" remains valid, with heightened vigor, especially in reference to coastal zone management! Similarly, these authors rued the transcendence of boundaries of institutions, in addition to the spatial scales of coastal processes that serve as an impediment to data collection and collation. This impediment also remains very much a dampener in coastal zone research. Despite much advancement in our understanding, there are many unanswered questions, and some are fundamental! For example, there is no consensus on the definition for the term "rock coasts," and data on their distribution and extent are not available (Naylor et al., 2010).

Mangroves are the most prominent tropical ecosystem where geomorphologic, sedimentary, and oceanographic processes have controlled landscape evolution. Mangrove vegetation can change over time as landforms can accrete or erode, which is a direct response to coastal sedimentary processes. This demonstrates that significant changes can occur on short time scales, and mangroves thus provide an excellent register of these modifications. Natural processes and human activities have extensively modified mangrove communities and, as a result, environmental, economic, and social impacts have increased over the past 2 decades (Filho et al., 2006). However, the extent of resilience, quantum of modification, and sustainability are yet to be documented from a local-regional scale, to be compiled on a global database. Coastal lagoons provide another challenge to management strategies. As these are isolated from the sea by a barrier, but connected to it, they play an important role, along with other critical transition zones, in the maintenance of the biogeochemical fluxes on the planet by allowing interchanges of mass, momentum, and energy between the sea, land, and atmosphere. They are also characterized by extreme fluctuations in salinity, temperature, water level, dissolved oxygen, and so forth, which means these far more sensitive and specific plans for their management and nourishment are yet to be developed (Moreno et al., 2010). According to Ericson et al. (2006), distinctiveness of deltas makes a comprehensive global assessment of the contemporary state and future sources of vulnerability in deltas difficult to establish, and the available data on apparent rates of contemporary sea-level rise for individual deltas remains problematic.

Despite the recent advances in sea level research, instrumental records augmented with high-resolution relative sea level (RSL) records derived from salt-marsh sedimentary sequences support the inference that modern rates of relative sea level rise (past 100 years) may be more rapid than the long-term rate of rise (between 800 and 1000 years ago), and that the timing of the 20th century acceleration may be indicative of a link with human-induced climate change (Rossi et al., 2011). The enigma remains pertaining to the differential rates of sea levels on a spatial scale, and variability in responses of different terrains on a spatial and temporal scale, that need to be resolved and need consensus. In addition, accurate estimates of sea level rise in the pre-satellite era are needed in order to provide a context for the 21st century-estimates and to calibrate climate models (Kopp et al., 2015). In this regard, while the limited number, distribution, and duration of tide-gauges precludes efforts to robustly test climate versus sea level hypotheses and establish the driving mechanisms responsible for such changes (Rossi et al., 2011), the absence of uniformity in proxies and methods of estimation thwarts establishment of an accurate long-term sea level database.

Although it is common knowledge that coastal regions are important in view of their productive ecosystems, concentration of population, and exploitation of renewable and non-renewable natural resources, sound management strategies depend on accurate and comprehensive scientific data on which policy decisions can be based. However, owing to the requirement of logistics, manpower, sophisticated analytical equipment, and time-consuming procedures of analyses and costs involved in generating such analytical data, such elaborate analyses were seldom in practice, much to the detriment of effective natural environmental management planning and implementation (Ramkumar et al., 2009). In addition, the time gap between data generation and publication is the major impediment that thwarts implementation of effective environmental management practices (Ramkumar et al., 2015a). Remote sensing and geographic information systems are routinely being used (Mumby et al., 1995) in order to obviate the time lag between data capture and utilization (Shanmugam et al., 2006).

During the past decade, significant advances in knowledge and understanding have been highlighted, such as the benefits of high and moderate spatial and spectral resolution data, which are better able to match the rich spatial and spectral diversity observed in coastal zones. Distinct from optical sensors (e.g., Landsat TM), which extract the target's information related to chemical composition and physical structure of the material from the reflected sunlight, backscattered microwave energy (e.g., RADARSAT-1) from the Earth's surface measured by a Synthetic Aperture Radar (SAR) provides information about the terrain's physical (macro-topography, slope, roughness) and electrical properties mainly controlled by the moisture content (Filho et al., 2006). While our understanding of the mainland coastal zones is appreciable, the coastal zones and associated marine regimes of island continents such as Australia or Sri Lanka, or archipelagos such as Andaman and the Nikobar group of islands, or the Maldives pose yet another complexity, and are poorly understood and need systematic documentation and understanding (e.g., Dawson and Smithers, 2010; Bandopadhyay and Carter, 2017).

The study of Hinkel et al. (2014) presented a global model on coastal flood risk, taking into consideration of a wide range of uncertainties in continental topography data, population data, protection strategies, socioeconomic development, and sea-level rise, emphasizing the need to improve the methodologies in using a variety of data. According to these authors, global scale models are yet to be developed against geohazards and adaptation strategies.

A compilation of land use-land cover data on a spatial-temporal scale attempted by Roy et al. (2015) has provided a critical link database for assessing the impacts of climate change on food production, health, urbanization, coastal zones, and so forth, which needs to be replicated on a global scale. Coastal habitat maps are a fundamental requirement in establishing coastal management plans. Coastal habitats are manageable units, and large-scale maps allow managers to visualize the spatial distribution of habitats, thus aiding the planning of networks of marine protected areas and allowing the degree of habitat fragmentation to be monitored (Mumby and Harborne, 1999). In a paradigm shift of how the geoscientific community and all other stakeholders started to look at the coastal zone is succinctly presented by Ramesh et al. (2015), who stated that the coastal zone is not a geographic boundary of interaction between the land and the sea, but a global compartment of special significance for biogeochemical cycling and processes, and ever increasingly for human habitation and economies. The current research and management plans, thus, need to reorient along these lines, recognizing the transcendence of physical boundaries and scales of processes.

4 Conclusions

From the review presented in this chapter, it is brought to light that consensus eludes the scientific community, despite the availability of sophisticated analytical instrumentation, establishment of reliable proxies, and creation of integration systems of a variety of data sets to monitor, assess, and manage coastal zones. Large gaps exist in creation of baseline databases, establishment of mapping units, development of methodology for classification schemes, and their adoptability in a variety of coastal settings, in establishing synergy between achievable goals from the perspectives of stakeholders, and establishment of a participatory approach for environmental protection and sustainable utilization, and so forth. Compounded with these gaps and lacuna are the effects of rising sea levels and the changing climate, superimposed on anthropogenic intervention into the natural dynamics. Together, these obviate implementation of management plans effectively. Topping all these are the boundaries between political, governmental, and societal institutions, across which the natural processes operate and require transcendence, for observation, data collection, and management program implementation. It requires a transgression of ideas between subject barriers and transcendence of sea level and coastal zone management research on a spatial and temporal scale for designing and implementing effective management strategies.

Acknowledgments

The review was made possible by the decades of work conducted by those authors whose papers are cited in this chapter. We thank those authors in the list and many others whose work influenced our thoughts and conceptualization.

References

Allen, J.C., Komar, P.D., 2006. Climate controls on US west coast erosion processes. J. Coast. Res. 22, 511–529.

Alongi, D.M., 2014. Carbon cycling and storage in mangrove forests. Annu. Rev. Mar. Sci. 6, 195–219.

Bandopadhyay, P.C., Carter, A., 2017. Introduction to the geography and geomorphology of the Andaman–Nicobar Islands. In: Bandopadhyay, P.C., Carter, A. (Eds.), The Andaman–Nicobar Accretionary Ridge: Geology, Tectonics and Hazards. In: vol. 47. Geological Society, Memoirs, London, pp. 9–18.

Baskaran, R., 2004. Coastal erosion. Curr. Sci. 86, 25.

Bouillon, S., Borges, A.V., Castañeda-Moya, E., Diele, K., Dittmar, T., Duke, N.C., Kristensen, E., Lee, S.Y., Marchand, C., Middelburg, J.J., Rivera-Monroy, V.H., Smith Iii, T.J., Twilley, R.R., 2008. Mangrove production and carbon sinks: a revision of global budget estimates. Glob. Biogeochem. Cycles. 22. https://doi.org/10.1029/2007GB003052.

Brommer, M.B., Bochev-van der Burgh, L.M., 2009. Sustainable coastal zone management: a concept for forecasting long-term and large-scale coastal evolution. J. Coast. Res. 25, 181–188.

Brown, D.R., Conrad, S., Akkerman, K., Fairfax, S., Fredericks, J., Hanrio, E., Sanders, L.M., Scott, E., Skillington, A., Tucker, J., van Santen, M.L., Sanders, C.J., 2016. Seagrass, mangrove and saltmarsh sedimentary carbon stocks in an urban estuary; Coffs Harbour, Australia. Reg. Stud. Mar. Sci. 8, 1–6.

Cicin-Sain, B., Belfiore, S., 2005. Linking marine protected areas to integrated coastal and ocean management: a review of theory and practice. Ocean Coast. Manag. 48, 847–868.

Conrad, S.R., Santos, I.R., Brown, D.R., Sanders, L.M., vanSanten, M.L., Sanders, C.J., 2017. Mangrove sediments reveal records of development during the previous century (Coffs Creek estuary, Australia). Mar. Pollut. Bull. 122, 441–445.

Coppola, H., 2011 Environmental Land Use Planning and Integrated Management at the River Basin Scale in Coastal North Carolina. Master's thesis submitted to the Duke University (unpublished).

Costanza, R., d'Arge, R., de Groot, R., Farber, S., Grasso, M., Hannon, B., Limburg, K., Naeem, S., O'Neill, R.V., Paruelo, J., Raskin, R.G., Sutton, P., van den Belt, M., 1997. The value of the world's ecosystem services and natural capital. Nature 387, 253–260.

Courtland, R., 2008. Polar bear numbers set to fall. Nature 453, 432–433.

Cowell, P.J., Stive, M.J.F., Niedoroda, A.W., Vriend, D.J., Swift, D.J.P., Kaminsky, G.M., Capobianco, M., 2003. The coastal-tract (Part 1): a conceptual approach to aggregated modeling of low-order coastal change. J. Coast. Res. 19, 812–827.

Cummins, V., Burkett, V., Day, J., Forbes, D., Glavovic, B., Glaser, M., Pelling, M., 2014. LOICZ Signpost: Consultation Document Signalling New Horizons for Future Earth—Coasts. LOICZ. http://www.loicz.org/cms02/about_us/FEcoasts/index.html.en.html.

Dawson, J.L., Smithers, S.G., 2010. Shoreline and beach volume change between 1967 and 2007 at Raine Island, Great Barrier Reef, Australia. Glob. Planet. Chang. 72, 141–154.

Dickert, T.G., Tuttle, A.E., 1985. Cumulative impact assessment in environmental planning: a coastal wetland watershed example. Environ. Impact Assess. Rev. 5, 37–64.

Diedrich, A., Tintoré, J., Navinés, F., 2010. Balancing science and society through establishing indicators for integrated coastal zone management in the Balearic Islands. Mar. Policy 34, 772–781.

Donato, D.C., Kauffman, J.B., Murdiyarso, D., Kurnianto, S., Stidham, M., Kanninen, M., 2011. Mangroves among the most carbon-rich forests in the tropics. Nat. Geosci. 4, 293–297.

Douvere, F., 2008. The importance of marine spatial planning in advancing ecosystem-based sea use management. Mar. Policy 32, 762–771.

Duarte, C.M., Losada, I.J., Hendriks, I.E., Mazarrasa, I., Marbà, N., 2013. The role of coastal plant communities for climate change mitigation and adaptation. Nat. Clim. Chang. 3, 961–968.

Ericson, J.P., Vörösmarty, C.J., Dingman, S.L., Ward, L.G., Meybeck, M., 2006. Effective sea-level rise and deltas: causes of change and human dimension implications. Glob. Planet. Chang. 50, 63–82.

Filho, P.W.M.S., Martins, E.S.F., da Costa, F.R., 2006. Using mangroves as a geological indicator of coastal changes in the Braganc-a macrotidal flat, Brazilian Amazon: a remote sensing data approach. Ocean Coast. Manag. 49, 462–475.

Foucat, V.S.A., 2002. Community-based ecotourism management moving towards sustainability, in Ventanilla, Oaxaca, Mexico. Ocean Coast. Manag. 45, 511–529.

Ghosh, P.K., Datta, D., 2012. Coastal tourism and beach sustainability—an assessment of community perceptions in Kovalam, India. J. Soc. Space 8, 75–87.

Hall, C.M., 2001. Trends in ocean and coastal tourism: the end of the last frontier? Ocean Coast. Manag. 44, 601–618.

Hema Malini, B., Rao, K.N., 2004. Coastal erosion and habitat loss along the Godavari delta front—a fallout of dam construction (?). Curr. Sci. 87, 1232–1236.

Hinkel, J., Lincke, D., Vafeidis, A.T., Perrette, M., Nicholls, R.J., To, R.S.J., Marzeion, B., Fettweis, X., Ionescu, C., Levermann, A., 2014. Coastal flood damage and adaptation costs under 21st century sea-level rise. PNAS 111, 3292–3297.

Hoyle, B., 2008. Accounting for climate ills. Nat. Rep. Clim. Change. https://doi.org/10.1038/climate.2008.43.

IPCC, 2007. Summary for policymakers. In: Soloman, S.D., Manning, Q.M., Chen, Z., Miller, H.L. (Eds.), Climate Change 2007: The Physical Science Basis. Contribution of Working Group I to the Fourth Assessment Report of the Intergovernmental Panel on Climate Change. Cambridge University Press, Cambridge, pp. 1–18.

Jayanthi, M., 2009. Mitigation of geohazards in coastal areas and environmental policies of India. In: Ramkumar, M. (Ed.), Geological Hazards: Causes, Consequences and Methods of Containment. New India Publishers, New Delhi, pp. 141–147.

Kallepalli, A., Rao, K.N., James, D.B., Richardson, M.A., 2017. Digital shoreline analysis system-based change detection along the highly eroding Krishna–Godavari delta front. J. Appl. Remote. Sens. 11. https://doi.org/10.1117/1.JRS.11.036018.

Kopp, R.E., Horton, B.P., Kemp, A.C., Tebaldi, C., 2015. Past and future sea-level rise along the coast of North Carolina, USA. Climate Change 132, 693–707.

Lenoir, J., Gegout, J.C., Marquet, P.A., de Ruffray, P., Brisse, H., 2008. A significant upward shift in plant species optimum elevation during the 20th century. Scientist 320, 1768–1771.

Leopold, A., Marchand, C., Deborde, J., Chaduteau, C., Allenbach, M., 2013. Influence of mangrove zonation on CO2 fluxes at the sediment-air interface (New Caledonia). Geoderma 202-203, 62–70.

Lewsey, C., Cid, G., Kruse, E., 2004. Assessing climate change impacts on coastal infrastructure in the Eastern Caribbean. Mar. Policy 28, 393–409.

Lichter, M., Zviely, D., Klein, M., 2010. Morphological patterns of southeastern Mediterranean river mouths: the topographic setting of the beach as a forcing factor. Geomorphology 123, 1–12.

McKenna, J., Coopera, J., O'Haganb, A.M., 2008. Managing by principle: a critical analysis of the European principles of Integrated Coastal Zone Management (ICZM). Mar. Policy 32, 941–955.

McLeod, E., Chmura, G.L., Bouillon, S., Salm, R., Björk, M., Duarte, C.M., Lovelock, C.E., Schlesinger, W.H., Silliman, B.R., 2011. A blueprint for blue carbon: toward an improved understanding of the role of vegetated coastal habitats in sequestering CO_2. Front. Ecol. Environ. 9, 552–560.

Moreno, I.M., Ávila, A., Losada, M.A., 2010. Morphodynamics of intermittent coastal lagoons in Southern Spain: Zahara de los Atunes. Geomorphology 121, 305–316.

Mumby, P.J., Harborne, A.R., 1999. Development of a systematic classification scheme of marine habitats to facilitate regional management and mapping of Caribbean coral reefs. Biol. Conserv. 88, 155–163.

Mumby, P.J., Gray, D.A., Gibson, J.P., Raines, P.S., 1995. Geographical information systems: a tool for integrated coastal zone management in Belize. Coast. Manag. 23, 111–121.

Naylor, L.A., Stephenson, W.J., Trenhaile, A.S., 2010. Rock coast geomorphology: recent advances and future research directions. Geomorphology 114, 3–11.

Parry, M., Palutikof, J., Hanson, C., Lowe, J., 2008. Squaring up to reality. Nat. Rep. Clim. Change 2, 68. https://doi.org/10.1038/climate.2008.50.

Ramesh, R., Chen, Z., Cummins, V., Day, J., D'Elia, C., Dennison, B., Forbes, D.L., Glaeser, B., Glaser, M., Glavovic, B., Kremer, H., Lange, M., Larsen, J.N., Le Tissier, M., Newton, A., Pelling, M., Purvaja, R., Wolanski, E., 2015. Land–ocean interactions in the coastal zone: past, present and future. Anthropocene 12, 85–98.

Ramkumar, M., 2000. Recent changes in the Kakinada spit, Godavari delta. J. Geol. Soc. India 55, 183–188.

Ramkumar, M., 2003. Geochemical and sedimentary processes of mangroves of Godavari delta: implications on estuarine and coastal biological ecosystem. Ind. J. Geochem. 18, 95–112.

Ramkumar, M., 2004a. Dynamics of moderately well mixed tropical estuarine system, Krishna estuary, India: part III. Nature of property-salinity relationships and quantity of exchange. Ind. J. Geochem. 19, 219–244.

Ramkumar, M., 2004b. Dynamics of moderately well mixed tropical estuarine system, Krishna estuary, India: part IV. Zones of active exchange of physico-chemical properties. Ind. J. Geochem. 19, 245–269.

Ramkumar, M., 2007. Spatio-temporal variations of sediment texture and their influence on organic carbon distribution in the Krishna estuary. Ind. J. Geochem. 22, 143–154.

Ramkumar, M., Pattabhi Ramayya, M., Swamy, A.S.R., 1999. Changing landuse/land cover pattern of coastal region: Gautami sector, Godavari delta, India. J. A.P. Acad. Sci. 3, 11–20.

Ramkumar, M., Anbarasu, K., Sathish, G., Suresh, R., Venkateswaran, S., 2009. Hyperspectral remote sensing of coastal sedimentary terrains: a solution in sight for faster, reliable and cost effective environmental impact assessment? In: Rajendran, S., Aravindan, S., Jayavel Rajkumar, T., Murali Mohan, K.R. (Eds.), Hyperspectral Remote Sensing and Spectral Signature Applications. New India Publishers, New Delhi, pp. 183–198.

Ramkumar, M., Kumaraswamy, K., Mohan Raj, R., 2015a. Land use dynamics and environmental management of river basins with emphasis on deltaic ecosystems: need for integrated study based development and nourishment programs and institutionalizing the management strategies. In: Ramkumar, M., Kumaraswamy, K., Mohan Raj, R. (Eds.), Environmental Management of River Basin Ecosystems. Springer-Verlag. https://doi.org/10.1007/978-3-319-13425-3-1.

Ramkumar, M., Kumaraswamy, K., Arthur James, R., Suresh, M., Sugantha, T., Jayaraj, L., Mathiyalagan, A., Saraswathi, M., Shyamala, J., 2015b. Sand mining, channel bar dynamics and sediment textural properties of the Kaveri River, South India: implications on flooding hazard and sustainability of the natural fluvial system. In: Ramkumar, M., Kumaraswamy, K., Mohanraj, R. (Eds.), Environmental Management of River Basin Ecosystems. Springer-Verlag, Heidelberg. https://doi.org/10.1007/978-3-319-13425-3-14.

Rao, K.N., Subraelu, P., Rao, T., Hema Malini, B., Ratheesh, R., Bhattacharya, S., Ajai, A.S.R., 2008. Sea-level rise and coastal vulnerability: an assessment of Andhra Pradesh coast, India through remote sensing and GIS. J. Coast. Conserv. 12, 195–207.

Rossi, V., Horton, B.P., Corbett, D.R., Leorri, E., Perez-Belmonte, L., Douglas, B.C., 2011. The application of foraminifera to reconstruct the rate of 20th century sea level rise, Morbihan Golfe, Brittany, France. Quat. Res. 75, 24–35.

Roy, P.S., Roy, A., Joshi, P.K., Kale, M.P., Srivastava, V.K., Srivastava, S.K., Dwevidi, R.S., Joshi, C., Behera, M.D., Meiyappan, P., Sharma, Y., Jain, A.K., Singh, J.S., Palchowdhuri, Y., Ramachandran, R.M., Pinjarla, B., Chakravarthi, V., Babu, N., Gowsalya, M.S., Thiruvengadam, P., Kotteeswaran, M., Priya, V., Yelishetty, K.V.N., Maithani, S., Talukdar, G., Mondal, I., Rajan, K.S., Narendra, P.S., Biswal, S., Chakraborty, A., Padalia, H., Chavan, M., Pardeshi, S.N., Chaudhari, S.A., Anand, A., Vyas, A., Reddy, M.K., Ramalingam, M., Manonmani, R., Behera, P., Das, P., Tripathi, P., Matin, S., Khan, M.L., Tripathi, O.P., Deka, J., Kumar, P., Kushwaha, D., 2015. Development of decadal (1985–1995–2005) land use and land cover database for India. Remote Sens. 7 (7), 2401–2430.

Sanders, C.J., Smoak, J.M., Waters, M.N., Sanders, L.M., Brandini, N., Patchineelam, S.R., 2012. Organic matter content and particle size modifications in mangrove sediments as responses to sea level rise. Mar. Environ. Res. 77, 150–155.

Shanmugam, P., Ahn, Y., Sanjeevi, S., 2006. A comparison of the classification of wetland characteristics by linear spectral mixture modelling and traditional hard classifiers on multispectral remotely sensed imagery in southern India. Ecol. Model. 194, 379–394.

Sreenivasulu, G., Jayaraju, N., Reddy, B.C.S.R., Lakshmanna, B., Prasad, T.L., 2018. Influence of coastal morphology on the distribution of heavy metals in the coastal waters of Tupilipalem coast, Southeast coast of India. Remote Sens. Appl. Soc. Environ. 10, 190–197.

Traini, C., Proust, J.N., Menier, D., Mathew, M.J., 2015. Distinguishing natural evolution and human impact on estuarine morpho-sedimentary development: a case study from the Vilaine Estuary, France. Estuar. Coast. Shelf Sci. 163, 143–155.

Vorosmarty, C.J., Syvitski, J.P.M., Day, J., de Sherbinin, A., Giosan, L., Paola, C., 2009. Battling to save the world's river deltas. Bull. At. Sci. 65, 31–43.

Weng, Q., 2000. Human-environment interactions in agricultural land use in a South China's wetland region: a study on the Zhujiang delta in the Holocene. Geol. J. 51, 191–202.

PART 1
Sealevel Cycles and Oceanography

CHAPTER 2

Development of Ideas and New Trends in Modern Sea Level Research: The Pre-Quaternary, Quaternary, Present, and Future

Nils-Axel Mörner
Paleogeophysics & Geodynamics, Stockholm, Sweden

1 Introduction

Humans, their habitation, and life activities have always been associated with the shores and coasts all over the globe, and the migration from one land area to another has been by crossing land bridges, shallow water, and sometimes even open water. This implies a very close link between the distribution of the human species and sea level changes. Therefore, it is not surprising when we read early stories and myths about sea level changes. We have the Gilgamesh epos, about five millennia old, telling about "the flooding" of Mesopotamia (recorded in stratigraphy by Wooley, 1929, and recently addressed as to age and sea level change by Mörner, 2015a); later to become the famous legend of The Great Flood and Noah and his Ark in the Bible. The general story in ancient myths is about a rising sea; that is what we today know as the main postglacial rise in sea level. In a few areas such as northern North America and Fennoscandia the story is the opposite; namely a general fall in sea level, or what we today know as glacial isostatic uplift. The Indians in northern North America talked about their living on the top of a huge turtle, which continually rose out of the sea. In Scandinavia, the old legend speaks about the habituated world living on top of a huge giant named Ymer, who slowly rose out of the sea.

The first more "scientific" note comes from Ovid (43 BCE to 17 CE), who in his remarkable book *Metamorphoses XV* wrote (Ovidius, n.d. (1 AD); translation: Mörner 1979):

> vidi ego, quod fuerat quondam solidissima tellus.
>   esse fretum, vidi factas exaequorure terras.

> I have myself seen what was once most solid ground.
> disappear into the sea, and I have heard of land risen out of the sea.

It seems reasonable to assume that Ovid himself had witnessed temples and buildings below sea level. Evidence of uplift, however, he had only heard about; could it have been the uplift recorded in Sweden (Mörner, 1979)? Regardless, Ovid's observations of changes (metamorphoses) in nature mark the beginning not only of geography and geology, but also of the modern concept of "global change." Therefore, at the 32nd International Geological Congress in Florence in 2004, we celebrated "the 2000 year anniversary of Global Change" with a special symposium in honor of Ovid (viz. symposium T28).

In this chapter, I have chosen to focus on my own views on the multifaceted issue of sea level changes. This will lead to bias in references with respect to my own papers (documenting this development of ideas, and providing the necessary background material for the reader who wants to dive deeper into the documentation). Still, I hope all different aspects of sea level changes will be covered in an adequate and truly open-minded way.

2 Development of Ideas

The development of ideas and recording of stratigraphic facts is a quite fascinating story. I have addressed it before (Mörner, 1979, 1987a). In this section I will highlight some of the main steps toward modern understanding (Fig. 1, Table 1).

2.1 Earth's Rotation

Newton (1687) was the first to realize that the Earth must be a globe strongly flattened at the poles and bulging at the equator. This implies the dynamic effects of rotation. Today, we know that the difference between the equatorial and polar diameters is 21,385 m. A century later, this was picked up by Frisi (1785), who used it to explain differential sea level changes between high and low latitudes as a function of changes in the speed of rotation. He proposed that this might explain the fall in sea level in Sweden and the rise of sea level in the Mediterranean (Fig. 2C; Mörner, 1987a, Fig. 11.1). In the late 17th century, an interest in recording and measuring changes in nature began; the first tide gauge was installed in Amsterdam in 1682, in 1694 Urban Hjärne in Sweden sent out a questionnaire about natural phenomena, including two questions about the changes in the land/sea relation, and the first water marks were established in Sweden in 1700 and 1704 (Mörner, 1979).

2.2 Discussions in the 18th Century

By observational facts it became obvious that the sea level changed with time; in some areas rising in over land, flooding old buildings and land surfaces (e.g., the Temple of Serapisat Pozzuoli outside Rome), in other areas receding and exposing more land (this was especially

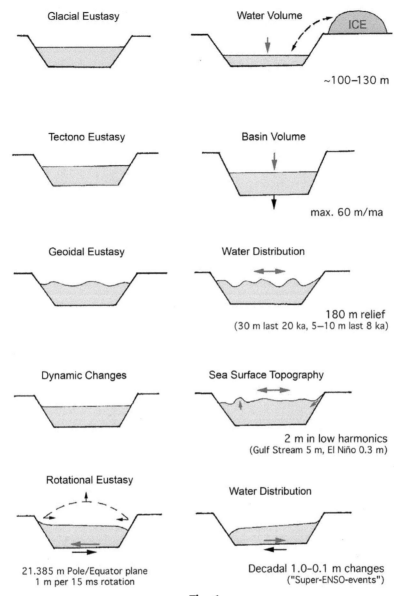

Fig. 1
Main eustatic variables controlling the water volume, the basin volume, the water distribution, the dynamic topography, and their rates and amplitudes. *From Mörner, N.-A., 2005. Sea level changes and crustal movements with special reference to the East Mediterranean. Z. Geomorph. Suppl. 137, 91–102.*

the case in Sweden, where a lively debate occurred; see Mörner, 1979). Something was changing; either it was the oceans themselves, or it was the land (as illustrated in Fig. 2). Proponents for both sides gathered under the names of "neptunists" (claiming that it was the ocean level that changed) and "plutonists" (claiming that it was the land level that changed).

Table 1 Ocean level variables and their maximum quantities in metres (Mörner, 1994, 1996a, 2000, 2013a)

Ocean Variables	Max. Amplitude in Metres
Present tides	
• Geoid tides	0.78
• Ocean tides	up to ~18
Present geoid relief	
• Maximum difference	180
Present dynamic sea surface	
• Low harmonics	up to max ~2
• Major currents	up to max ~5
Earth's radius differences	
• Equator vs. pole	21,385
Presently stored glacial volume ≈ metres sea level	
• Antarctica	~60
• Greenland	~6
• All mountain glaciers	0.5–0.4
Glacial eustasy	
• Last Ice Age amplitude	120–130
Geoid deformation	
• Last 150 ka	30–90
• Last 20 ka	30–60
• Last 8 ka	5–8
Rotational eustasy and ocean circulation	
• Holocene Gulf Stream beat	1.0–0.1
• Holocene Equatorial E–W pulses	~1
• North-South changes (grand solar maxima/minima)	0.7–0.3
• Super-ENSO events	~1
• El Niño/ENSO events	0.3
Thermal expansion	
• Open ocean	<1.0
• At 100 m water depth	<0.035
• At 10 m water depth	<0.0035
• At the shore	Always ±0.0
Coastal run-off	
• At river outlets	1–2
Evaporation/precipitation	
• Indian ocean evaporation low	0.3–0.4
Air pressure	
• 1 millibar	0.01
• Monsoonal range at Bangladesh	0.15
Rotation rate and sea level	
• ~15 ms rotation (LOD)	1.0

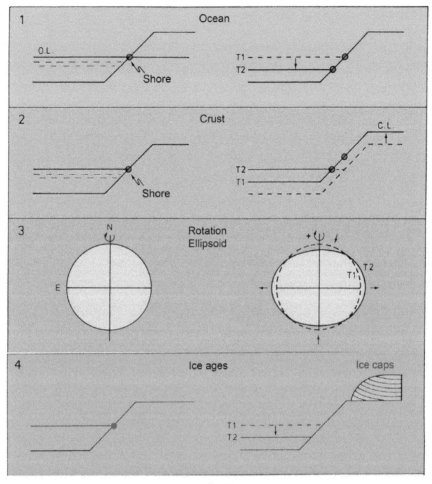

Fig. 2
Models of sea level changes as proposed in the 18th and 19th centuries. (A) ocean water lowering, the "water diminution" theory of the "neptunists" (e.g., Celsius, 1743), (B) crustal rise of the "plutonists" (e.g., Runeberg 1765; Lyell 1830), (C) the rotational theory of Frisi (1785) and (D) the glacial eustatic theory of Maclaren (1841). Modified from Mörner, N.-A., 1987a. Models of global sea-level changes. In: Tooley, M.J., Shennan, I. (Eds.), Sea Level Changes. Blackwell, p. 333-355; Mörner, N.-A., 1987b. Pre-Quaternary long-term changes in sea level. In: Devoy, R.J.N. (Ed.), Sea Surface Studies: A Global View, Croom Helm, Beckenham, p. 233-241 (Chapter 8a).

In Sweden, the continual withdrawal of the sea was easy to observe, even over the time of one generation (as reviewed in Mörner, 1979); generally believed to be a "water diminution" after the Great Flood (Fig. 2A; e.g., Celsius, 1743; von Linné, 1745). This would come back a century later in what we today call tectono-eustasy (Darwin 1842; Dana 1849; Suess 1888). In opposition to this, Runeberg (1765) claimed that it was the Swedish land level that rose (Fig. 2B); a proposition later to become adopted in the "elevation theory" by Lyell (1830) and the concept of postglacial land uplift, or glacial isostasy (e.g., De Geer, 1888).

2.3 Observations in the 19th Century

The 19th century is characterized by improved observations and measurements. A number of new concepts were formulated, all fundamental in sea level research (as reviewed in Mörner, 1979).

2.3.1 The elevation theory of the 19th century

The elevation theory evolved out of the idea by the "plutonists," claiming that it must be the land level that changes up or down with time (e.g., Playfair, 1802; von Buch, 1810; Lyell, 1835; Berzelius, 1843). Lyell had heard about the rise of land in Sweden. In 1834, he went there to see by himself, and indeed he observed and documented clear evidence of an ongoing uplift; that is, he got the final evidence for his "elevation theory" (Lyell, 1835). Consequently, we now had observational evidence for the Fig. 2B-model (i.e., crustal uplift) of the observed sea level regression in Sweden.

2.3.2 Stratigraphy: Multiple marine events

Lyell is known as the father of stratigraphy, and by that, geology (Lyell, 1830). The Great Flood of Noah (the deluge) rapidly came in as something actually having occurred. It even became a marker in stratigraphy, dividing up geology in what was before (antediluvian) and after (postdiluvian) the layer of the deluge. This generated a lot of misconceptions, both in geology and paleontology. Cuvier, a famous French zoologist and paleontologist, observed the stratigraphy of the Paris Basin, and noted a number of marine beds alternating with continental land periods. He also noted that the fossils became more and more advanced per new marine bed (Cuvier, 1825). He interpreted this as multiple floods ("revolutions"), where God at each event became a little more clever and created better and better animals up to the last one, when Man was created in his own image. With such stratigraphic records, the evolution theory began to evolve and was finally firmly established by Darwin (1859). As a matter of fact, it is easy to identify a logical evolutionary chain of understanding from Cuvier (1825) and Lyell (1830) to Darwin (1859). It is also interesting to note that we already, in the early 18th century, had the two main principles in stratigraphy established; that is, catastrophism by Cuvier and uniformitarianism by Lyell; principles that are vividly discussed even today. It must be noted, however, that the Danish geologist and philosopher Nicolò Steno (1638–1686) had already clearly explained the principles of stratigraphy and postdepositional deformations in the 17th century (Hansen, 2009a, b, Fig. 6).

2.3.3 The concept of Tectono-eustasy

Darwin's theory of the formation of coral reefs in the Pacific implied a gradual sinking of the ocean floor (Darwin, 1842). This subsidence was confirmed by the Dana's expedition to the Pacific (Dana, 1849). Chambers (1848) suggested that the widening of the ocean basin in the Pacific was the reason for the water diminution in Sweden; that is, an origin of sea level changes that we today term "tectono-eustasy" (or ocean basin volume changes, as further discussed in Mörner, 1987a). Consequently, we now saw observational evidence of the Fig. 2A-model (i.e., ocean basin volume changes or tectono-eustasy) of sea level changes.

2.3.4 The concept of ice ages and glacial eustasy

The notion that glaciers have had a much larger extension in the past began with the findings of erratic blocks in northern Germany (Bernhardi, 1832; Torell, 1872) and in the Swiss Alps (Charpentier, 1835), later transformed to the Ice Age theory (Agassiz, 1840; Hitchcock, 1841). Rapidly it was understood that this implied a novel cause of sea level changes as a function of storing water masses in ice volumes during Ice Ages, and returning the water to the sea at the postglacial melting. The concept of glacial eustasy was thus created (Maclaren, 1841; Tylor, 1868; Whittlesey, 1868). Consequently, we now saw observational evidence and a firm theoretical basis for changes in the ocean water volume according to the Fig. 2D-model of sea level changes.

2.3.5 The concept of geoid changes

In 1846, von Bruckhausen sent a manuscript to von Humbolt proposing that, because the mountain masses cause a considerable deviation between the plumb line and the geocentric line, this should lead to a rise of sea level along the continents due to the mass attraction of water. Fischer (1868) calculated that this "continental wave" would give rise to sea level differences in the order of 560–640 m (or 8 m per second of plumb-line deviation). In fact, this was the beginning of understanding the concept today named "geoidal eustasy" (Mörner, 1976). Listing (1873) introduced the word "geoid" to denote this gravity-controlled ocean surface. Penck (1882) realized that the ice masses of the Ice Ages must attract the ocean water and raise the ocean level in the vicinity of the ice caps. Von Drygalski (1887) and Woodward (1888) made detailed mathematical calculations of the effects of the glacial masses upon the geoid. Hergsell (1887) showed that Penck's (1882) figures of the geoid wave along the ice margin were greatly exaggerated.

2.3.6 The concept of glacial isostasy

De Geer (1888) showed that the geoidal attraction was by no means enough to explain the amount and duration of the relative sea level changes in Fennoscandia. De Geer concluded that the postglacial shore level displacement observed and documented in Sweden "primarily was caused by subsidence and uplift of the Earth's crust." This implies the introduction of the concept of glacial isostasy. Jamieson (1865, 1882) understood that the Earth is not rigid, and that the load of an ice cap had to deform the bedrock beneath, causing downwarping and uplift in response to the glacial advance and recession. He saw the Scottish and Fennoscandian uplift as evidence of this effect; that is, "glacial isostasy." Gilbert (1882, 1890) showed that the principle of isostasy also applied to the history of Lake Bonneville in the United States. The full evidence and description of glacial isostasy was given by De Geer (1888), however, and with this paper a new epoch began in the study of Fennoscandian uplift (Mörner, 1979).

2.3.7 Evolution of ideas

By the end of the 18th century, we had good holdings of changes in sea level, whether caused by changes in ocean level (neptunists) or land level (plutonists). By the end of the 19th century, most of the factors controlling sea level changes were known; that is: glacial eustasy, ocean basin subsidence (tectono-eustasy), geoid deformations, and glacial isostasy. In stratigraphy, both catastrophism and uniformitarianism have been addressed. The first half of the 20th century saw significant improvements and documentations of sea level changes. Daly (1910) established the approximate amplitude of the sea level lowering at Ice Ages at about 100m. Lidén (1938) presented an annually varve-dated sea level curve from the center of uplift in Sweden, a curve that for a long time was our best way of getting information on the rheological parameters of the upper Earth (e.g., Gutenberg, 1941).

2.3.8 Late 20th century achievements

The post-Second World War period meant enormous progress because of completely new and revolutionary technology and dating facilities. The enormous oceanic records suddenly became accessible.

3 Handling Sea Level Research

Our understanding of sea level changes must always be firmly anchored in observational facts. This is why we experience a quite interesting and sometimes impressive development of ideas through time as more and more observational facts become available, and our own ability to correctly to interpret what we record improves. It must be realized, however, that a sea level event in the past only leaves fragmentary records in the field. Those fragments we must record (document and date). Then we must be able to interpret the records correctly. Here enters the need of professional skill, but also imagination. It is thanks to this imagination that we have been able to advance our understanding so that we today can talk about the development of ideas. This is illustrated in Fig. 3. How can we expect complete records, when we, in fact, only are dealing with fragments (arrow 1 in Fig. 3)? How can we trust the interpretations given from the fragments retrieved (arrow 2 in Fig. 3)?

Far too often, there is a major drawback linked to the interpretational phase in simple human psychology. Too often we note interpretations not governed by the observations at hand, but by the general ideas of the time, or the inspiration of a strong guiding personality. If so, we have become slaves under a paradigm (Mörner, 2006). Sometimes, it leads to group behavior far from normal scientific deduction and curiosity. One example of this group behavior is the

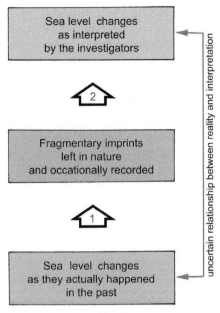

Fig. 3
Relationship between a sea level event in nature and the interpretation by the investigator. Arrow 1 refers to the highly selective processes of fragmentary imprints being left in nature and the chances of later finding them. Arrow 2 refers to the skill and psychology in the phase of interpretation. This implies that the interpretation might be very different from the event actually happening in the past.

general recognition of the instantaneous regression the Peter Vail (Vail et al., 1977) eustatic curve of the last 200 Ma. Another is the extensive application of the Fairbridge (1961) eustatic curve within global geomorphology. A third is the present idea of an anthropogenic, CO_2-driven, global warming with extensive coastal flooding.

4 Quaternary Sea Level Changes

Thanks to the piston corer (Kullenberg, 1947), the sedimentary records of the ocean floors were opened (e.g., Ericson and Wollin, 1964). The role of absolute dating thanks to the radiocarbon dating method (Arnold and Libby, 1949) cannot be underestimated. Suddenly, sea level changes could be assigned proper ages and compared among sites all over the globe (Fairbridge, 1961; Mörner, 1969). The oxygen isotope stratigraphy (e.g., Emiliani, 1955; Shackleton, 1969) provided glacial/interglacial records throughout the entire Quaternary Period. Uplifted coral reef sequences gave records and ages of former interglacial high stands (e.g., Bloom, 1980). Submarine coral reef beds were drilled and dated, providing new insight into the mode of sea level rise after the last glaciation maximum (e.g., Fairbanks, 1989).

4.1 The Postglacial Period and Definition of Eustasy

The word *eustasy* was used to denote the changes in the oceanic level in opposite to various factors affecting the land level. Often "eustasy" was made a synonym to ocean water volume changes. Its definition was clear: *"worldwide simultaneous changes in global sea level."* This implied that if we could just reconstruct the eustatic changes in sea level at one place, this record would have global significance. Indeed, this was exactly what Fairbridge (1961) attempted. As a basic standard curve, he used a relative sea level record from Sweden (Florin, 1944), subtracted a plausible istostatic factor, and arrived at an inferred global eustatic curve. By comparing and modulating this curve against available global sea level data, he proposed a master curve: *the Fairbridge eustatic curve*. It is a high-amplitude oscillation curve reaching maxima above the present level in the Mid- to Late Holocene era (Fig. 4). Shepard (1963) objected to this, and claimed that the sea level had risen continually and smoothly according to *the Shepard eustatic curve* (Fig. 4).

Several methods were applied to resolve this and the details are briefed herein. In the border zone between glacial isostatic uplift and forebulge subsidence—the Kattegatt Sea and the Swedish West coast—40 synchronous shorelines were identified, dated by >100 C^{14}-dates, and documented over a distance of 200–250 km in the direction of isostatic tilting (Mörner, 1969, 1971a). The isostatic and eustatic factors were identified and separated. The eustatic component is a low-amplitude oscillations curve: *the Mörner eustatic curve* (Fig. 4). This curve was tested against other sea level records (Mörner, 1969), against other northwest European records (Shennan, 1987), and against south Baltic data (Harff et al., 2001). A separate test was made for the past 300 years between absolute isostasy, tide-gauge records, and repeated leveling of benchmarks in solid rock (Mörner, 1973). It identified a eustatic rise of 11 cm, or 1.1 mm/year

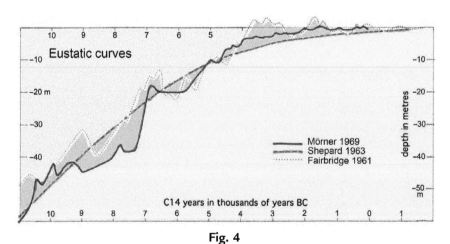

Fig. 4
The eustatic curves of Fairbridge (1961) Shepard (1963) and Mörner (1969). Purple field gives differences between the three curves.

from 1850 to 1950, later found to be in excellent agreement with the deceleration in Earth's rotation over the same period (Mörner, 2013a, Fig. 10). The study was updated later to find similar results (Mörner, 2014a). These three proposed eustatic curves are compared in Fig. 4. Despite differences, there are also similarities. One peculiarity is that all curves lie close together at around 7000–8500 C14-yrs BP, as illustrated in Fig. 5, which implies that the sea level must have been unequally distributed in the period in between (Mörner, 1971b).

With the publication of the satellite images of the irregular distribution of the geoid surface over the globe (e.g., Gaposchkin, 1974; Marsh and Vincent, 1974), it became obvious that the geoid surface must have changed significantly over time, and the concept of geoidal eustasy was coined (Mörner, 1976).

The *geoid* is an equipotential surface matching the mean density of water. Consequently, the geoid approximates the mean sea level. The geoid surface is highly irregular, with a maximum relief of 180 m between the high in New Guinea and the low in the Maldives. It is this irregular surface that changes – vertically as well as horizontally – with time (Mörner, 1976). The term *eustasy* was meant to denote changes in the ocean level (in opposite to changes in land level). Therefore, the old definition (worldwide simultaneous changes in global sea level) had to be redefined (Mörner, 1986). The new definition reads: *"ocean level changes or any absolute sea level changes regardless of causation and including both the vertical and the horizontal changes of the geoid surface, as well as changes of the dynamic sea surface topography."* The old and new concepts of eustasy are illustrated in Fig. 6 (Mörner, 1983a, 1986, 1987a, 2013a). A new era in sea level research commenced with this paradigm shift. We can no longer talk

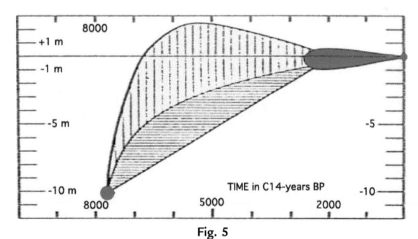

Fig. 5
Range of sea level positions of a large number of sea level curves from all around the globe. Instead of diverging back in time due to crustal differentiation, they all converge at around 8000 C^{14}-yrs BP. This was "the real sea level problem" (Mörner 1971b), later to become understood in terms of geoid changes (Mörner 1976).

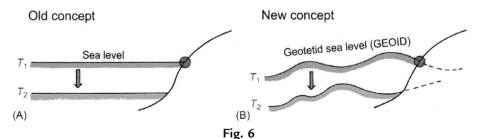

Fig. 6
The old (A) and new (B) concept of eustasy (Mörner, 1983a, 1987a). The old concept assumed total parallelism, and hence, global synchronism. The new concept realizes that sea levels are never fully parallel, and that gravitational and distributional changes of water masses must be globally compensated, and are irregular and not parallel to each other.

about a global eustatic curve, only regional eustatic curves. The first claim of a regional eustatic curve was the one for the Northwest European region (Mörner, 1980a).

There was one more factor to decode, however. In view of the new concept of changing geoid topography with time, one would expect to find a rather complex sea level picture over the globe. Indeed, this was one major novel fact. A closer examination of global sea level data revealed an additional factor, however. There occurred both positive and negative correlations that had to be explained in new terms. These changes are related to decadal and multidecadal changes in ocean circulation (Mörner, 1984, 1985, 1988, 1995, 1996b, 2010a, 2013b). These changes seem to be driven by an interchange of angular momentum between the hydrosphere and the solid Earth (*op.cit.*). Partly we see a centennial east–west flushing back and forth of equatorial water masses balancing a sea level high in East Africa with a sea level low in Peru, and a sea level low in East Africa with a sea level high in Peru (Mörner, 1992a, 1993a, 2000), similar to longer-term flushes back and forth between West Africa and Brazil (Mörner, 1993b, 1995, 1996c). Partly there is a SE–NW pulse in the Gulf Stream beat (Stommel, 1976; Mörner, 1984, 1995, 1996c, 2010a, 2015b) that seems synchronized with a similar beat of the Kuroshio Current in the North Pacific (Mörner, 1996c, 2015c). The Gulf Stream pulse exhibits a main cyclicity of ~60 years (Mörner, 1996b, 2010a, 2013b, 2015b).

Rotational eustasy with interchange of water masses between high latitudes and the equatorial region has recently been documented for the past 500 years (Mörner, 2017a, b; Mörner and Matlack-Klein, 2017). The amplitude is in the order of 70–30 cm. The pulses follow the accelerations at the Grand Solar Minima, and the decelerations at the Grand Solar Maxima (Mörner, 2010a, 2013a, 2015b, 2017b). The 60-year cycle in sea level and climate is driven by planetary influence on the Sun and the Earth-Moon system (Mörner, 2015b). The Grand Solar Cycles (GSC) are now shown to generate an oscillation in the oceanic system: the rotational eustatic GSC Oscillation (Mörner, 2015b, 2017b; Mörner and Matlack-Klein, 2017). This implies that the 60-year and GSC sea level oscillation are both ultimately driven by planetary beat on the Sun and Earth-Moon system (Mörner, 2015b).

4.2 The Glacial/Interglacial Cycles

Daly (1910) estimated the glacial eustatic lowering during the Ice Age at about 100 m, which has turned out to be quite a good estimate. One of the first and best came from Bassa Versilia in Italy, where Blanc (1937) obtained stratigraphic evidence of a regression during the Last Ice Age down to about −90 m. For a longer time, this was one of the best and clearest examples of the maximum glacial eustatic lowering (e.g., Mörner, 1971a). Fig. 7 illustrates the three main ways of obtaining a value of the glacial eustatic changes in oceanic water volume (from Mörner, 1983b); namely (1) sea level changes, (2) oxygen isotope changes in marine species, and (3) volumetric estimates of the ice extent. Each of these techniques has their sources of uncertainty, however.

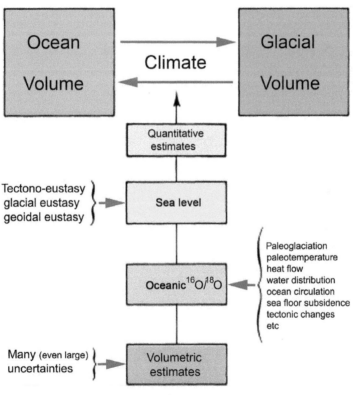

Fig. 7
The water budget of the oceans and cryosphere showing the major linkages (each of which includes multiple variables), demonstrating that neither eustatic sea levels nor oceanic oxygen-isotope records are a direct measure of changes in glacial volume *From Mörner, N.-A., 1983b. Illusions in water-budget synthesis. In: Street-Perrott, A., Beran, M., Ratcliffe, R., (Eds), Variations in Global Water Budget, Reidel Publ. Co., p. 419–423.*

Shepard (1963) found a sea level lowering of 132 m, and Milliman and Emery (1968) a lowering of 130 m at the Last Glaciation Maximum. The first high-quality record was the continually dated Barbados coral reef record by Fairbanks (1989), where the glacial lowstand was established at −120 m. Later, there were a number of additional continual coral reef records (e.g., Lambeck et al., 2014, Fig. 1) that established the regression maximum at about minus 110–130 m. The deep-sea cores provided long-term records of the temporal changes in ^{18}O, which was thought to be a good proxy of global paleoglaciation (e.g., Shackleton and Opdyke, 1973). As suggested in Fig. 7, there are several complications, however (cf. Mörner, 1983b). It was proposed that ∼2.0‰ ^{18}O corresponded to ∼120 m glacial eustatic sea level change. Therefore, it became customary to convert ^{18}O records into sea level curves (Chappell and Shackleton, 1986; Waelbroeck et al., 2002; Rohling et al. 2014). Personally, I have never been convinced that this relation is trustworthy and time consistent (Mörner, 1981a, 1983a,b, 1994). A third way of estimating the sea level lowering is to estimate the amount of water stored in the ice caps. This was done by Flint (1971). According to him, the Last Glaciation Maximum (LGM) at about 20 ka BP had built up ice caps, which, all taken together, corresponded to an ocean water volume of 120–132 m. All the three main methods of estimating the sea level lowering at the LGM at about 20,000 C^{14}-yrs BP are linked to problems (Fig. 7). At the same time, there is a surprisingly good agreement between the three methods; a maximum sea level lowering of 120–130 m at the LGM.

Flint (1971) also estimated the water masses still stored in Antarctica at ∼60 m, in Greenland at ∼6 m, and in all other continental glaciers at 0.5 m (cf. Mörner, 1996a, Table 4). What was remarkable with the ^{18}O records was that we now obtained long continual records of the changes in climate as illustrated in, for example, Fig. 8 (Shackleton and Opdyke, 1976). This implied a major step forward in Quaternary science (e.g., Berger et al., 1984). It was later followed up by similar long-term records from ice cores (Fig. 9; EPICA Community Members, 2004).

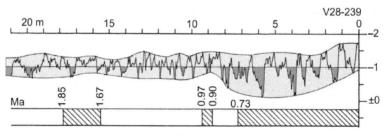

Fig. 8

Oxygen isotope record and palaeomagnetic chronology of the equatorial Pacific core 228–289. Both the amplitude and the frequency vary considerably through time (thin lines give amplitude range changes). The Ice Age/Interglacial cyclicity in the past 0.9 Ma is well recorded (cold down, warm up). *Modified from Shackleton, N.J., Opdyke, D., 1976. Oxygen isotope and paleomagnetic stratigraphy of Equatorial Pacific core 228–239, Late Pliocene to Latest Pleistocene. Geol. Soc. Am. Memoir. 145, 449–464.*

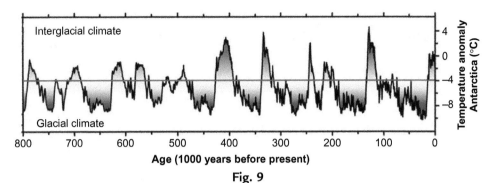

Fig. 9
Temperature record from the Vostok core in Antarctica spanning 800,000 years *Modified from EPICA Community Members, 2004. Eight glacial cycles from an Antarctic ice core. Nature 429, 623–628.*

4.3 The Problem With Global Glacial Isostasy

Walcott (1972, cf. 1980) called attention to the fact that the postglacial glacial isostatic compensation after the vanishing of the huge ice caps in Fennoscandia and North America (and elsewhere, too) would affect the crustal compensation regionally, if it took place in via a low-viscosity channel flow, while it would generate global compensational crustal motions if the viscosity had a linear profile (i.e., no channel flow). The two concepts are illustrated in Fig. 10 (from Mörner, 2005).

Model A (Fig. 10A), that is, the linear viscosity model (e.g., O'Connell, 1971; Cathles, 1975; Peltier, 1976), would imply that the glacial isostatic compensation had a direct global compensation (as first proposed by Bloom in 1967) and hence would lack a peripheral bulge surrounding the glacially depressed region. Clark calculated that the globe could be divided into six zones of different relative sea level changes (Clark, 1980, his Fig. 2). This theory was carried further by Peltier (1998) and Lambeck (1998). It has even become a general correction factor for present-day sea level records (e.g., Peltier, 2004).

Model B (Fig. 10B) implies that the glacial loading and deloading is compensated by lateral flow in a low viscosity channel, and that the glacially downloaded area was surrounded by a compensational forebulge. This is the classical theory of glacial isostasy in Fennoscandian (e.g., Nansen, 1928; van Bemmelen and Berlage, 1935; Mörner, 1969, 1979, 1980b 1990; Fjeldskaar and Cathles, 1991). All data available from Fennoscandia are in favor of a low-viscosity channel flow (Fig. 10B). The mass in the cone of absolute postglacial uplift is the same as the mass in the surrounding peripheral subsidence trough, indicating a horizontal mass flow from the collapsing forebulge (subsidence trough) to the rising uplift cone as quantified in 500-year steps (Mörner, 1979, 1980b, 1991, 2015c).

The determining factor in the discrimination between the two models (Fig. 10) is the shape of local and global sea level curves. Neither the near-field nor the far-field sea level records seem

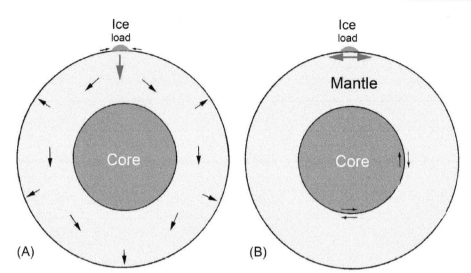

Fig. 10
Global versus regional loading. (A) In the global loading models, the glacial loading/unloading will be transferred through the mantle and affect the coasts and sea floors all around the globe. This requires a linear viscosity in the upper mantle. (B) In the regional loading model, the glacial loading/unloading is fully compensated in the region of glacial isostatic deformation via lateral mass flow in a low-viscosity asthenosphere channel in the upper mantle. Observational data from Fennoscandia and surrounding areas are consistent only with a regional loading model where compensation took place via a low-viscosity channel flow (Mörner, 1979, 1980b, 2015c; Fjeldskaar and Cathles 1991).

to support model A of Fig. 10, however (Mörner, 2005, 2015c). This conclusion is strongly supported by Houston and Dean (2012), who compared site specific values of global adjustment as predicted by the Peltier model (Peltier, 2004) and actual tide gauge records at 147 far-field sites, and found "remarkably little correlation."

This is fundamental for present day sea level analyses (Fig. 11). Available tide gauges spread all over the globe give a strong peak of mean relative sea level rise of +1.6 mm/yr (Mörner, 2014a, b) or, better, 1.14 mm/yr (Mörner, 2015c, 2017c), corresponding to an absolute sea level rise varying between ±0.0 and +1.0 mm/yr. The regional eustatic value (see Mörner, 2014b, 2016b) is set at +1.0 ± 0.1 mm/yr for the NE Atlantic, the Kattegatt, and the Baltic; at about +1.2 mm/yr for the American East Coast; and at about ±0.0 mm/yr for the Indian Ocean (Mörner, 2017a), the Fiji Islands (Mörner, 2017b), the Mediterranean (Mörner, 2014b), Pacific Islands such as Tuvalu, Vanuatu, and Kiribati (Mörner, 2014b, 2016a), and the Guyana-Surinam area (Mörner, 2010b). In most of these areas, we have firmly established long-term rates of the crustal components involved. Therefore, the values cited can be used as standards in the evaluation of regional eustasy. In addition, Parker and Ollier (2015) gave a mean value of +0.25 ± 0.19 mm/yr for 170 global tide gage stations with records of >60 years, and the revised satellite altimetry records of +0.55 ± 0.10 mm/yr according to Mörner (2015c).

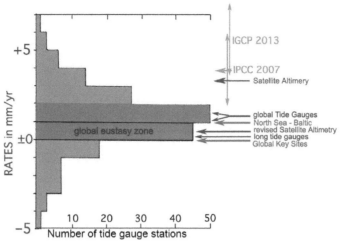

Fig. 11
Spectrum of sea level rate estimates (Mörner, 2013a, 2014b, 2015c, 2017c). The big differences indicate errors and mistakes. Satellite altimetry includes a subjective correction factor, which was removed by Mörner (2015c, 2017c). Removing this correction component, all records become compatible within a "global eustasy zone" of ±0.0 to +1.0 mm/yr.

Satellite altimetry gives a mean sea level rise of +3.2 mm/yr (NOAA 2014). This value is incompatible with available observational data (Mörner, 2013a, 2014b, 2017c); being 100% higher than the mean relative sea level rise from tide gauge data, and 2.0–2.3 mm/yr higher than European benchmark records (Fig. 11; Mörner, 2014b, 2015c, 2016b). The reason for this mismatch seems obvious; the satellite altimetry records include global isostatic adjustment components from the Peltier (2004) model of 2.3 to 2.0 mm/yr. Removing this factor (Mörner, 2011a, 2013a, 2015c, 2017c), the revised satellite altimetry value of +0.55 ± 0.10 mm/yr becomes compatible with the data on regional eustasy, and the mean rate of tide gauges, as given in Fig. 11. This is a very strong (maybe even conclusive) indication that the global loading model (Fig. 10A) is wrong. This view is reinforced by new analyses indicating that, indeed, there is a low-viscosity layer in the upper mantle/lithosphere region (Karato, 2012; Naif et al., 2013), in full agreement with model B of Fig. 10.

Therefore, it seems high time for a general revision of the global applicability of the Fig. 10A model of isostasy (as proposed in Mörner, 2015c).

4.4 Summary of Quaternary Sea Level Variables

Fig. 6 illustrates the old and new concept of eustasy (Mörner, 1983a, 1987a). In his paper proposing a redefinition of the word of "eustasy," Mörner (1986) recognized four main variables; namely, ocean basin volume, ocean water volume, ocean mass/level distribution, and dynamic sea level changes, as illustrated in Fig. 12.

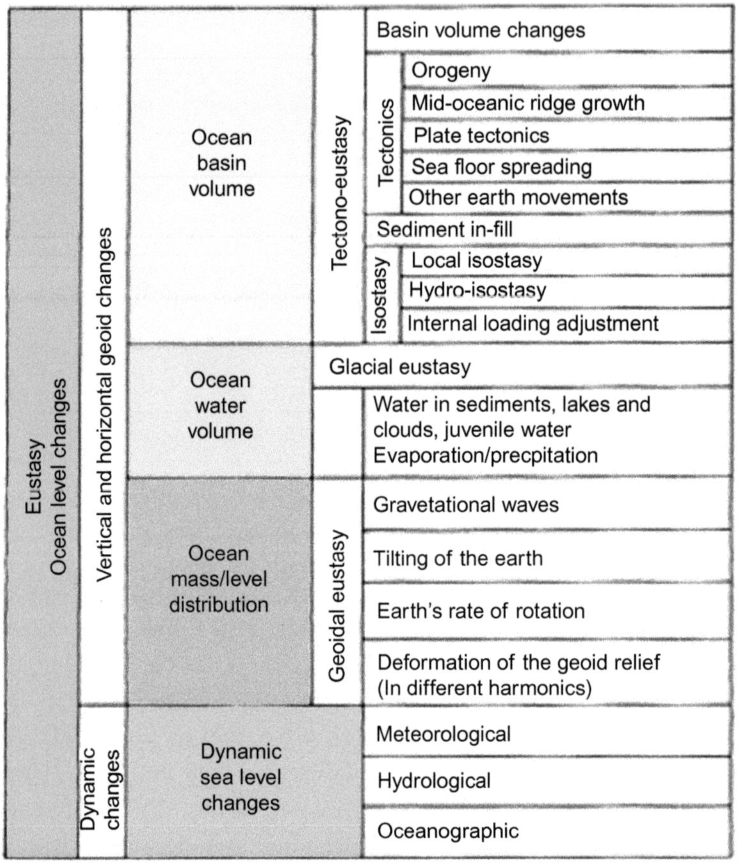

Fig. 12
Eustatic variables as recognized by Mörner in his redefinition of the concept of eustasy (Mörner, 1986).

Fig. 13 gives dynamic interaction (from Mörner, 1981b, 1983a); variables controlling the present sea level, variables that changes the sea level and land level, respectively, with time back into the past, as well as forward into the future.

Fig. 14 provides a simplified integrated scheme of the variables controlling the oceanic sea level. Fig. 15 provides an alternative illustration (from Mörner, 1980d).

This implies a quite complete understanding of sea level changes established already in the 1980s.

5 Pre-Quaternary Sea Level Changes—An Alternative View

In 1977, Vail et al. published a sea level graph of the past 200 million years, which came to be known as "*the Exxon Eustatic Curve*" (Vail et al., 1977, 1980). This curve (Fig. 16) provides a record covering 200 million years. It is characterized by 32 instantaneous regression episodes.

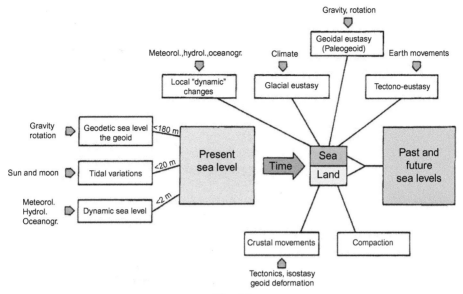

Fig. 13

Factors controlling and affecting sea level changes in the past, at present, and in the future *From Mörner, N.-A., 1981b. Space geodesy, paleogeodesy and paleogeophysics. Ann. Géophys. 37 (1), 69–76.*

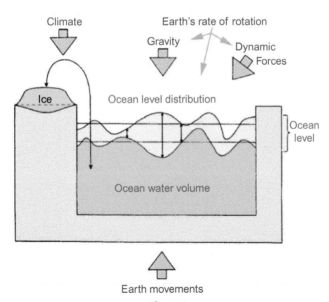

Fig. 14

Variables controlling the oceanic sea level position; that is, the eustatic factors.

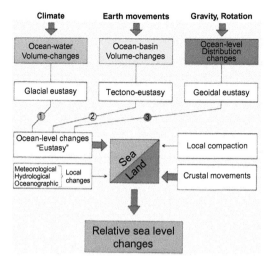

Fig. 15

Factors controlling sea level and land level, and their combined product of relative sea level changes as observed in nature. *From Mörner, N.-A., 1980d. Eustasy and geoid changes as a function of core/mantle changes. In: Mörner, N.-A. (Ed.), Earth Rheology, Isostasy and Eustasy, John Wiley & Sons, p. 535–553.*

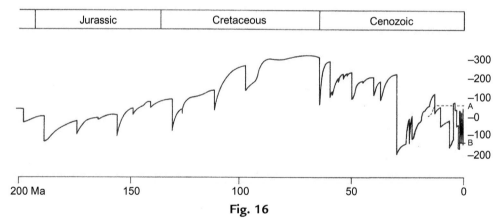

Fig. 16

"The Exxon Eustatic Curve" by Vail et al. (1977) composed of 32 sudden regressions in the past 200 Ma. This curve refers to the interpretation of sequence stratigraphy. It was a general mistake to claim that it was a eustatic curve; neither shape, amplitude, nor synchronicity can be of global validity. *Redrawn in Mörner, N.-A., 1987b. Pre-Quaternary long-term changes in sea level. In: Devoy, R.J.N. (Ed.), Sea Surface Studies: A Global View, Croom Helm, Beckenham, p. 233-241 (Chapter 8a).*

The amplitudes of those regressions are very large: 1 of 360 m, 10 of 140–160 m, 4 of 100–130 m, 7 of 70–90 m, 3 of 40–60 m, and 7 of 10–30 m.

We are facing two problems: one being the rapid rate (almost instantaneous drops in sea level) and the other being the high amplitudes (where 22 events are of the sizes of normal Quaternary

Ice Ages or even larger). There was simply no way of explaining the regressions in the Exxon Eustatic Curve. Changes in the ocean basin volume, that is, tectono-eustasy, were far too slow a process. Changes in the water volume, that is, glacial eustasy, were hard to advocate with the generally warm climatic conditions of the Mesozoic Period. Changes in the distribution of the water masses, that is, geoidal eustasy, could not be advocated, partly because of the rates and partly because of the proposed global synchronicity. There seemed only one way out; namely, to dismiss the curve as incorrect with respect to shape, amplitude, and synchronicity (Mörner, 1980c, 1981c, 1983a, 1992b, 1994). The curve emanated from new seismic stratigraphy and deep-sea drilling records. So the material was both excellent and novel. Still, there was something basically wrong in the interpretations (cf. Fig. 16).

5.1 The Paris 1980 Illusory Consensus

At the 26th International Geological Congress in Paris, 1980, there was a special symposium on "Geology of Continental Margins" devoted to seismic stratigraphy and long-term sea level changes. Speaker after speaker appeared and testified to the glory of the Exxon Eustatic Curve (Vail et al., 1977, 1980). Only those unconformities and regressions were found in place after place around the world. No exceptions, only confirmation. The picture appeared solid. It was claimed that they had a time resolution of 0.5 Ma, excluding any stratigraphic mistakes. There seemed to be a general and solid consensus (cf. above "Handling sea level research").

This is strange, and from the point of characteristics of different sea level variables (Table 1), even impossible. According to Kerr (1980), the offshore people now had "a new sea level index" and general consensus. For me it was all a misunderstanding (Mörner, 1981a, 1982, 1983a, b, 1992b).

5.2 The New Synthesis

The Exxon Eustatic Curve of 32 sudden regression events became the global standard; at least for several years. But then came the great disclosure by Haq et al. (1987): The number of events and the shape of the sea level changes were totally revised (Fig. 17)!

Instead of 32 sharp regressional events in the past 200 years, we now got 110 sinusoidal events in the past 200 Ma. This was fine, suddenly the record became much more geologically sound. It takes time for the sea level to fall as well as to rise again; that is, the true change must form a wave (regular sinusoidal or irregular does not matter; but certainly not "instantaneous" as was the case in the Vail et al. curve of 1977). The new curve met that criterion. The number of events was another total change; now 110 events, previously 32. Now, *a posterior*, one may, indeed, wonder how all the people at the Paris 1980 meeting could so firmly claim that they, in region after region, had verified the global applicability of the Exxon Eustatic Curve. As a matter of fact, it was a shameful mass psychosis, as further discussed in section "Handling Sea Level

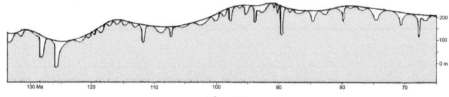

Fig. 17

The "eustatic" sea level curve of the Cretaceous as established by Haq et al. (1987). The shape, rates, and amplitudes are now totally different from the old "Exxon eustatic curve" (Fig. 16). But the global synchronicity remains a prime problem; there is no such thing as a global eustatic curve, only regional curves that may differ significantly (Mörner 1976, 1986).

Research" (Fig. 3). The new curve (Haq et al., 1987) has 58 events in the Cretaceous Period, where the old curve (Vail et al., 1977) had only four events!

There is one psychological factor left in the new curve, however. The transgressional peaks are lined up into "long-term curve," implying that the short term events represent regressional events or anomalies departing from this long term standard (this is the case in the curve of Haq et al., 1987, as well as in the recent curves of Haq, 2014, and Cloetingh and Haq, 2015). An alternative would be to line up the base levels into a long-term trend from which the short-term sea level changes sticks out as transgression spikes. The difference is of great significance for the interpretation in terms of forcing functions.

The Late Phanerozoic sea level curve of Haq et al. (1987) implies that the authors subscribed to the old definition of eustasy (Fig. 5); that is, "simultaneous changes in global sea level" (the same applies for the Paleozoic sea level curve of Haq and Schutter, 2008, and the "Cretaceous eustasy revisited" curve of Haq, 2014). As this definition no longer applies in Quaternary geology (Mörner, 1986), that is, in the period of time where we have the best documentation and understanding of the complex issue of sea level changes, this should also be the case in the pre-Quaternary Period, including the Cretaceous. It seems the "eustatic" curves of Haq (Haq et al., 1987; Haq and Schutter, 2008; Haq, 2014; Cloetingh and Haq, 2015) represent some sort of mean of local and regional "relative sea level changes" (cf. Fig. 15; cf. Mörner, 1980c, 1981c, 1982, 1983a).

5.3 Lessons From Field Studies in the 1970s

The sea level changes during the Cretaceous period have been the target of much research, and in the 1970s a special IGCP project (No. 58) was devoted to Mid-Cretaceous Events. Meetings were held in Hokkaido, Japan, in 1976, and in London, United Kingdom, in 1979. This provided an excellent opportunity to obtain sea level data from a large number of sites scattered all over the globe; in a form sprung directly out of observational facts. A priori, it was to be expected that the sea level changes were dominated by tectono-eustasy, geoidal eustasy, and quite active local tectonics (Mörner, 1980c,e, 1981c; Reyment and Mörner, 1977).

5.3.1 The land/sea distribution and geoid relief

The Cretaceous is a period of rapid opening of the Atlantic, and strong activity along the mid-oceanic ridges. By necessity, this also implies strong tectonic activity along the coasts and large-scale deformations of the geoid relief (Mörner, 1980c). According to Lubimova (1980), the asthenosphere may have evolved around 200 Ma ago, and then undergone an oscillatory evolution through time. The implication of this for crustal movements, geoid deformations, and sea level changes has been discussed by Mörner (1980c, Fig. 6). Fig. 18A gives an old version of the global land/sea distribution in the Mid-Cretaceous period (from Matsumoto, 1977).

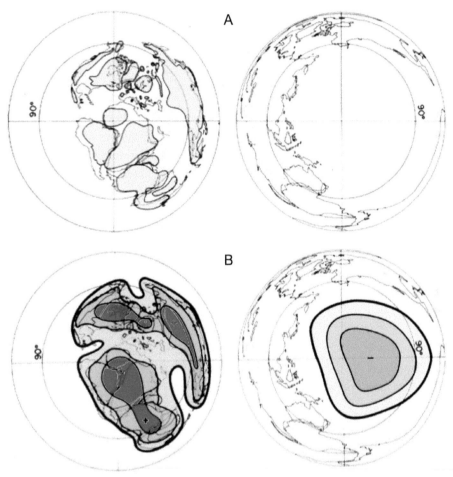

Fig. 18
Schematic illustration of the relation between land/sea distribution in the Mid-Cretaceous and the corresponding geoid topography. (A) the land/sea distribution according to Matsumoto (1977). (B) Corresponding geoid anomalies, a high-relief high over the continental half-sphere and a deep and large depression over the Pacific half-sphere. *Modified from Mörner, N.-A., 1980c. Relative sea-level, tectono-eustasy, geoidal-eustasy and geodynamics during the Cretaceous. Cretaceous Res. 1, 329-340.*

This distribution of mass must have caused a heavy geoid relief and topography, as illustrated in Fig. 18B (from Mörner, 1980c). At the further continental drift and redistribution of mass, this geoid relief must have undergone significant large-scale deformations. Therefore, it is impossible that the observed sea level changes could be expressed in one "eustatic" curve (e.g., Vail et al., 1977, 1980; Haq et al., 1987; Haq, 2014), but must, in fact, be expressed as different (sometimes even opposed) local to regional relative sea level curves (Mörner, 1980c, 1981c, 1983a).

5.3.2 The sea level changes observed

The material synthesized from the Mid-Cretaceous Event project consists of data compiled at the London 1979 meeting (L-79) and at the Japan 1976 meeting as reported by Matsumoto (1977, 1979; i.e., M-77, M-79 in Fig. 19). This extensive, fully observation-based, database was analyzed by Mörner (1980c, 1981c), further discussed as follows (Figs. 19–21).

The sea level graphs of Fig. 19 and the global distribution of transgression (T) and regression (R) event over the globe (Fig. 20) provide a clear and irrefutable indication that the Cretaceous sea level varied significantly both in time and space, which indicates that we are dealing with local to regional changes in relative sea level; that is, not in absolute "eustatic" sea level as claimed by Vail et al. (1977, 1980), Haq et al. (1987), and Haq (2014). On a further test (Mörner, 1981c), plot of the global latitudinal trends per time unit (Fig. 21), a novel phenomenon appeared; the wave-like transgressional/regressional motions up and down the globe. This was a revolutionary finding, because it brutally invalidated all talk about global synchronous sea level oscillations.

The sea level changes recorded in Figs. 20 and 21 provide quite different patterns throughout the Cretaceous, as follows: latitudinal waves of about 15 Ma duration during the Berriasian to Berremian, quite chaotic changes in the Aptian, quite similar change over the globe in the Albian to Cenomanian, latitudinal waves of about 4 Ma duration from the Turronian to the end of the Maastrichtian, broken by more globally consistent changes in the Coniacian and Campanian. The wave-like sea level pattern revealed in Fig. 21 was interpreted in terms of some kind of "gravitational drop motions" (Mörner, 1987b), as illustrated in Fig. 22.

This is what the Mid-Cretaceous Event project taught us; there is no—and cannot be any—global synchronicity in observed sea level changes. Consequently, neither the old Exxon Eustatic curve of Vail et al. (1977) nor the updated version by Haq et al. (1987) can record true global eustatic changes. The problem of understanding the possible mechanisms behind the Vail et al. (1977) global sea level curve has been very much debated (e.g., Pitman III and Golovchenko, 1983; Thorne and Watts, 1984; Schlanger, 1986). Schlanger (1986) realized the problems. He noted 6 cycles of relative sea level changes in the Albian to Maastrichtian, with amplitudes in the order of 150–250 m and durations ~5.5 Ma, implying rates of sea level rises and falls in the order of 50–100 m per Ma; that is, 0.1–0.05 mm/yr. This was in total

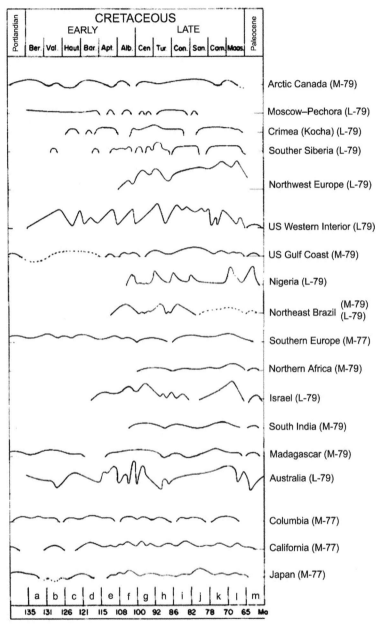

Fig. 19

Spectrum of sea level graphs as established within the Mid-Cretaceous project; M-77 and M-79 refers to Matsumoto (1977, 1979) and L-79 to the London 1979 meeting. The curves record quite different trends, indicating the dominance of local to regional factors.

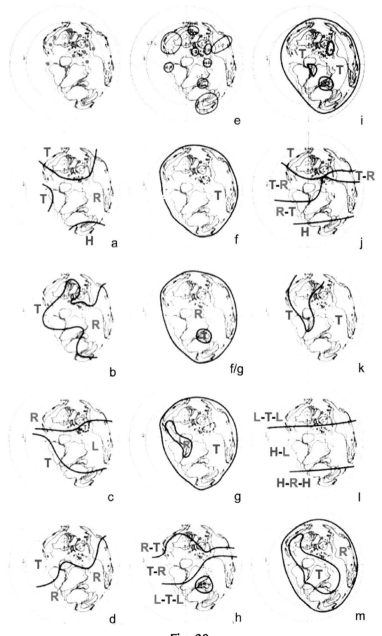

Fig. 20

Location of site (red dots in top left map) referring to the Fig. 19 sea level graphs. Maps (A) to (M) give the global distribution of transgressions (T) and regressions (R) and high (H) and low (L) sea level with respect to the 13 time units given at the base of Fig. 19. The maps indicate quite different trends over the globe that are impossible to be expressed in terms of one global eustatic curve. *From Mörner, N.-A., 1980c. Relative sea-level, tectono-eustasy, geoidal-eustasy and geodynamics during the Cretaceous. Cretaceous Res. 1, 329-340.*

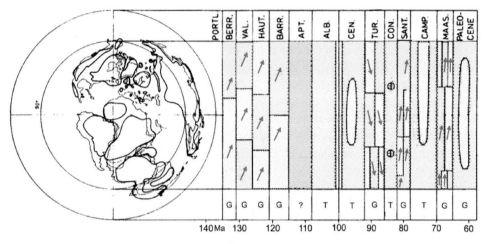

Fig. 21

Transgressions (*blue*) and regressions (*yellow*) events throughout the Cretaceous with respect to latitudinal occurrence per time unit. It documents waves (*red*) migrating up and down the globe; at a rate of ~15 Ma prior to 100 Ma, and about 2–4 Ma after 100 Ma. The dominant eustatic factor for each time unit is given at the base as G for geoidal eustasy and T for tectono-eustasy, with crustal movements always affecting local to regional relative changes in sea level. Glacial eustasy is an additional effect to be considered during periods of predominantly consistent sea level records. *From Mörner, N.-A., 1981c. Revolution in Cretaceous sea-level analysis. Geology 9, 344–346.*

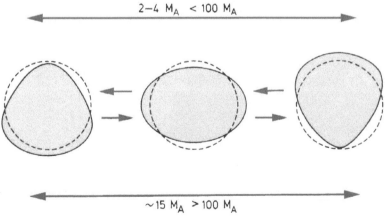

Fig. 22

"Gravitational drop motions" or waves of geoid deformation displaced up and down the globe as identified by Mörner (1981c, 1987b). The rate of wave displacement changed from ~15 Ma to 2–4 Ma in association with the rapid sea-floor spreading and opening of the Atlantic around 100 Ma ago.

disagreement with the sudden regressions of the Exxon Eustatic Curve (Vail et al., 1977). In the revised "eustatic" curve of Haq et al. (1987), there are about 25–28 sea level cycles during the same period of time, implying persisting problems.

5.4 Deep-Sea Hiatus

A stratigraphic gap is known as a hiatus. Stratigraphic hiatuses have also been found in the deep-sea; known as "deep-sea hiatuses." Usually, their origin was interpreted in terms of sea level changes (e.g., Moore et al., 1978; Thide, 1981; Barron and Keller, 1982). There is no physical reason for such an interpretation, however. Therefore, another mechanism was proposed (Mörner, 1988, 1995). The deep-sea hiatus must represent major changes in the depositional dynamics at the sediment/water interface. Differential rotation might have the potential to generate a deep-sea hiatus, at the same time as they generate the interchange of water with the Arctic basin (with another ^{18}O composition), and minor sea level adjustments, as illustrated in Fig. 23 (from Mörner, 1988, 1995).

A lowering of sea level is an ocean surface process, and will have no direct effect at the floor of the deep sea. This can only be generated by changes in the water circulation (Mörner, 1988); that is, friction at the sediment/water interface due to changes in bottom currents and/or differential rotations (Fig. 23).

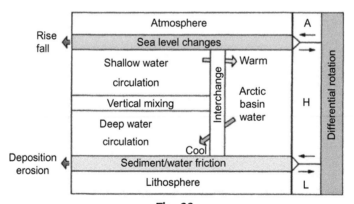

Fig. 23
Differential rotation changing the bottom currents and friction at the interface between water and sediments may offer an explanation to the occurrence of deep-sea hiatuses and their correlation with sea level changes and shelf unconformities, as well as oxygen isotope variations (yellow and blue arrows). *From Mörner, N.-A., 1988. Terrestrial variations within given energy, mass and momentum: paleoclimate, sea level, paleomagnetism, differential rotation and geodynamics. In: Secular Solar and Geomagnetic Variations in the Last 10,000 Years, FR Stephenson & AW Wolfendale, eds, p. 455478, Kluwer Acad. Press.*

5.5 Rates of Sea Level Changes

The rates —especially maximum rates— of eustatic sea level changes and crustal movements may offer a means of discrimination between different causation mechanisms, setting the frames of physical possibilities of proposed changes (as illustrated in the case of glacial eustasy and thermal expansion by Mörner, 2011a, 2016a). Therefore, this question was addressed repeatedly (Table 1, Fig. 1; Mörner, 1983a, 1987a, 1994, 1996a,d, 2013a) and others have, of course, done so too (e.g., Fairbridge, 1961; Rona 1973; Pitman III, 1978; Schlanger 1986; Cloetingh and Haq 2015). The main variables can be summarized as follows (Fig. 24). Glacial eustasy is global, with regional differentiations, and has a maximum rate frame of 10.0 ± 0.1 mm/yr (cf. Mörner, 2011a). Tectono-eustasy is a slow process of global significance, but regional differentiation, and has a maximum rate in the order of 0.06 mm/yr. Geoidal eustasy is regional, and compensational, with maximum rates in the order of 1.4 ± 0.1 mm/yr. Thermal expansion (steric effects) is local to regional, with changes in multidecadal cycles at maximum rates in the order of 1.0 mm/yr (cf. Mörner, 2011a, 2017d).

Fig. 24 (from Mörner, 1996d) gives the amplitude versus time (i.e., rate) distribution of sea level variables. The lines may be used to discriminate between different possible forcing functions, and also evaluate whether a claimed change in sea level is anchored in realistic rates or not; that is, inside or outside the frames set by facts (e.g., Mörner, 2011a).

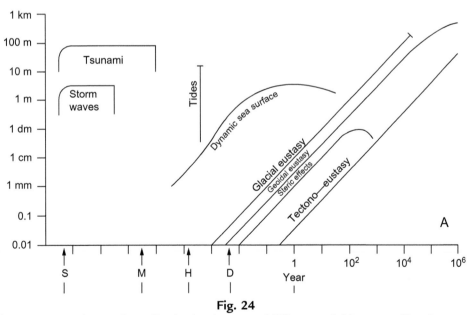

Fig. 24
The relations among time and amplitude; that is, rates of different variables controlling the ocean level (A) and the land level (B), respectively. The lines give maximum values. Scales are logarithmic; D=1 day, H=1 h, M=1 min, S=1 s. From Mörner, N.-A., 1996d. Rapid changes in coastal sea level. J. Coastal Res. 12, 797–800.

5.6 The "State of the Art"

Cloetingh and Haq (2015) have recently published a paper on the probable origin of the sea level changes in the Cretaceous era. Therefore, it is considered the latest "state of the art," and will be inspected with respect to the reconstruction and interpretation of sea level changes.

5.6.1 The sea level curve

Fig. 25 gives a sea level curve claimed to be "an updated eustatic curve for the Cretaceous Period" (Cloetingh and Haq, 2015). This curve by no means represents eustatic changes in sea level. Like the Fairbridge "eustatic" curve of the postglacial changes in sea level (Fairbridge, 1961), this curve represents an old view of sea level changes; that is, "simultaneous and global" instead of the novel concept "irregular and regional changes in sea level" (Mörner, 1986). All talk about one single eustatic curve of global applicability is untenable (Fig. 5) and only multiple regional curves can hold true to geological archives. With respect to the Cretaceous, this is obvious from Figs. 19–21.

The curve (Fig. 25) includes 58 sea level cycles in total (including the uncertain, dashed ones) from the entire Cretaceous period (66–145 Ma). This is a mean of one event every 1.36 Ma. The mean rate of changes (the total amplitude of oscillations divided by the total time) is 0.87 mm/yr.

During the Cretaceous period, the glacial eustatic factor was previously held to be minute to negligible. Steric changes are not applicable (too short in amplitude and time). The mean rate value of 0.87 mm/yr is more than one tenfold larger than the proposed maximum rate of tectono-eustasy, but well below the maximum values of geoidal eustasy (Fig. 24). This leaves us with a first general analysis: (1) tectono-eustasy cannot explain the pattern recorded, and (2)

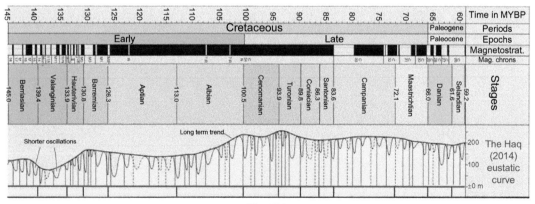

Fig. 25
The "eustatic curve" of Haq. There are 58 sinusoidal sea level cycles within the Cretaceous period. The mean rate of sea level changes over the entire period is 0.87 mm/yr. *Modified from Cloethingh, S., Haq, B.U., 2015. Inherited landscapes and sea level change. Science 347, 393–404.*

geoidal eustasy might, at least theoretically do so, but in that case, the oscillations would not be global. This leaves us with the most likely possibility that the curve does not represent global synchronous events, but a compilation of relative sea level changes, strongly influenced by local to regional geodynamic factors; that is, tectonics, isostasy, and geoid deformations (which the graphs in Figs. 19–21 seem to strongly indicate).

The Fig. 19 graph includes some 25 oscillations of irregular to opposed character; that is, changes that seem to be dominated by local tectonics and geoidal eustasy (Mörner, 1980c, 1981c). Schlanger (1986) saw 6 main cycles where Haq (2014) has 28 cycles. Similarly, Reyment and Mörner (1977) described 5 main sea level cycles, where Haq has 36 cycles. The sea level cycles in Haq's proposed "eustatic" curve (Fig. 25; Haq, 2014) refer to the documentation and interpretation of "sequence boundaries" in the local sequence stratigraphy. In Haq et al. (1987), these boundaries were termed "relative changes of coastal on-lap," and, indeed, this is what they represent: local sedimentary structures referring to local changes in the sedimentary dynamics. In terms of Fig. 3, these sequence boundaries are fragments left from past events, and it is upon the investigator to try to interpret the records as logically and correct as possible in the efforts of reconstructing what actually happened tens of millions of years ago (i.e., the problems illustrated in Fig. 3).

Haq's excellent palaeoceanographic work led to a very fine documentation of the occurrence of the "sequence boundaries" (Haq, 2014, Appendix). It is the conversion of this record into sea level changes that is to be questioned; however, especially the claim that the cycles represent eustatic changes in sea level (Fig. 25). Accordingly, Cloetingh and Haq (2015) admitted that "it has become apparent that a measure of the amplitude of eustatic events cannot be gleaned from any one location," but depends on averaging and interpretations. In line with this, they also state "quantitative assessments are fraught with difficulties."

Available oxygen-isotope records from the Contessa section in Italy (Haq, 2014, Fig. 3; Stoll and Schrag, 2000) seem too preliminary to enable the interpretation of scatter, origin, and shape. Regardless, it seems far too early to allow for an interpretation in terms of high-frequency glacial eustatic fluctuations. Occurrence of continental ice caps in the Antarctic region during parts of the Cretaceous period has also been advocated by Miller (Miller et al., 2003; Miller, 2009). Therefore, it seems reasonable to question the old idea of nonglacial conditions during the Cretaceous period. The number of possible cold events and occurrences of sea level lowering are far fewer (Miller et al., 2003) than the number of "eustatic" cycles in Haq's curve (2014), at a ratio of about 7:28.

It seems significant that Cloetingh and Haq (2015) themselves state: "we remain far from resolving the causes for third-order quasicyclic sea level changes ($\sim$500,000 to 3 million years in duration)." Having made the general examination of the sea level changes in the Haq (2014) eustatic curve, a detailed analysis of the database behind the curve, as provided in the Appendix in Haq (2014), was attempted. In this database, there are 58 sequence boundary events (interpreted as eustatic sea level oscillations) listed. Their occurrence is recorded in 13 main field regions, with the following distribution of events per sites: 51 (88%) in the West European Basin, 28 (48%) in the Arabic Platform, 18 (31%) in the Russian Platform, 12 in the Eastern US, 9 in India, 8 in the Gulf Coast region, 8 in New Zealand, 7 in Western Australia, 7 in the Pacific, 5 in northwest Africa, 4 in Africa, 3 in reef-carbonate sections, and 1 in the US interior. This means an exclusion of a massive amount of data (cf. Figs. 19–21). At the same time, the West European Basin emerges as the major site of recording (88%). The number of sites in which the individual events are observed is as follows: 1 in 6 sites, 8 in 5 sites, 6 in 4 sites, 16 in 3 sites, 15 in 2 sites, 9 in 1 site, and 3 in no site at all. This means that 53% of the events are only recorded in 2–3 sites. This is, of course, far too few sites per event to merit an interpretation in terms of global eustasy!

The classification of eustatic amplitude ranges is as follows: 7 events (KBa5, KAp1, KAp7, KAl8, KCe4, KTu4, KMa5) of >75 m, 33 events of 75–25 m, and 18 of <25 m. The conversion of sequence boundary registration (relative changes of coastal on-lap) into sea level changes is, of course, highly questionable. In fact, they rather seem to refer to local changes in sedimentary dynamics, which can be generated by many different factors; not just sea level changes alone. Finally, 42 events (72%) of the events are correlated with swings in the preliminary ^{18}O curve of the Contessa Section (Haq, 2014, Fig. 3). Obviously, this has influenced Haq (op. cit.) in his interpretation in terms of eustatic sea level changes. In three cases, there is not even a single stratigraphic event to support the event in question, and in two other cases, there are only geological data in one site. This is, of course, not good enough for the establishment of a global eustatic event. In conclusion, it must be realized that we are, in fact, still lacking a sea level solution for the Cretaceous period; just as the data in Figs. 19–21 suggested.

5.6.2 Geophysical explanations

Cloetingh and Haq (2015) elegantly reviewed different processes that may generate changes in sea level. However, all the processes discussed are the same as those discussed 40 years ago, and illustrated in Fig. 12 (Mörner, 1980e, 1983a, 1986, 1987a). So the talk about "substantial progress has been made" sounds somewhat relative. Undoubtedly, however, remarkable progress has been achieved in general geophysics and solid-Earth sciences.

The possible larger exchange of water with the mantle (i.e., what we previously termed "juvenile water") is new. With respect to the "dynamic topography" and their completely correct statement that warping "can dampen and enhance any underlying global signal" (Cloetingh and Haq, 2015), I find reasons to illustrate these relations by an old diagram (Fig. 26) of the deformation of an eustatic signal by uplift and subsidence over time. Only by understanding these relations, it can be possible to convert relative sea level changes into absolute (eustatic) sea level changes (Mörner, 1971a, 2013a). It seems significant that Cloetingh and Haq (2015) themselves admitted that "the reasons for shorter-term (third order) cyclicity remain elusive."

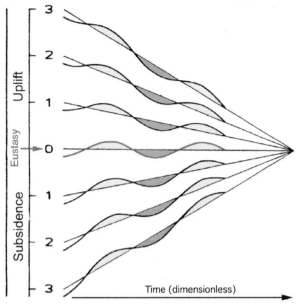

Fig. 26
Illustration of the mode of deformation of an eustatic signal (red) by increasing rates of uplift and subsidence *Redrawn from Mörner, N.-A., 1969. The Late Quaternary history of the Kattegatt Sea and the Swedish West Coast: deglaciation, shorelevel displacement, chronology, isostasy and eustasy. Sveriges Geol. Undersök. C460, 1–487, Mörner, N.-A., 1971a. Eustatic changes during the last 20,000 years and a method of separating the isostatic and eustatic factors in an uplifted area. Palaeogeogr. Palaeoclim. Palaeoecol. 9, 153–181, Mörner, N.-A., 2013a. Sea level changes: Past records and future expectations. Energy Environ. 24, 509–536.*

Geophysical research has advanced quite rapidly in modern times. Eppelbaum et al. (2018) have shown how, by using modern satellite gravity, big data can be used to decode tectonic-geodynamic changes with time back to the Mesozoic period in an area even as complicated as the Arabian-African region. From this type of analysis, it seems quite clear that we cannot, by simple stratigraphy, reconstruct any eustatic component. Grosheny et al. (2017) has shown how very complicated and illusive a bio- or lithostratigraphic search for an eustatic component can be, even in a limited area and time window. The satellite gravity big data methodology (Eppelbaum, 2017; Eppelbaum et al., 2018) seems to be a new and much more successful tool for summarizing the changes in ocean and land level in geological records in a realistic way, than the application of illusive "global eustatic" curves.

5.7 Concluding Comments on Cretaceous Sea Level Changes

The present standard sea level curve of the Cretaceous period is the one presented by Haq (2014), here reproduced as Fig. 25. It is a high-frequency oscillations curve, including 58 cycles of a mean duration of 1.36 Ma/cycle, a mean amplitude of 60 m, and a mean rate of 0.87 mm/yr.

The curve is claimed to be an eustatic curve (Haq, 2014; Cloetingh and Haq, 2015). There are good reasons to question both the number of events and their eustatic character, and there is no way the Haq curve (Fig. 25) can be a true global eustatic curve. Changes in sea level during the Cretaceous period should be dominated by tectono-eustatic factors (or what Cloetingh and Haq, 2015, calls "container effects"), and changes in geoidal eustasy (Figs. 19–22). This would suggest sea level records of irregular type, implying that there is no global eustatic curve, only regional eustatic curves. The possibility of the existence of ice caps—at least intermittently—on the landmasses in the Antarctic position opens the window for minor glacial eustatic changes. This remains hypothetical, however. The Cretaceous period is a very well-studied geological era (e.g., Cloetingh and Haq, 2015; Ramkumar and Menier, 2017); probably the best after the Holocene. Still, there is a long way to go in sea level research. The material is there (as recorded in, for example, the Appendix to Haq, 2014, as in all the old continental records; e.g., Fig. 19), but it needs to be organized and analyzed in a manner that takes into consideration the progress in Quaternary sea level research; that is, the Fig. 6B scenario. It can be said that at present, we still lack a theoretically and practically applicable synthesis of Cretaceous sea level changes.

6 Additional Notes on Tertiary Sea Level Changes

Both the Vail et al. (1977) and Haq et al. (1987) sea level curves also cover the Cenozoic. The Vail et al. curve includes 21 sudden regressions, and with a superregression of 370 m at 30 Ma. The Haq et al. curve includes some 26–30 sinusoidal oscillations, with deep regressions of 125 m at 58 Ma, of 115 m at 55 Ma, of 125 m at 49.5 Ma, of 170 m at 30 Ma, of 130 m at 10.5 Ma and three major events within the Quaternary. Chappell (1983) undertook an interesting analysis of the proposed eustatic sea level changes of Vail et al. (1977) with respect to tectonic and isostatic processes. He was well aware that the sea level curve must represent relative sea level changes, not global eustasy. He concluded that "many alternative scenarios" may be involved where "each of the main factors can be debated, or, at least qualified." His final statement reads: "only through careful analysis of separate regional records, each considered in their own plate-tectonic context, can we resolve the intriguing questions of gauge sea level changes." This is in full agreement with a previous study by Mörner (1983a).

Along the passive continental coast of South America, there is no evidence for all the sea level events recorded by either Vail et al. (1977) or Haq et al. (1987); instead, only four major sea level pulses within the Tertiary and two Quaternary highstands of the Last Interglacial and the Mid-Holocene periods (Mörner et al., 2009) can be recognized. The Tertiary changes are a regressive Early Tertiary marine event, an Eocene to Oligocene lowstand (Toba con Mamiferos), a 23–18 Ma marine highstand (Patagoniense), a 18–12 Ma lowstand (Santa Cruzian), a 12–10 Ma marine highstand (Paranense), a general lowstand of dry climate in the Late Miocene to Pliocene, a sharp sea level peak at around 3 Ma ago, and then the sea level

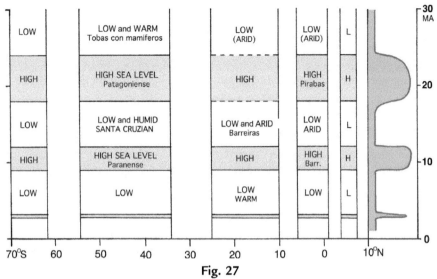

Fig. 27
Sea level changes in the past 30 Ma as recorded along the passive continental margin of Eastern South America from Lat. 10°N to Lat. 55°S, plus extension to the Antarctic Peninsula down to Lat. 70°S (Mörner, 2016b; Mörner et al., 2009). Three main periods of sea level highstand are recorded (right-hand column), occurring at ~20, ~10 and ~3 Ma.

highstands of the Last Interglacial and of the Mid-Holocene (Mörner, 2016b). Fig. 27 gives the regional sea level changes as observed in the field down the South American east coast from Surinam to the Antarctic Peninsula (Mörner, 2016b) for the past 30 Ma.

6.1 The Big Drop at Around 30 Ma

The huge and sudden regression of 370 m in the Vail et al. (1977) curve cannot be understood by means of any natural processes. It exceeds the total water-masses stored in the Antarctic ice cap by a factor of six. Its rapidity exceeds any known processes. Simply, it must be wrong (cf. Mörner, 1982, 1983a). Significantly, the amplitude of this regression is decreased by 50% to about 170 m by Haq et al., 1987). This amplitude exceeds the stored water masses in the Antarctic by about a factor of three. So, glacial eustasy is excluded at least as the main factor. It is interesting that there is an absence of both extinction events and changes in faunal assemblages in the period 35–25 Ma spanning the superregression event at 30 Ma according to the detailed analysis by Corliss and Lloyd Jr (1986). This seems to be a revealing fact indicating that there was neither any major climatic change nor any major change in palaeoceanography to be associated with the proposed superregression. This provides strong additional material speaking against the reality of the superregression at 30 Ma.

6.2 The Messinian "Salinity Crisis" and Sea Level Changes

The Messinian Salinity Crisis is dated at 5.96–5.33 Ma (Krijgsman et al., 1999). Extensive evaporates (salt) were formed in the Mediterranean in association with a sea level lowering. The volume of water changes in the Mediterranean is estimated at about $3.7 \times 10^6 \, km^2$, a volume equivalent to a 10 m global sea level lowering (Mörner, 2016b). The loss of water (i.e., evaporation) in the Mediterranean coincides with the trapping (i.e., precipitation) of huge quantities of water in the Amazonian Rain Forest system at about 6–5 Ma, and a causal connection was proposed leading to a general decrease in global water volume, generating a general sea level lowering (Mörner, 2016b). Recently, Leroux et al. (2018) have proposed that the Messinain Salinity Crisis was initiated by a worldwide tectonic revolution.

6.3 The Pliocene 3 Ma Highstand

Fig. 27 includes a short sea level peak at around 3 Ma, which can be followed all along the coast of Eastern South America (Mörner and Mijuca, 1989; Mörner, 1991, 2016b; Bidegain, 1991). Its stratigraphic position is well constrained by paleomagnetic data. It occurs at the lower transition from the Gauss Normal Polarity zone into the Kaena Reversed Polarity zone, that is, approximately at around 3 Ma. From the palaeomagnetic type-site in Argentina (Bidegain, 1991), it has become known as the Aldea Brasilera transgression. Its amplitude has to be in the order of 20–30 m. In East Africa, there is a very distinct palaeo-shore at +30 m, which can be tied to the same time period (Somi, 1993). Therefore, the sea level transgression at ~3 Ma is likely to have reached a sea level position ~30 m higher than today. If this transgression represents a glacial eustatic rise in sea level, it must represent a period of very strong ice melting of Antarctica (where 60 m of water is stored today) and Greenland (where 6 m of water is stored today). In order to raise sea level by 30 m, we need to have had a period of very extensive ice melting lasting for a short period of about 100,000 years. The age of the Kaena reversal event was dated at 2.9–3.0 Ma by Mankinen and Dalrymple (1979), 2.8–2.9 Ma by McFadden (1980), and 2.0–3.1 Ma by Hilgen (1991). The transgression began shortly below the onset of the Kaena event, and ended shortly before its end, implying that the entire transgression/regression episode lasted for about 100,000 years. While the rate of rise and fall in glacial eustatic sea level (in the order of 1.0 mm/yr) offers no problems with respect to the rate diagram of Fig. 27, the amplitude of possible melting seems far too large.

At about 3 Ma, however, there are observational facts of the occurrence of a boreal forest in northeast Greenland (the Cap Køpenhavn Formation), and in Antarctica there is the boreal mollusk fauna in the Cockburn Island Formation, which, indeed, suggests very strong ice melting (Mörner, 1993c; Jonkers and Kelley, 1998). It may be significant that the astronomical eccentricity cycle experienced a maximum (Hilgen, 1991) precisely where the transgression took place. Consequently, the 3 Ma sea level peak may, after all, correlate with a period of extensive glacial melting. This warm event has been observed and discussed by others, too (e.g., Jonkers and Kelley, 1998; Dowset et al., 2005; Cronin et al., 2005; Robinson et al., 2011; Brigham-Grette et al., 2013; Chandler et al., 2013). It indicates that, indeed, there was a Pliocene warm phase of considerable amplitude. According to Panitz et al. (2015), the climate may have been "6–14.5°C higher than present." This may explain the occurrence of a boreal forest in the Cap København Formation of NE Greenland. This has a bearing on our understanding of ice melting and sea level changes. But there is no straightforward implication to our present sea level situation. First of all, the rate of sea level rise seems to have been well below the maximum rate line of glacial eustasy in Fig. 24 (maybe on the order of 1.0–3.0 mm/yr), and hence offers no problem. Second, the time it would take to melt ice and raise the sea level by 30 m is on the order of 10,000–30,000 year. Third, we are now, since about 2.5 Ma ago (Mörner, 1993c), in a period dominated by the repeated alternation between Ice Ages and Interglacials, a trend that can hardly be broken in the present time.

7 The Future as Seen From the Past

The studies of sea level changes within the Late Pleistocene and the Holocene eras have taken us a long way in successively improved understanding of the processes and forcing-functions behind sea level changes. This database is the natural source of knowledge to be used for meaningful assessments of the present to future changes in sea level and climate.

In the review for the IGCP-609 project on "Cretaceous Sea-Level Changes" (UNESCO 2013–2017), the main reason for our studies of past sea level changes is explained as "to predict future sea levels, we need a better understanding of the record of past sea level change." Though there may not be any need for a motivation for Cretaceous sea level studies in a possible applicability to the present to future sea level debate. The sea level changes of the Cretaceous are far too important by themselves for a large number of fundamental questions about ocean dynamic processes (e.g., Hsü, 1986), crustal and mantle processes (e.g., Conrad, 2013; Cloetingh and Haq, 2015), and general planet Earth questions.

The study of pre-Quaternary sea level changes has not followed the same evolution as the study of Quaternary sea level changes. Instead, the main evolution has been devoted to the improvement in stratigraphy and, especially, the methods for detailed seismic stratigraphy of the shelf environments for hydrocarbon exploration (e.g., Payton, 1977). With respect to a firm sea level reconstruction, much progress seems to remain, however. For today's sea level scenarios, implications seem still to be virtually lacking. The possible occurrence of Antarctic ice caps, and, if so, their waxing and vanishing processes with time, may have relevance for our understanding of general climatic conditions on Earth.

In the Pliocene era, there was a short period of warm climate conditions, extensive ice melting, and sea level rise. This event has a direct bearing on present day warming/melting discussions. Still, no direct worries seem possible to associate with event in view of future scenarios;

the rates of associated sea level changes (~1.0 mm/yr) are well within normal conditions for the past 300 years (Mörner, 2004, 2011a, 2014b, 2017b); the time of melting must be in the order of a tens of kilo-years (maybe ~30,000 years), and Earth had not yet come into the stage of glacial/interglacial changes characterizing the last 2.5 Ma (Mörner, 1993c). This means that meaningful understanding and prediction of future sea level changes must primarily be based on our observations of present to subrecent sea level changes. The main variables operating over this time period are: glacial eustasy, thermal expansion and horizontal redistribution of water masses in association with ENSO events, the 18.6 yrs. main tidal cycle, the 60-yrs ocean oscillation, and the Grand Solar Cycle oscillation (e.g., Mörner, 2015b, 2017b, 2017d).

Fig. 11 gives a spectrum of present day values of regional eustasy. After removing the so-called "calibration" of the satellite altimetry data and considering local subsidence of some of the tide-gauge stations, all values fall with a zone of ±0.0 to 1.0 mm/yr sea level "rise" in present times (Fig. 11). Instead of a subjective selection of sites recording present day sea level changes (IPCC, 2007, 2013), one should consult "test areas" where the crustal component is known. Table 2 gives a summary of 25 such sites. Two things are obvious: (1) the satellite altimetry data (NOAA, 2014) very strongly overestimate true sea level changes, and (2) all data of present day local eustatic changes in sea level fit quite nicely within a zone ranging from ±0.0 to +1.0 mm/yr.

This implies negligible real threats of a future sea level rise with disastrous effects on low-lying coasts as claimed by the IPCC (2007, 2013). Besides, there are no records anywhere on the globe documenting any sea level acceleration. Whatever models may seem to predict, it must be true observational facts that counts (e.g., Mörner et al., 2017). This implies that the proposed threat of coastal flooding is very strongly exaggerated. The best estimate of sea level changes up to year 2100 seems to the +5 cm ±15 cm; that is, not exceeding 20 cm by 2100 (Mörner, 2004, 2014b, 2017e). The dominant factor in present day sea level changes seems to be rotational eustasy (Mörner, 2017b). Glacial eustasy plays a subordinate role, and there is no true indication of increased glacial melting (Easterbrook, 2016a, b; Mörner et al., 2017). Thermal expansion has limited effects on sea level, and is always zero at the coast (Mörner, 2017d). The 60-yrs sea level cycle or "oscillation" is important. It is manifested in the Gulf Stream beat (Mörner, 1996b, 2010a), in the Pacific Decadal Oscillation (PDO), North Atlantic Oscillation (NAO), etc. It can be linked to planetary beat on the Sun, the Earth and the Earth-Moon system (Mörner, 2015b).

The dominance of rotational eustasy during the past 500 years is a novel finding (Mörner, 2017b). It follows the alternations between rotational accelerations at the Grand Solar Minima, and rotational decelerations at the Grand Solar Maxima (Fig. 28). This means that it represents another oceanic oscillation driven by planetary beat. If we are facing a new Grand Solar Minimum at around 2030–2050 (Mörner, 2015d), it would be logical to foresee a significant

Table 2 Summary of sites with known subsidence/uplift rates, regional eustasy in tide gauges and mean satellite values of NOAA (2014)

Locality	Stability subm(+), em(−)	Regional Eustasy (mm/yr)	Reference	Satellite Altimetry (NOAA (2014))
Indian Ocean:				
• The Maldives	±0.0 in 40 yrs	Small to <1.0	1, 2	2.5–3.7
• Bangladesh	±0.0 in 50 yrs	Small to <1.0	3	3.7–5.0
• Goa, India	±0.0 in 50 yrs	∼0.0 in 50 yrs	4	2.5–3.7
• Minicoy	±0.0 in 40 yrs		5	2.5–3.7
• Perth	+1.4	∼0.0 in 15 yrs	6	3.7–5.0
• St Paul	stable	No rise in 135 yrs	7, 4	2.5–3.7
• Qatar	stable	No rise	8	2.5–3.7
Pacific:				
• Fiji, Yasawa	Stable	±0.0 in 50 yrs	9	2.5–3.7
• Kiribati		±0.0 in 20 yrs	11, 12	5.0–6.2
• Tuvalu		±0.0 in 30 yrs	12, 13	5.0–6.2
• Vanuatu		±0.0 in 20 yrs	14	5.0–6.2
• Galapagos		±0.0 in 30 yrs	15	0.0–2.5
S Atlantic:				
• Guyana	Stable	±0.0	16	3.7–5.0
NW Atlantic				
• Connecticut	+1.1	∼1.1 in 100 yrs	12, 16	0–2.5
Mediterranean:				
• Venice	+2.3 in 300 yrs	±0.0 in 140 yrs	12, 14	2.5–3.7
North Sea:				
• Brest	∼0.0 in 9500 yrs	∼1.0	15	1.2–2.5
• Amsterdam	+0.4 in >5000 yrs	∼1.1 in 100 yrs	15, 12	1.2–2.5
• Ijmuiden	+0.4 in >5000 yrs	∼1.2	15	1.2–2.5
• Cuxhaven	+1.4 in >170 yrs	∼1.1 in 160 yrs	15,	1.2–2.5
Kattegatt:				
• Korsör	±0.00 in 8000 yrs	0.81±0.18 in 125 yrs	17, 15	2.5–3.7
• Sliphaven	+0.10 in 8000 yrs	∼0.9 in 125 yrs	17, 15	2.5–3.7
• Aarhus	−0.28 in >3000 yrs	∼0.9 in 125 yrs	17, 15	2.5–3.7
• Varberg	−1.75 in >1000 yrs	∼0.9	17, 15	2.5–3.7
• Klagshamn	−0.30 in 8000 yrs	∼0.9	17, 15	2.5–3.7
The Baltic				
• Stockholm	−4.9 in >3000 yrs	1.1 in 200 yrs	17, 18	2.5–3.7

References: (1) Mörner 2007a, 2011b, (2) Mörner et al. 2004, (3) Mörner 2010c, (4) Mörner 2017a, (5) Mörner, 2011b, (6) Mörner & Parker 2013, (7) Testut et al. 2010, (8) Mörner 2015e, (9) Mörner & Matlack-Klein, 2017, (10) Mörner 2010a, (11) Parker 2016, (12) Mörner 2016a, (13) Parker 2017, (14) Mörner 2007b, (15) Mörner 2014b, (16) Mörner 2010b, (17) Mörner 2014a, (18) Mörner 1973.

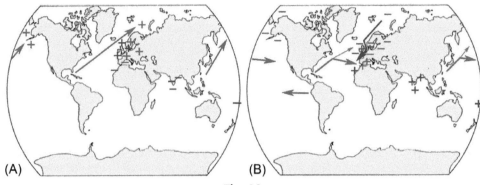

Fig. 28
Ocean circulation (arrows), temperature changes (red), and sea level changes (blue) in the equatorial Indian Ocean and Pacific, and in the NE Atlantic and Scandinavian regions during Grand Solar Maxima (A) with rotational slowing-down and Grand Solar Minima (B) with rotational speeding-up. Modified from Mörner, N.-A., 2015b. Multiple planetary influences on the Earth. In: Mörner, N.-A., (Ed.), Planetary-Solar-Terrestrial Influences on the Sun and the Earth and a Modern Book-burning. NovaScience Publishers. p. 39–49 (Chapter 4), Mörner, N.-A., 2017b. Our Oceans—Our Future: New evidence-basted sea level records from the Fiji Islands for the last 500 years indicating rotational eustasy and absence of a present rise in sea level. Int. J. Earth Environ. Sci. 2: 137. doi: 10.15344/2456-351X/2017/137.

cooling (maybe a Little Ice Age) with drastically changed ocean circulation, lowered sea levels in the high latitudes, and rising sea levels in the equatorial regions, as illustrated in Fig. 27B.

8 Conclusions

Forty years ago, it was proposed that continual deformations of the geoid surface relief implied that the sea level does not deform in a parallel manner over the globe (Mörner, 1976), and that the concept of eustasy therefore had to be redefined (Mörner, 1986), as illustrated in Fig. 6. This was later followed up by the IGCP-200 project, which, indeed, confirmed this concept (e.g., Pirazzoli, 1991). A new paradigm was born. It seems quite surprising that this redefinition and its confirmation in Holocene sea level records have not yet been adopted in pre-Quaternary sea level research. Therefore, the proposed eustatic curves of the Cretaceous period (Fig. 25 and related discussion) cannot be used to reconstruct eustatic changes in a correct way.

The Pliocene era sea level highstand, at about 3 Ma, has interesting general aspects related to climate and sea level changes. In view of the present sea level situation, there seem to be no relevant applications, however. Meaningful predictions of the present situation into the future must be based on observations and physical realities. Models not anchored in observations and physical laws are bound to fail. Our available data indicate that present day sea level changes are primarily driven by the redistribution of water masses – rotational eustasy – at interannual ENSO events, the main 18.6 yrs tidal cycle of the Earth-Moon system, the 60-yrs oscillation ultimately driven by planetary beat (Mörner, 2015b), and the Grand Solar Cycle oscillation driven by planetary beat on the Sun, the Earth, and the Earth-Moon system (Mörner, 2015b, 2017b). The observation-based prediction of future sea levels in the year 2100, therefore, is a modest rise of +5 cm ±15 cm (e.g., Mörner, 2004, 2014b, 2015c, 2017c). With respect to climate change, we are more likely facing an approaching Grand Solar Minimum with cold climatic conditions (Landscheidt, 2003; Mörner, 2015d) than an increased warming (IPCC, 2007, 2013).

Acknowledgment

I was kindly invited by Professor Ramkumar to contribute to this book with a paper on my views and experience with respect to sea level changes. I can now see that the paper nearly took the form of a review of my lifelong devotion to the science of sea level changes (Mörner, 2015f). Still, I hope it keeps a useful focus on what we know, need to know, and have to do.

References

Agassiz, L., 1840. Etudes sur les glaciers. (Neuchatel).
Arnold, J.R., Libby, W.F., 1949. Age determination by radiocarbon content: check with samples of known age. Science 110, 678–680.
Barron, J.A., Keller, G., 1982. Widespread Miocene deep-sea hiatuses: coincidence with periods of glacial cooling. Geology 10, 577–581.
van Bemmelen, R.W., Berlage, H.P., 1935. Versuch einer mathematiscen Behandlung geotektonischer Bewegungen unter besondererBerücksichtigung der Undattionteorie. Gerl. Beitr. Geophys. 43, 19–55.
Berger, A., Imbrie, J., Hays, J., Kukla, G., Saltzman, B., 1984. Milankovitch and Climate. Reidel Publ. Co., Dordrecht
Bernhardi, A., 1832. Wie kamen die aus dem Norden stammenden Felsbruchstücke und Geschiebe, welche man in Norddeutschland und den benachbarten Ländern findet, an ihre gegenwärtigen Fundorte? Jahrb. Mineralogie Geognosie Petrefaktenkünde 3, 257–267.
Berzelius, J., 1843. In: Några ord on den Skandinaviska vallens höjning öfver ytan af omkringliggande haf och om afslipningen och refflingen af dess berg.Förhandl. de skandinaviske naturforskarnes tredje möte i Stockholm. 1842.
Bidegain, J.C., 1991. Sedimentary Development, magnetostratigraphy and sequence of events of the Late Cenozoic of the Entre Rios and surrounding areas in Argentina. Ph.D. thesis, Stockholm University, Paleomagnetism & Geodynamics, p. 5.
Blanc, A.C., 1937. Low levels of the Mediterranean Sea during the Pleistocene glaciation. Q. J. Geol. Soc. Lond. 93, 621–651.
Bloom, A.L., 1967. Plesitocene shorelines. A new test of isostasy. Bull. Geol. Soc. Am. 78, 1477–1498.
Bloom, A.L., 1980. Late Quaternary sea level change on South Pacific coasts: a study in tectonic Diversity. In: Mörner, N.-A. (Ed.), Earth Rheology, Isostasy and Eustasy. John Wiley & Sons, Chichester, pp. 505–516.
Brigham-Grette, J., et al., 2013. Pliocene warmth, polar amplification, and stepped Pleistocene cooling recorded in NE Arctic Russia. Science 340, 1421–1427.
von Buch, L., 1810. Reise durch Norwegen und Lappland 2. Berlin, 1810.
Cathles, L.M., 1975. The Viscosity of the Earth's Mantle. Princeton Univ. Press, p. 386.
Celsius, A., 1743. Anmärkningar om vatnets förminskande så i östersjön som Vesterhafvet. Kungl. Vet. Akad, Handlingar.

Chambers, R., 1848. Ancient Sea-Margins, as Memorials of Change of the Relative Level of Land and Sea. Chambers, Orr &Co., Edinburgh.

Chandler, M.A., Sohl, L.E., Jonas, J.A., Dowsett, H.J., Kelley, M., 2013. Simulations of the mid-Pliocene warm period using two versions of the NASA/GISS ModelE2-R coupled model. Geosci. Model Dev. 6, 517–531.

Chappell, J., 1983. Aspects of Sea Levels, Tectonics and Isostasy since the Cretaceous. In: Gardner, R., Scoging, H. (Eds.), Mega-Geomorphology. Oxford Univ. Press, pp. 56–72 (Chapter 4).

Chappell, J., Shackleton, N.J., 1986. Oxygen isotopes and sea level. Nature 324, 137–140.

de Charpentier J (1835) Notice sur la cause probable du transport des blocs erratique de la Suisse. Annales des Mines, 3ms ser.

Clark JA (1980) A numerical model of worldwide sea level changes on a viscoelastic Earth. In: Earth Rheology, Isostasy and Eustasy, N-A Mörner, ed., p. 525–534. John Wiley & Sons, Chichester.

Cloetingh, S., Haq, B.U., 2015. Inherited landscapes and sea level change. Science 347, 393–404.

Conrad, C.P., 2013. The solid Earth'sinfluence on sea level. Goel. Soc. Am. Bull. 125, 1027–1052.

Corliss, B.H., Lloyd Jr., D.K., 1986. Eocene-Oligocene paleoceanography. In: Hsü, K.J. (Ed.), Mesozoic and Cenozoic Oceans. In: AGU/GSA, Geodynamic Series, 15GSA Publ, pp. 101–118.

Cronin, T.M., Dowsett, H.J., Dwyer, G.S., Baker, P.A., Chandler, M.A., 2005. Mid-Pliocene deep-sea bottom-water temperatures based on ostracode Mg/Ca ratios. Mar. Micropaleontol. 54, 249–261.

Cuvier, G., 1825. Discours sur les Révolutions de la surface du Globe, et sur les changements qu'elles ont produits dans le règne animal. Dufour et d'Ocagne, Paris.

Daly, R.A., 1910. Pleistocene glaciation and the coral reef problem. Am. J. Sci. 30, 297–308.

Dana, J.D., 1849. Geology. In: U.S. Exploring Expeditions.vol. 10, pp. 1–756. New York.

Darwin, C., 1842. The Structure and Distribution of Coral Reefs. Smith, Elder &Co., London.

Darwin, C., 1859. On the Origin of Species by Means of Natural Selection. John Murray, London.

De Geer, G., 1888. Om Skandinaviens nivåförändringar under Quartärperioden. Geol. Fören. Stockholm Förhandl. 10, 366–379 (1888) & 12: 61–110 (1890).

Dowset, H.J., Chandler, M.A., Cronin, T.M., Dwyer, G.S., 2005. Middle Pliocene sea surface temperature variability. Paleoceanography 20, 1–8. https://doi.org/10.1029/2005PA001133. PA2014.

von Drygalski, E., 1887. Die Geoiddeformation des Eiszeit. Zeitschrift Gesell, Erdkunde Berlin 22, 169–208.

Easterbrook, D.J., 2016a. Evidence that Antarctica is cooling, not warming. In: Easterbrook, D.J. (Ed.), Evidence-Based Climate Science. Elsevier, Amsterdam, pp. 123–136.

Easterbrook, D.J., 2016b. Temperature fluctuations in Greenland and Antarctica. In: Easterbrook, D.J. (Ed.), Evidence-Based Climate Science. Elsevier, Amsterdam, pp. 137–160.

Emiliani, C., 1955. Pleisticene temperatures. J. Goel. 63, 149–158.

EPICA Community Members, 2004. Eight glacial cycles from an Antarctic ice core. Nature 429, 623–628.

Eppelbaum, L.V., 2017. Satellite Gravimetry ('Big Data´) – A Powerful Tool in Regional Tectonic Examination and Reconstructions. In: Veress, B., Szigety, J. (Eds.), Horizons in Earth Science Research. In: Vol. 17. Nova Science Publishers, New York, pp. 115–148.

Eppelbaum, L., Katz, Y., Klokocnik, J., Kostelecky, J., Ben-Avraham, Z., Zheludev, V., 2018. Tectonic insights into the Arabian-African region inferred from a comprehensive examination of satellite gravity big data. Glob. Planet. Chang. 159, 1–24.

Ericson, D.B., Wollin, G., 1964. The Deep and the Past. Knopf, New York, p. 292.

Fairbanks, R.G., 1989. A 17,000-year glacio-eustatic sea-level record influence of glacial melting rates of the younger Dryas event and deep-ocean circulation. Nature 342, 637–642.

Fairbridge, R.W., 1961. Eustatic changes in sea level. Phys. Chem. Earth 4, 99–185.

Fischer, P., 1868. Untersuchungen über die Gestalt der Erde. Diehl, Darmstadt.

Fjeldskaar, W., Cathles, L., 1991. The present rate of uplift of Fennoscandia implies a low-viscosity asthenosphere. Terra Nova 3, 393–400.

Flint, R.F., 1971. Glacial and Quaternary Geology. John Wiley & Sons, Chichester, p. 892.

Florin, S., 1944. Havsstrandens förskjutningar och bebyggelseutvecklingen i östra Mellan-sverige under senkvartär tid. Geol. Fören. Stockh. Förh. 66, 531–634.

Frisi, P., 1785. Opera 3. (Mediolani).
Gaposchkin, E.M., 1974. Earth's gravity field to the eighteenth degree and geocentric coordinates for 104 stations from satellite and terrestrial data. J. Geophysical Res. 79, 53377–53411.
Gilbert, K.G., 1890. Lake Bonneville. U.S. Geol. Survey, Memoir 1, 1–438.
Gilbert, K.G., 1882. Contribution to the history of Lake Bonneville. In: U.S. Geol. Survey, 2nd Ann. Rep. pp. 169–200.
Grosheny, D., Ferry, S., Lecuyer, C., Merran, Y., Mroueh, M., Granier, B., 2017. The Cenomanian-Turonian boundary event (CTBE) in northern Lebanon as compared to regional data—another set of evidences supporting short-loved tectonic pulse coincidental with the event? Palaeogeogr. Palaeoclim. Palaeoecol. 64 (1), 168–185.
Gutenberg, B., 1941. Changes in sea level, postglacial uplift, and mobility of the Earth's interior. Bull. Geol. Soc. Am. 52, 721–772.
Hansen, J.M., 2009a. Naturhistoriens logiske grundlag: Fra Steno til geologerne Lyell, Darwin og Wegerer og filosofferne Kant, Peirce og Popper. Geol. Tidskr. 1, 46–60.
Hansen, J.M., 2009b. On the origin of natural history: Steno's modern, but forgotten philosophy of science. Bull. Geol. Soc. Denmark 57, 1–24.
Haq, B.U., 2014. Cretaceous eustasy revisited. Glob. Planet. Change 113, 44–58.
Haq, B.U., Schutter, S.R., 2008. A chronology of Paleozoic Sea-level changes. Science 322, 64–68.
Haq, B.U., Hardenbol, J., Veil, P.R., 1987. Chronology of fluctuating sea levels since the Triassic. Science 235, 1156–1167.
Harff, J., Frischbutter, A., Lampe, R., Meyer, M., 2001. Sea Level Changes in the Baltic: Interaction of Climate and Geological Processes. In: Gerhard, L.C., Harrison, W.E., Hanson, B.M. (Eds.), Geological Perspectives of Global Climate Change, pp. 231–250.
Hergsell, H., 1887. Über die Änderung der Gleichgewichtsflächen der Erde durch Bildung polarer Eismassen und die dadurch verursachten Schwankungen des Meresniveaus. vol. 1 Beitrag Geophys., Abhandl. Geogr. Sem. Univ., Strassburg, pp. 59–114.
Hilgen, F.L., 1991. Astronomical calibration of gauss to Matuyama sapropels in the Mediterranean and implication fort he geomagnetic polarity time scale. Earth Planet. Sci. Lett. 104, 226–244.
Hitchcock, E., 1841. First anniversary address before the Association of American Geologists. Am. J. Sci. 41, 232–275.
Houston, J.R., Dean, R.G., 2012. Comparisons at tide-gauge locations of glacial isostatic adjustment predictions with global positioning system measurements. J. Coastal Res. 28, 739–744.
Hsü, K.J. (Ed.), 1986. Mesozoic and Cenozoic Oceans AGU/GSA. In: Geodynamic Series, vol. 15. GSA Publ, Washington, DC, pp. 1–153.
IPCC, 2007. Fourth Assment Repost. Climate Change. Cambridge University Press, Cambridge.
IPCC, 2013. Fifth Assessment Report. Climate Change. Cambridge University Press, Cambridge.
Jamieson, T.F., 1865. On the history of the last geological changes in Scotland. Geol. Soc. Lond. Q. J. 21, 161–203.
Jamieson, T.F., 1882. On the cause of the depression and re-elevation of the land during the glacial period. Geol. Mag. N.S. 9, 400–407.
Jonkers, H.A., Kelley, S.P., 1998. A reassessment of the age of the Cockburn Island formation, northern Antarctic peninsula, and its plaeoclimatic implications. J. Geol. Soc. Lond. 155, 737–740.
Karato, S., 2012. On the origin of the asthenosphere. Earth Planet. Sci. Lett. 321 (22), 95–103.
Kerr, R.A., 1980. Changing global sea level as a geologic index. Science 209, 483–486.
Krijgsman, W., Hilgren, F.J., Raffi, I., Sierro, F., Wilson, D.S., 1999. Chronology, crisis and progression of the Messinian salinity crisis. Nature 4000, 652–655.
Kullenberg, B., 1947. The piston core samples. In: Svenska Hydr.-Biol.Komm. Skrift, Ser. 3, vol. 1(2). p. 1.
Lambeck, C., 1998. On the choice of timescale in glacial rebound modelling: mantle viscosity estimates and the radiocarbon timescale. Geophys. J. Int. 134, 647–651.
Lambeck, K., Rouby, H., Purcell, A., Sun, Y., Sambridge, M., 2014. Sea level and glacial volumes from the Last Glacial Maximum to the Holocene. PANAS. https://doi.org/10.1073/pnas.1411762111.
Landscheidt, T., 2003. New little ice age instead of global warming. Energy Environ. 14, 327–350.

Leroux, E., Aslanian, D., Rabineau, M., Pellen, R., Moulin, M., 2018. The late Messinian event: a worldwide tectonic revolution. Terra Nova 2018, 1–8. https://doi.org/10.1111/ter.12327.

Lidén, R., 1938. Den senkvartära strandförskjutningens förlopp och kronologi i Ånegermanland. Geol. Fören. Stockh. Förhandl. 60, 397–404.

von Linné, C., 1745. Öländskaoch Gothländskaresa på Riksens högloflige Ständers befallning förrättad 1741. (Stockholm).

Listing, J.B., 1873. Über unsere jetziger Keenntnisse der Gestalt und Grösse der Erde. Königl. Ges. Wiss. Göttingen, Nachrichten 3, 33–100.

Lubimova, E.A., 1980. Heat flow, epeirogeny and seismicity for the East European Platform. In: Mörner, N.-A. (Ed.), Earth Rheology, Isostasy and Eustasy. John Wiley & Sons, Chichester, pp. 91–109.

Lyell, C., 1830. Principles of Geology. John Murray, London.

Lyell, C., 1835. On the proof of a gradual risingof the land in certain partsof Sweden. Philos. Trans. R. Soc. London 1–38.

Maclaren, C., 1841. The glacial theory of Professor Agassiz of Neuchatel. The Scotsman Office, Edingburgh (also: Am. J. Sci. 42:346–365 (1842).

Mankinen, E.A., Dalrymple, G.B., 1979. Revised geomagnetic polarity time scale for the interval 0-5 m.y. B.P. J. Geophys. Res. 84, 615–626.

Marsh, J.G., Vincent, S., 1974. Global detailed geoid computation and model analysis. Geophys. Surv. 1, 481–511.

Matsumoto, T., 1977. On the so-called cretaceous transgressions. Special Papers of the Palaeontological Society of Japan 2pp. 75–84.

Matsumoto, T., 1979. In: Inter-Regional Correlation of Transgressions and Regressions in the Cretaceous Period. Abstracts, IGCP-58 Working Group 1 Discussion Meeting, London, September 1979.

McFadden, P.L., 1980. An overview of palaeomagnetic chronology with special reference to the south African hominid site. Palaeonto Afr. 23, 35–40.

Miller, K.G., 2009. Sea Level Changes, Last 250 Million Years. Springer, pp. 879–887.

Miller, K.G., et al., 2003. Late cretaceous chronology of large, rapid sea-level changes: Glacioeustasy during the greenhouse world. Geology 31, 585–588.

Milliman, J.D., Emery, K.O., 1968. Sea levels during the past 35,000 years. Science 162, 1121–1123.

Moore, J.R., van Andel, T.H., Sancetta, S., Psias, N., 1978. Cenozoic hiatuses in marine sediments. Micropaleontology 24, 113–138.

Mörner, N.-A., 1969. The late quaternary history of the Kattegatt Sea and the Swedish west coast: Deglaciation, shorelevel displacement, chronology, isostasy and eustasy. Sveriges Geol. Undersök. C460, 1–487.

Mörner, N.-A., 1971a. Eustatic changes during the last 20,000 years and a method of separating the isostatic and eustatic factors in an uplifted area. Palaeogeogr. Palaeoclim. Palaeoecol. 9, 153–181.

Mörner, N.-A., 1971b. The Holocene eustatic sea level problem. Geol. en Mijnbouw 50, 699–702.

Mörner, N.-A., 1973. Eustatic changes during the last 300 years. Palaeogeogr. Palaeoclim. Palaeoecol. 13, 1–14.

Mörner, N.-A., 1976. Eustasy and geoid changes. J. Geol. 84, 123–151.

Mörner, N.-A., 1979. The Fennoscandian uplift and late Cenozoic geodynamics: Geological evidence. GeoJournal 3, 287–318.

Mörner, N.-A., 1980a. The northwest European "sea-level laboratory" and regional Holocene eustasy. Palaeogeogr. Palaeoclim. Palaeoecol. 29, 181–300.

Mörner, N.-A., 1980b. The Fennoscandian uplift: geological data and their geodynamical implication.
In: Mörner, N.-A. (Ed.), Earth Rheology, Isostasy and Eustasy. John Wiley & Sons, Chichester, pp. 251–284.

Mörner, N.-A., 1980c. Relative Sea-level, tectono-eustasy, geoidal-eustasy and geodynamics during the cretaceous. Cretac. Res. 1, 329–340.

Mörner, N.-A., 1980d. Eustasy and geoid changes as a function of core/mantle changes. In: Mörner, N.-A. (Ed.), Earth Rheology, Isostasy and Eustasy. John Wiley & Sons, Chichester, pp. 535–553.

Mörner, N.-A., 1980e. Paleogeoid changes and their possible impact on the formation of natural resources in Africa. Palaeogeogr. Palaeoclim. Palaeoecol. 20, 225–232.

Mörner, N.-A., 1981a. Eustasy, paleoglaciation and paleoclimatology. Geol. Rundsch. 70, 691–702.

Mörner, N.-A., 1981b. Space geodesy, paleogeodesy and paleogeophysics. Ann. Géophys. 37 (1), 69–76.

Mörner, N.-A., 1981c. Revolution in cretaceous sea-level analysis. Geology 9, 344–346.

Mörner, N.-A., 1982. Sea level changes as an illusive "geological index" Neotectonics Bull. 5, 55–64.

Mörner, N.-A., 1983a. Sea levels. In: Gardner, R., Scoging, H. (Eds.), Mega-Geomorphology. Oxford Univ. Press, Oxford, pp. 73–91 (Chapter 5).

Mörner, N.-A., 1983b. Illusions in water-budget synthesis. In: Street-Perrott, A., Beran, M., Ratcliffe, R. (Eds.), Variations in Global Water Budget. Reidel Publ. Co., Dordrecht, pp. 419–423

Mörner, N.-A., 1984. Planetary, solar, atmospheric, hydrospheric and endogene processes as origin of climatic changes on the Earth. In: Mörner, N.-A., Karlén, W. (Eds.), Climatic Changes on a Yearly to Millennial Basis. Reidel Publ. Co., Dordrecht/Boston/Lancaster, pp. 483–507.

Mörner N-A (1985) Possible "Super-ENSO" in the past (p. 31), Earth/Ocean Interchange of momentum and major climatic changes on a decadal basis (p. 44), Paleo-interchanges of energy, mass and momentum (p. 68), The role of rotational and gravitational changes for climate change on the Earth (p. 94), Dynamic sea surface changes in the past and redistribution of mass and energy (p. 118). Abstracts IAMAP/IAPSO Joint Assembly, Honolulu 1985.

Mörner, N.-A., 1986. The concept of eustasy: a redefinition. J. Coastal Res. S1 (1), 49–51.

Mörner, N.-A., 1987a. Models of Global Sea-Level Changes. In: Tooley, M.J., Shennan, I. (Eds.), Sea Level Changes. Blackwell, Oxford, pp. 333–355.

Mörner, N.-A., 1987b. Pre-Quaternary long-term changes in sea level. In: Devoy, R.J.N. (Ed.), Sea Surface Studies: A Global View. Croom Helm, Beckenham, pp. 233–241 (Chapter 8a).

Mörner, N.-A., 1988. Terrestrial Variations within Given Energy, Mass and Momentum: Paleoclimate, Sea Level, Paleomagnetism, Differential Rotation and Geodynamics. In: Stephenson, F.R., Wolfendale, A.W. (Eds.), Secular Solar and Geomagnetic Variations in the Last 10,000 Years. Kluwer Acad. Press, Dordrecht, pp. 455–478.

Mörner, N.-A., 1990. Glacial isostasy and long-term crustal movements in Fennoscandia with respect to lithospheric and asthenospheric processes and properties. Tectonophysics 176, 13–24.

Mörner, N.-A., 1991. Course and origin of the Fennoscandian uplift: The case for two separate mechanisms. Terra Nova 3, 408–413.

Mörner, N.-A., 1992a. Ocean circulation, sea level changes and East African coastal settlement. In: Sinclair, P. (Ed.), Urban Origins in East Africa. Board Antiquity, Uppsala, pp. 258–263.

Mörner, N.-A., 1992b. Eustatic changes in level and topography. Past long- and short-term changes. Present and future trends. Neotectonics Bull. 15, 53.

Mörner, N.-A., 1993a. Global change: The last millennia. Glob. Planet. Change 7, 211–217.

Mörner, N.-A., 1993b. Global change: the high-amplitude changes 13-10 Ka ago – Novel aspects. Glob. Planet. Change 7, 243–250.

Mörner N-A (1993c) Neotectonics. The global tectonic regiment during the last 3 ma and the initiation of ice ages. Ann. Brazilian Acad. Sci., 65 (Supl. 2): 295–301. Also (available at): http://tallbloke.wordpress.com/2012/12/08/nils-axel-morner-neotectonics-and-the-recent-ice-ages/#comment-37682.

Mörner, N.-A., 1994. Internal response to orbital forcing and external cyclic sedimentary sequences. In: de Boer, P.L., Smith, D.G. (Eds.), Orbital Forcing and Cyclic Sequences. Spec. Publ. Int. Ass. Sedimentology, vol. 19, pp. 25–33.

Mörner, N.-A., 1995. Earth rotation, ocean circulation and paleoclimate. GeoJournal 37, 419–430.

Mörner, N.-A., 1996a. Sea level variability. Z. Geomorph. N.F, Suppl-Bd. 102, 223–232.

Mörner, N.-A., 1996b. Global Change and interaction of Earth rotation, ocean circulation and paleoclimate. An. Brazilian Acad. Sci. 68 (Suppl. 1), 77–94.

Mörner, N.-A., 1996c. Earth rotation, ocean circulation and paleoclimate: The North Atlantic–European case. In: Andrews, J.T., Austin, W.E.N., Bergsten, H., Jennings, A.E. (Eds.), Late Quaternary Palaeoceanography of the North Atlantic Margins. Geol. Soc. Spec. Publ., vol. 111, pp. 359–370.

Mörner, N.-A., 1996d. Rapid changes in coastal sea level. J. Coastal Res. 12, 797–800.

Mörner, N.-A., 2000. Sea level changes and coastal dynamics in the Indian ocean. Integr. Coast. Zone Manag. 1, 17–20.

Mörner, N.-A., 2004. Estimating future sea level changes. Glob. Planet. Chang. 40, 49–50.

Mörner, N.-A., 2005. Sea level changes and crustal movements with special reference to the East Mediterranean. Z. Geomorph. Suppl. 137, 91–102.

Mörner, N.-A., 2006. The danger of ruling models. In: Burdyuzha, V. (Ed.), The Future of Life and the Future of our Civilization. Springer, pp. 105–114.

Mörner, N.-A., 2007a. Sea level changes and tsunamis, environmental stress and migration over the seas. Internationales Asienforum 38, 353–374.

Mörner, N.-A., 2007b. The Greatest Lie Ever Told. P&G print, 20 pp (1st ed 2007, 2nd ed 2009, 3rd ed 2010).

Mörner, N.-A., 2010a. Solar minima, Earth's rotation and little ice ages in the past and in the future. The North Atlantic—European case. Global Planet. Change 72, 282–293.

Mörner, N.-A., 2010b. Some problems in the reconstruction of mean sea leveland its changes with time. Quat. Int. 221, 3–8.

Mörner, N.-A., 2010c. Sea level changes in Bangladesh: New observational facts. Energy Environ. 21, 235–249.

Mörner, N.-A., 2011a. Setting the Frames of Expected Future Sea Level Changes by Exploring Past Geological Sea Level Records. In: Easterbrook, D.J. (Ed.), Evidence-Based Climate Science. Elsevier, Amsterdam, pp. 185–196 (Chapter 6).

Mörner, N.-A., 2011b. The Maldives as a Measure of Sea Level and Sea Level Ethics. In: Easterbrook, D.J. (Ed.), Evidence-Based Climate Science. Elsevier, Amsterdam, pp. 197–209 (Chapter 7).

Mörner, N.-A., 2013a. Sea level changes: Past records and future expectations. Energy Environ. 24, 509–536.

Mörner, N.-A., 2013b. Solar wind, Earth's rotation and changes in terrestrial climate. Phys. Rev. Res. Int. 3, 117–136.

Mörner, N.-A., 2014a. Deriving the eustatic component in the Kattegatt Sea. Glob. Perspect. Geogr. 2, 16–21.

Mörner, N.-A., 2014b. Sea level changes in the 1920th and 21st centuries. Coordinates X (10), 15–21.

Mörner, N.-A., 2015a. The flooding of Ur in Mesopotamia in new perspectives. Archaeol. Discov. 3, 26–31.

Mörner, N.-A., 2015b. Multiple planetary influences on the Earth. In: Mörner, N.-A. (Ed.), Planetary-Solar-Terrestrial Influences on the Sun and the Earth and a Modern Book-Burning. NovaScience Publishers, New York, pp. 39–49 (Chapter 4).

Mörner, N.-A., 2015c. Glacial isostasy: regional—not global. Int. J. Geosci. 6, 577–592.

Mörner, N.-A., 2015d. The approaching new grand solar minimum and little ice age climate conditions. Nat. Sci. 7, 510–518.

Mörner, N.-A., 2015e. Coastal Erosion and Coastal Stability. In: Barens, D. (Ed.), Coastal and Beach Erosion. Processes, Adaptation Strategies and Environmental Impacts. Nova Science Publishers, NY, pp. 69–82.

Mörner, N.-A., 2015f. Into the unknown. In: Guttman, M. (Ed.), We discover. Amazon, pp. 165–178, ISBN 978-0-9849802-3-9, 2015f.

Mörner, N.-A., 2016a. Sea level as observed in nature. In: Easterbrook, D.J. (Ed.), Evidence-based Climate Science. second ed. Elsevier, Amsterdam, pp. 185–196 (Chapter 6).

Mörner, N.-A., 2016b. The origin of the Amazonian rainforest. Int. J. Geosci. 4 (4), 470–478.

Mörner, N.-A., 2017a. Coastal morphology and sea-level changes in Goa, India, during the last 500 years. J. Coast. Res. 33 (2), 421–434.

Mörner, N.-A., 2017b. Our oceans—our future: New evidence-basted sea level records from the Fiji Islands for the last 500 years indicating rotational eustasy and absence of a present rise in sea level. Int. J. Earth Environ. Sci. 2, 137. https://doi.org/10.15344/2456-351X/2017/137.

Mörner, N.-A., 2017c. Sea level manipulation. Int. J. Eng. Sci. Invent. 6 (8), 48–51.

Mörner, N.-A., 2017d. Thermal Expansion. In: Finkle, C.W., Makowski, C. (Eds.), Encyclopedia of Coastal Science. Springer. https://doi.org/10.1007/978-3-319-48657-4_375-1.

Mörner, N.-A., 2017e. Changing Sea Levels. In: Finkle, C.W., Makowski, C. (Eds.), Encyclopedia of Coastal Science. Springer. https://doi.org/10.1007/978-3-319-48657-4_66-2.

Mörner, N.-A., Mijuca, A., 1989. General Cenozoic Sea-level changes and paleoenvironmental changes in Antarctica and Argentina. 28th IGC, Washington DC. Dent. Abstr. 2, 463.

Mörner, N.-A., Tooley, M., Possnert, G., 2004. New perspectives for the future of the Maldives. Glob. Planet. Chang. 40, 177–182.

Mörner, N.-A., Rossetti, D., de Toledo, P.M., 2009. The Amazonian Rain forest only some 6–5 million years old. In: Vieira, I.C.G. et al., (Ed.), Biological and Cultural Diversity of Amazonia. Museu Goeldi, Parana, pp. 3–18.

Mörner, N.-A., Parker, A., 2013. Present-To-Future Sea level changes: The Australian case. Environ. Sci. Ind. J. 8 (2), 43–51.

Mörner, N.-A., Matlack-Klein, P., 2017. New records of sea level changes in the Fiji Islands. Oceanogr. Fishery Open Access J. 5 (3), 1–20. https://doi.org/10.19080/OFOAJ.2017.05.555666.

Mörner, N.-A., Parker, A., Easterbrook, D., Matlack-Klein, P., 2017. Estimating sea level changes, assessing costal hazards, avoiding misguiding exaggerations, and recommending present day management. Int. Refereed J. Eng. Sci. 7 (4), 19–25.

Naif, S., Key, K., Constable, S., Evans, R.L., 2013. Melt-rich channel observed at the lithosphere-asthenosphere boundary. Nature 495, 356–359.

Nansen, F., 1928. The Earth's crust, its surface forms and isostatic adjustment. Norsk. Vid. Akad. Oslo, Mat. Nat. 12, 1–122 1927.

Newton, I., 1687. Principia (English Translation by a. Motte, 1845). Daniel Adee, New York.

NOAA, 2014. Laboratory for Satellite Altimetry/Sea Level Rise. http://www.star.nesdis.noaa.gov/sod/lsa/SeaLevelRise/.

O'Connell, R.J., 1971. Pleistocene glaciation and the viscosity of the lower mantle. Geophys. J. 23, 299–327.

Ovidius, n.d. (1 AD), Metamorphoseon libri (Methamorphoses).

Panitz, S., Salzmann, U., Risebrobakken, B., De Schepper, S., Pound, M.J., 2015. Climate variability and long-term expansion of peat lands in Arctic Norway during the late Pliocene (ODP site 642, Norwegian Sea). Clim. Past 11, 5755–5798.

Parker A (2016) Analysis of the sea levels in Kiribati. A rising sea of misconception sinks Kiribati. Nonlinear Eng., aop, 1–7, DOI https://doi.org/10.1515/nleng-2015-0031.

Parker, A., 2017. Tuvalu sea level rise. Land loss, mismanagement and overpopulation. NCGT. 5 (in press).

Parker, A., Ollier, C., 2015. Sea level rise for India since the start of tide gauge records. Arabian J. Geosci. 8 (9), 6483–6495.

Payton, C.E. (Ed.), 1977. Seismic stratigraphy. Applications to hydrocarbon exploration. Am. Assoc. Pet. Geol. Memoir. 26.

Peltier, W.R., 1976. Glacial isostatic adjustments. Geophys. J. 46 (605–646), 669–796.

Peltier, W.R., 1998. Postglacial variations in the level of the sea: implications for climate dynamics and soild-earth geophysics. Rev. Geophys. 36, 603–689.

Peltier, W.R., 2004. Global glacial isostasy and the surface of the ICE-age earth: The ICE-5G (VM2) model and GRACE. Annu. Rev. Earth Planet. Sci. 32, 111–149.

Penck, A., 1882. Schwankungen des Meeresspiegel. Geogr. Ges. München 7, 1–70.

Pirazzoli, P.A., 1991. World atlas of Holocene Sea-level changes. Elsevier Oceanogr. Ser. 58, 1–300.

Pitman III, W.C., 1978. Relashonship between eustasy and stratigraphic sequences of passive margins. Geol. Soc. Am. Bull. 89, 1389–1403.

Pitman III, W.C., Golovchenko, X., 1983. The effect of sea-level changes on the shelf edge and slope of passive margins. SEMP Spec. Pap. 33, 41–58.

Playfair, J., 1802. Illustrationsto the Huttonian Theoryof the Earth. Edinburgh.

Ramkumar, M., Menier, D., 2017. Eustasy: High-Frequency Sea Level Cycles and Habitat Heterogeneity. Elsevier, Amsterdam.

Reyment, R.A., Mörner, N.-A., 1977. Transgressions and regressions exemplified by the South Atlantic. Palaeont. Soc. Japan, Sp. Papers 21, 247–261.

Robinson, M.M., Valdes, P.J., Haywood, A.M., Dowset, H.J., Hill, D.J., Jones, S.M., 2011. Bathmetric controls on Pliocene North Atlantic and Arcric Sea surface temperature and Deepwater production. Palaeogeogr. Palaeoclim. Palaeoecol. 309, 92–97.

Rohling, E.J., Foster, G.L., Grant, K.M., Marino, G., Roberts, A.P., Tamisiea, M.E., Williams, F., 2014. Sea-level and deep-sea-temperature variability over the past 5.3 million years. Nature 508, 477–482.

Rona, P.A., 1973. Relations between rates of sediment accumulation on continental shelves, sea-floor spreading, and eustacy inferred from Central North Atlantic. Bull. Geol. Soc. Am. 84, 2851–2872.

Runeberg, E.O., 1765. Om några förändringar på jordytan i allmänhet och under kalla kimat i synnerhet. In: Kungl. Sv. Vet. Akad. Handlingar.

Schlanger, S.O., 1986. High frequency sea-level fluctuations in cretaceous time: an emerging geophysical problem. In: Hsü, K.J. (Ed.), Mesozoic and Cenozoic Oceans. AGU/GSA, Geodynamic Series, vol. 15. GSA Publ, Washington, DC, pp. 61–74.

Shackleton, N., 1969. The last interglacial in the marine and terrestrial records. Proc. R. Soc. London, B 174, 135–154.

Shackleton, N.J., Opdyke, D., 1973. Oxygen isotope and paleomagnetic stratigraphyof equatorial Pacific core 228-238: oxygen isotope temperatures and ice volume on a 20^5 and 10^6 year scale. Quat. Res. 3, 39–55.

Shackleton, N.J., Opdyke, D., 1976. Oxygen isotope and paleomagnetic stratigraphy of equatorial Pacific core 228–239, Late Pliocene to latest Pleistocene. Geol. Soc. Am. Memoir 145, 449–464.

Shennan, I., 1987. Holocene Sea-Level Changes in the North Sea Region. In: Tooley, M.J., Shennan, I. (Eds.), Sea Level Changes. Blackwell, Oxford, pp. 109–151.

Shepard, F.P., 1963. Thirty-five thousand years of sea level. In: Clements, T. (Ed.), Essays in Marine Geology in Honor of K.O. Emery. Univ. S. California Press, Los Angeles, pp. 1–10.

Somi, E.J., 1993. Paleoenvironmantal changes in the central and coastal Tanzania during the upper Cenozoic. Magnetostratigraphy, sedimentary records and shorelevel changes. Ph.D. thesis, Stockholm University. Paleomagnetism & Geodynamics, No. 6.

Stoll, H.M., Schrag, D.P., 2000. High-resolution stable isotope records from the upper cretaceous rocks of Italy and Spain: Glacial episodes in a greenhouse planet? GSA Bull. 112, 308–319.

Stommel, H., 1976. The Gulf stream. A physical and dynamical description. Univ. California Press. 248 pp.

Suess, E., 1888. Das Anlitzder Erde, II: Die Meereder Erde. Wien.

Testut, L., Miguez, B.M., Wöppelmann, G., Tiphaneau, P., Poureau, N., Karpytchev, M., 2010. Sea level at Saint Paul Island, southern Indian Ocean, from 1874 to the present. J. Geophys. Res. 115. https://doi.org/10.1029/2010JC006404 C12028, 10 pp.

Thide, J., 1981. Reworked neritic fossils in upper Mesozoic and Cenozoic Central Pacific deep-sea sediments monitor sea-level changes. Science 211, 1422–1424.

Thorne, J., Watts, A.B., 1984. Seismic reflectors and unconformities at passive continental margins. Nature 311, 365–368.

Torell, O., 1872. Undersökningar öfver istiden. I & II. öfvers. Kongl. Sv. Vet. Akad. Förhandl. (I) 10, 25–66.

Tylor, A., 1868. On the Amiensgravel. Geol. Soc. Lond. Q. J. 24, 103–125.

Vail, P.R., Mitchum Jr., R.M., Shirley, T.H., Buffler, R.T., 1980. Unconformities of the North Atlantic. Phil. Trans. R. Soc. London A294, 137–155.

Waelbroeck, C., Labeyrie, L., Michel, E., Duplessy, J.C., McManus, J.F., Lambeck, K., Balnon, E., Labracherie, M., 2002. Sea-level and deep water temperature changes derived from bentic foraminifera isostopic records. Quat. Sci. Rev. 21, 295–305.

Walcott, R.I., 1972. Past sea level, eustasy and deformation of the earth. Quat. Res. 2, 1–14.

Walcott, R.I., 1980. Rheological models and observational data of glacio-isostatic rebound. In: Mörner, N.-A. (Ed.), Earth Rheology, Isostasy and Eustasy. John Wiley & Sons, Chichester, pp. 3–10.

Whittlesey, C., 1868. Depression of the ocean during the iceperiod. Am. Ass. Adv. Sci. Proc. 16, 92–97.

Woodward, R.S., 1888. On the form and position of the sea level. U.S. Geol. Surc. Bull. 48, 1–88.

Wooley, L., 1929. Ur in the Chaldees. Ernest Benn, London.

Vail PR, Mitchum RM Jr, Thompson S III, Todd RG, Sangree JB, Widmier JM, Bubb IN, Hatlelid. W G (1977) Seismic stratigraphv and global changes of sea level. Mem. of American Assoc. Pet. Geol., 26: 49–212.

PART 2

Shoreline and Coastal Changes

CHAPTER 3

Typology and Mechanisms of Coastal Erosion in Siliciclastic Rocks of the Northwest Borneo Coastline (Sarawak, Malaysia): A Field Approach

Dominique Dodge-Wan, R. Nagarajan
Department of Applied Geology, Curtin University, Miri, Malaysia

1 Introduction

Rocky coasts are dynamic, and respond to a number of environmental factors, at a range of spatial and temporal scales (Naylor and Stephenson, 2010). Many rocky coasts are, or are becoming, vulnerable to erosion, particularly in response to climate change-induced extreme weather events and sea level rise (IPCC, 2013). Among the environmental factors that influence the evolution of rocky coast landforms are climate and weather, waves, and tides (Sunamura, 1994). The ability of these factors to modify the coast will depend on a variety of geological conditions. Of particular significance are lithology and rock strength, as well as discontinuities within the rock mass (Dickson et al., 2004, Naylor and Stephenson, 2010). Among geological discontinuities are bedding planes, joints, faults, and changes in lithology and their orientation relative to the shoreline and waves. In addition, local ecology and conditions may lead to bioerosion from coastal plants and/or animals. Hydrogeology also plays a role, with the possibility of seepage contributing to increased erosion. Effective monitoring and management are pressing priorities in many developing countries, with increasing coastal populations and infrastructure (Paw and Thia-Eng, 1991; Johan et al., 2016; Cabral et al., 2017).

Rocky cliffs with arches and caves are potentially attractive geotourism sites, and, when located within a few kilometers of a major city, they may become popular tourism destinations. In recent years, the northwest coast of Borneo has seen an increase in visitor numbers due to improved access and the growing population. Additional visitors have also been attracted by an episodic blue bioluminescence phenomenon in the sea dubbed "Blue Tears" (Chong, 2016). Several scenic locations are now being developed with basic facilities, such as parking spaces, food, and drinks stalls, in order to cater to more visitors, and the cliffs at Tusan have been described as a regional tourism hotspot (Borneo Post Online, 2015). Improved roads have also led to new facilities at Peliau, a few kilometers south of Tusan.

As pointed out by Stephenson et al. (2013), the erodibility of rocky coasts is paradoxically what gives them their scenic value. The rock outcrops of the northwest Borneo coast have long been known to geologists as sites of geological interest, and high fossil content and their erosion potential has been pointed out (Wannier et al., 2011). Recently, there has been concern over the risks posed to visitors (Borneo Post Online, 2017). Similar coastal erosion issues have been documented in other parts of southeast Asia, which are also experiencing an increase in development and tourism (Sajinkumar et al., 2017). Despite this, to our knowledge, only limited research on the types, magnitude, and processes of coastal erosion has been carried out to date in this area (Anandkumar et al., 2018). This chapter aims to contribute by assessing and describing the types of erosion features, and providing an indication of their magnitude, as well as outlining the processes that may be responsible. Particular attention is given to geological discontinuities and their contribution to the type and location of erosional features.

2 Study Area

The study area is located on the northwest coast of Borneo, in the Malaysian state of Sarawak, south of Miri city (Fig. 1 and Table 1). Two areas of the coast are popular with visitors, and the rocky portions of these were selected to be examined in detail, and their erosion features mapped: the Tusan cliffs area, which is approximately 35 km SW of Miri, and the Peliau cliffs area, located 41 km SW of Miri. Three other individual sites were also observed: the south end of Taman Selera beach in Miri town, cliffs south of Beraya Beach, and a rocky portion of the Bungai Beach. All five sites are accessible by road and popular with visitors. Except for Taman Selera Beach, which is close to the Miri town center, the other sites are considered to have insignificant to low economic parameter scores on the Malaysian government's scale for evaluating economic parameters (National Coastal Erosion Study, 2015). However, there is evidence of some development in the past 10 years in the form of road upgradation from unsealed to sealed, new visitor facilities (covered market and toilets) at Bungai Beach, and drinks stalls, toilets, and parking areas at Tusan and Peliau.

Overall, this coast consists of relatively long stretches with straight beaches, with and without inland cliffs, interrupted by a small number of rocky headlands, most of which are fronted by beaches. At the Taman Selera Beach, the coastline is oriented 033 degrees, and to its immediate south lies the Tanjong Lobang headland, one of the few headlands without a beach. From Tanjong Lobang, the coast extending southwest to the Luak Bay Beach is the southern limit of Miri town. Further southwest from Luak Bay, the orientation of the coastline changes gradually from 025 to 043 degrees at Tusan cliffs, with a 10-km long central stretch oriented at 039 ± 1 degrees. Further southwest, the 9-km long stretch of coast that separates Tusan from Bungai Beach is marked by several erosional rocky headlands with minor beaches, including Tanjung

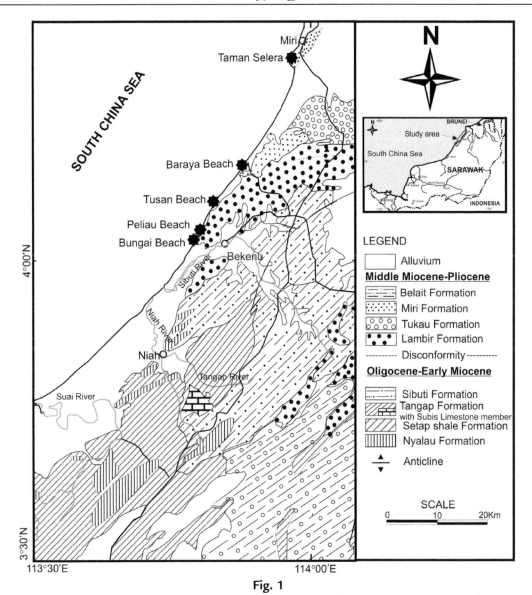

Fig. 1
Geological map of the study area showing the location of observation sites. *From Liechti, P.R., Roe, F.W., Haile, N.S., 1960. The geology of Sarawak, Brunei and the western part of North Borneo. British Borneo Geol. Surv. Bull. 3, 1–360.*

Batu, and a number of minor bays. Along this stretch, the orientation of the coast shows some irregularity, ranging from 012 to 034 degrees.

At the field areas of interest at Tusan and Peliau, the coastline is almost straight, and trends at an orientation of 043 to 031 degrees. The rock cliffs are generally more than ten meters high at the back of a sandy beach. The coast is microtidal, with a reported tidal range of 1.79 m, and the tide

Table 1 List of individual observation sites and mapped coastline sites with their coordinates

Site Name (In Order From NE to SW)	Investigation for This Study	NE Limit		SW Limit		Orientation of Coastline at Site
		Latitude	Longitude	Latitude	Longitude	
Taman Selera Beach (Miri)	Individual observation site	04.36775	113.96630	04.36775	113.96630	033 degrees
Beraya Beach South	Individual observation site	04.20394	113.88931	04.20394	113.88931	039 degrees
Tusan cliffs	Mapped coastline site	04.1294	113.8250	04.1256	113.8218	043 degrees
Peliau cliffs	Mapped coastline site	04.0786	113.7950	04.0784	113.7947	031 degrees
Bungai Beach	Individual observation site	04.06290	113.78391	04.06290	113.78391	034 degrees

type is classified as mixed (dominant diurnal) (Lim and Koh, 2010). In a number of places, the foot of the cliffs is immersed at medium and high tide, as well as during storm surges, and may be subject to direct wave action.

Climate in this region falls under the Koppen-Geiger climate classification category Af; that is, tropical rain forest climate. The average annual rainfall at the nearest rainfall station (Miri Airport) is 2694 mm (25 year average). The driest three consecutive months are generally July, August, and September, and the wettest are November, December, and January. Interannual and intermonthly variations in rainfall are high, and intense rainfall events are common.

3 Methodology

Several kilometers' length of rocky coast line with adjacent sandy beaches between Miri and Bungai Beach were examined (Table 1 and Fig. 1). Two sites, Tusan and Peliau, were selected for detailed observations, survey mapping, and photography. Coordinates were measured with Garmin 76CSx GPS. Key erosion features were surveyed using a tripod mounted Leica Disto D810 range meter. All macroscale erosion features were mapped, including cliff rim, cliff face, cliff toe, caves, scree, fallen blocks, and so forth. Measurements included map view surveys as well as profiles. All surveys were linked to UTM coordinates and sea level. Survey data were plotted with AutoCAD 2014 to produce maps and profiles. A Suunto MC2 compass clinometer was used to measure dip and strike of bedding, jointing, and rock faces. The orientation of geological surfaces is given in the usual expression, which is [strike/dip value and dip direction]. Recent photographs were compared with photographs by the authors dated 2007, to assess temporal changes. The eroded area was quantified using AutoCAD 2014 and calibrated photos. Structural data (strata and joint strikes and dips) were analyzed using Stereonet 9.5 software by Cardozo and Allmendinger (2013). A number of rocks were sampled and examined in thin sections with a Nikon Eclipse LV100NPOL polarizing microscope. Others were observed in the field using standard geological field methods. Elevations were computed by comparison of the surveyed water level at the time of each observation, with the corresponding tide levels at Miri. The datum is the 0 m of the tide chart, and all elevations are reported in "mad" (meters above datum). To analyze the density of isopod perforations on rock surfaces, sample areas of $25\,cm^2$ were photographed in the field for counting and surface estimation using AutoCAD.

4 Results

4.1 General

Previous research has documented decadal shoreline changes in a 74-km long coastal belt from Kuala Baram to Bungai Beach, with both accretion and erosion, and it has identified erosion hotspots at several sites, including Tusan and Peliau beaches (Anandkumar et al., 2018). This chapter is focused more specifically on the typology of erosion features at these sites.

The coastal region was observed to be dynamic, with evidence of recent recession and evolution, including landslides, active cave formation and collapse, eroding arches, and remnant sea stacks and debris piles. The processes involve both cliff top retreats, modifications of cliff faces and bases, as well as temporal changes in beach sand and scree deposits.

Typical beach width ranges from more than 40 to 25 m (from low tide line to base of cliff), and in several places, the high tide rises above the cliff-beach junction, with direct wave action on the rock surface. The slope of the beach face, below the wrack line, was measured at Tusan (five profiles) and at Peliau (one profile), and found to be approximately 2 degree at all sites. Higher slopes, up to 12 degree, were observed in places adjacent to the cliff, showing that there are areas of sediment supply from landslides and debris piles.

This coast is classified as Type—A Sloping Shore Platform in the classification proposed by Sunamura (2015), and this is the case at both sites studied in detail, irrespective of strata dip. At Beraya Beach, all cliffs are fronted by a beach, and there are no rocky platforms exposed. In limited parts of the beach at Tusan, Peliau, Bungai, and Taman Selera, there is a partially sand covered wave cut intertidal platform of less than ten meters' width. At Bungai Beach, the more resistant beds of the tilted bedrock form low rocky ridges, which protrude <0.5 m above the sand at low tide. The less resistant rock beds, such as mudstones, are eroded, recessed, and covered with sand. Cliff top vegetation is grasses, shrubs, the screw pine *Pandanus,* and small *Casuarina* trees.

4.2 Geology

The northern region of Sarawak is underlain by a thick sequence of Cenozoic sediments. On the whole, the sequence is younging toward the N and NNE, from Oligocene to Pliocene, with a number of lithostratigraphic formations (Nyalau, Setap, Tangap, Sibuti, Belait, Lambir, Tukau and Miri) formed under shallow to deep marine conditions. In the area of interest, the rocks are of the Miocene age, and belong to the Sibuti, Lambir, Tukau, and Miri formations (Nagarajan et al., 2015).

The early Miocene to late-middle Miocene age Sibuti Formation (Hutchison, 2005; Simon et al., 2014) is essentially shaley, with some coarser clastics and minor limestones. The Lambir Formation consists predominantly of sandstones, with minor shales and rare marl/limestone interbeds that were deposited in deltaic to fluvial environments (Nagarajan et al., 2017). It is dated at mid-Miocene on the basis of planktonic foraminifera, and contains sedimentary structures such as hummocky to planar cross-bedding (Hutchison, 2005). The Middle Miocene Miri Formation (Hutchison, 2005) consists of alternating sandstone and shale, with shales slightly more dominant in the lower part, and arenaceous rocks more dominant in the upper part. This formation is affected by a number of faults, and hosts the oil-bearing rocks that have been exploited in the oilfields around Miri, both onshore and offshore (Jai and Rahman, 2009). Table 2 summarizes the geological information for each of the outcrops discussed in this chapter, with information on the varying strikes and dips of these tilted formations. The geology at each site is briefly described herein. Additional information on the geology of the region is also available in Chapter 12 of this volume (Nagarajan et al., 2019).

At Taman Selera, the rocks belong to the "Pujut Shallow Sands" sequence, which is part of the Miri Formation, striking approximately 020 degrees and dipping 23°W (Wannier et al., 2011). They are massive sandstones with some *Ophiomorpha* burrows and minor mudstone laminae (Fig. 2A).

At the Beraya Beach southern cliff, the rocks belong to the Miri Formation, which consists predominantly of fossiliferous mudstones, siltstones, and sandstones dissected by a series of faults (Wannier et al., 2011). The sandstones contain quartz pebble, lignite, and amber clasts. Bed thickness is variable, with some rare sandstone beds more than 2 m thick. *Skolithos* and *Ophiomorpha* burrows are common, together with irregular mudstone and organic laminations. Resistant sandstone cliffs are present in only limited places due to the predominance of mudstones, siltstones, and faulting.

At Tusan, the sandstone beds of the Lambir Formation show variable thicknesses with mudstone units being frequently interbedded with the sandstones. The mudstone beds are generally thinner than the more massive sandstones. Ichnofossils are frequent, including *Ophiomorpha, Skolithos,* and *Dactyloidites*. The sandstones are commonly cross-laminated and show current bedding. Bedding planes often show ripple marks. Rock surfaces are coated in

Table 2 Rock formations, lithologies, and structural data at the five rocky coastal sites observed for this study

Site Name (In Order From NE to SW)	Rock Formation	Dominant Lithology	Average Strike of Strata	Average Dip of Strata	Photograph of the Sites
Taman Selera Beach	Miri Formation (Pujut Shallow Sands)	Massive sandstone, minor mudstone laminae	020 degrees	23°W	
Beraya Beach south	Pleistocene Sand, Miri Formation	Weak pebbly sandstone and mudstones	048 degrees	20°W	

(*Continued*)

Table 2 Rock formations, lithologies, and structural data at the five rocky coastal sites observed for this study—Cont'd

Site Name (In Order From NE to SW)	Rock Formation	Dominant Lithology	Average Strike of Strata	Average Dip of Strata	Photograph of the Sites
Tusan cliffs	Pleistocene Sand, Tukau Formation, Lambir Formation	Alternating sandstone (dominant) and mudstone beds	020–054 degrees	32° to 38°NW	
Peliau cliff	Pleistocene Sand, Lambir Formation	Alternating sandstone (dominant) and mudstone beds	050 degrees	86° to 88°NW	
Bungai Beach	Sibuti Formation	Alternating sandstone and mudstone beds	060–070 degrees (anticlinal fold)	60° to 80° to NW and to SE (fold)	

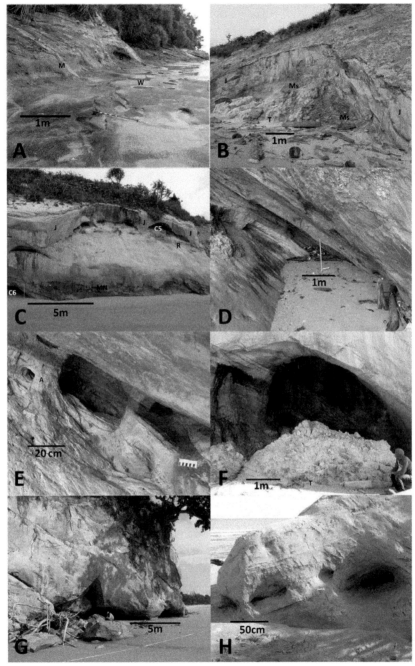

Fig. 2

(A–H) Field photographs of macro- and mesoscale erosion features. (A) *Wave cut platform* (W), *marine erosion notch* (M) and *bedding plane enlargement* (B) at Taman Selera during low water spring tide. Above the cliff, the vegetation is dominated by *Pandanus*. High tide level is approximately 1 m above the line of green algae visible to base of notch. Upper portion of cliff face is wet due to groundwater percolating through overlying soil. (B) *Mass wasting* feature EF1 at Tusan, showing translational to rotational slide

iron oxides, sometimes forming a fine brittle crust-like deposit. The coastline is oriented approximately 043 degrees, roughly parallel to the strata that show strike values with a range of 020–054 degrees ($n=13$). The dip ranges from 32° to 38°NW (Fig. 7). The formation is affected by faults whose alignments are generally perpendicular to the coast (Wannier et al., 2011). Sandstone porosity is variable from 9% to 28% (Ali et al., 2016). At the top of the rock cliff, there is a horizontally bedded lignitic sand deposit (Pleistocene Terrace). It is a few meters thick, and unconformably overlies the tilted Miocene strata. Tectono-geomorphic investigations indicate that the topography of northern Sarawak is evolving in response to recent uplift (Mathew et al., 2016). Lignite samples taken from the Pleistocene Terrace deposit at Tusan were dated by C^{14} analysis as being $15,550 \pm 80$ years BP, and their current elevation 22 m above sea level indicates very recent uplift (Kessler and Jong, 2014). These superficial deposits are permeable sands, and are also present at several other sites (Table 2).

At Peliau, the subvertical to steeply dipping beds of the Lambir Formation are aligned generally in a seaward orientation. Where these beds consist of relatively resistant sandstones, they form a

erosion of seaward dipping strata affecting cliff face and cliff toe, but not yet extending to cliff top. Oblique joints (marked J) and mudstone bed (Ms) have resulted in crescent shaped feature and debris pile. Present size of debris pile, and linear toe of debris pile (T) suggests that some debris has been removed by wave action. (C) *Mass wasting* feature EF2 at Tusan, showing role of joints (marked J) in promoting translation slide type erosion on seaward dipping sandstone strata and cave formation controlled by lithology at base of upper cliff portion. Below cave C5, flow on rock surface has formed shallow runnel in bedding cliff face (R). The lower cliff portion shows marine notch (MN) and small cave (C6) at beach level on extreme left. (D) *Tabular lithologically controlled cave* C3 at Tusan, with mud rich beds exposed in back wall (visible in top left corner of photo). The base of overlying resistant sandstone forms the cave ceiling (hanging wall top right). The cave has total length of 19 m, of which about half is visible in the photo, and it extends parallel to strata strike. (E) *Mud rich bed* visible in side walls of cave C2 at Tusan: notice formation of small arch top left (marked A). At bottom right, circular light colored patches in the mud rich bed are trace fossil burrows filled with sandstone. Sloping ceiling of cave (hanging wall) is bedding plane formed by base of resistant sandstone bed above the cave. This cave is classified as "tabular—lithologically controlled". (F) *Domed collapse controlled cave* C7 at Tusan showing arched ceiling caused by collapse and debris pile below. Plant rootlets visible in cave ceiling and evidence of meteoric water infiltration indicate that collapse of the pseudokarst dropout doline above may be imminent with only a few meters of rock thickness remaining above the domed cave ceiling. Strata dip is from top left to bottom right. Toe of debris pile is eroding during high water (marked T). (G) *Prismatic joint controlled caves* C13 (behind person) and C14 (approximately 5 m to right) in the cliff at Peliau. Strata are sub-vertical and bedding surface forms the cliff face. At least two joint directions are visible in the rock face of which the one dipping approximately 30 degrees to SW (to the right in photo) is the most dominant. In foreground, debris pile from rock fall mass wasting with blocks over 3 m length which are submerged at high tide. (H) *Bedding plane enlargements* associated with discontinuous mud rich beds, at site P2 (Tusan) where strata dip approximately 35°NW (towards the sea visible in background). Orange linear features are evidence of iron cementation in *Skolithos* and *Ophiomorpha* ichnofossils.

steep rock cliff at the back of the beach. Strike is consistently 050 degrees, and dip values are around 86° to 88°NW. The area of interest consists predominantly of sandstones, although further south toward Peliau Bay, the lithology is more variable with thick mudstones, siltstones, and some carbonate-rich clastic beds.

At Bungai Beach, folded Miocene rocks of the Sibuti Formation, including alternating thin beds of sandstones, siltstones, and mudstones, are visible at low tide in a wave-cut platform where these rocks form an anticlinal fold with subparallel limbs and strata dips exceeding 60 degrees. The presence of a few resistant sandstone beds in this location has created a small headland that is currently eroding.

At all the sites, except, to a lesser degree, in the Miri Formation at Taman Selera, the sandstones are heavily jointed. Jointing is discussed in the sections on macroscale and mesoscale erosion features.

4.3 Typology of Coastal Erosion

A number of erosion features have been observed at different locations. They also show varying controls of lithology, structure, and so forth. Based on these, a typology of these erosion features has been established and classified according to size, from macroscale (features more than 10 m in size) to mesoscale (features from a few meters to less than 1 m in size), and down to microscale (features less than 1 cm). Here they are listed and discussed, in order of decreasing size, as summarized in Table 3.

4.3.1 Macroscale erosion features

Mass wasting erosion features

Mass wasting erosion features exceeding $50 m^2$ in a map view were observed in sandstone bedrock at four locations along the coast at Tusan. Information on these locations, dimensions, and key characteristics of the features are presented in Table 4, and shown in photographs in Fig. 2B and C. The features are shown in a map view and a cross-section view in Fig. 3, and described briefly herein.

On rocky coastlines, mass wasting may contribute to recession of the cliff top, the cliff face, and/or the cliff base (Young et al., 2009). This type of erosion feature (EF) is often complex, and may combine several other types of erosion features, that is, bedding plane enlargements, bedding plane caves, joint enlargement, joint caves, and so forth. They also provide evidence of different stages of the process of coastal retreat.

Erosion features EF4 and EF3 are close to each other and display similar characteristics, but with EF4 being a larger feature (see Fig. 3). Both features involve coastal retreat, that is, a recess in the entire rocky cliff of 11 m (EF3) to 12 m (EF4) (measured perpendicular to coast and

Table 3 Typology of coastal erosion features observed and their relative importance in the surveyed areas of the different sites in NW Borneo coast

Scale	Typology	Taman Selera	Beraya Beach	Tusan Cliffs	Peliau Cliffs	Bungai Beach
Macro	Mass wasting (major features)	−	++	+++	++	NE
	Rock falls and topples	−	+	+	+++	NE
	Caves	+	−	+++	+	NE
	Pseudokarst dropout doline	−	−	+	−	NE
	Arches	−	+	+++	++	NE
Meso	Bedding plane enlargements	++	+	++	+	NE
	Joint enlargements	−	+	++	+++	NE
	Pockets	+	+	+	+	NE
	Subaerial gulleys and runnels	−	+	+	−	NE
	Tafoni	−	+	+	+	NE
	Wave cut notches	+	+	++	+	NE
	Near-horizontal platforms	+	NE	+	+	+
	Intertidal runnels	+	NE	+	−	−
	Intertidal rock pools	+	NE	+	−	−
Micro	*Isopod perforations*	+	NE	+	+	+

+++, very common feature and significant component of erosion; ++, common feature; +, observed but either not common or not significant component of erosion; −, not observed (in surveyed area); NE, not expected (see text for details). Features shown in italics are in the intertidal zone.

strike), and both features are backed by a subvertical rock wall forming the headscarp. At EF4, there is a 24-m long scree slope between the headscarp and the beach. At EF3 there is no scree, and the headscarp is a 10-m high subvertical rock face of relatively soft sandstone, suggesting that rock fall is the dominant mechanism in this case. Both recesses have side walls that are controlled by joints oriented approximately perpendicular to strike, and these give rise to planar subvertical rock walls forming the two sides that are perpendicular to the coast. Hence, in a map

Table 4 Details of large-scale mass wasting erosion features observed in Tusan cliffs

Site Reference (Listed From NE to SW)	Latitude N	Longitude E	Width (Parallel to Coast)	Depth of Recess, i.e., Coastal Retreat (Perpendicular to Coast)	Map View Shape	Associated Caves	Comments
EF3	4.12940	113.82504	9 m	11 m	Rectangular	C7 plus three smaller caves	Adjacent to EF4
EF4	4.12918	113.82481	20 m	12 m (plus additional 24 m scree slope)	Rectangular	C3 in SW	C3 is longest cave observed; scree slope on inland side
EF2	4.12873	113.82429	13 m	9 m	Semi-circular	C5 and other smaller caves, C6 at beach level (not connected)	Crescent shape, high up on cliff
EF1	4.12562	113.82184	16 m	7 m (from rim break of slope)	Semi-circular	None	Several mudstone failure surfaces

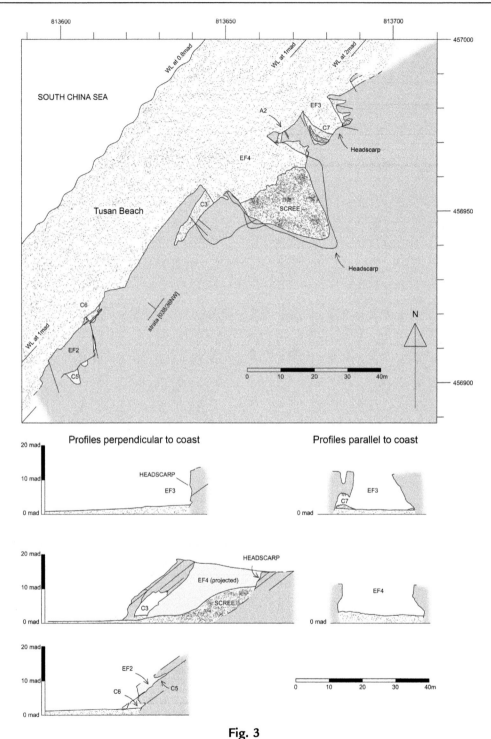

Fig. 3
Map and sections of coastline showing cliff features at Tusan. Diagonal lines top right and bottom left indicate position of water line at different tidal stages in meters above datum. Coordinates of map grid are UTM Zone 49N.

view, the features appear rectangular, with their long side parallel to the coast (Fig. 3). The cliff base in the side walls may be eroded and overhanging by several meters, and/or may host caves, such as cave C7 (Fig. 3). The toe of the rubble piles, both from rock-slide scree and from cave roof collapse (as in C7), is eroded during high tide, and this material contributes to beach sand accumulation.

EF1 is 16m wide and 3m high, and has an arcuate headscarp in soft sandstone (Fig. 2B). The lateral walls are formed by subvertical joints oriented obliquely to the coastline (154 and 074 degrees). This gives the feature a curved aspect in a map view, in contrast to the rectangular map view shapes of EF4 and EF3 (Fig. 3). There are several beds of mudstone that form failure surfaces, and this is more obvious at EF1 than at the other sites, where the lithology is more homogenous. In profile, the rupture surface at EF1 shows three characteristics (Fig. 2B): (1) an upper portion with a vertical headscarp below, (2) a small area of exposure of the underlying mudstone bed with the slip surface dipping 38 degrees seaward, and (3) a lower portion with the typical curved profile of a rotational slide. This suggests that the mechanism is mixed: initially translational, and followed by rotational slide. The rotational slide has cut into much of the mudstone bed, which formed the rupture surface of the initial translational slide.

EF2 differs from the other three large-scale features in being above the beach on the sloping rock cliff, beyond the reach of wave action (Fig. 2C). In a map view, the headscarp has a curved aspect, with dimensions similar to EF1. The feature is a translational slide. Several caves are located at the foot of the headscarp, and they are of the bedding plane enlargement type, with some collapse also evident in the largest cave (C5, see Fig. 3 profile). There is evidence of groundwater flow seeping out along the failure surface, and formation of a slight erosion runnel on the bedding surface and cliff below (Fig. 2C). For this mass wasting feature, there is no remaining talus debris or scree, suggesting that the material slipped to beach level and was subsequently eroded. This feature is an example of cliff face erosion (Tables 5 and 6).

In addition to the mass wasting features, tension cracks are present at the cliff top at Tusan. They are located close to the cliff edge and parking-utility area, where people gather to view the

Table 5 Details of dimensions and volume of material eroded in large mass wasting features at Tusan (excludes volume of associated caves)

Site Reference (Listed From NE to SW)	Area in Map View (m^2)	Average Height (m)	Approximate Volume (m^3)	Dominant Mass Movement Type
EF3	87	10.9	948	Rock fall
EF4	671 (297+320 scree)	9.3 (scree 7)	5002 (2760+2240 scree)	Rotational
EF2	113	2.5	282	Translational
EF1	82	2.6	215	Translational +rotational

Table 6 Components of coastal cliff retreat observed in mass wasting features at Tusan

Site Reference (Listed From NE to SW)	Component of Cliff Retreat			Headscarp	
	Cliff Top	Cliff Face	Cliff Toe	Map View	Profile
EF3	8 m	8 m	8 m	Linear/strike	Subvertical
EF4	12 m	12 m	12 m	Linear/strike	Subvertical in soil (base covered with scree)
EF2	None	9 m	1.3 m	Arcuate (radius: 10.8 m)	Irregular (with caves at base, including C5)
EF1	None	Translational: 2.2 m Rotational: 0.7 m	5 m	Arcuate (radius: 10.6 m)	Mixed (see text)

Cliff slope retreat is measured perpendicular to strata bedding.

coastline, particularly at sunset. This suggests that further mass wasting events will likely occur in the near future, with the potential of risk to infrastructure and visitors. Similar mass wasting is noted at other locations along the northwest Borneo coastline, with a number of large features, particularly in the Tusan to Peliau zone.

Rock falls

There is evidence of rock fall at a number of sites. The mass wasting feature EF3 at Tusan (see the previous section) shows evidence of former rock fall, with most of the debris pile having now been eroded from the foot of the subvertical headscarp cliff. At *Peliau*, where the strata dip is subvertical, rock fall is common, with two major debris piles observed along the 80 m length of surveyed coastline (see maps in Figs. 3 and 5, and photos in Figs. 2G and 6C). The Peliau rock falls contain blocks over 3 m diameter, and these are partially submerged at high tide. Fig. 6C shows pronounced jointing in vertical bedded rocks, suggesting that future rock falls are most likely at this site.

Caves

Caves are commonly defined as cavities large enough for human entry (Selley et al., 2005), and sea caves are a common feature of this coastline. A total of seven caves were observed at Tusan, and eight at Peliau, in addition to many smaller cavities (<human size), which are described under mesoscale erosion features.

Table 7 lists the caves and presents their key features, including dimensions and relationship to strata bedding direction and/or jointing. All the listed caves, except one, have their main orifice at or close to the elevation of the beach. So the majority of caves are reached by the high tide and/or by storm surge waves, and their formation is due to the erosive action of waves. Cave C5,

Table 7 List of caves at Tusan and Peliau sites (largest caves in bold)

Site	Cave Ref. No.	Dominant Shape and Control Mechanism			Dimensions			Approx. Volume (m³)	Map View Shape	Figure Number
		Tabular Lithological Control	Prismatic Joint Control	Domed Collapse Control	Parallel to Strike (m)	Perpendicular to Strike (m)	Height (m)			
Tusan	C1	+			5.1	0.71	0.3 to 1	<2	Semi-circular with 2 branches	
	C2	+			3	2 to 3.3	0.8 to 1.9	10	Irregular triangle	Figs. 2E and 8C
	C3	+			19	2.34 to 4.2	5.5	820	Elongated with subparallel walls	Figs. 2D and 3
	C4	+			11.6	2 to 2.7	1.8	20	Elongated with subparallel walls	Figs. 2C and 3
	C5	+			3.4	4.2	0.7	10	Semi-circular	
	C6	+			5.4	1.2 to 2.3	0.3 to 1.2	4	Rectangular with lateral branches	Fig. 2C
Peliau	C7			+	3.5	7.3	3.6 to 4.2	72	Semi-circular	Figs. 2F and 3
	C8		+		2.7	7.5	0.8	<1	Elongated triangle	
	C9		+		1.1	2.4	0.4	16	Elongated rectangle with subparallel walls	
	C8+9	+	+		6.6	12.4	4	110	Rectangle with two branches perpendicular to strike	Fig. 5
	C10		+		1.4	4.8	0.7	<5	Elongated subparallel walls	Fig. 5
	C11		+		1.0	5.1	1.4	<5	Elongated with subparallel walls	Fig. 5
	C12		+		2.6	7.0	0.9	8	Elongated oval	Fig. 5
	C13		+		4.8	6.3	1.1	17	Elongate oval	Figs. 2G and 5
	C14		+		3.1	9.6	0.9	13	Elongated with subparallel walls	Figs. 2G and 5
	C15		+		1.6	>7.3	1.4	8	Elongated with subparallel walls	Fig. 5

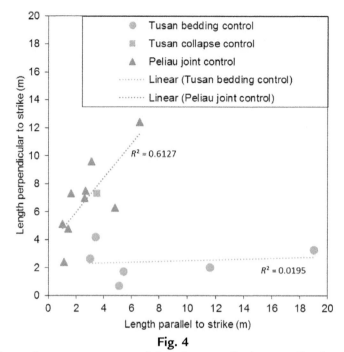

Fig. 4

Principal dimensions of caves measure parallel and perpendicular to strike showing the difference between the three types of caves: tabular-lithological control *(blue)*, prismatic-joint control *(orange)*, and domed-collapse control *(grey)*.

and several other similar small caves that are not listed, but are visible in Fig. 2C, are located on dip slope 10 m above datum, beyond the reach of high tide.

Among the seven caves observed at *Tusan*, four caves have their orifice positioned on a joint in the rock, and three caves have their orifice in the rock bedding surface. All of the Tusan caves, except C7, have their maximum extension (length) parallel to the strike of bedding, and this is particularly clear at C3, which extends 19 m parallel to strike, but is less than 5 m wide throughout. Fig. 4 shows the principal dimensions of the caves measured parallel and perpendicular to strike. The Tusan caves are elongated parallel to strike. The majority of the Tusan caves can be considered to have a tabular cavity shape inclined parallel to the strata dip, and a mud-rich bed, or a series of mud-rich laminations, are evident in the deepest recesses of the cave walls (Fig. 2D and E). In several cases, a more resistant sandstone bed is evident in the cave's ceiling, that is, the hanging wall. These observations indicate a strong lithological control of cave formation and morphology. In the classification proposed here, these caves are described as "*tabular-lithologically controlled.*" Cave C5, which is above the reach of marine erosion, is also lithologically controlled, with evidence of groundwater percolation through the cave caused by the presence of a lower permeability layer (Fig. 2C). The erosion of C5, and several other small caves in a similar situation, appear to have been facilitated by the translation slide type mass movement on a tilted bedding surface, discussed in the previous section (EF2). This mass movement has exposed a crescent-shaped headscarp, and led to recession at the foot of it, where the caves are located. Cave C7, shown in Fig. 2F, differs from the other caves at Tusan, as it has a dome shape with an arched ceiling. There is a debris pile below, suggesting that collapse is the more dominant control factor in the cave's formation and morphology. It is classified as "*domed—collapse controlled.*" On the land surface above C7 there is a linear shallow valley that runs approximately SE to NW, that is, perpendicular to the coastline. In the valley directly above the cave there is a 2-m diameter circular depression, similar to a doline. The bottom of it is approximately 3 m below the average ground surface, and the pit walls are vertical. At the time of observation (20/7/2016), the valley was dry, but there was stagnant water ponding in the pit. The ponding water feeds the percolation drips in the cave ceiling below. The pit can be considered as a pseudokarst dropout doline (Waltham et al., 2005) formed in semi-cohesive sand and weak sandstone. It is likely that with further erosion it will evolve into a collapse doline. *Pandanus* plants grow in the valley and within the pit, and their roots appear to be providing some temporary stability to the domed cave roof.

In comparison, at *Peliau*, the strata bedding is subvertical, and the cliff face is parallel to strike. All the caves at Peliau show different configurations relative to bedding compared with those at Tusan, as shown on the Peliau map in Fig. 5. The smaller Peliau caves are prism-shaped cavities with triangular entrances located at the foot of the cliff (Fig. 2G). Their sloping side walls are a result of widening of a subvertical joint oriented roughly perpendicular to the coastline. They are classified as "*prismatic—joint controlled.*" The larger caves have more varied morphologies, but all have a maximum extension (length) approximately perpendicular to the strike of the bedding. In all Peliau caves, the length perpendicular to strike exceeds that parallel to strike, as shown in Fig. 4. The cave lengths perpendicular to bedding are generally between 5 and 10 m. Within these caves, small-scale erosion of softer, mud-rich beds may also be evident with minor lateral passages parallel to bedding. On the whole, at Peliau, cave formation does not appear to be predominantly controlled by lithological variations. In one case, two caves, C8 and C9, have developed along joints located less than 20 m apart from each other, and have now coalesced into a single wider cave in which the joint control is still obvious, but in which bedding control is also indicated. Several Peliau caves are connected to the surface by subvertical shafts with joint and/or lithological control. In cave C9 at Peliau, a recent iron oxide-rich conglomerate mass measuring several meters long and approximately 0.5 m thick was observed, where pebbles and clasts were embedded, forming a cave beach rock. Less than a year later, this material was no longer present, indicating that mechanical marine erosion is active within the cave at a short time scale. It should be noted that at Peliau, the foot of the cliff and the base of all the caves are submerged at high tide.

The comparison between the two sites shows that in addition to lithology, strata tilt has a critical role in determining the type of cave formation. In moderately tilted strata such as at Tusan,

84   Chapter 3

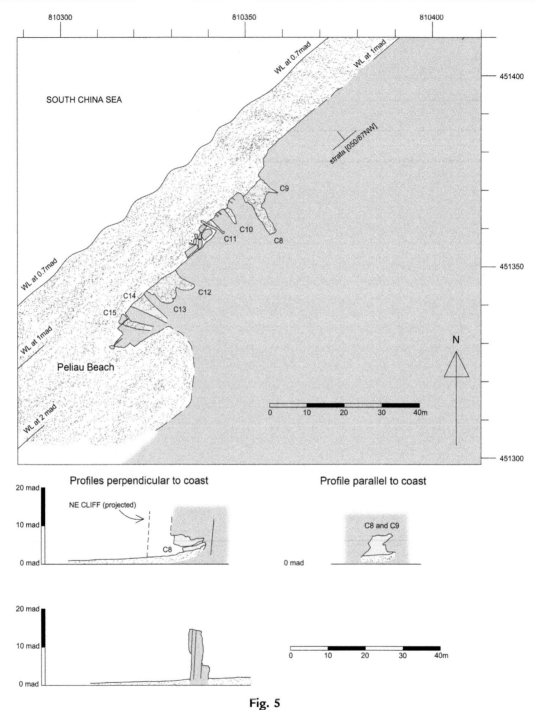

Fig. 5
Map and sections of coastline showing cliff features at Peliau. Diagonal lines *top right* and *bottom left* indicate position of water line at different tidal stages in meters above datum. Coordinates of map grid are UTM Zone 49N.

caves have developed preferentially along lithological weaknesses (bedding control). In more steeply tilted to subvertical strata, such as at Peliau, they have developed preferentially along joints.

Arches

Wannier et al. (2011) have shown before and after photographs of an arch collapse feature that occurred in the spring of 1999 in the Tusan cliffs. While that arch has collapsed, several others are still surviving. For this study, three other arches at Tusan were examined, as listed in Table 8.

Arch A0 at Tusan is the largest remaining arch in the region. It is frequently visited by local photographers, and features as an attraction in the Miri Resort City website (Miri Resort City, 2017). We have estimated its dimensions based on scaled photos taken recently and in 2007. This shows that in the past 10 years, there was an increase in span height from <4 m to more than 8 m, and an increase in up dip length from 5 m to approximately 15 m. Overall, the area of the orifice in profile has increased from approximately $14\,m^2$ to more than $70\,m^2$. This amounts to a five-fold increase in volume, with the greatest increase being in the up-dip direction, indicating a strong lithological control. The remaining keystone has a stratigraphic thickness of 24 m. The side walls of the pillar appear to be joint-controlled, and oriented approximately perpendicular to the coastline. There are relatively large landslides on either side of the arch where weaker mudstones have been exposed.

Arch A1 at Tusan is much smaller, only approximately 1.4 m high, and is due to the erosion of a mudstone layer approximately 0.5 m thick. The keystone is a sandstone bed less than 1 m thick. The arch is particular in having two pillars (legs) that extend seaward of the main cliff face by 6 m, and with a horizontal span of less than 4 m. In a map view, the pillars form the two sides of a triangle, and each is less than 2 m wide. The lateral faces of the pillars are controlled by two sets of subvertical joints trending respectively at 148–156 and 172 degrees.

Arch A2 at Tusan is a small arch located between mass wasting features EF3 and EF4. The pillar is 3 m diameter, and the arch appears to be predominantly controlled by lithology, while the side faces of the pillar are joints oriented roughly perpendicular to the coast. The keystone is a 2.5-m thick sandstone bed.

All three arches show lithological control of the cavity that formed where a weaker bed or beds were more readily eroded. There is joint control of the lateral faces of the pillar(s). It is unclear if the Tusan arches result from the breakdown of former caves, but that is certainly likely. In that case, the precursor caves fall under the "tabular—lithologically controlled" cave type. This phenomenon is observed at small scale too, in the presence of small arches (<1 m in any dimension) inside lithologically controlled caves such as C2 (see Fig. 2E). Knowing that elongated lithologically controlled caves may evolve into arches can assist in predicting the site of future arches and eventual collapse locations.

Table 8 Location and dimensions of three arches at Tusan (numbers refer to 2017 situation)

Site	Ref. No.	Latitude	Longitude	Key Dimensions					Note
				Horizontal Span		Keystone			
				Parallel to Strike (m)	Perpendicular to Strike (m)	Height of Cavity Below (m)	Stratigraphic Thickness (m)		
Tusan	A0	04.12167	113.81830	<10	8	8	24		Large single arch
	A1	04.12685	113.82275	4.3	4.2	1.4	1		Small double arch, collapse likely
	A2	04.1293	113.825	2.7	1.1	1	2.5		Small single arch

4.3.2 Mesoscale erosion features

Bedding plane enlargements

Bedding plane enlargements are small-scale features, below cave size, that is, they are not large enough for human entry, but their formation is similar to that of "tabular—lithologically controlled" caves. They may be considered an initial step in the formation of this type of cave. They are typically more than a meter deep, extending parallel to strike, but less than a meter high and/or wide. They are commonly located in the lowest 2 m of the cliffs, although similar features are also found at the base of the headscarp in translational mass wasting forms (such as EF2, see Fig. 2C). Bedding plane enlargements are related to mechanical erosion of softer mud rich beds.

Joint enlargements

At *Tusan*, where the strata dip 32° to 38°NW, much of the cliff face has a similar slope. The rocks are jointed, and many of the joints are enlarged close to the intertidal zone, as shown in Fig. 6A. Strike and dip of joints measured ($n=41$) were plotted on stereonet (Fig. 7). The figure shows a clear clustering, with one set of subvertical joints striking between 120 and 160 degrees. The joints are typically enlarged to approximately 30 cm width at most, and the cavities are 1–2 m long (perpendicular to cliff face). They are typically 1–2 m high, closing to a tight joint above the zone of marine erosion (Fig. 6A). At Tusan, although there are many enlarged joints, none are large enough to be considered caves.

At *Peliau*, where the strata dip is subvertical, the situation is different, and erosion of joints is a major process, with several joint sets playing a major role. Many of the joints at the toe of the cliff are enlarged to several meters width, height, and/or length; and hence, are considered caves, as discussed previously (see Table 7 and the previous section). There are also numerous joints showing only minor enlargement, comparable to those observed at Tusan. Two sets of joints with the following strikes and dips were notable: set 1 [136–152 degrees/approx. 30°SW] and set 2 [108–110 degrees/approx. 88°NE]. There are also other, less-frequent joints with different orientations [108°/60°SW] and [140°/45°E], as shown on the stereonet (Fig. 7).

Pockets

Pockets are mesoscale features formed by mechanical erosion around a cylindrical-shaped rock weakness, such as trace fossil burrow. The trace fossil *Ophiomorpha* is common, with straight branching burrows up to a meter in depth. The trace fossil burrows are approximately 2 cm diameter, but erosion may widen them to pockets of up to 15 cm diameter (Fig. 8A). The length of the erosion pockets ranges from a few centimeters to several decimeters. Although the pockets are common and found at numerous sites, volumetrically they do not account for much of the overall erosion.

Fig. 6

(A–D) Field photographs of mesoscale erosion features. (A) *Joint enlargement* (JE) at foot of Tusan cliff. Cliff face is almost parallel to bedding with shallow wave cut notch. Rock face shows several small pocket erosion features, the largest is marked (P). Walls of pocket as well as other areas adjacent to joint enlargement show *perforations by isopod Sphaeroma triste*. There are also minor bedding plane enlargements. (B) *Subaerial gulley* in soft sandstone at Beraya Beach cliff. Note white sand that has been washed off the cliff and now deposited on beach surface. The strata are dipping toward sea (located to left, off the photo) and bedding surfaces are visible dipping in top left corner of photo.

Orange surface is iron oxide coating of joint. (C) *Tafoni* (marked T) on surface of steep sandstone cliff at Peliau with *rock fall* at foot of cliff. Note that strata are subvertical with strike parallel to photograph and some bedding surfaces have fossil ripple marks (R). Several sets of joints are visible. (D) *Intertidal runnel* on near-horizontal platform at Taman Selera, as observed during low water spring tide. The gulley is eroded in sandstone and has meandriform pattern with two rock pools where pebbles have accumulated. Note the differential encrustation of various rock faces with barnacles and algae and the lack of encrustation in the runnel.

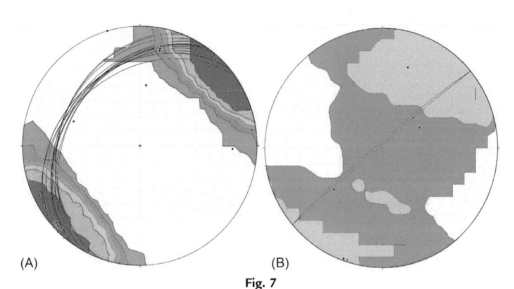

Fig. 7

Stereogram of bedding and jointing at Tusan (A) and Peliau (B). (A) Stereogram for Tusan: great circles ($n=13$) show bedding, black dots are poles of joints ($n=41$) and contours for joints show one dominant cluster. (B) Stereogram for Peliau: great circles ($n=2$) show bedding which is subvertical [050/86 to 88 NW], black dots are poles of joints ($n=9$) and these show two clusters as described in text. *Plotting software by Cardozo, N., Allmendinger, R.W., 2013. Spherical projections with OSXStereonet. Comput. Geosci. 51, 193–205. doi: 10.1016/jcageo.2012.07.021.*

Subaerial gulleys and shallow runnels

Clear evidence of soft rock gully formation in sandstones was only observed at the Beraya site where the pebbly sandstones appear to be weaker than at the other sites. At Beraya, several irregular gullies have formed on a steeply dipping joint controlled cliff face in soft sandstone (Fig. 6B). The gullies are stepped, and approximately 5–15 cm wide, generally widening downstream. Each step contains deposited loose sand, and sand is washed out onto the beach surface at the toe of the cliff.

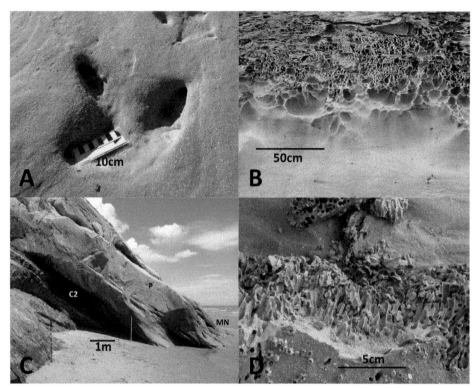

Fig. 8

(A–D) Field photographs of meso- and micro-scale erosion features. (A) *Erosion pockets* at Tusan. The pockets have developed due to widening by erosion of a branching trace fossil *Ophiomorpha*. The characteristic wall of *Ophiomorpha* is visible inside the top pocket, which is oriented perpendicular to the line formed by the other two pockets. (B) *Honeycomb weathering (tafoni)* on sloping bedding plane with iron oxide crust at Tusan. Note that the tafoni appear to be protected by remnant crust and several tafoni may coalesce as they erode once the crust is itself eroded. (C) *Marine wave-cut notch* (MN) and *cave* C2 at Tusan that has formed on mudstone bed (visible top left). The rock face is joint controlled and consists of cross-bedded sandstone of Lambir Formation. There are several bedding plane enlargements. On the jointed rock face are small erosion pockets (near P for example). Photo taken at low tide. Interior details of cave C2 are shown on Fig. 2E. (D) *Isopod perforations* on freshly broken sandstone surface in the intertidal zone at Tusan. Note the relative absence of perforations where more resistant iron oxide rich coating has not been breached (bottom of photo). The perforations are by the isopod *Sphaeroma triste* many of which can be seen in their burrows and on the freshly fractured rock surface, which is lighter color.

At Tusan cave C5, in the cliff face several meters above the beach, there is evidence of groundwater flow seeping out along the failure surface, and formation of a shallow erosion runnel on the bedding surface and cliff below (Fig. 2C). Occurrence of gullies and runnels in areas of visible evidence of groundwater flow suggests their formation is by mechanical erosion due to meteoric water flow.

Tafoni

Clusters of tafoni, a form of microcavernous weathering, are observed between 2 and 6 m above the high tide line at both sites (Figs. 6C and 8B). Most are located above the reach of storm waves. They are found on massive sandstone beds, and can be considered as sidewall tafoni because they are found on subvertical rock surfaces that lack discontinuities (Goudie, 2004). They vary in diameter, from a few centimeters, to approximately 10 cm, and are generally less than 5 cm deep. Their morphologies are well rounded with no sharp edges. In several places, the tafoni are associated with a capping of mineral crust that appears to offer some protection from further erosion. Where the crust is eroded, the tafoni are larger, coalesced, and less pronounced, indicating a more advanced stage of erosion. There is eventually full removal/disappearance some distance from the protective crust. This indicates that the tafoni are ephemeral, in other words, not long lasting, an indication of the general soft nature of the sandstone rock prone to surficial weathering. No secondary tafoni (forming inside other tafoni cavities) are observed.

Wave cut notches

Minor wave cut notches are present at several sites where tilted relatively homogenous sandstone beds are exposed to wave action, and associated with bioerosion in the intertidal zone (Figs. 2A and 8C). The most pronounced notch is at Tusan, near caves C2 and C3. Here the notch is 2.8 m high, and the concavity has a radius of 1.4 m (Fig. 8C). The base of the notch where the toe of the cliff is covered by beach sand was at 1.1 mad at the time of observation (15/6/16). However, this may be expected to change with seasonal sand mobility. The top of the notch is at 4 mad. The height of the notch exceeds the tidal range of 1.79 m, and this indicates that erosive action above high tide level is likely due to wave run up and storm surges. The deepest part of the notch concavity is situated at 2.7 mad, which is approximately 1 m above high tide.

Near-horizontal platforms

In the intertidal zone at Taman Selera, the sandstone is both massive and relatively gently dipping (see Table 2). There is a wave-cut platform approximately 4 m wide, and relatively flat (Fig. 2A). At the other sites (Tusan, Peliau, and Bungai) where lithology is less homogenous with alternating harder sandstone beds and softer siltstone and mudstone beds, platforms are also observed, but are less well developed on the whole. Where rocks are visible on the beach, they consist of more resistant sandstone beds protruding in small ridges parallel to strike. The softer rock beds form depressions, or furrows, and are partially or totally covered with sand and pebbles. Hence, the intertidal platform is not smooth, but can be considered a corrugated surface with hard layers projecting (Sunamura, 1994).

Intertidal runnels

Runnels are observed on the near-horizontal sandstone platform in the intertidal zone at Taman Selera. They were not observed at other sites. The runnels are around 3 m long, and their width is 40–20 cm, narrowing down at their base. The overall plan shape is relatively straight and perpendicular to the cliff face. Within the runnels, the deepest part is meandriform (see Fig. 6D). The convex faces of the meanders form a pointed rocky protrusion, whereas the concave face is a smooth curve with a radius of 5–10 cm. Subcircular rock pools are noted at several places on runnels, and the presence of pebbles in these pools suggests increased abrasion and/or increased water flow turbulence. No evidence of jointing is observed in the runnels, and their meandriform nature suggests that the dominant control is not jointing. Furthermore, when followed up slope, they do not relate to rock joints. While the near-horizontal platform is covered in encrusting algae of various types and barnacles, the walls of the runnels are clear of encrustations (Fig. 6D).

4.3.3 Microscale erosion features

Isopod perforations

The isopod *Sphaeroma triste* (Heller, 1865) is reported here for the first time in Sarawak boring into the soft sandstone at numerous locations in the intertidal zone (Dodge-Wan and Nagarajan, 2017, personal communication). Previous reports of boring by Sphaeromatids in and around Malaysia are of wood boring by other species in brackish and marine environments (Singh and Sasekumar, 1994; Hossain and Bamber, 2013).

In this study, the species *Sphaeroma triste* was observed perforating rocks at the four sites where soft sandstones are found in the intertidal zone (Taman Selera, Tusan, Peliau, and Bungai). The highest density of infestation is on subhorizontal surfaces at Peliau and Tusan. The density of perforations exceeds 8 holes/cm^2 in areas of very high infestation, as shown in Fig. 8D. The density commonly exceeds 2 holes/cm^2 in areas of high infestation ($n=19$). The holes average approximately 3 mm diameter and 2 cm in length. Secondary enlargement of holes is also common, especially in the upper half of the intertidal zone, and coalescing of two or more adjacent holes may occur if the fine wall between holes is eroded.

The isopod perforations are located in soft rocks, and preferentially clustered and in sheltered positions, such as within the rim of a rock pool, on the side walls of an enlarged joint, or on the concavities of the rock surface (Fig. 6A). Sphaeroma perforations were found both on soft sandstone and siltstone, but not on mudstones. They appear to be limited by rock hardness, and do not occur where hard iron oxide-rich crusts have formed on the rock surface. High infestations are found, for example, where iron oxide crusts have been breached by erosion, exposing a softer rock beneath (Fig. 8D). Some rare abandoned uninhabited perforations are present above, but close to the high tide line, but the greatest concentration is in the upper half of the intertidal zone on in situ rocks, large immobile fallen blocks, and intertidal platforms. In the upper intertidal zone and in the splash zone above it, the rock surfaces may also host clusters of barnacles and small gastropods. There is some overlap between the barnacle zone (generally above) and the isopod zone (generally below).

5 Discussion

Moore et al. (1999) list some of the factors leading to coastal erosion citing "wave exposure, terrestrial processes (runoff and groundwater seepage, rainfall induced sliding, seismicity) as well as the lithology of the cliff, and in particular, structural weaknesses and the presence/absence of protective beach." All of these factors appear to affect the study area, although seismicity can be expected to have a lesser contribution, being generally low in this region (Hall, 2013). Lim et al. (2010) stressed that cliff failures should not be seen as "events," but rather as part of an ongoing process with multiple interconnected processes and mechanisms. This preliminary study, which is largely observational, is not able to identify all processes and mechanisms. However, it can provide indications of the dominant processes and the resulting features, and guide future investigations.

Emery and Kuhn (1982) stressed that homogeneity (or lack of it) is one of the most significant characteristics of rocks influencing sea cliff erosion, particularly with regard to rock hardness. In addition, the discontinuities of rock structure, such as strata of erodible rocks and joints, also play a role, particularly when dipping oceanward (Emery and Kuhn, 1982; Naylor and Stephenson, 2010). In the area of interest, the sedimentary formations are commonly interbedded, alternating sandstone and mudstone layers; and all dip oceanward, with subvertical beds at Peliau. Interbedded lithology leads to the presence of weaker mudstone strata with high potential for erosion. Even where mudstones are relatively thin (less than 30 cm thick), tabular lithologically controlled caves have formed, with resistant hanging walls of sandstone. These caves may be >10 m long, and eventually may evolve into arches if a second opening is formed. Lower permeability in mudstone interbeds can cause an increase in groundwater flow above them, with associated enlargement of bedding planes, resulting in the formation of caves, even without the added effect of the mechanical action of the sea. This is observed at Tusan in EF2 (Fig. 2C).

Meteoric waters percolating on cliff faces may carve gulleys in very soft rocks, or shallow runnels in more resistant rocks, but the effect of meteoric water is greatest when infiltrating into the rock mass, and eventually leading to mass movements. Groundwater percolation may swell clays leading to instability. Groundwater may also lubricate sloping surfaces, making rotational and translational slides more likely. On tilted rocks, infiltration occurs preferentially where there are structural discontinuities, such as joint caves at Peliau, which are connected to the ground surface several meters above. The presence of the superficial capping layer of permeable Pleistocene terrace material may further lead to cliff top instability by encouraging increased infiltration. This situation is similar to that reported in Dorset under the permeable head deposits (Conway, 1979). At Peliau, there is a pseudokarst dropout doline above a domed collapse cave, which is another indication of the role of infiltration.

Sunamura (2015) emphasized the episodic nature of soft rock coastal erosion, with single storm events being responsible for recession rates several orders of magnitude higher than the long-term averages. We note from our observations and based on Wannier et al. (2011) that sudden events such as storms cause significant erosion over short periods of time. In addition, the deposition of sand and coarser rock clasts on the beach is episodic, and dynamic with changes in time. An example of this is the modification, burial, and removal of a recent iron cemented conglomeratic beach rock observed at Tusan and Peliau both on the beach, at the foot of the cliff and in a cave, over time spans of several months. This supports the findings of Anandkumar et al. (2018), who observed cyclic decadal changes in erosion and accretion along this shoreline.

The observations at Tusan (Fig. 3) indicate strong geological control of erosion, with beds dipping coastward in headlands, and joints perpendicular to them. Interbedded resistant sandstones with minor mudstones and less resistant sandstones led to a step wise recession of the cliff from one resistant sandstone bed to the next. Initially, joints allowed marine erosion to excavate weaker mudstone beds at the cliff toe, forming tabular caves oriented parallel to the coast. Sunamura (1994) showed that cave formation by wave action on joints or other rock discontinuities operates by impulsive pressure (wedge action), and there is a positive feedback mechanism by which small openings form, and are then widened and deepened, leading to cave formation. With ongoing erosion, mesoscale bedding enlargements may become macroscale caves, and these caves will continue to be excavated following the weakest rocks. If the cave's hanging wall bed is sufficiently resistant, it may eventually become the keystone of an arch, if the cave opens on both sides of a headland or rock mass between two joints. If the keystone bed is not resistant, it will be readily eroded, exposing the former foot wall of the cave, which then becomes the new cliff face. In this situation, further erosion of the cliff face will extend up dip and down the stratigraphic sequence, until another resistant sandstone bed is reached. In a map view, the coastline shows the progressive stepping back of ongoing recession with a series of linear cliffs corresponding to successive resistant sandstone beds (as shown in Fig. 3). This scenario may operate even where the resistant beds are not prone to undercutting. Wave cut notches, enhanced by bioerosion in the intertidal zone, will favor undercutting. Undercutting, particularly in relatively soft rocks, may destabilize slopes, further accelerating the mass wasting process (Trenhaile, 2015).

At Peliau, where the strata are subvertical, discontinuities within sandstone beds, such as joints, appear to control where erosion is the greatest, and joint controlled caves are common. There is a potential for undercutting and then rock fall. Joint controlled caves may favor marine erosion, both into the rock mass, and, if weaker beds are present, along strike. With subvertical strata, there is also the potential for opening up of skylights as caves erode up dip and meteoric water, and plant roots penetrate between resistant beds. Several of the Peliau caves have such characteristics.

Where the stratigraphic sequence includes thick mudstones or thick weak sandstones, the potential for mass movement is high. This is particularly true where high tide is able to erode the foot of the cliff (Emery and Kuhn, 1982), as in many places observed in this study. High tide, wave action, and storm surges eroding the softer materials, including fallen rocks and collapse debris, provide sediment to the beach, which is prone to variations and modifications as sediment is deposited and eroded.

The NW Sarawak coast is considered to be *active* in the definition of Emery and Kuhn (1982) because of the fact that the bedrock is exposed to continuous retreat due to marine and subaerial processes. The evidence for this is abundant: lack of base mantling (protection), talus slopes commonly >30 degree, talus slopes lacking vegetation, rock falls, slumps, and other mass movements, as well as sediment mobilization. The cliff profiles are variable, as would be expected seeing the stratigraphic and structural features of the rock formations (Emery and Kuhn, 1982). Here the geology commonly includes rocks of differing lithology, hardness, and permeability.

6 Conclusions and Implications

This study has examined several kilometers of the NW Sarawak coastline to the south of Miri city where Miocene clastic sedimentary rocks outcrop. The research has identified, mapped, and described 14 different types of coastal erosion features, ranging from macro- to meso- and microscale.

The largest features are mass wasting, and these range in volume from several hundred to several thousand cubic meters. They involve rock falls, and rotational and translational mass movement, and cause retreat of cliff toe, cliff face, or cliff top, or a combination of these. Mass movement features are episodic in frequency, accelerated by a retreat of cliff toe and other erosion processes, including the formation of caves and arches. The potential for further mass movement is high, including in areas of high visitor access where tensional cracks have been observed.

Caves are the second-largest feature, and their morphology and dimensions are strongly controlled by geology, including the presence of easily erodible mudstones interbedded within more resistant sandstones, as well as joints in the sandstones. Three types of caves were identified: (1) tabular caves dominantly controlled by lithology and elongated parallel to strike, (2) prismatic caves dominantly controlled by erosion of joints and elongated perpendicular to strike, and (3) domed caves that show evidence of collapse processes and less elongation. Strata dip value determines which type of cave is the most common at any one site, with shallower dips favoring tabular caves (as at Tusan), and subvertical dips favoring prismatic caves (as at Peliau). Caves may be linked to the ground surface with the potential for erosive effect of water percolation, and one pseudokarst dropout doline was observed, indicating the location of a probable future collapse feature.

A number of mesoscale and microscale erosion features have been identified and classified, including bedding enlargements, joint enlargements, erosion pockets, tafoni, gulleys, and intertidal runnels with rock pools. The most significant microscale features observed are perforations into soft rock by the isopod *Sphaeroma triste,* reported here for the first time in Sarawak. Although isopod perforations are microscale features, and have a relatively small contribution to overall coastal erosion, they would accelerate the formation of marine notches, and the widening of joints and pools in the intertidal zone.

The sites are in a region with a low economic parameter score (National Coastal Erosion Study, 2015). However, the coastal cliffs, with their unique and attractive erosional features, such as rock beds, caves, and arches, warrant special consideration in coastal zone management plans, as they are sensitive geotourism areas. Cliff top hazard indexes have been proposed in similar geological settings in California (Young, 2018), and could be applied in this region to identify cliff steepening that may pose an immediate risk of failure in the future. It is recommended, in the short term, that warning signage should be installed to indicate the high erosion spots identified in this study. In addition, the magnitude, location, and likely consequences of coastal erosion should be considered in all future development planning. Further research is warranted, and should be targeted to provide more quantitative data, and in particular, an analysis of the rates of coastal recession along this active coastline, particularly with anticipated increasing numbers of visitors and development of new facilities and concern for public safety.

Acknowledgments

The authors wish to thank Curtin University, Malaysia, for funding, facilities, and equipment for this research. Our thanks also go to Chu Wai Hsin for his help during field surveys, as well as to the anonymous reviewers and editors for their comments and suggestions that have greatly helped to improve the manuscript.

References

Ali, A.M., Padmanabhan, E., Baioumy, H., 2016. Petrographic and microtextural analyses of Miocene sandstones of onshore West Baram Delta Province, Sarawak Basin: Implications for porosity and reservoir rock quality. Petrol. Coal 58 (2), 162–184.

Anandkumar, A., Vijith, H., Nagarajan, R., Jonathan, M.P., 2018. Evaluation of decadal shoreline changes in the coastal region of Miri, Sarawak, Malaysia. In: Krishnamurthy, R.R., Jonathan, M.P., Srinivasalu, S., Glaeser, B. (Eds.), Coastal Management: Global Challenges and Innovations. Elsevier. https://doi.org/10.1016/B978-0-12-810473-6.00008-X.

Borneo Post Online, 2015. Tusan Beach a Stunning Open Secret in Miri. Borneo Post Online, 21 September 2015. http://www.theborneopost.com/2015/09/21/tusan-beach-a-stunning-open-secret-in-miri/ (Accessed 21 November 2017).

Borneo Post Online, 2017. Over 200 Join Tusan Community Project. Borneo post Online, 5 November 2017. http://www.theborneopost.com/2017/11/05/over-200-join-tusan-community-project/ (Accessed 21 November 2017).

Cabral, P., Augusto, G., Akande, A., Costa, A., Amade, N., Niquisse, S., Atumane, A., Cuna, A., Kazemi, K., Mlucasse, R., Santha, R., 2017. Assessing Mozambique's exposure to coastal climate hazards and erosion. Int. J. Disaster Risk Reduct. 23, 45–52.

Cardozo, N., Allmendinger, R.W., 2013. Spherical projections with OSXStereonet. Comput. Geosci. 51, 193–205. https://doi.org/10.1016/jcageo.2012.07.021.

Chong, E. V., 2016. Blue Tears and Bioluminescence Phenomenon Back Again in Miri! Borneo Post Online, 22 November 2016. http://www.theborneopost.com/2016/11/22/blue-tears-bioluminescence-phenomenon-back-again-in-miri/ (Accessed 21 November 2017).

Conway, B.W., 1979. The contribution made to cliff instability by Head deposits in the west Dorset area. Q. J. Eng. Geol. 12 (4), 267–275.

Dickson, M.E., Kennedy, D.M., Woodroffe, C.D., 2004. The influence of rock resistance on coastal morphology around Lord Howe Island, Southwest Pacific. Earth Surf. Process. Landf. 29, 629–643.

Dodge-Wan D., Nagarajan R., 2017. Personal Communication on Field Observations of *Sphaeroma triste* Bioerosion Between Miri and Bungai Beach.

Emery, K.O., Kuhn, G.G., 1982. Sea cliffs: their processes, profiles, and classification. Geol. Soc. Am. Bull. 93, 644–654.

Goudie, A., 2004. Encyclopedia of Geomorphology. Routledge, New York. 1200 p. ISBN: 0-415-27298-X.

Hall, R., 2013. Contraction and extension in northern Borneo driven by subduction rollback. J. Asian Earth Sci. 76, 399–411. https://doi.org/10.1016/j.seaes.2013.04.010.

Heller, C. 1865. Crustaceen. In: Reiseder Osterreichischen Fregatte "Novara" um die Erde in den Jahren 1857, 1858, 1859. Zoologischer Theil 2(3), 1–280, pls 1–25.

Hossain, M.B., Bamber, R.N., 2013. New record of a wood-boring isopod, *Sphaeroma terebrans* (Crustacea: Sphaeromatidae) from Sungai Brunei estuary, Brunei Darussalam. Mar. Biodivers. Records 6, 1–3.

Hutchison, C.S., 2005. Geology of North-West Borneo: Sarawak, Brunei and Sabah. Elsevier, Amsterdam. 421 p. ISBN: 978-0 444-51998-6.

IPCC, 2013. Climate Change 2013: The Physical Science Basis. In Contribution of Working Group I to the Fifth Assessment Report of the Intergovernmental Panel on Climate Change. IPCC. http://www.ipcc.ch/report/ar5/wg1/ (Accessed 9 April 2018).

Jai, T.Y., Rahman, A.H.A., 2009. Comparative analysis of facies and reservoir characteristics of Miri Formation (Miri) and Nyalau Formation (Bintulu), Sarawak. Bull. Geol. Soc. Malaysia 55, 39–45.

Johan, F.E., Boateng, I., Osman, A., Shimba, M.J., Mensah, E.A., Adu-Boahen, K., Chuku, E.O., Effah, E., 2016. Shoreline change analysis using end point rate and net shoreline movement statistics: An application to Elmina, Cape Coast and Moree section of Ghana's coast. Reg. Stud. Mar. Sci. 7, 19–31.

Kessler, F.L., Jong, J., 2014. Habitat and C-14 ages of lignitic terrace deposits along the northern Sarawak Coastline. Bull. Geol. Soc. Malaysia 60, 27–34.

Lim, Y.S., Koh, S.L., 2010. Analytical assessments on the potential of harnessing tidal currents for energy generation in Malaysia. Renew. Energy 35 (5), 1024–1032. https://doi.org/10.1016/j.renene.2009.10.016.

Lim, M., Rosser, N.J., Allison, R.J., Petley, D.N., 2010. Erosional processes in the hard rock coastal cliffs at Staithes, North Yorkshire. Geomorphology 114, 12–21.

Mathew, M.J., Menier, D., Siddiqui, N., Kumar, S.G., Authemayou, C., 2016. Active tectonic deformation along rejuvenated faults in tropical Borneo: Inferences obtained from tectono-geomorphic evaluation. Geomorphology 267, 1–15.

Miri Resort City, 2017. http://www.miriresortcity.com/content/tusan-beach (Accessed 30 November 2017).

Moore, L.J., Benumof, B.T., Griggs, G.B., 1999. Coastal erosion hazards in Santa Cruz and San Diego Counties, California. J. Coast. Res. Spec. Issue 28, 121–139.

Nagarajan, R., Armstrong-Altrin, J.S., Kessler, F.L., Hidalgo-Moral, E.L., Dodge-Wan, D., Taib, N.I., 2015. Provenance and tectonic setting of Miocene siliciclastic sediments, Sibuti formation, northwestern Borneo. Arab. J. Geosci. 8, 8549–8565. https://doi.org/10.1007/s12517-015-1833-4.

Nagarajan, R., Armstrong-Altrin, J.S., Kessler, F.L., Jong, J., 2017. Petrological and geochemical constraints on provenance, paleo-weathering and tectonic setting of clastic sediments from the Neogene Lambir and Sibuti Formations NW Borneo (Chapter 7). In: Mazumder, R. (Ed.), Sediment Provenance: Influences on Compositional Change From Source to Sink. Elsevier, Amsterdam, Netherlands, pp. 123–153. https://doi.org/10.1016/B978-0-12-803386-9.00007-1.

Nagarajan, R., Anandkumar, A., Hussain, S.M., Jonathan, M.P., Ramkumar, M., Eswaramoorthi, S., Saptoro, A., Chua, H.B., 2019. Geochemical characterization of beach sediments from NW Borneo: emphasis on provenance, weathering and environmental status (Chapter 7). In: Ramkumar, M., Arthur James, R., Menier, D., Kumaraswamy, K. (Eds.), Coastal Zone Management: Global Perspectives, Regional Processes, Local Issues. Elsevier. https://doi.org/10.1016/B978-0-12-814350-6.00012-4 (This volume).

National Coastal Erosion Study, 2015. NCES Erosion Categorisation, by Malaysian Department of Irrigation and Drainage. Retrieved from http://nces.water.gov.my/nces/About (Accessed 3 April 2018).

Naylor, L.A., Stephenson, W.J., 2010. On the role of discontinuities in mediating shore platform erosion. Geomorphology 114, 89–100.

Paw, J.N., Thia-Eng, C., 1991. Climate changes and sea level rise: implications on coastal areas utilization and management in South-East Asia. Ocean Shoreline Manag. 15, 205–232.

Sajinkumar, K.S., Kannan, J.P., Indu, G.K., Muraleedharan, C., Rani, V.R., 2017. A composite fall-slippage model for cliff recession in the sedimentary coastal cliffs. Geosci. Front. 8, 903–914. https://doi.org/10.1016/j.gsf.2016.08.006.

Selley, R.C., Cocks, L., Plimer, R.M., Ian, R., 2005. Encyclopedia of Geology. vols. 1–5. Elsevier.http://app.knovel.com/hotlink/toc/id:kpEGV00001/encyclopedia. Accessed 21 November 2017.

Simon, K., Bin Amir Hassen, M.H., Barbeito, M.P.J., 2014. Sedimentology and stratigraphy of the Miocene Kampung Opak limestone (Sibuti Formation), Bekenu, Sarawak. Bull. Geol. Soc. Malaysia 60, 45–53.

Singh, H.R., Sasekumar, A., 1994. Distribution and abundance of marine wood borers on the west coast of Peninsular Malaysia. Hydrobiologia 285 (1), 111–121.

Stephenson, W.J., Dickson, M.E., Trenhaile, A.S., 2013. Rock coasts. In: Shroder, J.F. (Ed.), Treatise on Geomorphology. In: 10, Elsevier, Amsterdam, pp. 289–307.

Sunamura, T., 1994. Rock control in coastal geomorphic processes. Trans. Jap. Geomorphol. Union 15 (3), 253–272.

Sunamura, T., 2015. Rocky coast processes: with special reference to the recession of soft rock cliffs. Proc. Japan Acad. Sci. Ser. B 91 (9), 481–500.

Trenhaile, A.S., 2015. Coatal notches: Their morphology, formation and function. Earth Sci. Rev. 150, 285–304.

Waltham, A., Bell, F., Culshaw, M., 2005. Sinkholes and Subsidence: Karst and Cavernous Rocks in Engineering. Springer, Chichester. ISBN: 3-540-20725-2.

Wannier, M., Lesslar, P., Lee, C., Raven, H., Sorkhabi, R., Ibrahim, A., 2011. Geological excursions around Miri, Sarawak: Celebrating the 100th Anniversary of the discovery of the Miri oil field. EcoMedia Software, Miri. 279 p. ISBN: 978-9834216030.

Young, A.P., 2018. Decadal-scale coastal cliff retreat in southern and central California. Geomorphology 300, 164–175.

Young, A.P., Flick, R.E., Gutierrrez, R., Guza, R.T., 2009. Comparison of short-term seacliff retreat measurement methods in Del Mar, California. Geomorphology 112, 318–323. https://doi.org/10.1016/j.geomorph.2009.06.018.

Further Reading

Gutierrez, M., 2012. Geomorphology. CRC Press, Boca Raton. 1014 p. ISBN: 9780415595339.

Leitchi, P.R., Roe, F.W., Haile, N.S., 1960. The geology of Sarawak, Brunei and the western part of North Borneo. Brit. Borneo Geol. Survey Bull. 3, 1–360.

CHAPTER 4

Monitoring Spatial and Temporal Scales of Shoreline Changes in the Cuddalore Region, India

Subbarayan Saravanan*, Parthasarathy K.S.S.[†], S.R. Vishnuprasath[‡]

*Department of Civil Engineering, NIT Trichy, Trichirappalli, India [†]Department of Applied Mechanics and Hydraulics, NIT Surathkal, Mangalore, India [‡]Bharathidasan University, Tiruchirappalli, India

1 Introduction

The coastal area is the interface between land and sea. The shoreline is one of the most rapidly changing landforms of the coastal zone. Climate change and development of infrastructure put pressure on the coastal environment, and this leads to various coastal hazards, such as coastal erosion, flooding, sea level rise, seawater intrusion, and bioresource degradation (Kaliraj et al., 2013). Moreover, the global mean sea level is projected to rise by 0.09 to 0.88 m over the same period, as a result of the thermal expansion of the oceans, and the melting of glaciers and polar ice sheets (Chand and Acharya, 2010). The shoreline is one of the most dynamic landforms in the coastal environment, and it tends to change due to erosion and accretion processes induced by wave action, sea level rise, and sediment transportation. Further anthropogenic activities along the coast, changes in river catchments, offshore developments such as land reclamation in coastal areas, port and harbor construction, river damming, diversion of rivers, sand mining, dredging, and so forth, also contribute to shoreline changes (Kankara et al., 2014). Therefore, coastal zone monitoring is an important task in sustainable development and environmental protection.

The coastal areas are also the places where natural disasters are experienced. The tsunami that occurred on December 26, 2004, was one of the most unexpected catastrophes to occur along the Indian coast, causing severe damage to life and property. In the Bay of Bengal, cyclonic disturbances (at an average of 5–6 per year) and storms (at an average of 6 per year) are quite frequent. Healthy coastal ecosystems cannot completely protect the coast from impacts of storms and floods, but they do play an important role in stabilizing shorelines and buffering coastal development from the impact of the storm (Kumaravel et al., 2012).

The application of geospatial technology is a promising tool for providing the synoptic coverage multitemporal satellite images of the coastal area in different resolutions to estimate periodic shoreline changes. Detection, extraction, and monitoring the shoreline are regarded as important tasks in safe navigation, coastal zone management, environmental protection, and sustainable coastal development and planning (Gens, 2010). A variety of data sources have been used to detect the shoreline, such as historical land-based photographs, coastal maps and charts, aerial photography, GPS field surveys, and remote sensing imagery (Bouchahma and Yan, 2012). Among such data, various remote sensing images have been used (Mujabar and Chandrasekar, 2013), including multispectral and hyperspectral images (Feng and Han, 2012), airborne light photography (Liu et al., 2007), and microwave images (Ashar and Nobuhiro, 2009), to name a few. Remotely sensed imagery from Landsat, SPOT, SAR, QuickBird, IKONOS, and LiDAR have been widely used to map shorelines and detect shoreline changes (Feng and Han, 2012). However, the selection of data for a study area is generally determined by the availability of data (Boak and Turner, 2005). Long-term shoreline changes and predictions are estimated by linear regression in DSAS (Addo et al., 2008). Saravanan et al. (2015) studied an automatic shoreline extraction and coastal vulnerability study by using automatic water body extraction techniques for segment water and land regions for Chennai, on the east coast of India.

There are many change detection techniques currently in use, including image enhancement, multitemporal data classification, density slicing using single or multiple bands, and supervised and unsupervised multispectral classification. In India, shoreline change studies have been successfully investigated using remote sensing and GIS by many researchers at different times (Alesheikh et al., 2007; Chandrasekar et al., 2011; Srinivasa Kumar et al., 2008; Mani Murali et al., 2013; Kaliraj et al., 2013; Chenthamil Selvan et al., 2014; Mageswaran et al., 2015). Arun and Kunte (2012) studied coastal vulnerable index (CVI) mapping for the Chennai coast to assess the coastal erosion impact. Aedlaa et al. (2015) developed an automatic shoreline extraction method using clipped histogram equalization-based contrast enhancement for enhancing coastal pixels and thresholding techniques for segmenting water and land regions for the Netravati-Gurpur River mouth area, on the Mangalore coast, on the west coast of India.

In this study, the modification of normalized difference water index (MNDWI) method to extract the raster shoreline-based contrast value of coastal pixels and thresholding techniques for segmenting water and land regions was attempted. DSAS software and reference digitized shoreline boundary data were used for the analysis of shoreline changes of the Cuddalore Coast from the Gadilam River to the Vellar River on the east coast of India for the period of 2000–2015.

2 Study Area

India has a 7500 km long shoreline interspersed with deltaic, rocky, cliff coasts (Ramkumar, 2003), in which the Tamil Nadu state constitutes a 750-km shoreline. The study area selected is a coastal zone of Cuddalore city (Fig. 1), in the Tamil Nadu state, in India. This analysis was carried out from the Gadilam River to the Vellar River for a stretch of 33 km distance. The study area is located between 11°46′N-11°28′N latitudes and 79°44′E-79°49′E longitudes. The region receives annual rainfall of 1104 mm/year. It includes both the south west (373 mm) and north east (731 mm) monsoons. The maximum annual temperature is 36.8°C and the minimum temperature is 19.9°C. The geomorphology of the Cuddalore coastal stretch includes the coastal plain, with an average width of 6 km. Its coastal landforms include strand-lines, raised beaches, sand dunes, mangrove swamps, and tidal flats with predominantly sandy beaches on the northern side, and mangrove swamps to the south. The coastal towns of Cuddalore in the north and Porto Novo (Parangipettai) in the south are densely populated. There are 19 fishing habitations along the coastline in the study area. Fishing and allied activities are the main sources of livelihood in the area, with the minimum extent under cultivation. Northeast monsoons occur during the wettest period of the year, when cyclonic activities in the Bay of Bengal bring in the rainfall. The coastline, which includes tourist resorts, ports, hotels, fishing villages, industries, and towns, has experienced threats from many disasters such as storms, cyclones, floods, tsunamis, and erosion. This was one of the affected areas during the 2004 Indian Ocean tsunami, and the various cyclones that passed included: Fanoos (22.12.2005), Nisha (29.11.2008), Jal (10.11.2010), Thane (31.12.2011), Nilam (1.11.2013), and Madi (13.12.2013).

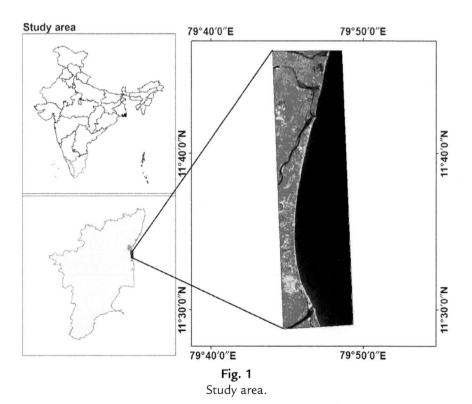

Fig. 1
Study area.

3 Methodology

Changes in the shoreline through accretion and erosion can be analyzed in a geographic information system (GIS) by measuring differences in past and present shoreline locations. In this study, we used the Modified Normalized Difference Water Index (MNDWI), as it can efficiently suppress the noise from built-up land, vegetation, and soil (Xu et al., 2013). The digital shorelines were processed in DSAS for change detection analysis. The choice of DSAS statistical parameters in this study enabled an exploration of the temporal and spatial dynamics of the coastal change, and the geomorphic variability along the beach because of their ability to use all shoreline positions, the cumulative shoreline movement, and the time variations; which encapsulate the rate range of the historical dataset (Moussaid et al., 2015). The detailed methodology is shown in Fig. 2. The estimation of erosion and accretion along the Cuddalore coast was performed by using satellite images in ENVI using the automatic delineation method. The spatial variability of shoreline changes were studied by using geometrically corrected and ortho-rectified satellite images of the years 2000, 2005, 2010, and 2015. The shorelines were delineated using the MNDWI technique. Using these two sets of results, NSM and EPR were analyzed. These data were processed under a GIS environment to determine the areas of high and low erosion, and accretion regions.

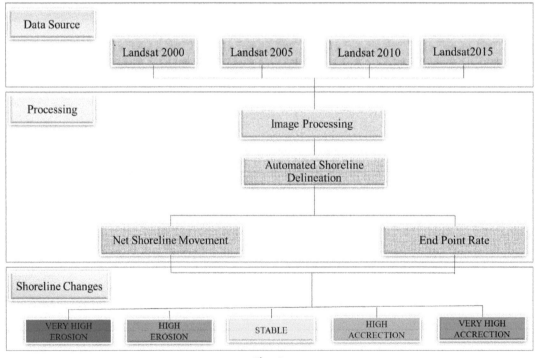

Fig. 2
Flow chart of methodology.

3.1 Shoreline Delineation and Image Processing Techniques

The dynamic nature of the coast introduces uncertainty in accurate mapping of instantaneous shoreline positions. This is due to the fact that at any given time, the position of the shoreline is influenced by the short-term effect of the tide and the long-term effect of the sea level's rise.

Image processing is a technique using computer algorithms that are applied to digital images for processing. The extraction of water information from multispectral satellite images can be done by analyzing the signatures of each ground target among the different spectral bands, and using these signatures as input for a specific classification method. Another way is using a band ratio approach of two spectral bands. Ratio images are calculated by dividing DN values of each pixel in one spectral band by the corresponding pixel value in another band. The advantage of this method is that the environment-induced variations in the DN values of a single spectral band are reduced by considering the ratio between two bands (Tran and Trinh, 2009). In this chapter, we used the Modified Normalized Difference Water Index (MNDWI), as it can efficiently suppress the noise coming from built-up land, vegetation, and soil (Xu et al., 2013) from the water information. It is calculated using

$$MNDWI = (\rho 2 - \rho 6)/(\rho 2 + \rho 6) \qquad (1)$$

where ρ represents the reflectance from the Landsat 8 for Bands 2 (blue), 3 (green), 4 (red), 5 (near-infrared), 6 (SWIR1) and 7 (SWIR2), and 10 (TIR). After calculating the MNDWI for each image, the K-means clustering method is used to separate the water from land, as the mean values of the water cluster are always greater than zero.

3.2 Shoreline Change Analysis

Multiple shoreline positions, along with a fictitious baseline, are the basic requirements for analyzing the shoreline. A continuous shoreline position with a regular time interval was delineated in ENVI software for the years 2000 (Fig. 3), 2005 (Fig. 4), 2010 (Fig. 5), and 2015 (Fig. 6).

3.2.1 Net shoreline movement

The net shoreline movement (NSM) reports a distance. The NSM is associated with the dates of only two shorelines. It reports the distance between the oldest and youngest shorelines for each transect. When the distance so calculated is divided by the number of years elapsed between the two shoreline positions, the result is the end point rate (EPR).

3.2.2 End point rate

The EPR was calculated by dividing the distance of shoreline movement by the time elapsed between the earliest and latest measurements at each transect (Fig. 8). The major advantages of the EPR are the ease of computation and minimal requirement of only two shoreline dates. The

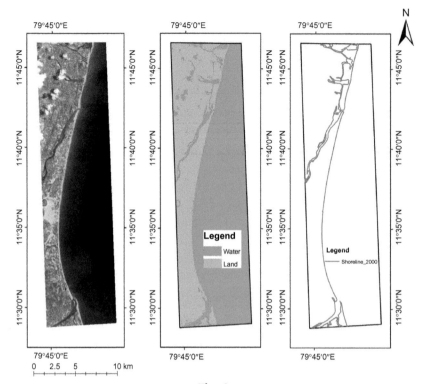

Fig. 3
Shoreline delineation for the year 2000.

major disadvantage is that in cases where more data are available, the additional information is ignored. Changes in sign (e.g., accretion to erosion), magnitude, or cyclical trends may be missed (Crowell et al., 1991). Based on the EPR data, the shoreline changes are classified as very high erosion, high erosion, stable, high accretion, and very high accretion.

DSAS casts a number of transects perpendicular to a baseline, and records the intersection position between the transect and each shoreline. DSAS automatically generates several statistical methods, such as the shoreline change envelope (SCE), NSM, EPR, and so forth. In the present study, shoreline changes were estimated using two statistical approaches, the NSM and EPR.

3.2.3 Shoreline change detection

The NSM and EPR depict the number of shoreline changes in a region, and classify them as very high erosion, high erosion, stable, high accretion, or very high accretion (Fig. 8). Standard color coding of the intensity of erosion/accretion includes red box—very high erosion, brown

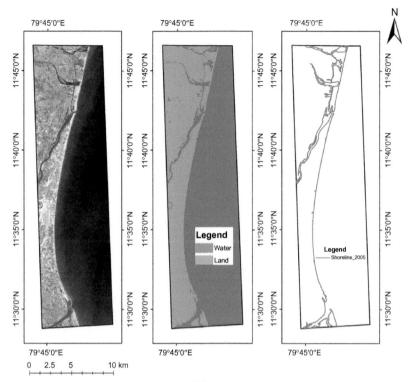

Fig. 4
Shoreline delineation for the year 2005.

box—high erosion, yellow box—stable region, light green box—high accretion, and thick green box—very high accretion.

4 Results

Shoreline change analysis was attempted in order to calculate the rate of change of the shoreline from the time series of multiple shoreline positions. These shoreline delineation images are presented in Figs. 3–6. The results of NSM are presented in Fig. 7. Fig. 8 presents the classification of the shoreline according to erosion/accretion. The coastal region of Cuddalore is undergoing a rapid change due to the presence of the river mouth in both the northern and southern part of the study area, where two major rivers have confluences. A maximum accretion rate of 1.2 m/year has been observed in the vicinity of the Uppanar River, causing sedimentation of around 450 m below the Uppanar River mouth. The study region was divided into five segments according to the coastal morphology. The average erosion is −1.66 m/year (Fig. 8), and the accretion trend is around +2.5 m/year. On the other hand, the region of the Northern Vellar River mouth experiences accretion of at a rate of 4.2 m/year during the study period.

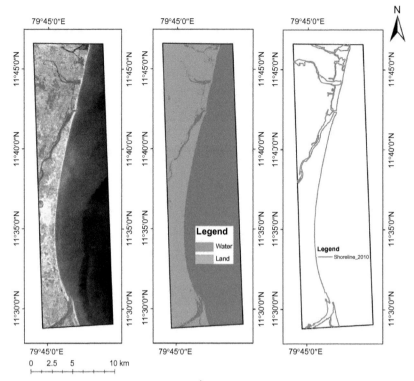

Fig. 5
Shoreline delineation for the year 2010.

5 Discussion

The intense coastal erosion in the northern region shows a diminishing trend toward the middle part, which in turn changes to accretion toward the southern part (Fig. 9). The trends could be due to the heavy influx of sediments from the Vellar River, which is located in the southern part of the study area. These sediments are redistributed toward the north due to intense littoral currents directed northerly. Despite the presence of the confluence of the Gadilam River in the northern part, the observed significant erosion (compare Figs. 7 and 8) could have been due to the presence of groins near the Uppanar confluence. These structures might act as dampeners in sediment movement toward the north under the influence of a littoral current, causing settling of the sediments brought by the river in the southern side of the confluence. The interplay between sediment influx, coastal current, and artificial structures could be gauged by the very high variability of beach and coastal profiles and beach morphology in this region.

Between the years 2000 and 2015, it seems that larger tracts of the coastal regions were eroded, especially due to the occurrences of multiple storms and cyclones, namely, Fanoos (22.12.2005), Nisha (29.11.2008), Jal (10.11.2010), Thane (31.12.2011), Nilam (1.11.2013), and Madi (13.12.2013). The region between the Gadilam River confluence to the southern

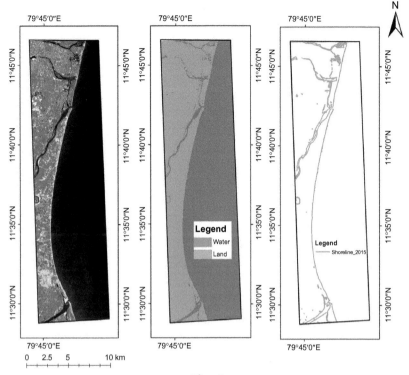

Fig. 6
Shoreline delineation for the year 2015.

extremity had a recorded maximum erosion of −3.8 m/year, and a maximum accretion of +7.6 m/year had been recorded by the region north of the Vellar River confluence.

Present observations also suggest the roles played by dwindling sediment influx, and construction of artificial structures such as groins that deflect coastal currents in shoreline stability and dynamics. Occurrences of enhanced dynamism of shoreline positions in a region historically known for recurrent storms and cyclones would naturally aggravate the repercussions, often detrimentally to the infrastructure and coastal ecosystems. The rates computed through the present analysis are, in this regard, a cause to worry, and need suitable remedial measures.

6 Conclusions

Applications of the modified normalized differential water index automated shoreline extraction from satellite images using contrast enhancement and thresholding-based techniques, and significantly improved the contrast between coastal edges and coastal objects for clear recognition and delineation. EPR and NSM statistical methods have shown substantial shoreline changes on the Cuddalore Coast. Analysis of the computed quantitative data,

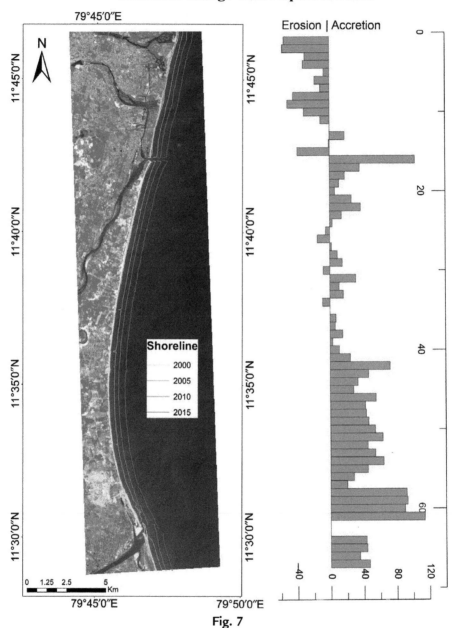

Fig. 7
Net shoreline movement in comparison with the study area.

classification of the coastline according to erosion/accretion statistics, and corroboration with field information have suggested fragility of the study area, which is already reeling under the recurrence of cyclones and storms. The results have also suggested the roles played by artificial structures in modification of shoreline positions, and active sites of erosion/deposition.

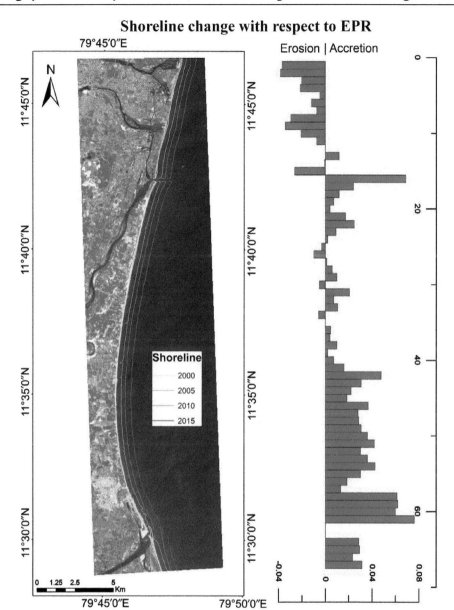

Fig. 8
End point rate in comparison with the study area.

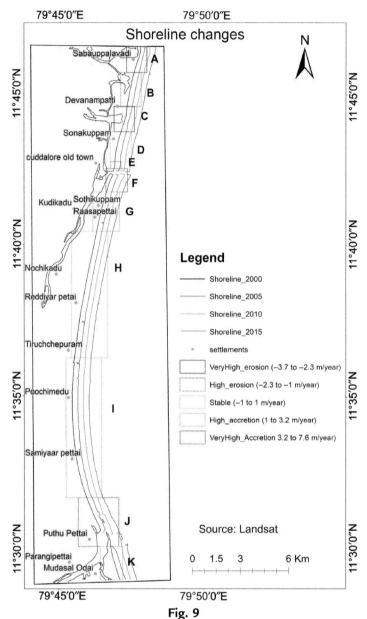

Fig. 9
Shoreline change detection of Cuddalore coast.

References

Addo, A.K., Walkden, M., Mills, J.P., 2008. Detection, measurement and prediction of shoreline recession in Accra, Ghana. J. Photogramm. Remote Sens. 63, 543–558.

Aedlaa, R., Dwarakish, G.S., Venkat Reddy, D., 2015. Automatic shoreline detection and change detection analysis of Netravati-Gurpur river mouth using histogram equalization and adaptive thresholding technique. Aquatic Proc. 4, 563–570.

Alesheikh, A.A., Ghorbanali, A., Nouri, N., 2007. Coastline change detection using remote sensing. Int. J. Environ. Sci. Technol. 4 (1), 61–66.

Arun, A., Kunte, D., 2012. Coastal vulnerability assessment for Chennai, east coast of India using geospatial techniques. Nat. Hazards 64, 853–872.

Ashar, M.L., Nobuhiro, I., 2009. Shoreline changes and vertical displacement of the 2 April 2007 Solomon Islands earthquake Mw 8.1 revealed by ALOS PALSAR images. Phys. Chem. Earth 34 (6), 409–415.

Boak, E.H., Turner, I.L., 2005. Shoreline definition and detection: a review. J. Coast. Res. 21 (4), 688–703.

Bouchahma, M., Yan, W., 2012. Automatic measurement of shoreline change on Djeba Island of Tunisia. Comput. Inf. Sci. 5 (5), 17–24.

Chand, P., Acharya, P., 2010. Shoreline change and sea level rise along coast of Bhitarkanika wildlife sanctuary, Orissa: an analytical approach of remote sensing and statistical techniques. Int. J. Geomat. Geosci. 1 (3), 436–455. ISSN 0976–4380.

Chandrasekar, N., Joevivek, V., Soundaranayagam, J.P., Divya, C., 2011. In: Geospatial analysis of coastal geomorphological vulnerability along southern Tamilnadu coast.GeoSpatial World Forum, pp. 18–21.

Chenthamil Selvan, S., Kankara, R.S., Rajan, B., 2014. An adaptive approach to monitor the Shoreline changes in ICZM framework: a case study of Chennai coast. Ind. J. Mar. Sci. 43(7).

Crowell, M., Leatherman, S.P., Buckle, M.K., 1991. Historical shoreline change error analysis and mapping accuracy. J. Coast. Res. 7 (3), 839–852.

Feng, Y., Han, Z., 2012. Cellular automata approach to extract shoreline from remote sensing imageries and its application. J. Image Graph. 17 (3), 441–446.

Gens, R., 2010. Remote sensing of coastlines: detection, extraction and monitoring. Int. J. Remote Sens. 31 (7), 1819–1836.

Kaliraj, S., Chandrasekar, N., Magesh, N.S., 2013. Impacts of wave energy and littoral currents on shoreline erosion/accretion along the south-west coast of Kanyakumari, Tamil Nadu using DSAS and geospatial technology. Environ. Earth Sci. 71 (10), 4523–4542.

Kankara, R.S., Selvan, S., Rajan, B., Arockiaraj, S., 2014. An adaptive approach to monitor the Shoreline changes in ICZM framework: a case study of Chennai coast. Ind. J. Mar. Sci. 43(7).

Kumaravel, S., Ramkumar, T., Gurunanam, B., Suresh, M., 2012. Quantitative estimation of shoreline changes using remote sensing and GIS: a case study in the parts of Cuddalore district, East coast of Tamil Nadu, India. Int. J. Environ. Sci. 2 (4). ISSN 0976–4402.

Liu, H., Sherman, D., Gu, S., 2007. Automated extraction of shorelines from airborne light detection and ranging data and accuracy assessment based on Monte Carlo simulation. J. Coast. Res. 23, 1359–1369.

Mageswaran, T., Ram Mohan, V., Chenthamil Selvan, S., Arumugam, T., Usha, T., Kankara, R.S., 2015. Assessment of shoreline changes along Nagapattinam coast using geospatial techniques. Int. J. Geomat. Geosci. 5 (4), 555–563.

Mani Murali, R., Ankita, M., Amrita, S., Vethamony, P., 2013. Coastal vulnerability assessment of Puducherry coast, India, using the analytical hierarchical process. Natural Hazards Earth Syst. Sci. 13 (12), 3291–3311. https://doi.org/10.5194/nhess-13-3291-2013.

Moussaid, J., Fora, A.A., Zourarah, B., Maanan, M., Maanan, M., 2015. Using automatic computation to analyse the rate of shoreline change on the Kenitra coast, Morocco. Ocean Eng. 102, 71–77.

Mujabar, P.S., Chandrasekar, N., 2013. Shoreline change analysis along the coast between Kanyakumari and Tuticorin of India using remote sensing and GIS. Arab. J. Geosci. 6 (3), 647–664.

Ramkumar, M., 2003. Progradation of the Godavari delta: a fact or empirical artifice? Insights from coastal landforms. J. Geol. Soc. India 62, 290–304.

Saravanan, S., Parthasarathy, K.S.S., Kumaresan, P.R., Vishnu Prasath, S.R., Vasanth Kumar, T., 2015. Shoreline change detection for chennai coast using geospatial techniques, Presented to HYDRO 2015 International, IIT Roorkee, India, 17–19 December.

Srinivasa Kumar, T., Mahendra, R.S., Nayak, S., Radhakrishnan, K., Sahu, K.C., 2008. Coastal vulnerability assessment for Orissa state, east coast of India. J. Coast. Res. 26 (3), 523–534.

Tran, T.V., Trinh, T.B., 2009. Application of remote sensing for shoreline change detection in cuu long estuary. VNU J. Sci. Earth Sci. 25 (4), 217–222.

Xu, J.Y., Zhang, Z.X., Zhao, X.L., Wen, Q.K., Zuo, L.J., Wang, X., Yi, L., 2013. Spatial-temporal analysis of coastline changes in northern China from 2000 to 2012 (in Chinese). Acta Geogr. Sin. 68, 651–660.

Further Reading

Thi, V.T., Xuan, A.T.T., Nguyen, H.P., Dahdouh-Guebas, F., Koedam, N., 2014. Application of remote sensing and GIS for detection of long-term mangrove shoreline changes in Mui Ca Mau, Vietnam. Biogeosciences 11, 3781–3795.

CHAPTER 5

Shoreline Evolution Under the Influence of Oceanographic and Monsoon Dynamics: The Case of Terengganu, Malaysia

Effi Helmy Ariffin[*,†,‡], Mouncef Sedrati[*], Nurul Rabitah Daud[‡,§],
Manoj Joseph Mathew[‡,¶], Mohd Fadzil Akhir[‡], Nor Aslinda Awang[‖],
Rosnan Yaacob[†], Numair Ahmed Siddiqui[#,**], Mohd Lokman Husain[†,‡]

[*]*Géosciences Océan Laboratory UMR CNRS 6538, University of South Brittany, Vannes Cedex, France,* [†]*School of Marine and Environmental Science, Universiti Malaysia Terengganu, Kuala Terengganu, Malaysia,* [‡]*Institute of Oceanography and Environment, Universiti Malaysia Terengganu, Kuala Terengganu, Malaysia,* [§]*Universiti Teknologi MARA, Shah Alam, Malaysia,* [¶]*South-East Asia Carbonate Research Laboratory, Universiti Teknologi PETRONAS, Tronoh, Malaysia,* [‖]*National Hydraulic Research Institute of Malaysia, Ministry of Natural Resources and Environment, Seri Kembangan, Malaysia,* [#]*Department of Geosciences, Universiti Teknologi PETRONAS, Tronoh, Malaysia,* [**]*Institute of Hydrocarbon Recovery, Universiti Teknologi PETRONAS, Tronoh, Malaysia*

1 Introduction

Global coastal zones are dynamic areas of great importance because they house a large number of residential properties worldwide. These areas provide an essential environment for flora and fauna, and food production. In the context of economic and social values, the coastal zone constitutes places that are necessary for human development. However, over-development can create major problems in coastal zones (i.e., pollution, extinction of flora and fauna species, flooding, coastal erosion, etc.).

The main problem along coastlines is coastal erosion. This issue is of concern globally, as it is estimated that at least 60% of sandy beaches across the globe are currently being eroded (Durgappa, 2008). The main causes of erosion worldwide are clearly related to natural, as well as anthropogenic factors. Natural phenomena such as storms (monsoonal storms), with their associated strong winds, waves, and currents, are among the obvious factors that contribute to erosion along the coasts of many countries (Ariffin et al., 2016; Ashraful Islam et al., 2016; Awang et al., 2014; Rajawat et al., 2015).

Erosion also arises from various anthropogenic modifications of the coasts, including structures being built along coastlines. Although most of these structures (e.g., revetments, ripraps, seawalls, groynes, and breakwaters) are designed to control erosion, it is interesting to note that they can actually cause erosion in many cases. For example, at Agadir Bay, Morocco, breakwaters were constructed to address erosion problems after the establishment of Agadir Port in 1978 (Aouiche et al., 2016).

Furthermore, in the case of Gopalpur Harbour, India (which was built in 1987), similar problems necessitated the construction of a groyne. This structure juts out into the sea, and is designed to trap longshore sediment for the conservation of a busy protected beach. In this way, the groyne serves to reduce erosion, as well as prevent longshore drift from reaching downdrift areas in nearby ports or inlets (Mohanty et al., 2012). In this regard, groynes are able to modify the longshore transport of sediment, resulting in the accumulation of sand, mostly on the updrift side, but causing the erosion of sand on the downdrift side (Aouiche et al., 2016; Mohanty et al., 2012).

Prior to constructing coastal defense systems, the lack of comprehensive knowledge pertaining to coastal physical parameters and coastal processes was the main problem in coastal management, especially the lack of proper documentation (Anfuso et al., 2012; Mulder et al., 2011; Portman et al., 2012). For example, in India, coastlines have been relentlessly modified as a result of mounting development activities that have not been properly managed (Rajawat et al., 2015). In Ragusa, Italy, ineffective erosion management at ports and harbors produced critical erosion problems (Anfuso et al., 2012).

According to Anker et al. (2004), over the past several decades, policy-makers have indicated the lack of integration as a barrier to successful coastal management. Integrated coastal zone management (ICZM) is a commonly-used term in the documentation on coastal zone management. Additional documentation includes an Environmental Impact Assessment (EIA) that has been applied in Thailand for every breakwater project. The objective of EIA is to protect people and vulnerable environmental resources from being affected (Saengsupavanich, 2012).

In Malaysia, the government conducted a National Coastal Erosion Study in 1985 that was funded by the Asian Development Bank to estimate coastal erosion, and the results indicated that 30% of the Malaysian coastal areas were threatened by erosion (Nor Hisham, 2006). They continued to carry out further assessments and management plans in order to ensure sustainable development continues along the Malaysian coasts. The current plan is an Integrated Shoreline Management Plan (ISMP) that addresses the major issues and problems faced by the shorelines. This integrated approach takes into account factors such as economic, social, environmental, and ecological issues. However, it is focused on selected states (i.e., Pulau Pinang, Negeri Sembilan, Melaka, Pahang), based on the criteria of critical erosion (Apat, 2009).

According to Ariffin et al. (2016), currently, Kuala Terengganu, the capital city of the Malaysian state of Terengganu facing the South China Sea, tends to experience critical erosion in the areas undergoing development. According to comprehensively documented studies, only 12% of the overall Terengganu coastline is under critical erosion (Nor Hisham, 2006). This chapter aims to identify the current rate of coastal erosion along the Terengganu coastline, describe the evolution of the shoreline, and document the causes of changes along 244 km of coastline.

1.1 Regional Geological Setting

The regional geology of the land adjacent to the coast is relevant to the investigation, as well as the interpretation of the results. This is because sediment is transported and deposited in a given area, depending on the nature of the bordering terrain. The geology of Peninsular Malaysia records a Phanerozoic history from the late Cambrian to the late Triassic periods. During the Triassic period, sedimentation took place mainly in a continental environment, while older deposits took place primarily in a marine environment. The Triassic period is also characterized by the eruption of the Pahang Volcanic Series, which is found mainly in Kelantan, Terengganu, and Pahang (Ghani, 1980; JMG, 1985).

However, in terms of Quaternary geological formations, (1 to 4) are composed of continental and marine deposits, with unconsolidated sands of mainly marine origins. In Garba et al. (2015), they define the Quaternary geological formations as:

i. Quaternary (1): continental and marine deposits with mainly unconsolidated marine sand.
ii. Quaternary (2): continental and marine deposits with unconsolidated silt and clay (mainly).
iii. Quaternary (3): continental and marine deposits with unconsolidated humic clay, peat, and silt.
iv. Quaternary (4): marine and continental deposits with unconsolidated clay, silt, sand, and gravel, undifferentiated.

Presently, the Terengganu coast is covered almost entirely by sand (90%), and is constantly enriched by ample sediment loads from several major rivers, such as the Terengganu River (Nor Hisham, 2006; Sultan et al., 2011). Strong coastal currents operating in the South China Sea bring only sand into the Terengganu coast, with the weathered granitic products as the main source of sediment (Garba et al., 2015). According to Yaacob and Hussain (2005), erosion of beaches is the other major source that provides sediment to the Terengganu coast by a longshore transport system.

The coast of Terengganu faces the South China Sea (in the east coast of the Peninsular Malaysia province), with the coastline extending for 244 km. The Terengganu coastline commences from Kuala Besut in the Besut region in the north, near the state of Kelantan, and terminates at Kuala Kemaman in the Kemaman region near the southern boundary of the state of Pahang. This study distinguishes Besut, Setiu, and Kuala Terengganu as northern regions, and Marang, Dungun, and Kemaman as southern regions. The Terengganu International Airport is located on the Tok Jembal coastline (SD 4), where the runway extension was planned (Fig. 1).

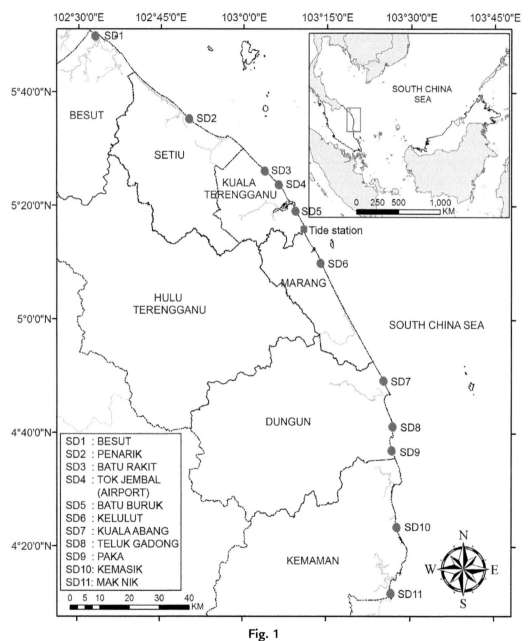

Fig. 1
Terengganu coast with tide station and scattered sediment stations at all regions (SD).

According to Ariffin et al. (2016), the construction of the runway extension would change the coastal processes. The coastal processes on the Terengganu coastline are greatly influenced by the monsoons. The monsoon seasons bring about a greater intensity of related physical phenomenona, including high frequency rainfall, winds, current velocities, and waves. Numerous studies have been conducted to demonstrate that the monsoon period depends on rainfall distribution (e.g., Bagyaraj et al., 2015; Chang et al., 2005).

Fig. 2 shows there are two monsoon conditions on the Terengganu coast based on rainfall distribution and wind parameters, namely the southwest monsoon and the northeast monsoon. The dry season (or less rainfall) is observed during southwest monsoon season, and increases to heavy rainfall during the northeast monsoon season. Moreover, during the southwest monsoon season, the winds are southwesterly between May and September. The northeast

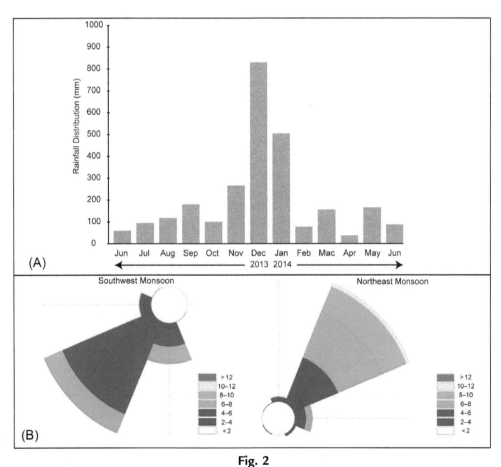

Fig. 2
Monsoon conditions along the Terengganu coast in 2013–2014 with (A) rainfall distribution and (B) wind distribution.

monsoon season occurs between November and March, with northeasterly winds blowing over the South China Sea (Kok et al., 2015).

The same pattern is observed in the wind parameters during the southwest monsoon season, with calm conditions in the currents and waves. However, during the northeast monsoon season, the coast is exposed to strong currents and waves that lead to beach erosion (Ariffin et al., 2016). The waves are higher than usual during this season due to the strong onshore winds, and thus can cause comparatively more damage (Mohd Lokman et al., 1995). Overwash deposits are thrown to higher elevation surfaces and further inland than the backshore. However, the beaches will build up or recover during the southwest monsoon season (Ariffin et al., 2016). However, the tides on Terengganu coast are semi-diurnal and micro- to meso-tidal.

2 Methodology

Multiple methods were used for assessing coastal erosion along the Terengganu coastline in this study. Shorelines changes were indicated by a Digital Shoreline Analysis System (DSAS), with only one tide evolution station at Chendering Port, Kuala Terengganu. To observe the mean grain size distribution, eleven sediment stations (SD) were selected along the Terengganu coast. Physical parameters (currents and waves) were determined by using a Regional Ocean Modelling System (ROMS) and Mike 21-SW (Spectral Wave).

2.1 Shoreline Evolution

This section briefly discusses shoreline evolution based on the results obtained from a Digital Shoreline Analysis System (DSAS) (Thieler et al., 2009), using aerial photographs taken between 1995 and 2016. The images were first processed by geometric correction using a UTM 48-WGS 1984 projection. 100 m interval of transects were generated, with the results incorporating only the End Point Rate (EPR) method and Lineal Regression (LRR) approaches, which presented negative and positive values for erosion and accretion rates, respectively. However, the retreat or accretion rates at all regions were calculated and grouped into four categories of coastal evolution trends based on the Rangel-Buitrago et al. (2015) study. The groups are "high erosion" (≥ -1.5 m/year), "erosion" (between -0.2 and -1.5 m/year), "stability" (between -0.2 and $+0.2$ m/year), and "accretion" (≥ 0.2 m/year).

2.2 Sediment Mean Size Measurement

In this study, eleven sediment stations (SD) were selected along the Terengganu coast to identify the mean grain size distribution during the southwest season (June 2011) and northeast season (December 2011) monsoons. The mean grain size (D50) of the sediment was determined using the GRADISTAT V4.0 program.

2.3 Physical Setting

To assess the impact of monsoons on the Terengganu coast, the extracted tidal elevation was observed based on in situ data at Chendering Port from 1984 to 2014 to quantify the sea level rise. We then compares the hydrodynamic models on current and wave parameters during the southwest and northeast monsoon seasons. A three-dimensional model created by a Regional Ocean Modelling System (ROMS) was used to produce a surface circulation current pattern in the southern part of the South China Sea in order to show the controlling processes around the Terengganu coast. The model was forced by the ECMWF surface winds, heat fluxes, TOPEX tidal and HYCOM temperature, and salinity. The current model was simulated by daily output from 2013 to 2014 for monsoon season comparison.

The current and wave parameters were simulated in a similar fashion by daily output from 2013 to 2014 for monsoon season comparison; however, the wave model was forced only by the wind (source from ECMWF) and tidal observation at Chendering Port. It should be noted that the wave models were used by Mike 21-SW. Furthermore, the wave model for 2013–2014 was validated with in situ field data based on the significant wave height (H_s) from November 2013 to February 2014. The validation presented a Root Mean Square (RMSE) of 0.29.

3 Result

3.1 Magnitude of Coastal Evolution

The analysis of aerial photographs from 1995 to 2016 revealed that 55% of the Terengganu coast undergoes erosion (Fig. 3). Specifically, the southern part of the Terengganu coast revealed the highest erosion trends, as recorded by the maximum erosion rates in Marang (−2.2 m/year), Dungun (−4.2 m/year) and Kemaman (−2.8 m/year). On the northern part, Kuala Terengganu showed an accretion trend with the maximum accretion rates of +0.4 m/year; while the maximum erosion rates can be observed in Tok Jembal (SD4) at 10.8 m/year. However, other regions in the northern part revealed that the maximum erosion rates were below 1.5 m/year in the Besut and Setiu regions.

3.2 Grain Size Distributions

Grain size analysis of two selected sediment samples during the southwest and northeast monsoon seasons showed a decreasing trend from north to south along the Terengganu coast (Table 1). During the southwest monsoon season, a considerable quantity of medium-grain sand was found in the southern part, located at Marang, and continued to the northern part, from Kuala Terengganu to Besut. A decreasing trend toward more coarse sand is observed during the northeast monsoon season in these areas. In contrast, other regions in the southern part (Kemaman to Dungun) revealed coarse sand during both monsoon seasons.

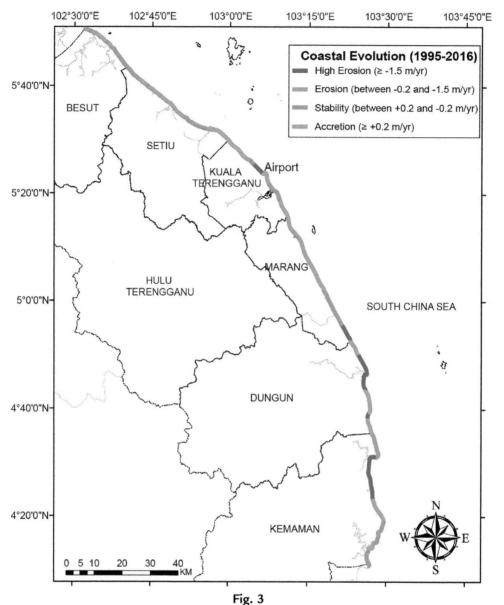

Fig. 3
Coastal evolution from 1995 to 2016 along Terengganu coast.

3.3 Main Causes of Erosion Along the Terengganu Coast

Three main causes of erosion along the Terengganu coast were determined to be hydrodynamic factors. The main hydrodynamic factor was relative sea level rise, followed by currents and waves. The current and wave parameters are describe in detail by comparing the factors between the southwest and northeast monsoon seasons.

Table 1 Particle size along Terengganu coast during southwest and northeast monsoons

Station	Mean (phi)			
	Southwest Monsoon Season		Northeast Monsoon Season	
Besut				
SD1 (Besut)	1.31	Medium sand	0.56	Coarse sand
Setiu				
SD2 (Penarik)	1.27	Medium sand	0.80	Coarse sand
Kuala Terengganu				
SD3 (Batu Rakit)	0.94	Coarse sand	0.35	Coarse sand
SD4 (Tok Jembal)	1.06	Medium sand	0.37	Coarse sand
SD5 (Batu Buruk)	1.06	Medium sand	0.30	Coarse sand
Marang				
SD6 (Kelulut)	1.05	Medium sand	0.34	Coarse sand
Dungun				
SD7 (Kuala Abang)	0.93	Coarse sand	0.64	Coarse sand
SD8 (Teluk Gadong)	−0.15	Very coarse sand	−0.39	Very coarse sand
SD9 (Paka)	0.93	Coarse sand	0.47	Coarse sand
Kemaman				
SD10 (Kemasik)	0.43	Coarse sand	−0.03	Very coarse sand
SD11 (Maknik)	0.56	Coarse sand	0.55	Coarse sand

3.3.1 Relative sea level rise

The tidal elevation at Chendering station showed increasing levels of 0.03 m of water over the past 28 years (Fig. 4). During the southwest monsoon season, the water is carried away with the southwesterly wind (Fig. 2) toward the center of the South China Sea, thus lowering the water level at the coast. By comparing the tidal cycles between the southwest and northeast monsoon seasons, the northeast monsoon season has a higher water level because of stronger winds in the South China Sea, thus uplifting the water at the coast. Furthermore, during the El Nino years (1997, 2010, and 2015), lower water levels were observed, compared with normal years. On the other hand, the water level increased during the La Nina years (1994, 2000, and 2011), which implies that the water level is effected by seasonal changes and El Nino-Southern Oscillation phenomena.

3.3.2 Monsoon seasons

Seasonal surface current in the southern South China Sea region

The southwest monsoon season commences in May, when the southwesterly wind blows over the Southeast Asian region. It drives the northward current circulation in the southern South China Sea region. The northward current is stronger, and flows parallel to the coastline

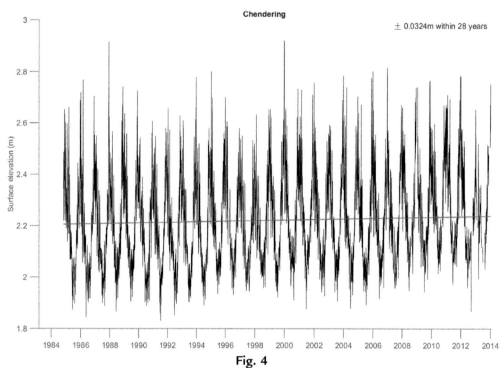

Fig. 4

Time series of tidal elevation in Chendering station in the Kuala Terengganu region (the only one along the Terengganu coast).

along the east coast of Peninsular Malaysia (at 2°N to 5°N latitude). Fig. 5 presents the southern part of the Terengganu coast, especially at Dungun and Kemaman, demonstrating stronger currents. Furthermore, the current was deflected, and shows a decreasing trend to the east at 6°N latitude off the Terengganu coast (toward Kuala Terengganu, Setiu, and Besut). The gulf circulation redirected currents to the Vietnam coast, and due to this mechanism, there was an eddy spotted in the region center at 105°E longitude and 5°N latitude.

Meanwhile, the northeast monsoon season generally develops in November, when the winter season in East Asia blows up strong northeasterly winds over the South China Sea. This generates a strong southwestward current in the southern South China Sea. A strong southward current is observed along the east coast of Peninsular Malaysia at 2°N to 5°N latitude (especially Dungun and Kemaman). There is no eddy spotted during northeast monsoon season in the southern South China Sea region, as the current is driven by strong northeasterly winds, and the gulf area is sheltered by this strong wind.

Seasonal wave distribution in the Terengganu coast

Fig. 6 shows the significant wave height and direction during the southwest and northeast monsoon seasons. The southwest monsoon season commences in May, and the significant wave height (H_s) reveals a calm pattern on the Terengganu coast. H_s decreases gradually northward,

Shoreline Evolution Under the Influence of Oceanographic and Monsoon Dynamics 123

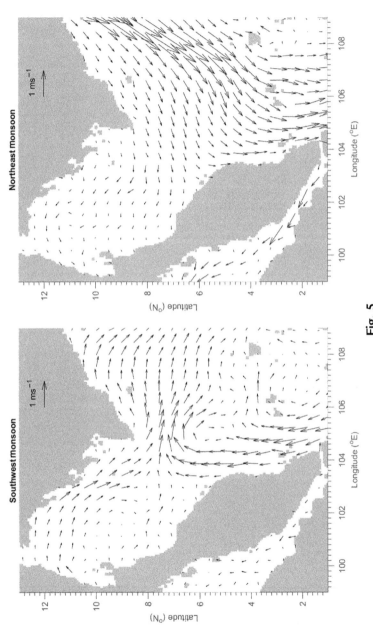

Fig. 5

Current speed and direction pattern along Terengganu coast with observation at Peninsular Malaysia during southwest and northeast monsoon seasons.

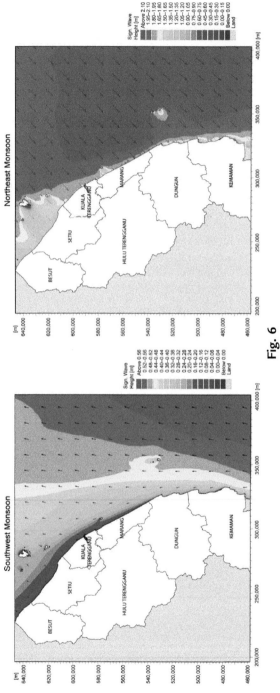

Fig. 6

Significant wave height and direction along the Terengganu coast during southwest and northeast monsoon seasons.

where the values are relatively lower, with ranges from 0.08 to 0.20 m on the Kuala Terengganu, Setiu, and Besut coasts. The lower value of H_s in these regions might be due to a shadowing effect caused by the many islands in front of the coast. Furthermore, this monsoon period did not bring any storm events along the Terengganu coast. Conversely, during the northeast monsoon season, signs of storms can be observed along the Terengganu coast, with the maximum value of H_s up to 1.5–2.0 m. However, these values reduce to 1.2–1.5 m in Setiu and Besut because of the obstruction effects of several islands in these regions.

4 Discussions

4.1 Coastal Processes

The Terengganu coast along the South China Sea is multi-dynamic, and can be divided into the southern and northern parts (see Fig. 1). Specifically, the southern part of the Terengganu coast, including Marang, Dungun, and Kemaman, revealed the highest erosion trends, which were mainly affected by topographic (not covered by islands) and monsoon factors. This is supported with physical (current and wave) and sediment dynamics that indicate that the southern part is more dynamic based on strong physical parameters, and the coarser sediment in this area. According to Dora et al. (2014), slower rates of recovery can be observed, based on coarser sediment during the annual cycle of monsoons. In addition, the recovery process along the coast is highly dependent on topographic features.

The trend is similar in the Ashraful Islam et al. (2016) study of Bangladesh, which experiences various coastal hazards, especially monsoons. In Brittany and Normandy (located in western France with a geomorphological point of view facing the Atlantic Ocean, which is more dynamic), a severe storm hit the coastline, which, along with the high spring tide, caused major damage on the coast (Suanez et al., 2012).

However, different cases in the northern part of Terengganu, which is covered by many islands, reveal a more stable coastline. The coast in this area still shows erosion, especially at Setiu and Besut, because of impacts from natural dynamics, such as monsoon storms and the morphodynamics of lagoons and estuaries (Rosnan and Mohd Zaini, 2009). Normally, during the northeast monsoon season, signs of storms can be observed along the Terengganu coast, with maximum values of H_s up to 1.5–2.0 m (Ariffin et al., 2016; Mirzaei et al., 2013). However, these values reduce to 1.2–1.5 m in Setiu and Besut because of the obstruction effect of several islands in these regions. The existence of multiple islands in these regions obstructs wave propagation, resulting in a reduction of H_s (Mirzaei et al., 2013).

Owing to the natural dynamics in the northern part of Terengganu, critical areas located at Tok Jembal (SD4) in the Kuala Terengganu province reveal unfavorable erosion along the coastline due to the extension of the Terengganu airport runway (Ariffin et al., 2016;

Muslim et al., 2011). Naturally, the coastal processes will be disturbed if there is any environmental stress along the coastlines (Cooper and Mckenna, 2008; Mulder et al., 2011). For example, in Morocco and India, the authorities built ports (resulting in erosion), and a series of coastal defenses to address erosion problems (Aouiche et al., 2016; Mohanty et al., 2012).

4.2 Coastal Defense Structures and Erosion Control

In terms of coastal defense or protection, different people can interpret the impact of structures differently. According to Cooper and Mckenna (2008), the environmentalists and ecologists expected the coastal defense system to protect the ecosystem. Additionally, a number of engineers and landowners wanted the construction of structures to protect property. However, this defense is satisfactory only if its impacts are overlooked (Anfuso et al., 2012; Cooper and Mckenna, 2008; Portman et al., 2012).

Coastal defense is a common application in coastal management around the world (Bunicontro et al., 2015; Mohanty et al., 2012; Rangel-Buitrago et al., 2018; Saengsupavanich, 2012), especially in Malaysia (Ariffin et al., 2016; Awang et al., 2014). For almost 30 years, the first option (almost always) is the construction of hard structures; whereas, seawalls, groynes, and breakwaters are favorite choices in Malaysia. This coastal defense is the most common engineering practice for defense against erosion problems at the coast (Bunicontro et al., 2015; Vaidya et al., 2015).

The current coastal erosion management in Malaysia is only for action-reaction or post-disaster scenarios, with solutions only for emergencies, and not for risk prevention (Ariffin et al., 2016; Awang et al., 2014; Mokhtar and Ghani Aziz, 2003; Nor Hisham, 2006). In similar studies conducted in Thailand (close to Terengganu coast), authorities have taken action only after erosion problems. However, they take action according to management policies, and response to local stakeholders' pressure. The action and management applied were adapted to meet the requirements of the environmental impacts assessment and preliminary potential risk assessments (Saengsupavanich, 2013a, b). This is necessary to implement good management strategies (Lucrezi et al., 2016; Rangel-Buitrago et al., 2018; Saengsupavanich, 2012). Furthermore, implementation of good and timely management strategies can save millions of dollars of the public's investment, and reduce the time frame required to mitigate erosion (Rangel-Buitrago et al., 2018).

However, from our assessment, coastal management is required to determine beach morphodynamics based on the knowledge of erosion and accretion patterns. Hence, the processes of sediment transport also need to be identified before constructing a coastal protection plan. It is necessary to construct a coastal protection plan, or build other protection options, such as beach nourishment, or other solutions.

Furthermore, the coastal defense system has altered the natural environment of the Terengganu coast (Ariffin et al., 2016) and other regions (Aouiche et al., 2016; Bunicontro et al., 2015; Mohanty et al., 2012; Vaidya et al., 2015), causing:

- Sediment erosion in one site and accumulation in another site;
- Significant reduction of sediment supply, especially from the river mouth;
- Changes in hydrodynamic conditions, especially to the current and wave parameters.

Fig. 7 shows the example of the coastal defense along the Terengganu coast to mitigate erosion close to properties and coastal communities. On the Terengganu coast, the main problem is mitigating the erosion from two monsoon seasons (southwest and northeast monsoons). While the monsoons affect the coast in a drastic way, the government requires time to approve the budget for appropriate erosion-mitigation programs.

Hence, while waiting for budget approval processes, the areas are being further eroded due to the northeast monsoon storms. Concurrently, the contractors of coastal defense structures cannot mitigate the erosion-risk during the storm conditions of the northeast monsoon season. The task of mitigating the erosion problem should be prioritized, and undertaken by every individual involved. These include the Malaysian government, to help the state government mitigate erosion, and to assist the local authorities and coastal communities. Hence, the

Fig. 7
Examples of structures used along the study area in order to protect the coast from erosion; (A) breakwater and seawall at Besut, (B) groyen at Kuala Terengganu, (C) breakwaters at Kuala Terengganu, (D) expansion block at Marang, (E) groyens at Dungun, and (F) breakwater and revetments at Kemaman.

cooperation of all these stakeholders together can help avoid overlapping efforts, reduce coastal erosion related risks, and improve the quality of life for coastal communities across the Terengganu region.

5 Conclusions

For almost 20 years, the Terengganu coast has undergone a serious erosion trend in the southern area, owing to topographic and monsoon conditions. Evidence indicates that the hazard is closely related to the northeast monsoon storms that produce strong currents and waves on the coast, especially in the southern part of Terengganu. Furthermore, the negative impacts of the implantation of hard structures has direct and indirect effects on the Terengganu coast, especially in the Kuala Terengganu province. Many examples prove that this coastal defense, in fact, has created more of a problem than a solution.

Poor understanding of the coastal processes system can result in negative impacts along the coast. This problem must be confronted and solved sustainably through high coastal management regulations, which are currently lacking in the coastal management policies in Terengganu. Recently, improvement has been seen due to the collaboration between stakeholders to mitigate erosion. Simultaneously, time and money can be saved.

However, to improve coastal zone management, the Terengganu authorities could adopt appropriate methods and practices from other states in Malaysia. Perhaps studies on coastal processes can bridge the gap of documentation on coastal management in Terengganu. Engaging in international programs with neighboring countries can have an impact on integration for coastal management, and help stakeholders continue to learn from countries that have more experience, specifically in coastal management tools.

Acknowledgments

The Institute of Oceanography and Environment (INOS), Universiti Malaysia Terengganu, Malaysia, funded this research. We would like to thank the Laboratory of Physical Oceanography and Geology, Universiti Malaysia Terengganu, Malaysia, for providing the physical data. We are also most grateful to Mr. Mohd Raihan Shah Abdullah for his support in developing our project.

References

Anfuso, G., Martı́nez-del-Pozo, J., Rangel-Buitrago, N., 2012. Bad practice in erosion management: the Southern Sicily case study. In: Pilkey, O., Cooper, A. (Eds.), Pitfalls of Shoeline Stabilization, first ed. Springer, New York, pp. 215–233. https://doi.org/10.1007/978-94-007-4123-2.

Anker, H.T., Nellemann, V., Sverdrup-Jensen, S., 2004. Coastal zone management in Denmark: ways and means for further integration. Ocean Coast. Manag. 47 (9–10), 495–513. https://doi.org/10.1016/j.ocecoaman.2004.09.003.

Aouiche, I., Daoudi, L., Anthony, E.J., Sedrati, M., Ziane, E., Harti, A., Dussouillez, P., 2016. Anthropogenic effects on shoreface and shoreline changes: input from a multi-method analysis, Agadir Bay, Morocco. Geomorphology 254, 16–31. https://doi.org/10.1016/j.geomorph.2015.11.013.

Apat, F., 2009. Pengurusan Pantai. JPS, Kuala Lumpur.

Ariffin, E.H., Sedrati, M., Akhir, M.F., Yaacob, R., Husain, M.L., 2016. Open sandy beach morphology and morphodynamic as response to seasonal monsoon in Kuala Terengganu, Malaysia. J. Coast. Res. Spec. Issue (75), 1032–1036. https://doi.org/10.2112/SI75-207.1.

Ashraful Islam, M., Mitra, D., Dewan, A., Akhter, S.H., 2016. Coastal multi-hazard vulnerability assessment along the Ganges deltaic coast of Bangladesh—a geospatial approach. Ocean Coast. Manag. 127, 1–15. https://doi.org/10.1016/j.ocecoaman.2016.03.012.

Awang, N.A., Jusoh, W.H.W., Hamid, M.R.A., 2014. Coastal erosion at Tanjong Piai, Johor, Malaysia. J. Coast. Res. Spec. Issue 71, 122–130. https://doi.org/10.2112/SI71-015.1.

Bagyaraj, M., Bhuvaneswari, M., Priya, B.N., 2015. The high/low rainfall fluctuation mapping through GIS technique in Kodaikanal Taluk, Dindigul, District, Tamil Nadu. Int. J. Adv. Res. 3 (11), 1606–1613.

Bunicontro, M.P., Marcomini, S.C., López, R.A., 2015. The effect of coastal defense structures (mounds) on southeast coast of Buenos Aires province, Argentine. Ocean Coast. Manag. 116, 404–413. https://doi.org/10.1016/j.ocecoaman.2015.08.016.

Chang, C.P., Wang, Z., McBride, J., Liu, C.H., 2005. Annual cycle of Southeast Asia—maritime continent rainfall and the asymmetric monsoon transition. J. Clim. 18 (2), 287–301. https://doi.org/10.1175/JCLI-3257.1.

Cooper, J.A.G., Mckenna, J., 2008. Working with natural processes: the challenge for coastal protection strategies. Geogr. J. 174 (4), 315–331.

Dora, G.U., Kumar, V.S., Vinayaraj, P., Philip, C.S., Johnson, G., 2014. Quantitative estimation of sediment erosion and accretion processes in a micro-tidal coast. Int. J. Sediment Res. 29 (2), 218–231. https://doi.org/10.1016/S1001-6279(14)60038-X.

Durgappa, R., 2008. Coastal protection works. In: Seventh International Conference of Coastal and Port Engineering in Developing Countries, Dubai.

Garba, N.N., Ramli, A.T., Saleh, M.A., Sanusi, M.S., Gabdo, H.T., 2015. Terrestrial gamma radiation dose rates and radiological mapping of Terengganu state, Malaysia. J. Radioanal. Nucl. Chem. 303, 1785–1792. https://doi.org/10.1007/s10967-014-3818-2.

Ghani, A. C. (1980). Preliminary Investigation of the Quaternary Deposit in the Lowland Area of Kuantan, Pahang: Annual Report of Geological Survey for 1981.

JMG, 1985. Geology of Terengganu: Anuall Report on Geology of Terengganu 1985. JMG, Malaysia.

Kok, P.H., Akhir, M.F., Tangang, F.T., 2015. Thermal frontal zone along the east coast of Peninsular Malaysia. Cont. Shelf Res. 110, 1–15. https://doi.org/10.1016/j.csr.2015.09.010.

Lucrezi, S., Saayman, M., Van der Merwe, P., 2016. An assessment tool for sandy beaches: a case study for integrating beach description, human dimension, and economic factors to identify priority management issues. Ocean Coast. Manag. 121, 1–22. https://doi.org/10.1016/j.ocecoaman.2015.12.003.

Mirzaei, A., Tangang, F., Juneng, L., Mustapha, M.A., Husain, M.L., Akhir, M.F., 2013. Wave climate simulation for southern region of the South China Sea. Ocean Dyn. 63 (8), 961–977. https://doi.org/10.1007/s10236-013-0640-2.

Mohanty, P.K., Patra, S.K., Bramha, S., Seth, B., Pradhan, U., Behera, B., Mishra, P., Panda, U.S., 2012. Impact of groins on beach morphology: a case study near Gopalpur Port, East Coast of India. J. Coast. Res. 28 (1), 132–142. https://doi.org/10.2112/JCOASTRES-D-10-00045.1.

Mohd Lokman, H., Rosnan, Y., Shahbudin, S., 1995. Beach erosion variabiltiy during a northeast monsoon: the kuala setiu. Pertanika J. Sci. Technol. 3 (2), 337–348.

Mokhtar, M., Ghani Aziz, S.A., 2003. Integrated coastal zone management using the ecosystems approach, some perspectives in Malaysia. Ocean Coast. Manag. 46 (5), 407–419. https://doi.org/10.1016/S0964-5691(03)00015-2.

Mulder, J.P.M., Hommes, S., Horstman, E.M., 2011. Implementation of coastal erosion management in the Netherlands. Ocean Coast. Manag. 54 (12), 888–897. https://doi.org/10.1016/j.ocecoaman.2011.06.009.

Muslim, A.M., Ismail, K.I., Razman, N., Zain, K., Khalil, I., 2011. Detection of shoreline changes at Kuala Terengganu, Malaysia from multi-temporal satellite sensor imagery. In: 34th International Symposium on Remote Sensing of Environment—The GEOSS Era: Towards Operational Environmental Monitoring, Sydney, Australia.

Nor Hisham, M.G., 2006. Coastal erosion and reclamation in Malaysia. Aquat. Ecosyst. Health Manag. 9 (2), 237–274. https://doi.org/10.1080/14634980600721474.

Portman, M.E., Esteves, L.S., Le, X.Q., Khan, A.Z., 2012. Improving integration for integrated coastal zone management: an eight country study. Sci. Total Environ. 439, 194–201. https://doi.org/10.1016/j.scitotenv.2012.09.016.

Rajawat, A.S., Chauhan, H.B., Ratheesh, R., Rode, S., Bhanderi, R.J., Mahapatra, M., Kumar, M., Yadav, R., Abraham, S.P., Singh, S.S., Keshri, K.N., Ajai, 2015. Assessment of coastal erosion along the Indian coast on 1: 25,000 scale using satellite data of 1989–1991 and 2004–2006 time frames. J. Curr. Sci. 109 (2), 347–353.

Rangel-Buitrago, N.G., Anfuso, G., Williams, A.T., 2015. Coastal erosion along the Caribbean coast of Colombia: magnitudes, causes and management. Ocean Coast. Manag. 114, 129–144. https://doi.org/10.1016/j.ocecoaman.2015.06.024.

Rangel-Buitrago, N., Williams, A., Anfuso, G., 2018. Hard protection structures as a principal coastal erosion management strategy along the Caribbean coast of Colombia. A chronicle of pitfalls. Ocean Coast. Manag. 156, 43–58. https://doi.org/10.1016/j.ocecoaman.2017.04.006.

Yaacob, R., Hussain, M.L., 2005. The relationship of sediment texture with coastal environments along the Kuala Terengganu Coast, Malaysia. Environ. Geol. 48 (4–5), 639–645. https://doi.org/10.1007/s00254-005-1321-3.

Rosnan, Y., Mohd Zaini, M., 2009. Grain-size distribution and subsurface mapping at the Setiu wetlands, Setiu, Terengganu. Environ. Earth Sci. 60 (5), 975–984. https://doi.org/10.1007/s12665-009-0236-9.

Saengsupavanich, C., 2012. Unwelcome environmental impact assessment for coastal protection along a 7-km shoreline in Southern Thailand. Ocean Coast. Manag. 61, 20–29. https://doi.org/10.1016/j.ocecoaman.2012.02.008.

Saengsupavanich, C., 2013a. Detached breakwaters: communities' preferences for sustainable coastal protection. J. Environ. Manag. 115, 106–113. https://doi.org/10.1016/j.jenvman.2012.11.029.

Saengsupavanich, C., 2013b. Erosion protection options of a muddy coastline in Thailand: stakeholders' shared responsibilities. Ocean Coast. Manag. 83, 81–90. https://doi.org/10.1016/j.ocecoaman.2013.02.002.

Suanez, S., Cariolet, J.M., Cancouët, R., Ardhuin, F., Delacourt, C., 2012. Dune recovery after storm erosion on a high-energy beach: Vougot Beach, Brittany (France). Geomorphology *139–140*, 16–33. https://doi.org/10.1016/j.geomorph.2011.10.014.

Sultan, K., Shazili, N.A., Peiffer, S., 2011. Distribution of Pb, As, Cd, Sn and Hg in soil, sediment and surface water of the tropical river watershed, Terengganu (Malaysia). J. Hydro Environ. Res. 5 (3), 169–176. https://doi.org/10.1016/j.jher.2011.03.001.

Thieler, E. R., Himmelstoss, E. A., Zichichi, J. L., & Ergul, A. (2009). Digital Shoreline Analysis System (DSAS) Version 4.3-An ArcGIS Extension for Calculating Shoreline Change: U.S. Geological Survey Open-File Report 2008-1278.

Vaidya, A.M., Kori, S.K., Kudale, M.D., 2015. Shoreline response to coastal structures. Aquat. Proc. 4, 333–340. https://doi.org/10.1016/j.aqpro.2015.02.045.

CHAPTER 6

Spatial and Statistical Analyses of Clifftop Retreat in the Gulf of Morbihan and Quiberon Peninsula, France: Implications on Cliff Evolution and Coastal Zone Management

Soazig Pian, David Menier

Géosciences Océan Laboratory UMR CNRS 6538, University of South Brittany, Vannes Cedex, France

1 Introduction

Most coastal research studies deal with the geomorphological and morphodynamic behavior of beach systems, while a few others focus on cliff evolution (Naylor et al., 2010). However, cliff retreat processes play an important role in the behavior of coastal systems by providing sediment sources (Dornbusch et al., 2008). Cliff failure and coastline retreat rates are assessed using a wide range of methods, based on both field and remote sensing data (Quinn et al., 2010). Long-term cliff retreat is often assessed using historical topographic maps or aerial photographs (Moore and Giggs, 2002; Dornbusch et al., 2008). Cliff retreat can be related to a wide range of processes, as reviewed by Sunamura (1992) and Dong (2009), including numerous factors such as cliff lithology, geological constraints, susceptibility of rocks to wave impacts and freshwater processes, soil layer width, vegetation cover, cliff exposure to prevailing winds and waves, protective beach width, sea-level rise rates, and also anthropogenic pressures such as footpath location or an increase of artificial areas (French, 2001). In sheltered areas, clifftop retreats are mainly driven by subaerial weathering processes (Robinson and Jerwood, 1987; Sallenger et al., 2002). Subaerial mechanisms can outweigh marine processes, in which case cliff failures are mainly related to rainfall and rock resistance (Young et al., 2009). Anthropogenic pressures exerted on clifftop areas also interact with weathering processes, leading to considerable coastline recession (Adriani and Walsh, 2007; Kumar et al., 2009). At the same time, the recession of coastal cliffs leads to significant risks to persons and property (Lee et al., 2001): coastline retreat contributes to triggering human activities, thus affecting infrastructure and tourist resorts, and could represent significant losses of land area. Recent studies (Del Rio and Gracia, 2009; Nunes et al., 2009) used Geographical Information Systems (GIS) to assess the risks of cliff retreat. These studies highlight the need to achieve a global understanding of cliff retreat processes to improve the efficiency of cliff retreat management (Fall, 2009).

This chapter combines the application of GIS with spatial and statistical analysis to assess the role of both natural and anthropogenic factors likely to drive clifftop retreat. This approach aims at identifying the main factors associated with differing rates of clifftop retreat in order to produce an effective set of data for coastal managers. The study focuses on two cliff systems located in South Brittany (France): the sheltered and weathered low cliffs of the Gulf of Morbihan, and the rocky cliffs of the Quiberon Peninsula. From the 1950s, these systems have been subject to subaerial weathering processes leading to clifftop retreat rates of more than 0.03 m/year (Pian, 2010). Both of these cliff systems are associated with highly important consequences for tourism, and thus an understanding of their evolution over a long time scale (50 years) may help plan coastal management for the study areas and elsewhere.

2 Study Area

The South Brittany coast is located in the western part of France (Fig. 1), which experiences westerly and southerly winds (Lemasson, 1999). The dominant waves are characterized by a height ranging from 0.5 to 2.5 m with a period of 5 to 9 seconds, and directed mainly from the northwest and west (Tessier, 2006).

2.1 The Gulf of Morbihan Cliffs

The Gulf of Morbihan is a sheltered coastal system where hydrodynamic conditions are predominately influenced by tidal currents. The waters in this Gulf interact with the Atlantic Ocean only though a pass at Port Navalo, which does not exceed 900 m in width. Numerous small islands are scattered in the Gulf. The two largest islands are the Ile aux Moines and the Ile d'Arz. The propagation of the tidal pulse is controlled, to a large extent, by the general shape of the Gulf, the location of the islands, and the bathymetry. In the west of the Gulf, around the Port Navalo pass, the maximum water depth reaches 30 m in tidal creeks. Toward the east, the water depths decrease, ranging from 10 to 1 m east of the Ile d'Arz. As a consequence, tidal currents are stronger at the Port Navalo pass, where they reach 9 *knots* during high spring tides (Pian and Regnauld, 2007). Tidal speeds of currents decrease between the Ile aux Moines and Ile d'Arz, and both flood and ebb currents reach their lowest values east of the Ile d'Arz. Due to these low-energy hydrodynamic conditions, tidal flats and salt marshes are well developed in the Gulf, especially in the eastern part. Most of the Gulf coastline is composed of small weathered cliffs no more than 10 m high. Cliff heights range from less than 0.50 m to a maximum of 10 m, with a mean height ranging between 1.5 and 4 m. The cliffs are composed of highly weathered

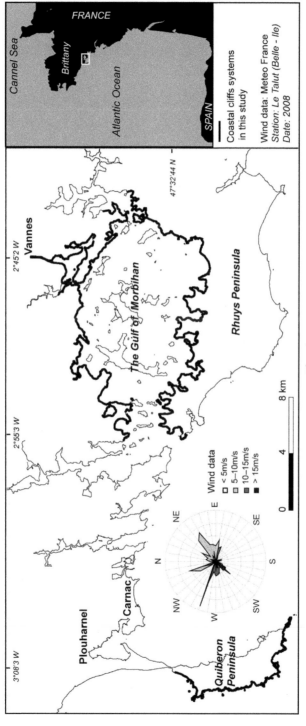

Fig. 1
Study area.

Fig. 2
Various types of soft rock cliffs within the Gulf of Morbihan.

materials derived from granite rocks (Fig. 2). A fine soil layer, which does not exceed 0.50 m in thickness, overlies the weathered layer and is colonized by different vegetation types such as grass, shrubs, or trees (Pian and Regnauld, 2007). Anthropogenic pressures on the cliffs are mainly linked to the presence of a footpath running along the top of the cliffs. In addition, numerous private properties associated with residential areas have been developed. Rigid structures such as sea walls are constructed to protect these properties from cliff recession.

2.2 The Quiberon Peninsula Cliffs

The Quiberon Peninsula is mainly composed of weathered rocky cliffs, orientated toward the west and exposed to prevailing winds and waves. This cliff system is located in a much higher energy environment than the Gulf of Morbihan. At the northern and southern extremities of the peninsula, cliffs with heights not exceeding 4 m are developed in highly weathered materials and are based on a hard rock shore platform. On the central part of the peninsula, cliffs are cut in granite and metamorphic rocks. These hard rock cliffs are capped by a weathered layer (Fig. 3). Cliff heights range from 15 to 5 m, and decrease from the north to the south. A former highly eroded climbing dune is established on the clifftop. A few small pocket beaches partly fed by cliff retreat occur at the foot of the cliffs. These beaches are composed of material of heterogeneous grain size ranging from pebbles to medium sands (Pian, 2010).

Relatively strong anthropogenic pressures are exerted on the cliffs tops: the Quiberon peninsula is one of the most important tourist resorts of South Brittany. It records the highest variations in population during the summer months. A large number of footpaths run along the clifftops, and there are numerous car parks in the vicinity. The cliffs are visited throughout the year.

3 Materials and Methods

To map a clifftop retreat, a combined set of field and aerial photograph data are utilized and integrated within a spatial database. The spatial database was constructed using ESRI ArcCatalog (Pirot and Saint-Gérand, 2005). A personal geodatabase was set up to store and

Fig. 3
Hard rock cliffs of the Quiberon Peninsula.

analyze different layers of spatial data dealing with the clifftop retreat, along with the characteristics of natural and anthropogenic factors.

3.1 Measuring Clifftop Retreat From Analysis of Aerial Photographs

Cliff retreat on a regional scale was measured from aerial photographs. The use of vertical aerial photographs in coastal studies has been widely discussed (Williams et al., 2001; Moore and Giggs, 2002; Graham et al., 2003; Fletcher et al., 2003). The oldest aerial photographs used in the present study date back to 1952. They were scanned with a spatial resolution of 12,000 dpi and geo-rectified, using ArcGis9.2 software. For each geo-rectified photograph, the RMS error was around 4, and the cell sizes varied from 0.68 to 0.8 m (Table 1). Once the photographs were geo-rectified, all the aerial photographs covering each site for 1952 were assembled into a mosaic using Envi3.0 software. Parallax was taken into account by conserving only the central parts of the aerial photographs and by measuring the offset between two fixed features (car parks) located on the clifftops, which were plotted onto the 1952 mosaic as well as the 2004 orthophotography.

The clifftop position was then determined by means of a geomorphological indicator whose migration during the considered time interval is assumed to represent the evolution of the shoreline (Parker, 2003). In this study, the clifftop position was assigned to the break of slope at the top of the cliff (Moore and Giggs, 2002). It was plotted on the orthophotographs of 2004, as

Table 1 Characteristics of the aerial photographs

	Cell Size (m)	Root Mean Square Error (RMS) (m)
Orthophotography 2004	0.50	
Aerial photographs 1952 (Gulf of Morbihan)	0.80	4.50
Aerial photographs 1952 (Quiberon Peninsula)	0.69	4.05

well as the on the aerial photographs dating from 1952, as a polyline using the ArcGis Editor tool. Then, for each site, polylines representing the shoreline positions in 1952 and 2004 were merged into a new layer using ArcGis toolbox functions and converted into a polygon layer (Levin and Benn Dor, 2004; Rodriguez et al., 2009). The values of the error margins associated with the geo-rectification and digitalization processes were extracted from the polygons (Table 1). To estimate the maximum amount of cliff retreat (outside error margins), the width of a rectangle was measured whose length was parallel to the coastline, and which included the whole polygon, using the polygon area and perimeter. Because cliff failure is constrained by faults and fractures orthogonal to the coastline, the polygon shape is a good approximation of the rectangle shape, so this method allowed an automatic quantification of clifftop retreats from aerial photographs. Cliff retreat rates were then estimated using the effective end point rate method (Doland et al., 1991), by dividing clifftop retreat values for each site by the number of years of the studied time interval.

3.2 Construction of the Spatial Database

The compilation of the spatial database (Fig. 4) was carried out in several steps, including the identification, classification, and digitization of natural and anthropogenic factors, and spatial data layer intersection.

As mechanical resistance of rocks, soil depth, or vegetation cover represent natural factors likely to control subaerial weathering processes (Robinson and Jerwood, 1987; French, 2001; Sallenger et al., 2002), we focus here on these factors to explain cliff retreat in both the Gulf of Morbihan and Quiberon Peninsula. For the more exposed Quiberon Peninsula cliffs, the orientation of the coastline to prevailing winds and waves was also taken into account, because erosion by wave action also controls cliff recession (Sunamura, 1977; Hapke and Richmond, 2002).

First, different categories of cliff and vegetation cover were mapped from field data and aerial photographs to characterize each stretch of the studied coastlines. The cliff classification used here is based on the previous studies of Emery and Kuln (1982) and Claytone and Shamonn (1998), and uses the following criteria: cliff orientation to prevailing winds and waves, cliff

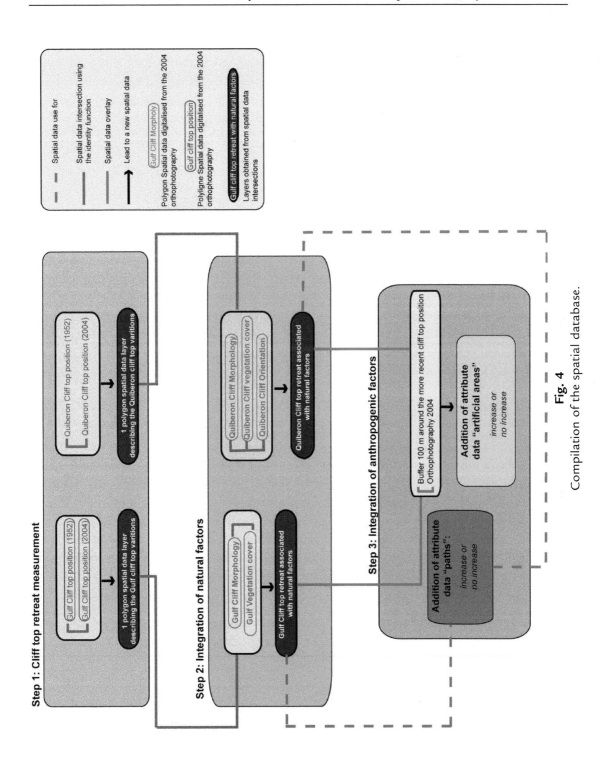

Fig. 4
Compilation of the spatial database.

height, presence/depth of weathered layer, presence/depth of soil layer, and resistance of the rock to weathering processes (soft rock vs. hard rock). From the aerial orthophotographs of 2004, a number of sites were selected and visited at least twice, at low and high tides. Between June 2006 and July 2009, nine sites were visited along the cliffs of the Quiberon Peninsula. A total of 46 sites were visited in the Gulf, where two field trips were carried out by boat. The choice of sites was based on screening and visual analysis of aerial photographs, which provided an overview of the different types of cliff morphology. For each site, a field index card was drawn up containing a picture of both the 2004 orthophotography and the survey map, as well as a description of the cliff and vegetation cover obtained by visual analysis of aerial photographs. For each site, this description was completed in the field: the resistance of the rock was assessed, the depths of the weathered and soil layers were measured, clifftop vegetation cover was checked, and cliff height was measured using D-GPS data.

The first cliff category includes microcliffs, no more than 1 m high, that are cut in weathered materials or salt marshes. They are widespread throughout the Gulf, especially in the east where low-energy hydrodynamic conditions favor the development of mudflats and saltmarshes. At high tide, the retreat of microcliffs could be partly related to tidal currents; however, their evolution is mainly controlled by subaerial weathering processes (Fig. 2). The second cliff category corresponds to low weathered cliffs, whose height ranges from 1 to 4–5 m (Fig. 2). The weathered layer makes up at least one half of the total cliff height. Sometimes, it lies on top of hard metamorphic or highly fractured granitic rocks. The weathered layer was found to be capped by a soil layer not exceeding a depth of 0.50 m. Because these types of cliffs are located in sheltered areas, the main processes controlling their retreat are driven by subaerial weathering. Such cliffs are found along the Gulf coastline and the south-eastern extremity of the Quiberon Peninsula. The third cliff category (Fig. 2) is composed of high weathered cliffs that can reach a height of 10 m. Unlike the previous type, the weathered layer represents less than half of the total cliff height, but overlies the same fractured metamorphic or granite rocks. These cliffs are located in sheltered areas within the Gulf, as well as along the more exposed coastlines, such as the north of the Quiberon Peninsula. Thus, their retreat is the result of both subaerial processes and wave action. The fourth cliff category concerns the exposed rocky cliffs of the Quiberon Peninsula (Fig. 3), which range in height from 10 to 15 m. The weathered layer is much thinner than in the previous types. The soil layer is very thin and often absent.

Three categories of vegetation cover were recognized on the orthophotographs of 2004, and then validated in the field: grass, scrubs, and tree vegetation. A fourth category was added to describe areas where soil layers and vegetation are absent.

To store these factors as spatial data layers within the geodatabase, the cliff coastlines were divided into different segments according to cliff and vegetation cover categories (Nunes et al., 2009). These segments were digitalized as polygons from the aerial photographs taken in 2004. Nominative attributes describing the four categories of cliffs and vegetation cover were assigned to each polygon. The Quiberon cliff coastline was also divided according to its orientation. Five orientations were recognized along the coastline: southeast, south, southwest, west, and northwest. Each segment was digitalized as a polygon at a scale of 1:25,000. An attribute value was assigned to each polygon in order to store the information related to the orientation.

Then, these spatial data layers were intersected with the polygon layers describing clifftop variations using the *identity* function provided by the ESRI ArcGis9.3 software (Fig. 4). As a result, for each stretch of the coastline in the Gulf of Morbihan and the Quiberon Peninsula, a spatial data layer defining a value of clifftop retreat in meters, as well as the location of a set of natural factors controlling cliff weathering and erosion was obtained. Then, two anthropogenic factors affecting clifftops and likely to interact with subaerial erosion processes were added to the spatial database (Fig. 4). The first factor refers to the addition of a pathway between 1952 and 2004. It was assessed by screening and visual analysis of aerial photographs. When a pathway was added on the clifftop, and intersected a polygon describing clifftop retreat, it was recorded within the spatial database in terms of nominative attribute values (*0 = no footpath increase or 1 = footpath increase*). The second anthropogenic factor takes into account the increase of artificial areas—that is, car parks—established on the clifftops within the same time interval. To identify this last factor, a 100 m wide buffer zone was created around the polygons representing clifftop retreat. Any increase of artificial area between 1952 and 2004 within this buffer zone was also recorded within the spatial database in terms of nominative attribute values (*0 = no artificial area increase or 1 = artificial area increase*).

3.3 Spatial and Statistical Analysis

To explain the spatial variations of clifftop retreat, a series of spatial and statistical analyses were carried out and are detailed herein.

3.3.1 Spatial analysis

First, spatial analysis was carried out to assess the spatial relationships between the distributions of clifftop retreat values along the coastlines, as well as the characteristics of the analyzed factors likely to control these changes (Pian, 2010). Such analyses rely on the assumption that the spatial distribution of geographic objects is related to their characteristics (O' Sullivan and Unwin, 2003), owing to the spatial auto-correlation principle that allows us to quantify the relationship between the proximity of different places and their degree of resemblance (Pumain and Saint-Julien, 2004). In the scope of this study, the geographic objects refer to the polygon representing clifftop retreat, and their characteristics are described within the spatial database as attribute values. A nearest-neighbor analysis was performed with the aim of assessing the distance between the density of different clifftop retreat values recorded along the coastline, and the density of certain factor characteristics contained within the spatial database. Indeed, this analysis attempts to identify certain spatial associations that group together a given factor (i.e. tree vegetation cover) with a specific clifftop behavior (i.e., high clifftop retreats values).

The density of the different factor characteristics was mapped using the ESRI *create vector grid tool* to create regular grids from spatial features contained in a spatial data layer source, while taking into account specific attribute values (Fig. 5). Cell sizes were determined according to the areal extent of spatial data layers and contain at least two polygons. Cell resolution is 1000 m for the Gulf of Morbihan, but decreases to 500 m for the Quiberon Peninsula due to the smaller spatial extent. Polygon attribute values contained in a given cell were assigned to the cell as new attribute data. Selected attribute values were added, and the sum weighted by the polygon areas. For each cliff coastal system, a grid was created for each attribute value related to each studied factor. Finally, the different grids were superimposed onto the spatial data layer describing the spatial variations of clifftop retreat values.

3.3.2 Statistical analysis

To complete the spatial analysis, the statistical relationships between the different factors contained within the spatial database were assessed through multivariate statistical analysis using XlStat software. First, a multiple component analysis (MCA) was carried out. The MCA allowed us to assess the statistical relationships between all the modalities of the set of nominative variables (Minvielle and Souiah, 2003). Clifftop retreat values are expressed in m and refer to quantitative data. To be incorporated within the MCA, these values were grouped

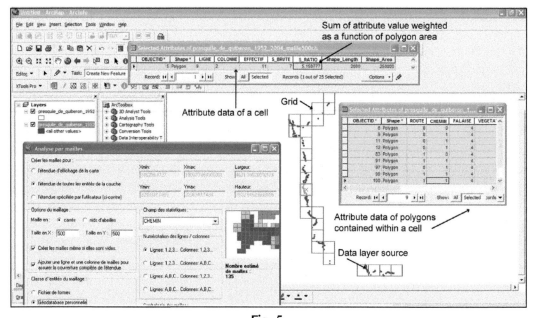

Fig. 5
Mapping density.

into five classes using a natural break classification that allowed us to define classes of similar numbers of data, and which does not distort the statistical analysis (Escoffier and Pagès, 1998).

It was followed by a hierarchical clustering analysis (HCA) on the coordinates of modalities provided by the MCA. The HCA aims at establishing a typology of the studied objects (clifftop retreat sites) by clustering them on the basis of their resemblance. Resemblance is defined as the sharing of a large number of the same modalities (Escoffier and Pagès, 1998), and can be assessed using Euclidian distance.

4 Results

4.1 Clifftop Retreat Mapping

Fig. 6 presents the spatial distribution of clifftop retreat values along the Gulf of Morbihan coastline between the years 1952 and 2004. While retreat rates higher than 0.02 m/year are less common, the four other clifftop retreat classes are well represented, and their distribution is relatively homogeneous. This inference implies that spatial variations of hydrodynamic conditions do not control the spatial distribution of clifftop retreat, or, alternatively, hydrodynamic conditions play a lesser role in clifftop retreat. At this juncture, the dominant role of other factors, such as subaerial weathering, can be assumed. This presumption is affirmed as detailed herein.

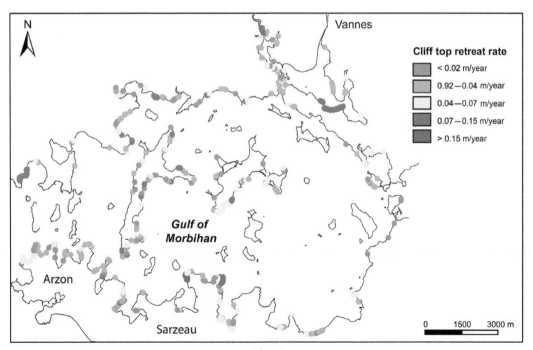

Fig. 6
Clifftop retreat along the Gulf of Morbihan coastline (1952–2004).

The clifftop retreat values along the Quiberon Peninsula display a similar homogeneous spatial distribution (Fig. 7). Cliff retreat rates ranging from less than 0.02 to 0.07 m/year are more frequent, and are observed all along the coastline. The maximum retreats, faster than 0.15 m/year, are located on the central part of the Quiberon Peninsula windward coast. They are associated with weathered material capping the cliff face.

4.2 Spatial Associations Inferred From the Nearest-Neighbor Analysis

Fig. 8 displays the spatial distribution of tree and scrub vegetation cover, as well as the distribution of clifftop retreat values along the Gulf of Morbihan coastline. Both the types of vegetation cover are homogeneously distributed along the Gulf coastline, except for tree cover, which is less dense toward the southeast. Higher clifftop retreat values are located in sectors with a denser tree cover. On the contrary, lower clifftop retreat values are located where scrub vegetation cover is denser. These results suggest a spatial relationship between the distributions of cliff vegetation cover and clifftop retreat values, which can be interpreted in terms of the spatial correlation. Previous authors have proposed that vegetation plays a role in controlling subaerial weathering (French, 2001). Based on this, the increased densities of specific classes of clifftop retreat located close to higher concentrations of certain types of vegetation cover can be assumed to suggest the catalytic role of vegetation on cliff retreat processes in the Gulf of Morbihan.

On the Quiberon Peninsula, most of the hard rock cliffs are capped either by an eroded soil layer where vegetation cover is lacking (category 4) or by grass vegetation (Fig. 9). Higher values of clifftop retreat are measured in sectors showing higher densities of category 4. In addition, these sectors broadly correspond to those associated with an enhanced development of anthropogenic features, especially footpaths. These observations are suggestive of positive relationships between the distribution of clifftop retreat values and the state of vegetation cover, as well as the intensity of human pressure.

4.3 Analysis of Statistical Relationships Between Clifftop Retreat Values and the Studied Factors

In the Gulf of Morbihan, MCA reveals two main axes that account for 57.66% of the total variance (Fig. 10). The first axis explains 40.92% of the total variance. It is mainly defined by variables referring to natural factors such as cliff and vegetation cover categories, or higher classes of clifftop retreat. Microcliffs are associated with grass vegetation cover, as well as both higher and lower values of cliff retreats. Weathered cliffs are associated with shrub vegetation and medium classes of cliff retreat. The second axis explains 16.47% of the total variance. It is defined by variables that refer to natural factors, such as vegetation cover, as well as anthropogenic pressures such as the addition of footpaths or increase of artificial areas. This latter axis allows us to link the high weathered cliffs (type 3) covered by shrub vegetation with

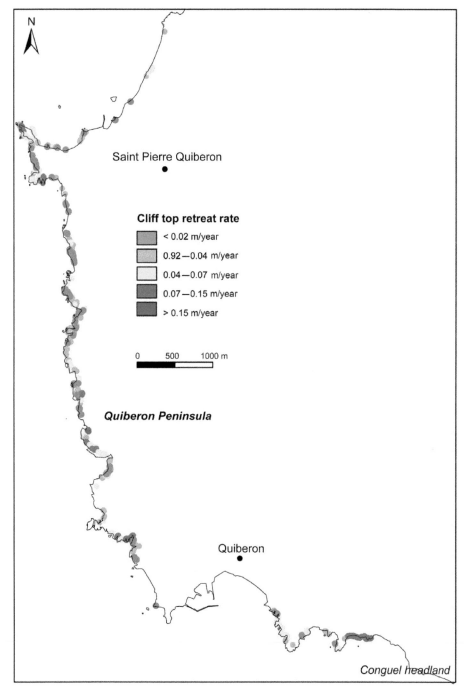

Fig. 7
Clifftop retreat along the Quiberon Peninsula coastline (1952–2004).

144   Chapter 6

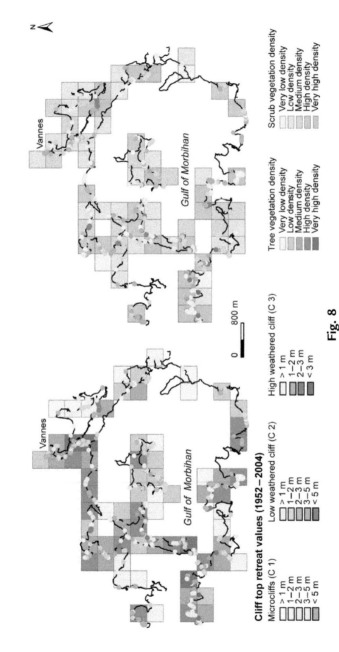

Fig. 8   Spatial distribution of clifftop retreat values and vegetation cover in the Gulf of Morbihan.

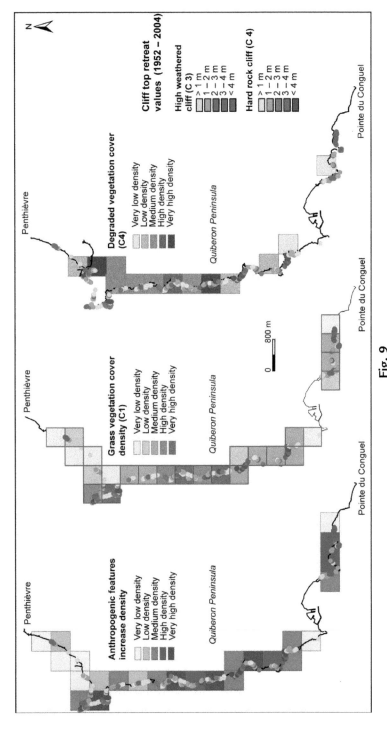

Fig. 9

Spatial distribution of clifftop retreat values, vegetation cover, and anthropogenic features on Quiberon Peninsula.

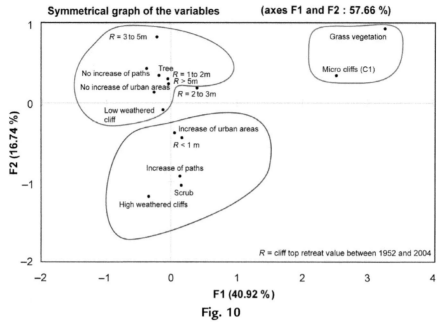

Fig. 10
Projection of modalities onto the first two axes obtained from MCA (Gulf of Morbihan—1952/2004).

an increase of anthropogenic pressures and lower values of clifftop retreat. On the contrary, higher values of clifftop retreat are associated with small weathered cliffs (type 4) and tree vegetation cover. According to the results displayed by these axes, clifftop retreat appears controlled by the type of vegetation cover, combined with the morphology of the cliffs (height, depth of weathered layer, and soil layer). Indeed, MCA leads to associate different values of clifftop retreat with each cliff category, in relation with their vegetation cover: microcliffs are likely to record both high and low values of cliff retreat. Weathered cliffs are associated with low to medium classes of clifftop retreat, except when the vegetation cover is composed of trees. The combination of tree vegetation cover with a deep weathered layer leads to higher cliff retreat values.

HCA highlights these results and emphasizes the role of vegetation cover. It allows identification of three clusters. Cluster 1 is composed of low retreat sites associated with a scrub or grass vegetation. These sites are also often characterized by an increase of anthropogenic pressure. Cluster 2 refers to high retreat sites recorded on cliffs belonging to category 2 and covered by tree vegetation. They are located in sectors with higher densities of tree vegetation (Fig. 11). Cluster 3 groups high and low values of clifftop retreat recorded on category-3 cliffs covered by grass vegetation.

For the cliffs of the Quiberon Peninsula, MCA results identify two axes that explain 52.77% of the total variance (Fig. 12). The first axis explains 40.73% of the total variance and is mainly defined from variables related to natural controlling factors such as cliff category, vegetation

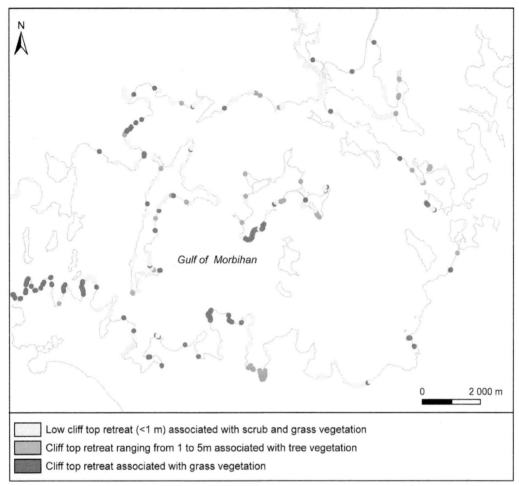

Fig. 11
Spatial distribution of clusters obtained from HCA (Gulf of Morbihan—1952/2004).

cover, and cliff orientation, and, to a lesser extent, the addition of footpaths on clifftops. Along this axis, higher classes of retreat are associated with the south and northwest oriented weathered cliffs covered by shrub and tree vegetation, as well as with the addition of footpaths. Rocky cliffs with no soil layer or vegetation cover are associated with lower classes of cliff retreat. The second axis explains 12.40% of the total variance, and is defined by variables related to the vegetation cover, the increase in anthropogenic pressure on clifftops, and the orientation of the coastline. It reveals a contrast between two groups of variables: on the one hand, the north, northeast, or east oriented weathered cliffs covered by grass vegetation associated with the addition of footpaths on clifftops, and on the other hand, the southeast oriented cliffs characterized either by grass vegetation cover or no soil layer or vegetation cover. Thus, it is safe to conclude that the evolution of the Quiberon Peninsula cliffs is likely to

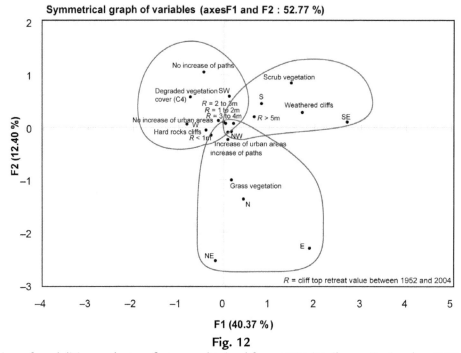

Fig. 12
Projection of modalities on the two first axes obtained from MCA (Quiberon Peninsula—1952/2004).

be controlled by their morphology, their exposure to prevailing winds and waves, and the vegetation cover, as well as the intensity of anthropogenic pressure, and especially footpaths running along the coastline: higher classes of clifftop retreat are associated with the location of footpaths on the clifftop.

Results obtained from HCA attribute the sites to three clusters (Fig. 13). Cluster 1 groups together high-retreat sites located on hard rocky cliffs associated with an increase of anthropogenic features established on clifftops. Cluster 2 refers to low-retreat sites located on hard rock cliffs. These two clusters reinforce the inferences drawn from MCA and spatial analysis. The third cluster is composed of high-retreat sites located on soft rock clifftops.

5 Discussion

The results presented in the previous section highlight the dominant role of vegetation cover in controlling clifftop retreat in the Gulf of Morbihan. Both spatial and statistical analysis show that higher classes of clifftop retreat are observed on low soft rock cliffs characterized by shallow soil, a thick weathered layer in relation to the cliff height, and the presence of tree vegetation cover. The higher the density of tree cover, the higher the values recorded for clifftop retreat. On the contrary, higher concentrations of shrub vegetation are associated with lower clifftop retreat values. Anthropogenic features established on clifftops do not lead to an increase

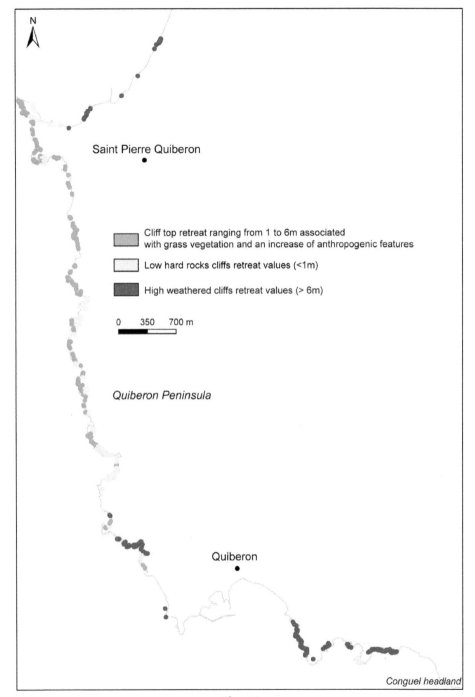

Fig. 13
Spatial distribution of clusters obtained from HCA (Quiberon Peninsula—1952/2004).

Fig. 14
Example of accelerated clifftop retreat due to growth of trees.

of clifftop retreat values. The retreat of cliffs covered by tree vegetation is accelerated by wind breakage, which is directly linked to the presence of tree vegetation on clifftops (Fig. 14). The growth of roots from trees on the clifftop contributes to destabilizing the thin soil layer and the weathered layer. Percolation is enhanced along the roots which, in turn, increases the local fracture networks and intensifies weathering. As a consequence of this destabilization and erosion of the weathered and soil layers, trees tend to be blown down while tearing up large areas of soil and weathered layers. In addition, within the Gulf, most of the coastline is occupied by private properties associated with residential use and hidden from view by numerous trees planted by the owners. In such a context, most of the trees established on the Gulf clifftop are present owing to human choice related to subjective landscape preferences and private property management practices (Roger, 1997; Salomé, 2000). Thus, the increase of clifftop retreat associated with the location of tree vegetation can be interpreted as a special form of morphological change on the coast brought about by human activities.

Along the Quiberon Peninsula coastline, spatial and statistical analyses point to the role of interactions between cliff morphology, vegetation cover, cliff exposure to wind and wave action, as well as human pressure on clifftops to account for the distribution of clifftop retreat values. Higher values of cliff retreat are associated with a high concentration of anthropogenic pressures, and especially the presence of footpaths running along clifftops. Sectors of the coastline with hard rock cliffs capped by a weathered layer are also characterized by a high density of eroded soil layers where vegetation cover is absent or composed of short vegetation cover. The exposure of the Quiberon cliffs to strong westerly winds contributes to limiting the development of vegetation and favors the establishment of short vegetation cover, providing little protection from weathering processes, and being easily disturbed by trampling. Interactions between the establishment of footpaths, trampling, and weathering have been previously discussed (Gallet and Roze, 2001). The establishment of footpaths facilitates the access to the clifftops, and thus increases trampling, which, in turn, contributes to disturbing the vegetation cover. The vegetation cover disturbance increases weathering by increasing runoff processes and favoring the formation of gullies. At the same time, trampling processes contribute to compacting the footpath soil, thus limiting infiltration while increasing runoff. It also hinders the development of the vegetation cover and, ultimately, reduces the water infiltration capacity of the clifftop (French, 2001). Water infiltration processes are thus concentrated at the borders of the footpaths, where weathering processes due to rain water action are thus focused (Van Waerbeke, 1999). In the case of the Quiberon cliffs, weathering processes are likely to be favored by natural factors, including geological and meteorological constraints. At the same time, natural clifftop weathering is likely to be enhanced by increased visiting of clifftops, especially in areas with denser footpath networks.

According to these observations, interactions between vegetation cover, the intensity of anthropogenic pressures, and cliff morphology can produce some specific combinations leading to a major local increase of weathering processes and clifftop retreat along the cliff coastline of South Brittany. Along the Gulf coastline, human-induced clifftop retreat is mainly related to the type of vegetation established on the clifftop. Where trees are planted on highly weathered low cliffs characterized by a thin soil layer, *wind breakage* leads to high values of clifftop retreat. Along the Quiberon Peninsula coastline, the evolution of the cliffs is likely to be strongly controlled by cliff morphology and exposure. Moreover, higher concentrations of footpaths and degradation of clifftop vegetation tend to favor an acceleration of clifftop retreat.

6 Conclusion

Spatial and statistical analyses allowed an overall analysis of clifftop retreat factors on a sub-regional scale. For each coastal cliff system, spatial and statistical analyses yield the same trend. Because these analyses are carried out using independent methods, the observations contribute to validating each of these approaches. The main results discussed here suggest that cliff retreat results from a combination of natural and anthropogenic factors. Human activities lead to an acceleration of clifftop retreat processes. However, such an increase could occur in at least two different ways in South Brittany: In the Gulf of Morbihan, the human impact on cliff evolution is characterized by a time lag due to the time required for tree growth. Human pressures can be related to the introduction of non-natural vegetation cover, which, in turn, lead to an intensification of subaerial processes. On the Quiberon Peninsula, human impacts due to trampling processes are associated with an acceleration of natural subaerial weathering. The approach adopted here, based on spatial data analysis, allowed us to map areas likely to be concerned by these changes, hence providing an efficient tool for coastal managers.

Acknowledgments

We thank Professor Hervé Regnauld for his valuable comments for improvement of the manuscript, Dr. M.S.N. Carpenter for linguistic corrections, and the Conseil Général of Morbihan for supporting this study.

References

Adriani, F.G., Walsh, N., 2007. Rocky coast geomorphology and erosional processes: a case study along the Murgia coastline South of Bari, Apulia—SE Italy. Geomorphology 87, 224–238.

Claytone, K.M., Shamonn, N., 1998. A New approach to the relief of Great Britain II. Classification of rocks based on relative to denudation. Geomorphology 25, 155–171.

Del Rio, L., Gracia, F.J., 2009. Erosion risk assessment of active coastal cliffs in temperate environments. Geomorphology 112, 82–95.

Doland, R., Fenster, M.S., Holme, S.J., 1991. Temporal analysis of shoreline recession and accretion. J. Coast. Res. 7, 723–744.

Dong, P., 2009. Cliff erosion—how much do we really know about it. Environ. Geol. 58, 815–832.

Dornbusch, U., Robinson, D.A., Moses, C.A., William, R.B.G., 2008. Temporal and spatial variations of chalk cliff retreat in East sussex, 1873 to 2001. Mar. Geol. 249, 271–282.

Emery, K.O., Kuln, G.G., 1982. Sea cliffs: their processes, profiles and classification. GSA Bull. 93, 644–654.

Escoffier, B., Pagès, J., 1998. Analyses factorielles simples et multiples, objectifs, méthodes et interprétation. Dunod.

Fall, M., 2009. A GIS-based mapping of historical coastal cliff recession. IAEG Bull. 68, 473–482.

Fletcher, C., Rooney, J., Barbee, M., Lim Siang, C., Richmond, B., 2003. Mapping shoreline change using digital orthophotogrammetry on Maui, Hawaii. J. Coast. Res. 38, 106–124.

French, P.W., 2001. Coastal Defenses: Processes, Problems and Solutions. Routledge, London.

Gallet, S., Roze, F., 2001. Resistance of Atlantic Heathlands to trampling in Brittany (France): influence of vegetation type, season and weather conditions. Biol. Conserv. 97, 189–198.

Graham, D., Sault, M., Bailey, J., 2003. National ocean service shoreline, past, present and future. J. Coast. Res. 38, 14–32.

Hapke, C., Richmond, B., 2002. The impact of climatic and seismic events on the short term evolution of seacliffs based on 3-D mapping: northern Monterey Bay, California. Mar. Geol. 187, 259–278.

Kumar, A., Seralathan, P., Jayappa, K.S., 2009. Distribution of coastal cliffs in Kerala, India: their mechanisms of failure and related human engineering response. Environ. Geol. 58, 815–832.

Lee, E., Hall, J.W., Meaddowcroft, I.C., 2001. Coastal cliff recession: the use of a probabilistic prediction method. Geomorphology 40, 253–269.

Lemasson L (1999) Vents et tempêtes sur le littoral de l'Ouest de la France: variabilité, variations et conséquences morphologiques. PhD Dissertation, University of Rennes 2.

Levin, N., Benn Dor, E., 2004. Monitoring sand dune stabilization along the coastal dunes of Ashdod-Nizanim, Israel, 1945-1999. J. Arid Environ. 58, 335–355.

Minvielle E, Souiah SA (2003) L'analyse statistique et spatiale: Statistiques, Cartographie, Télédétection. SIF, Edition du Temps, Nantes.

Moore, L.J., Giggs, G.B., 2002. Long term cliff retreat and erosion hotspots along the central shores of the Monterey Bay National Marine Sanctuary. Mar. Geol. 181, 265–283.

Naylor, L.A., Stephenson, W.J., Trenhaile, A.S., 2010. Rock coast geomorphology: recent advances and future research directions. Geomorphology 114, 3–11.

Nunes, M., Ferreira, O., Schaefer, M., Clifton, J., Baily, B., Moura, D., Loureiro, C., 2009. Hazard assessment in rock cliffs at Central Algarve (Portugal): A tool for coastal management. Ocean Coast. Manag. 52, 506–551.

O' Sullivan, D., Unwin, N., 2003. Geographic Information Analysis. John Wiley, London.

Parker, B., 2003. the difficulties in measuring a consistently defined shoreline: the problem of vertical referencing. J. Coast. Res. 38, 44–56.

Pian S., 2010. Analyse multiscalaire et multifactorielle de l'évolution et du comportement géomorphologique des systèmes côtiers sud bretons. PhD Dissertation, University of Rennes 2.

Pian S., Regnauld H., 2007. La carte qui change les concepts EspacesTemps.net (web archive link, 06 June 2007). fttp://espacestemps.net/document2466.html (Accessed 6 June 2007).

Pirot, F., Saint-Gérand, T., 2005. La Géodatabase sous ArcGis, des fondements conceptuels à l'implantation logicielle. In: Geomatic Expert.41/42, pp. 61–66.

Pumain, D., Saint-Julien, T., 2004. L'analyse spatiale: Localisations dans l'espace. Armand Colin—SEJER, Paris.

Quinn, J.D., Rosser, N.J., Murphy, W., Lawrence, J.A., 2010. Identifying the behavioural characteristics of clay cliffs using intensive monitoring and geotechnical numerical modelling. Geomorphology 120, 107–122.

Robinson, D.A., Jerwood, L.C., 1987. Sub-aerial weathering of chalk shore platforms during harsh winters in southeast England. Mar. Geol. 77, 1–14.

Rodriguez, I., Montoya, I., Sanchez, M.J., Carreno, F., 2009. Geographic information systems applied to integrated coastal zone management. Geomorphology 107, 100–105.

Roger, A., 1997. Court traité du paysage. Gallimard, Paris.

Sallenger, A.H.J., Krabill, W., Brock, J., Swift, R., Manizades, S., Stockdon, H., 2002. Sea cliff erosion as a function of beach changes and extreme wave runup during the 1997-1998 El Nino. Mar. Geol. 187, 279–297.

Salomé L., 2000. La Muse bretonne. Le musée des Beaux Arts de Bretagne, Rennes.

Sunamura, T., 1977. A relationship between wave-induced cliff erosion and erosive force of waves. J. Geol. 85, 613–618.

Sunamura, T., 1992. Geomorphology of Rocky Coasts. John Wiley, Chichester.

Tessier C., 2006. Caractérisation et dynamique des turbidités en zone côtière: l'exemple de la région marine Bretagne Sud. PhD Dissertation, University of Bordeaux.

Van Waerbeke, D., 1999. L'incidence des sentiers côtiers sur la dynamique des falaises en roche meuble du Petit Trégor (Bretagne Nord). In: Les documents de la MRSH Caen: Littoraux, entre environnement et aménagement.vol. 10, pp. 153–162.

Williams, A.T., Altveirinho-Dias, J., Novo, G., Garcia-Moro, M.R., Curr, R., Pereira, A., 2001. Integrated coastal dune management: checklists. Cont. Shelf Res. 21, 1937–1960.

Young, A.P., Guza, R.T., Flick, R.E., O'Reilly, W.C., Gutierrez, R., 2009. Rain, waves, and short-term evolution of composite seacliffs in southern California. Mar. Geol. 267, 1–7.

CHAPTER 7

Erosional Responses of Eastern and Western Coastal Regions of India, Under Global, Regional, and Local Scale Causes

K.Ch.V. Naga Kumar*, G. Demudu[†], V.P. Dinesan*, Girish Gopinath*, P.M. Deepak*, K. Lakshmanadinesh*, Kakani Nageswara Rao[‡]

*Geomatics Division, Centre for Water Resources Development and Management (CWRDM), Kozhikode, India, [†]Department of Geography, Andhra University, Visakhapatnam, India, [‡]Department of Geo-Engineering, Andhra University, Visakhapatnam, India

1 Introduction

Coastal erosion and/or submergence may be a temporary phenomenon due to tidal cycles, and/or seasonal changes in sea levels, or even episodic events such as storm surges and tsunamis. In all such cases, the coast usually regains equilibrium in several days or years. However, if the coastal erosion is permanent due to rise of the global sea level, or a decrease in sediment supply from rivers due to upstream dams, or construction of permanent coastal structures, there is little scope for the coast to regain its original position. Such permanent and irreversible coastal erosion can be identified by studying the shoreline behavior on a long-term basis.

As the global relative sea levels stabilized after ~7000 BP, coastal margin productivity increased, leading to the emergence of civilizations (Chen et al., 2008; Day et al., 2012). Although the coastal zones at the interface between the continents and oceans constitute a mere 5% of the total landmass, they sustain approximately one-third of the world's population (Lin and Pussella, 2017). This high concentration of population in the coastal zones is mainly due to the availability of natural resources such as rich soil, water, mineral and marine resources, and moderate climate, in addition to being conducive for marine trade and transportation, industrial activity, and tourism (Nageswara Rao et al., 2008a; Kummu et al., 2016). At the same time, the coastal zones, which are very low-lying, are highly dynamic and vulnerable to erosion. Generally, the coasts are vulnerable to various natural and man-made hazards. Storms and tsunamis are considered natural hazards, while man-made hazards are those caused by human activities at the global, regional, and local scales (Neumann et al., 2015).

Human activities at the global scale, such as air pollution and deforestation, lead to global warming, which in turn causes large-scale ice melt, thereby contributing to eustatic sea level rise. In addition to the global sea level rise, the relative sea level rise induced by human activities at the regional and local scales triggers even more severe coastal erosion (Kriauciuniene et al., 2013; Dornbusch, 2017; He-Qin and Ji-Yu, 2017; Lageweg and Slangen, 2017; Tessler et al., 2017). Human activities at the regional scale, such as construction of upstream dams in the river basins, is the main cause of coastal erosion, especially in the deltaic coasts, as the major source of sediment for the coast through the rivers is impounded at the dam sites, depriving the coastal zones of the material for delta building, while waves and tides continue to modify the coast (Day et al., 1995; Stanley and Warne, 1998; Anthony and Blivi, 1999; Hema Malini and Nageswara Rao, 2004; Syvitski et al., 2005; Biggs et al., 2007; Gamage and Smakhtin, 2009; Saito et al., 2007; Saito, 2008; Nageswara Rao et al., 2010, 2015, 2012; Li et al., 2017; Muzaffar et al., 2017; Wang et al., 2017). On local scale, construction of harbor structures such as groins, jetties, and breakwaters on the coastline permanently disturb the sediment drift patterns, resulting in accretion on the updrift side and erosion along the downdrift coast of such structures (WHOI, 2000; Kriauciuniene et al., 2013); whereas construction of seawalls parallel to the coast changes the near-shore process, triggering erosion (Jayappa et al., 2003; Gabriel and Terich, 2005; Kudale, 2010; Kumar and Ravinesh, 2011; Balaji et al., 2017). Similarly, sand mining in the beaches, and destruction of mangrove forests and salt marshes also leads to coastal erosion on a local scale (Thornton et al., 2006; Woodroffe et al., 2006; Masalu et al., 2010). The study described in this chapter is aimed at assessing the shoreline changes that occurred during the past several decades along the Indian coast, with the states of Andhra Pradesh (East Coast) and Kerala (West coast) as examples of human impact at the global level, the Krishna Delta front coast for regional activities, and the Visakhapatnam city coast for local activities.

2 Study Area

Bordered by the Bay of Bengal on the east, the Arabian Sea on the west, and the Indian Ocean on the south, the Indian coastal zones experience varied oceanographic conditions. The east coast region, in general, is a relatively low energy marine environment, with microtidal and low to moderate wave conditions under an easterly swell regime (Davies, 1957). However, strong longshore sediment transport is common along the southern parts of the east coast, while its northern sectors are strongly influenced by river discharges (Rajawat et al., 2015; Ramkumar et al., 2016). The east coast of India is also known for frequent tropical cyclones, and associated floods and tidal surges, causing loss of life and property in the region, apart from severe coastal erosion (Gopal and Chauhan, 2006). Along the west coast, a tide-dominated environment prevails in the northern parts; whereas high wave energy prevails along the southern parts, especially during the summer monsoon when the significant wave height (H_{m0}) reaches up to 6 m (Kumar, 2006).

The east coast of India bordering the Bay of Bengal includes four states: West Bengal (housing part of the Ganga Delta), Odisha (Mahanadi-Brahmani-Baitarani composite deltas), Andhra Pradesh (deltas of the Krishna-Godavari and Penner Rivers) and Tamil Nadu (Cauvery Delta) from north to south. All these deltaic sections of the east coast are extremely low-lying and densely populated with intense economic activity. The states on the west coast of India include Gujarat, Maharashtra, Goa, Karnataka, and Kerala, from north to south.

The east coast of India exhibits typical landforms, such as wide beaches, barrier spits, mangrove swamps, tidal channels, foredunes, and beachridges, mainly associated with the deltaic sections, and occasional patches of coral reefs in the extreme southern parts. However, erosional features such as sea cliffs, wave-cut benches, and sea stacks are also present at a few locations along the east coast, especially on both sides of the city of Visakhapatnam. The west coast of India, on the other hand, is essentially a narrow, nondeltaic coast with estuaries, cliffs, beaches, coastal sand dunes, mangroves, spits, mudflats, raised marine benches, wave-cut platforms, pocket beaches, lagoon-barrier complexes, and extensive saline encrusted lowlands configuring it in different parts (Ramkumar et al. 2016). For this study, the 1080 km-long AP coast from the east coast, including the Krishna Delta Coast and Visakhapatnam City Coast, and the 590 km- long Kerala State on the west coast of India, are examined for describing the impact of human activities on coastal zones.

AP State has a ~1080 km-long coastline facing the Bay of Bengal (Fig. 1). The Eastern Ghats mountain chain is aligned in a NE-SW direction, mostly parallel to, and, at variable distances, inland from the east coast of India, except in the Visakhapatnam region, where they swerve almost at right angles and run perpendicular to the coast. As a result, rock promontories jut into the Bay of Bengal at the eastern end of these *E-W* trending ridges. Thus, over a stretch of 65 km on both sides of the city of Visakhapatnam, the coastline presents a headland-and-bay configuration. The rest of the coastline in AP is a low-lying plain built of coastal/delta alluvium over varying widths in different parts. At a few locations, especially along the deltaic coasts (Krishna-Godavari deltas and Penner delta) where the land is low-lying and thus subjected to regular tidal inundation up to several kilometers inland, halophytic vegetation exists, mostly in the form of mangrove swamps. The AP coast experiences a microtidal and low wave energy environment. A tropical, semiarid climate with <1000 mm annual rainfall and mild winters and hot summers prevails over the region. The total ~11,550 km^2 coastal region in AP is densely populated (563 persons/km^2), with >6.5 million people (2001 census) living within a 5-m elevation above sea level, including in the port cities of Visakhapatnam, Kakinada, and Machilipatnam (Nageswara Rao et al., 2008b). The state

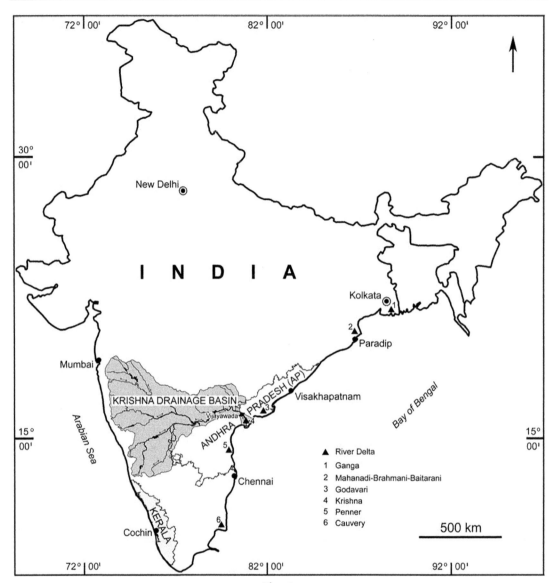

Fig. 1
Outline map of India showing the location of the states of Andhra Pradesh (AP) on the east coast and Kerala on the west coast, and drainage basin of the Krishna River. Mumbai, Cochin, Chennai, Visakhapatnam, and Paradip are the selected tide-gauge locations along the Indian Coast.

of Kerala, in the southwestern part of the Indian Peninsula, has a ~590 km-long narrow and remarkably straight coastline oriented in a general NNW- SSE direction. Kerala state experiences a humid, tropical, wet climate, with an average annual rainfall of about 3107 mm. Facing the Arabian Sea along the west coast of India, most parts of the Kerala coast is a submergent coast (Nair, 1974). The area is densely populated (2362 person/km^2), with several low-lying segments studded with wetlands, including estuaries and backwaters (Fig. 1).

The Krishna Delta, one of the two coastal segments selected in this study from the AP coast, is at the seaward end of a large, $2.5 \times 10^5 \, km^2$ drainage basin of the Krishna River in peninsular India (Fig. 1). Constituting about 7.9% of India's total landmass, the Krishna is the third largest river system in the country, after the Ganges and Godavari Rivers. The river originates on the eastern flank of the Western Ghats near Mahabaleswar at 1438 m elevation, just about 60 km east of the west coast of India, and flows eastward over a length of 1400 km across the Indian Peninsula before debauching into the Bay of Bengal along the east coast of India. The river basin is under the influence of a tropical, semiarid climate, with the average temperatures ranging between 25°C in January and more than 30°C in May. The average annual rainfall in the Krishna River Basin as a whole is 840 mm (Biggs et al., 2007). The city of Visakhapatnam, the second coastal segment from AP selected in this study, is one of the major port cities in India (Fig. 1). With a population of almost two million, the city is spread over a 540 km² area. The city experiences a semiarid climate, with relatively higher mean annual temperatures (26–28°C), and experiences an annual rainfall of about 1178 mm (Hema Malini et al., 2017). The most densely populated part of the city is bound by two *E*-W trending hill ranges, the Kailasa Range to the north and the Yarada Range to the south, which are about 10 km apart.

3 Methods of Study

The study is mainly based on the interpretation of time series satellite images for assessing the shoreline changes along different coastal regions selected for the study. Initially, all the topographic maps pertaining to the respective areas of study, used as base maps for rectifying the satellite images, were first converted into digital images and then geo-coded using digital image processing software (ERDAS Imagine). All the time-series satellite images covering the entire coastal zones of AP and Kerala (Table 1) were downloaded from the USGS website (http://edcsns17.cr.usgs.gov/EarthExplorer/). The individual band data in TIFF (tagged image file format) were stacked and converted into an image format using digital image processing software, and were rectified using the topographic map-based image, and brought to the polyconic projection system common to both the datasets.

The temporal changes in the shoreline on a decadal scale have been analyzed using satellite imagery. Two satellite datasets from 1990 (Landsat TM) and 2017 (Sentinal MSI) have been used for representing the shoreline positions for the AP and Kerala states, as well as the Krishna Delta front coast. For the Visakhapatnam coast, two sets of satellite data, that is, CORONA (1965) and Sentinel MSI (2017), were used. The position of the shoreline has been traced from all the respective datasets, and then overlaid to find out the temporal changes. The near-infrared band images were chosen (Band 4 of Landsat TM and Band 8 of the Sentinel MSI) because the land-water boundary is sharper (Nageswara Rao et al., 2010). The selected images were edge enhanced, and then resampled using the cubic convolution method in the image processing software to smoothen the pixels, so that the shoreline could

Table 1 Specifications of the satellite imagery used in the study. Last column shows the near infrared band used for differentiating land-water boundaries. https//earthexplorer.usgs.gov/

Satellite	Sensor	Path/Row	Image Date	Spatial Resolution (m)	Radiometric Resolution	Swath (km)	Near Infrared Band
			Andhra Pradesh State, India				
Landsat 5	TM	140/47	28-11-1990	30	8-bit	170 × 185	4
	TM	141/47	31-10-1989	30	8-bit	170 × 185	4
	TM	141/48	12-10-1988	30	8-bit	170 × 185	4
	TM	141/49	31-10-1989	30	8-bit	170 × 185	4
	TM	142/49	10-11-1990	30	8-bit	170 × 185	4
	TM	142/50	25-08-1991	30	8-bit	170 × 185	4
	TM	142/51	25-08-1991	30	8-bit	170 × 185	4
		Tile Number					
Sentinel 2A	MSI	T44QRG	17-02-2017	10	12-bit	100 × 100	8
		T44QRF	17-02-2017	10	12-bit	100 × 100	8
		T44QQF	17-02-2017	10	12-bit	100 × 100	8
		T44QQE	17-02-2017	10	12-bit	100 × 100	8
		T44QPE	02-03-2017	10	12-bit	100 × 100	8
		T44QPD	22-03-2017	10	12-bit	100 × 100	8
		T44QND	22-03-2017	10	12-bit	100 × 100	8
		T44PNC	10-02-2017	10	12-bit	100 × 100	8
		T44PMC	10-02-2017	10	12-bit	100 × 100	8
		T44PLB	10-02-2017	10	12-bit	100 × 100	8
		T44PMB	10-02-2017	10	12-bit	100 × 100	8
		T44PMV	10-02-2017	10	12-bit	100 × 100	8
		T44PMA	10-02-2017	10	12-bit	100 × 100	8
			Kerala State, India				
Landsat 5	TM	145/51	02-01-1991	30	8-bit	170 × 185	4
		145/52	15-01-1990	30	8-bit	170 × 185	4
		144/53	24-01-1990	30	8-bit	170 × 185	4
		144/54	25-02-1990	30	8-bit	170 × 185	4
		143/54	11-03-1992	30	8-bit	170 × 185	4
		Tile Number					
Sentinel 2A	MSI	T43PDQ	16-02-2017	10	12-bit	100 × 100	8
		T43PDP	16-02-2017	10	12-bit	100 × 100	8
		T43PEP	16-02-2017	10	12-bit	100 × 100	8
		T43PEN	16-02-2017	10	12-bit	100 × 100	8
		T43PFM	16-02-2017	10	12-bit	100 × 100	8
		T43PFL	03-02-2017	10	12-bit	100 × 100	8
		T43PFK	03-02-2017	10	12-bit	100 × 100	8
		T43PGK	03-02-2017	10	12-bit	100 × 100	8

be traced through on-screen digitization. The area lost by erosion and gained by accretion between two successive data periods was computed using the Union option in the overlay analysis in the GIS software (ArcGIS).

4 Results

4.1 Coastal Erosion in the AP and Kerala States

Towards understanding the nature of changes along AP and Kerala coasts, the two sets of geo-referenced images—Landsat TM images from 1990 and Sentinal MSI images from 2017 were overlaid for comparison using the GIS software. The stretches of coast that appeared landward in 2017, when compared with their positions in 1990, indicated erosion, while those coastal sectors that were advanced into the sea indicated zones of accretion. The comparison also showed that erosion prevailed along a cumulative length of 717 and 327 km on the AP and Kerala coasts, respectively. Similarly, deposition along 292 and 172 km were observed on the AP and Kerala coasts, respectively. The remaining 71 km-long coastal sector in AP and 91 km-long sector in Kerala remained without any noticeable change during the 27-year period (1990–2017). As a result, erosion accounted for the loss of 172 km^2 (AP) and 23 km^2 (Kerala) land, while deposition led to an addition of 70 km^2 (AP) and 10 km^2 (Kerala) area with a net loss of 102 km^2 (AP) and 13 km^2 (Kerala) of the coastal lands during the past three decades (Figs. 2 and 3, and Table 2).

4.2 Coastal Erosion in the Krishna Delta

Rivers play a link between the land and ocean by transporting material (Walling, 2006; Pozzato et al., 2017); they bring almost 95% of the sediment entering the ocean (Syvitski et al., 2003). And, these river-borne sediments play a significant role in the building of deltas and coastal zones (Anthony et al., 2015), with deltas alone constituting 50% of the total sediment flux (Syvitski and Saito, 2007). But, during the recent past, construction of dams in the upper reaches of the rivers for water diversion, hydro-power generation, and so forth, have led to coastal erosion, as the coastal areas adjacent to the river mouths are deprived of the sediment, while the marine processes of waves and tides continue to act up on the shoreline (Syvitski et al., 2005; Nageswara Rao et al., 2010; Syvitski and Kettner, 2011; Ramkumar, 2003, 2015; Besset et al., 2017; Li et al., 2017; Pietron, 2017; Wang et al., 2017; Wu et al., 2017).
The situation in India is no different, construction of dams has picked up momentum since the 1950s and 1960s in order to increase the irrigation potential of the country for meeting the growing food demands of the population. As a result of the impounding of water at the reservoirs, water discharges and sediment loads into the sea have decreased manifold (Nageswara Rao et al., 2010). With this background, it would be interesting to know if dam construction in the river basin has impacted the Krishna Delta-front shoreline.

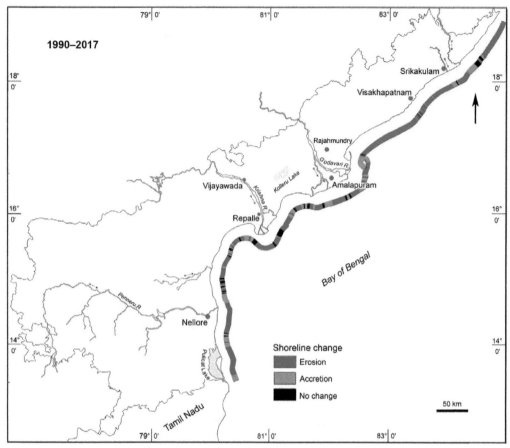

Fig. 2
Map showing the changes along the Andhra Pradesh state during 1990–2017. Green colored segments of the coast indicate shoreline advance into the sea by accretion, red colored segments indicate landward retreat of the shoreline by erosion and black colored segments indicate no change.

A comparison of the satellite imagery from 1990 and 2017 revealed pronounced erosion along the Krishna Delta front coast. Although deposition has occurred over a cumulative length of 39 km (~28%) of the 140 km-long Krishna Delta front coast, leading to accretion of 18.52 km^2 area, erosion has resulted in a loss of 46.04 km^2 area along a 92 km (~66%) stretch of the delta coast. As a result, the net loss was about 27.52 km^2 area in the Krishna Delta during the 27-year period (Fig. 4 and Table 2).

4.3 Coastal Erosion in Visakhapatnam

The city of Visakhapatnam on the east coast of India facing the Bay of Bengal is characterized by a headland-bay type of coast. The two hill ranges the northern Kailasa Range and the southern Yarada Range that encircle the main part of the city protrude into the sea as headlands.

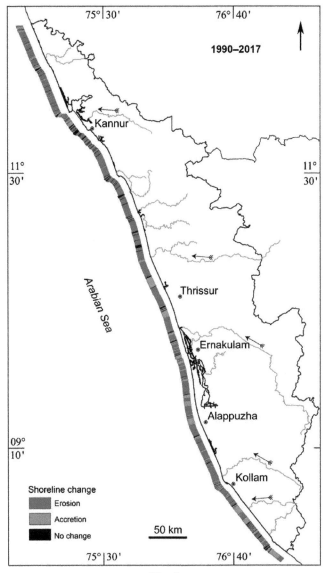

Fig. 3
Map showing the changes along the Kerala state during 1990–2017. Green colored segments of the coast indicate shoreline advance into the sea by accretion, red colored segments indicate landward retreat of the shoreline by erosion and black colored segments indicate no change.

The eastern tip (headland) of the Kailasa hill range is called Kailasagiri Hill, and the eastern tip (headland) of the southern hill range is called Dolphin's Nose. To the immediate north of the southern headland (Dolphin's Nose) is the mouth of a 12 km-long deep tidal channel that runs parallel to the Yarada Range and serves as a natural harbor. At about 7.5 km north of this harbor channel is a low, almost flat rocky projection into the sea, which is called the Jalaripeta Headland. The shoreline between the entrance point of the harbor channel

Table 2 Data on coastal changes along AP, Kerala, Krishna delta and Visakhapatnam city

Coastal Region	Time Interval	Erosion Length (km)	Accretion Length (km)	Stable Length (km)	Area lost by Erosion (km²)	Area gained by Deposition (km²)	Net Loss (km²)
Andhra Pradesh (AP State)	1990–2017	717	292	71	172	70	102
Kerala State	1990–2017	327	172	91	23	10	13
Krishna Delta	1990–2017	92	39	9	46.04	18.52	27.52
Visakhapatnam	1965–2017	7.34	0.78	2.15	0.62	0.08	0.54

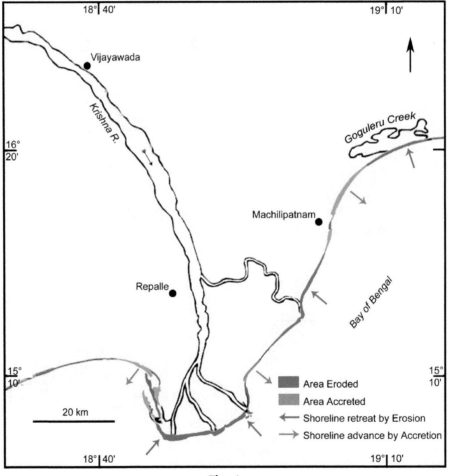

Fig. 4
Erosion-Accretion trend along the Krishna delta shoreline during 1990–2017. Green colored segments of the coast indicate shoreline advance into the sea by accretion and red colored segments indicate landward retreat of the shoreline by erosion.

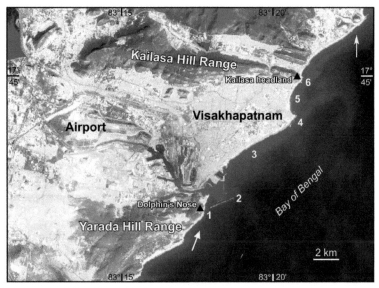

Fig. 5
Sentinel MSI image showing Visakhapatnam coastal zone. 1. Dolphin's Nose headland, 2. Outer Harbor, 3. Crenulate-shaped Ramakrishna Beach, 4. Jalaripeta headland, 5. Lawson's Bay, 6. Kailasagiri headland. Arrows indicate the direction of dominant longshore drift during the southwest monsoon season.

and the Jalaripeta Headland is crenulate-shaped. Further north, the shoreline between the Jalaripeta low-headland and the Kailasagiri is the beautiful Lawson's Bay, with a perfect crescent-shaped beach (Fig. 5). The Visakhapatnam Coast is known for its sylvan beaches, and especially the Ramakrishna (RK) Beach. The area between the harbor channel and the Jalaripeta headland is a famous tourist attraction. The gently sloping sandy RK Beach was almost 140 m wide until the 1960s. An outer harbor was built during the 1970s by constructing breakwaters, which projected into the sea beyond the seaward limit of the southern headland, the Dolphin's Nose. Further, the city administration has constructed several parks and one submarine museum at the sandy beach and fortified them with masonry walls on three sides, including the seaward side, while on the landward side of the beach is a shore-parallel road. These encroachments have disturbed the equilibrium conditions in the nearshore processes, leading to severe shoreline erosion and irreversible loss of most of the beach. A comparison of the CORONA satellite photograph pertaining to 1965 and the recent Sentinal MSI image from 2017 showed a net loss of $0.54\,km^2$ area of the beach along the $\sim$10 km-long Visakhapatnam Coast (Table 2). The average loss of the beach was about 5.4 ha per km along the 10 km stretch.

5 Discussion

5.1 Global-Scale Activities and Coastal Erosion—The AP and Kerala Coasts

Although no general conclusions can be drawn relating sea-level and shoreline changes on a global scale (Cozannet et al., 2014), a number of studies in different countries have showed persistent erosion of large sectors of the coasts. Previous studies showed widespread erosion in the United States, such as along the east coast barrier beaches (Galgano, 2004), the California Coast (Moore et al., 1999), and the Gulf of Mexico (Morton and McKenna, 1999). Studies also showed widespread erosion in China (Dongwing et al., 1993; Cai et al., 2009), the Yasawa Islands, Fiji (Mörner and Matlack-Klein, 2017), southern Italy (Barbaro, 2016), Portugal (Pinto, 2016), Scotland (Fitton et al., 2018), Trinidad (Darsan, 2013), northeast Bali (Husrin et al., 2016), Benin (Bahini, 2016), Subang, Indonesia (Kikuyama et al., 2017), Uruguay (Lopez et al., 2010), Ghana (Addo et al., 2011), Malaysia (Ghazali, 2006), Chile (Martinez et al., 2018), Vietnam (Tien et al., 2003), Kuwait (Neelamani, 2017), Greece (Petrakis et al., 2014), Thailand (Vongvisessomjai et al., 1996), Korea (Boer et al., 2017), Taiwan (Hsu et al., 2007), Darkar, Senegal (Sane and Yamagishi, 2004), Puerto Rico (Morelock and Barreto, 2003), Brazil (Gouveia Souza, 2009), Australia (Harvey and Caton, 2010), Mozambique (Palalane et al., 2015), and Bangladesh (Emran et al., 2016).

An extensive study on shoreline changes in India through a comparison of two sets of satellite data from 1989 to 1991 and 2004–06 by Rajawat et al. (2015) showed that about 45% of the 8412 km-long Indian coastline, which includes 1039 km (41%) of the 2514 km-long east coast, 1287 (39%) of the 3263 km-long west coast, and 1502 (57%) of the 2635 km coast in the island territories has been eroded. On the whole, at least 70% of the sandy beaches in the world are receding (Bird, 1985). There must be a worldwide cause for such pervasive shoreline erosion (Zhang et al., 2004). The most commonly proposed cause for such shoreline erosion globally is atmospheric warming and the attendant sea level rise (Leatherman et al., 2000).

Increased air pollution due to the release of industrial and automobile exhausts, and a build-up of greenhouse gasses, especially carbon dioxide in the atmosphere (Nageswara Rao et al., 2011), are leading to global warming, which in turn is resulting in the melting of ice and the rise of world ocean levels. As of April 2018, the atmospheric carbon dioxide levels are at 410.31 ppm (http://co2now.org/), an alarming increase of about 150 ppm from pre-industrial levels of 260–270 ppm (Wigley, 1983). Studies from Karl et al. (2015) showed that global average temperatures rose at 0.113°C per decade during 1950–99, and increased further to 0.116°C per decade during 2000–14, which has led to a considerable rise in the eustatic sea level (Cozannet et al., 2014). There is already evidence of a large-scale ice melt in the three major ice repositories of the world. Sasgen et al. (2012) reported that the average rate of mass loss of ice in Greenland increased from 34 Gt/yr during 1992–2001 to 215 Gt/yr during

2002–11, contributing to a sea level rise of 0.47±0.23 mm/yr during 1991–2015 (Tedstone et al., 2017). Since the middle of the past century, the extent of Arctic ice has been decreasing at almost 8% per decade (Stroeve et al., 2007), with the past two decades showing a thinner and smaller extent of ice cover (Comiso 2012, Stroeve et al., 2012a, b). The summertime Arctic Sea ice reaches its minimum during September. Satellite derived observations during 1979–2016 proved that September Arctic Sea ice is declining at an alarming rate of 13.2% per decade (NSIDC, 2007). Perhaps the most alarming is the widespread loss of ice in West Antarctica (Rignot et al., 2008), contributing to a global sea level rise of ∼0.36 mm/yr (Chen et al., 2008). Globally, the sea level is estimated to have risen by about 1.5–1.9 mm/yr between 1901 and 2010 (Church et al., 2013), and somewhat higher during 1993–2010, at an average rate of 3.2 mm/yr (Ablain et al., 2009; Church et al., 2013). Even if the global temperatures are leveled off at this stage, the sea level will continue to rise over the 21st century (Meehl et al. 2005), while the increased sea-surface temperature would also result in frequent and intensified cyclonic activity and associated storm surges affecting the coastal zones (Wu et al., 2002; Unnikrishnan et al., 2006). Analysis of satellite altimeter data over the north Indian Ocean also revealed a 3.2 mm/yr rise in sea level during 1993–2012, which is close to the estimated global mean sea-level rise (Unnikrishnan et al., 2015). Tide-gauge data of several decades are likely to provide reliable estimates of sea level changes. Based on 110 years of tide-gauge data from different parts of the world, the Intergovernmental Panel on Climate Change (IPCC) in its fifth assessment report estimated that the global mean sea level rose at 1.7 mm/yr during 1901–2010. Some other estimates showed a rise of 1.8 mm/yr between 1961 and 2003 (Church et al., 2001; Bindoff et al., 2007; Unnikrishnan et al., 2015). Similar trends have been observed from the Indian maritime states. Unnikrishnan et al. (2015) reported an average sea level rise of 1.3 mm/yr between 1901 and 2010 based on the analysis of past tide gauge measurements. Data on the net sea level rise at Mumbai and Cochin from the west coast, and Chennai, Visakhapatnam; and Paradip from the east coast compiled from the Global Linear Relative Mean Sea Level (MSL) trends at 95% Confidence Intervals (CI), with the Glacial Isostatic Adjustment values included, which are available from the Permanent Service for Mean Sea Level (PSMSL) web portal (http://www.psmsl.org/). Cochin, along the west coast of India in Kerala, showed a rise of 2.07 mm/yr during 1939–2007, while Visakhapatnam, which is located on the east coast of India in AP State, showed a rise of 1.03 mm/yr. From these data and the discussion, it is evident that global sea levels, including along the Indian Coast, are rising (Table 3).

Whatever the cause, the sea-level rise eventually leads to coastal erosion (Cozannet et al., 2014; Brammer 2014; Epanchin-Niell et al., 2017; Lin and Pussella, 2017). Therefore, the net coastal erosion and loss of 102 km^2 land along the 1080 km-long AP coast in a 27-year period at an average of about 9 ha per one km length is considered to be the result of sea level rise, which in turn can be the regional expression of the consequence of human activities at a global scale. Similarly, the net loss of 13 km^2 land at an average of about 2 ha per one km length of the 590 km-long Kerala coast, in spite of fortification by the construction of seawalls along

Table 3 Trends in Sea-level rise at selected tide gauge stations along the Indian Coast (Mumbai, Cochin, Chennai, Visakhapatnam, and Paradip)

Station	Period of Analysis	Number of Years of Data Availability	Trend in Relative Sea-Level Rise (mm/yr)	GIA Correction (mm/yr)	Net Sea-Level Rise (mm/yr)
Mumbai	1878–2008	130	0.79 ± 0.11	−0.31	1.10
Cochin	1939–2007	68	1.71 ± 0.36	−0.36	2.07
Chennai	1916–2008	92	0.32 ± 0.37	−0.28	0.60
Visakhapatnam	1937–2013	70	0.79 ± 0.45	−0.24	1.03
Paradip	1966–2008	42	0.77 ± 1.13	−0.23	1.00

Source of Data: Net Sea-level-rise at selected tide-gauge stations is derived from Global Linear Relative Mean Sea Level (MSL) trends and 95% Confidence Intervals (CI) in mm/yr (https://tidesandcurrents.noaa.gov/sltrends/msl GlobalTrendsTable.htm) and the Glacial Isostatic Adjustment (GIA) (http://www.atmosp.physics.utoronto.ca/~peltier/datasets/psmsl/drsl250. PSMSL. ICE5Gv1.3 _VM2_L90_2012b.txt) available at http://www.psmsl.org/.

310 km of the coast, is an indication of rising sea levels. Similar satellite-data-based studies made by Hema Malini and Nageswara Rao, 2004; Nageswara Rao et al., 2007, 2008a, b, 2010; Kankara et al., 2015; Kannan et al., 2016 also showed sustained coastal erosion along different sections of the AP Coast. The studies of Ramkumar et al. (2000), and Ramkumar and Gandhi (2000) based on beach rock exposures below the low-tide water lines of Godavari and Krishna deltas estimated a combined effect of 1.2 mm/year of sea-level rise and coastal submergence.

Several previous studies on the Kerala coast also reported persistent erosion along different sectors (Mallik et al., 1987; Jayappa et al., 2003; Sundar et al., 2007; Sachin Pavithran et al., 2014; Rajawat et al., 2015; Noujas and Thomas, 2015). The National Centre for Sustainable Coastal Management has reported (http://ncscm.res.in/cms/geo/pdf/research/kerala_fact_sheet.pdf) that erosion is prevalent along 370 -long sectors of the total 587 km Kerala Coast, including about a 310 km-long stretch (53%) that has been fortified with seawalls, which are also repeatedly undermined by waves and collapsed (Noujas and Thomas, 2015). Therefore, the persistent erosion along the AP State on the east coast of India, and Kerala State on the west coast, may be considered a consequence of global sea-level rise.

5.2 Regional Activities and Coastal Erosion—Krishna Delta Coast

Erosion along the Krishna Delta front coast is a standing example of the impact of regional-scale human activities on coasts. The Krishna Delta, one of the major depocenters along the east coast of India on a millennial scale, has prograded into the sea during the past 6000 years by the addition of 3539.25 km^2 land. Even on a decadal scale, the delta showed an overall progradation during the pre-dam period, as an area of about 40.5 km^2 was added to the delta by a net deposition between 1930 and 1965 (Nageswara Rao et al., 2010). But the trend was reversed after the inception of construction of dams across the river.

The post-Independence period coincided with major dam construction activity in the Krishna River Basin, especially from the 1960s onward (Nageswara Rao et al., 2010). During this period, numerous dams of different sizes have been built across the Krishna River and its tributaries, storing almost the total yield of the basin, as a result of which it has become an overexploited and almost a closed river basin (Biggs et al., 2007; Gaur et al., 2007). This has significantly affected the water discharges and suspended sediment loads through the river into the delta. Data on water discharges recorded during 1951–2008 at the delta apex near Vijayawada showed a distinct decrease from an annual average of $61.88\,km^3$ during 1951–59 to $11.82\,km^3$ in 2000–08 (Nageswara Rao et al., 2010). The sediment loads also indicated the distinct effect of dams. The available data recorded at Vijayawada showed an annual average of 9.0 million tons during 1966–69 and as low as 0.4 million tons during 2000–05 (Nageswara Rao et al., 2010), indicating a distinct decrease in sediment flux into the sea, apparently because a bulk of the sediment is being trapped at the dams. This is evident from a comparison of sediment loads recorded at Agraharam and Vijayawada, respectively, located upstream and downstream of the two large dams at Srisailam and Nagarjunasagar across the Krishna River, which showed higher sediment loads in the river at Agraharam than at Vijayawada, apparently due to reservoir trapping (Gamage and Smakhtin, 2009; Nageswara Rao et al., 2010). This is also corroborated by the bathymetry surveys conducted by the Central Water Commission, Government of India, which showed that the storage capacity of the Srisailam reservoir was reduced by 14% from 8.72 to $7.47\,km^3$ during 1984–2004, and that of the Nagarjunasagar by one-fourth from 11.55 to $8.16\,km^3$ during the four-decade period since its inception in 1967, clearly indicating the role of dams in sediment retention (Nageswara Rao et al., 2010).

Reduction in sediment supply not only diminishes the delta growth, but also leads to coastal erosion and shoreline recession, as is the case with many deltas around the world, such as, for example, the Mississippi, Rhone, and Ebro Deltas (Day et al., 1995); the Nile Delta (Stanley and Warne, 1998); the Volga Delta (Anthony and Blivi, 1999); the Chao Phraya, Huanghe, and Song Hong Deltas (Saito et al., 2007; Saito, 2008); the Mekong Delta (Saito et al., 2007; Saito, 2008; Li et al., 2017), the Yellow River Delta (Wang et al., 2017), and the Indus Delta (Muzaffar et al., 2017). Indian deltas are no exception, as indicated by pronounced erosion along the deltas of the Godavari (Hema Malini and Nageswara Rao, 2004, Nageswara Rao et al. 2010), and Krishna Rivers (Biggs et al., 2007; Gamage and Smakhtin, 2009; Nageswara Rao et al. 2010), located along the east coast of India, fringing the Bay of Bengal.

Because the net loss of land along the 140 km-long Krishna Delta front coast was $27.52\,km^2$ during a 27-year period (1990–2017), the average loss may be estimated to be 19.65 ha per km length of the coast. This is more than double the average loss of land of 9 ha per km length along the entire 1080 km-long AP coast. Therefore, the distinctly higher rate of coastal erosion along the Krishna delta part of the AP coast, when compared with the AP coast as a whole, points to the impact of human activities at the regional scale, that is, construction of dams and water diversion in the upper river basin, contributing to the increased rate of erosion at the delta, which is otherwise one of the major depocenters along the AP coast.

5.3 Local Activities and Coastal Erosion—Visakhapatnam City Coast

Coastal erosion at the local scale has become a major problem, but only after human activities began to occupy the coasts. Local activities tend to disturb the dynamic equilibrium of the beaches and shorelines. When human activities get too close to the coast, the problem of shoreline erosion appears hazardous. In attempting to protect the structures, emphasis is generally given to modifying the behavior of the beaches, rather than on people trying to learn to live with natural shoreline dynamics. Construction of parks/permanent structures, and jetties for development activities cause erosion. For example, severe beach erosion was triggered at surfside along the Texas coast due to construction of long jetties (Watson, 2003). In India, there are also several instances of human interference in the beaches that invariably resulted in coastal erosion, such as near Pondicherry, Kanyakumari, and Nagapattinam (Jena et al., 2001) and Ennore (Pandian et al., 2004) along the Tamil Nadu coast, and Dibbapalem Beach (Nageswara Rao et al., 2008a) along the AP coast. In this context, an attempt is made to find how permanent structures, such as harbor breakwaters along the coastal regions, effect the shoreline configuration by taking an example from Ramakrishna Beach, in Visakhapatnam city.

The satellite image of Sentinel MSI (2017) covering Visakhapatnam city shows that the shoreline, including RK Beach and adjacent beaches in between the Dolphin's Nose headland to the south, and the Jalaripeta low-headland to the north is crenulate (Fig. 5). Prior to the construction of the outer harbor, the material carried by the prevailing longshore current from the southwest was not available to the inner harbor area, which is on the immediate leeward of the Dolphin's Nose headland. But further northward, the drift material is accreted in the RK Beach area, and further north up to the Jalaripeta headland. Therefore, the shoreline in between these two headlands assumed a crenulate shape, with erosion in the immediate leeward part of the beach north of the Dolphin's Nose, and sand accretion in the RK Beach further north and beyond. But with the construction of breakwaters/jetties of the outer harbor about 45 years ago, the equilibrium in the shoreline was altered as a part of the RK Beach area has come into the leeward side of the extended artificial headland (harbor breakwaters), which has been built into the sea far eastward beyond the seaward limit of the Dolphin's Nose. The resultant increase of leeward stretch of the coast causes erosion along the RK Beach area and further northward. In fact, a comparative analysis of the Corona satellite images from 1965 and Sentinel MSI image from 2017 indicated more erosion than deposition along the city coast (Fig. 6). In addition to the construction of the harbor breakwaters by the Port Trust, the construction of tourist facilities such as lamp-posts, parks, and vehicle parking zones, a submarine museum, and so forth, on the beach has exacerbated the coastal erosion. These types of local activities, such as construction of harbor structures, and even seawalls, trigger coastal erosion. For example, the RK Beach was subjected to severe erosion, and with it,

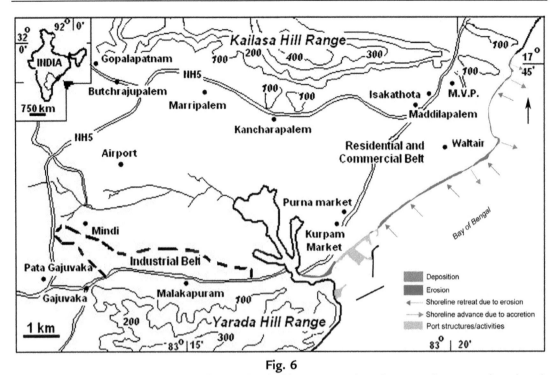

Fig. 6

Shoreline change between 1965 and 2017 in Visakhapatnam city. The coastal sectors where beach erosion occurred during the 52-year period are shown in red color, whereas the green colored areas indicate accretion. Beach erosion led to shoreline retreat marked by red arrows and beach accretion resulted in shoreline advance into the sea represented by green arrows.

the man-made structures were also destroyed. Moreover, the seawalls, though considered to protect the coast, as in the case of the Kerala (Shareef, 2007) and Mumbai coasts (Meher-Homji, 2007), may not serve the intended purpose in the long run. The case of the collapsed wall of the submarine museum in Ramakrishna Beach in Visakhapatnam city is an example of the failure of seawalls. When the submarine was hauled onto the beach and a wall was built in the foreshore zone in 2003, the width of the sandy beach between the wall and the shoreline was very wide (Fig. 7A). But the wall was undermined by waves and collapsed during the monsoon swell in 2005 (Fig. 7B). Although the wall was rebuilt and fortified with riprap on the seaside, the beach between the wall and the shoreline was lost permanently (Fig. 7B). In fact, the seawalls are known to compound the problem of beach erosion. The problem with the seawalls is that once they are built on the beach, waves directly impinge on them, especially during high-tide, and get reflected back into the sea dragging the beach material along. In addition, the seawalls prevent the sand from the dunes and cliffs behind them to reach the eroding beach, which would otherwise replenish the beaches. The net result is permanent loss of the beach, and even collapse of the seawalls. Moreover, waves refracted by these walls would impinge on the adjacent parts of the beach, leading to erosion, as happened on both sides of the submarine museum in RK Beach, Visakhapatnam. On several occasions, the road along RK

Fig. 7
Shoreline erosion along Visakhapatnam city coast triggered seawall construction. A wall was built in RK Beach in Visakhapatnam around the submarine in 2003 (A), which was collapsed by wave-undermining in 2005 (B). The beach between the rebuilt wall riprap support and the shoreline was completely lost by 2007 (C). The road along RK Beach to the south of the submarine museum is eroded by waves refracted by the museum wall (D).

Beach was undermined and collapsed during strong swell waves (Fig. 7D). Under such circumstances, an alternative method to protect the beach seems to be the construction of offshore barriers or artificial reef structures to protect the coastal regions from the effects of erosion, as in the case of Kaike Beach in Yumigahama Peninsula, western Japan (Nageswara Rao and Sadakata, 2006).

6 Conclusion

Extensive erosion along the coastal regions of the AP and Kerala coasts as a whole could be considered as the coastal response to global sea level rise that is induced by human activities such as deforestation, air pollution, and increased levels of greenhouse gasses in the atmosphere on a global scale. Regional factors, such as upstream dam construction across rivers, cause coastal erosion in the sediment-starved deltaic sections as evident from the Krishna delta front coast. Human activities on a local scale, such as the construction of breakwaters, seawalls and similar encroachments, trigger coastal erosion as in the case of the Visakhapatnam city coast. Erosion of this nature might continue in the future as well leading to the loss of more and more highly resourceful land along the AP and Kerala coasts, and elsewhere along the world coasts, unless concerted efforts are made through appropriate coastal management plans.

Acknowledgments

The authors are thankful to the Executive Director, CWRDM, an institution of Kerala State Council for Science Technology and Environment (KSCSTE) for constant encouragement and support for preparing this manuscript. KNR is grateful to the Indian Council of Social Science Research (ICSSR), New Delhi, for the award of Senior Fellowship (Award No. F. 2-21/2016-17/SF).

References

Ablain, M., Cazenave, A., Valladeau, G., Guinehut, S., 2009. A new assessment of the error budget of global mean sea level rate estimated by satellite altimetry over 1993–2008. Ocean Sci. 5, 193–201.

Addo, K.A., Jayson-Quashigah, P.N., Kufogbe, K.S., 2011. Quantitative analysis of shoreline change using medium resolution satellite imagery in Keta, Ghana. Mar. Sci. 1 (1), 1–9.

Anthony, E.J., Blivi, A.B., 1999. Morphosedimentary evolution of a delta-sourced, drift-aligned sand barrier–lagoon complex, western Bight of Benin. Mar. Geol. 158, 161–176.

Anthony, E.J., Brunier, G., Besset, M., Goichot, M., Dussouillez, P., Nguyen, V.P., 2015. Linking rapid erosion of the Mekong River delta to human activities. Sci. Rep. 5, 14745. https://doi.org/10.1038/srep14745.

Bahini, A., 2016. Coastal erosion and climate change in Benin. Journee de la Renaissance Scientifique de I'Afrique (JRSA), Cotonou, Benin.

Balaji, R., Sathish Kumar, S., Misra, A., 2017. Understanding the effects of seawall construction using a combination of analytical modeling and remote sensing techniques: case study of Fansa, Gujarat, India. Int. J. Ocean Climate Syst. 8 (3), 153–160.

Barbaro, G., 2016. Master Plan of solutions to mitigate the risk of coastal erosion in Calabria (Italy), a case study. Ocean Coast. Manag. 132, 24–35.

Besset, M., Anthony, E., Sabatier, F., 2017. River delta shoreline reworking and erosion in the Mediterranean and Black Seas: the potential roles of fluvial sediment starvation and other factors. Elementa Sci. Anthrop. 5(54). https://doi.org/10.1525/elementa.139.

Bird, E.C.F., 1985. Coastline Changes. In: A global Review. Wiley, New York, 219 pp.

Biggs, T.W., Gaur, A., Scott, C.A., Thenkabail, P., Rao, P.G., Gumma, M.K., Acharya, S., Turrel, H., 2007. Closing of the Krishna Basin: irrigation, stream flow depletion and macroscale hydrology. http://dlc.dlib.indiana.edu/dlc/bitstream/handle/10535/3678/RR111[1].pdf?sequence=1/ (Accessed 22 October 2017).

Bindoff, N.L., Willebrand, J., Artale, V., Cazenave, A., Gregory, J., Gulev, S., Hanawa, K., Le Quere, C., Levitus, S., Nojiri, Y., Shum, C.K., Talley, L.D., Unnikrishnan, A., 2007. Observations: Oceanic Climate Change and Sea Level. In: Solomon, S., Qin, D., Manning, M., Chen, Z., Arquis, M., Averyt, K.B., Tignor, M., Miller, H.L. (Eds.), Climate Change 2007: The PHYSICAL Science Basis. Contribution of Working Group I to the Fourth Assessment Report of the Intergovernmental Panel on Climate Change. Cambridge University, United Kingdom and Newyork, NY.

Boer, W., Huisman, B., Yoo, J., McCall, R., Scheel, F., Swinkels, C., Friedman, J., Luijendijk, A., Walstra, D., 2017. Understanding coastal erosion processes at the Korean East Coast. In: Aagaard, T., Deigaard, R., Fuhrman, D. (Eds.), Proceedings of Coastal Dynamics, Helsingor. Denmark, pp. 1336–1347.

Brammer, H., 2014. Bangladesh's dynamic coastal regions and sea-level rise. Climate Risk Manag. 1, 51–62.

Cai, F., Su, X., Liu, J., Li, B., Lei, G., 2009. Coastal erosion in China under the condition of global climate change and measures for its prevention. Prog. Nat. Sci. 19 (4), 415–426.

Chen, Z., Zong, Y., Wang, Z., Wang, H., Chen, J., 2008. Migration patterns of Neolithic settlements on the abandoned and Yellow and Yangtze River deltas of China. Quat. Res. 70, 301–314.

Church, J.A., Gregory, J.M., Huybrechts, P., Kuhn, M., Lambeck, K., Nhuan, M.T., Qin, D., Woodworth, P.L., 2001. Changes in Sea Level. In: Houghton, J.T., Ding, Y., Griggs, D.J., Noguer, M., Van der Linden, P.J., Dai, X., Maskell, K., Johnson, C.A. (Eds.), Climate Change 2001: The Scientific Basis: Contribution of Working Group I to the Third Assessment Report of the Intergovernmental Panel.

Church, J.A., Clark, P.U., Cazenave, A., Gregory, J.M., Jevrejeva, S., Levermann, A., Merrifield, M.A., Milne, G.A., Nerem, R.S., Nunn, P.D., Payne, A.J., Pfeffer, W.T., Comiso, J.C., 2013. Large Decadal Decline of the Arctic Multiyear Ice Cover. J. Clim. 25, 1176–1193.

Comiso, J.C., 2012. Large decadal decline of the Arctic multiyear ice cover. J. Clim. 25 (4), 1176–1193.

Cozannet, G.L., Garcin, M., Yates, M., Idier, D., Meyssignac, B., 2014. Approaches to evaluate the recent impacts of sea-level rise on shoreline changes. Earth-Sci. Rev. 138, 47–60.

Darsan, J., 2013. Beach morphological dynamics at Cocos bay (Manzanilla). Trinidad.

Davies, J.L., 1957. The importance of cut and fill in the development of sand beach ridges. Aust. J. Sci. 20, 105–111.

Day, J.W., Pont, D., Hensel, P.F., Ibanez, C., 1995. Impacts of sea-level rise on deltas of the Gulf of Mexico and the Mediterranean: the importance of pulsing events to sustainability. Estuaries 18, 636–647.

Day, J.M., Gunn, J.D., Folan, W.J., Yanez-Arancibia, A., Horton, B.P., 2012. The influence of enhanced post-glacial coastal margin productivity on the emergence of complex societies. J. Island Coast. Archaeol. 7, 23–52.

Dongwing, X., Wenhai, W., Guiqiu, W., Jinrui, C., Fulin, L., 1993. Coastal erosion in China. Acta Geographica Sinica 5, 468–476.

Dornbusch, U., 2017. Design requirement for mixed sand and gravel beach defences under scenarios of sea level rise. Coast. Eng. 124, 12–24.

Emran, A., Rob, M.A., Kabir, M.H., Islam, M.N., 2016. Modeling spatio-temporal shoreline and areal dynamics of coastal island using geospatial technique. Model. Earth Syst. Environ. 2 (1), 4.

Epanchin-Niell, R., Kousky, C., Thompson, A., Walls, M., 2017. Threateneed protection: sea level rise and coastal ptotected lands of the eastern United States. Ocean Coast. Manag. 137, 118–130.

Fitton, J.M., Hansom, J.D., Rennie, A.F., 2018. A method for modelling coastal erosion risk: The example of Scotland. Nat. Hazards 91 (3), 931–961.

Gabriel, A.O., Terich, T.A., 2005. Cumulative patterns and controls of seawall construction, Thurston County, Washington. J. Coast. Res. 430–440.

Galgano Jr., F.A., 2004. Long-term effectiveness of a groin and beach fill system: a case study using shoreline change maps. J. Coast. Res. Special Issue No. 33. In: Functioning and Design of Coastal Groins: The Interaction of Groins and the Beach-Process and Planning, pp. 3–18.

Gamage, N., Smakhtin, V., 2009. Do river deltas in East India retreat? A case study of the Krishna delta. Geomorphology 103, 533–540.

Ghazali, N.H.M., 2006. Coastal erosion and reclamation in Malaysia. Aquat. Ecosyst. Health Manag. 9 (2), 237–247.

Gopal, B., Chauhan, M., 2006. Biodiversity and its conservation in the Sundarban mangrove ecosystem. Aquat. Sci. Research Across Boundaries 68 (3), 338–354. https://doi.org/10.1007/s00027-006-0868-8.

Gouveia Souza, C.R., 2009. A erosãocosteira e osdesafios da gestãocosteira no Brasil. Revista de GestãoCosteiraIntegrada-J. Integr. Coast. Zone Manage. 9(1).

Harvey, N., Caton, B., 2010. Coastal Management in Australia. University of Adelaide Press, Adelaide.

He-Qin, C., Ji-Yu, C., 2017. Adapting cities to sea level rise: a perspective from Chinese deltas. Adv. Clim. Chang. Res. 8, 130–136.

Hema Malini, B., Nageswara Rao, K., 2004. Coastal erosion and habitat loss along the Godavari delta front—a fallout of dam construction (?). Curr. Sci. 87, 1232–1236.

Hema Malini, B., Visweswara Reddy, B., Gangaraju, M., Nageswara Rao, K., 2017. Malaria risk mapping: a study of Visakhapatnam district. Curr. Sci. 112 (3), 463–465.

Husrin, S., Pratama, R., Putra, A., Sofyan, H., Hasanah, N.N., Yuanita, N., Meilano, I., 2016. Jurnal Segara.

Hsu, T.W., Lin, T.Y., Tseng, I.F., 2007. Human impact on coastal erosion in Taiwan. J. Coast. Res. 961–973.

Jayappa, K.S., Kumar, G.V., Subrahmanya, K.R., 2003. Influence of coastal structures on the beaches of southern Karnataka, India. J. Coast. Res. 389–408.

Jena, B.K., Chandramohan, P., Sanil Kumar, V., 2001. Longshore transport based on directional waves along north Tamilnadu coast, India. J. Coast. Res. 17, 322–327.

Kankara, R.S., Selvan, S.C., Markose, V.J., Rajan, B., Arockiaraj, S., 2015. Estimation of long and short term shoreline changes along Andhra Pradesh coast using remote sensing and GIS techniques. Procedia Eng. 116, 855–862.

Kannan, R., Ramanamurthy, M.V., Kanungo, A., 2016. Shoreline change monitoring in Nellore coast at East Coast Andhra Pradesh District using remote sensing and GIS. J. Fish. Livest. Prod. 4, 161.

Karl, T.R., Arguez, A., Huang, B., Lawrimore, J.H., McMahon, J.R., Menne, M.J., Peterson, T.C., Vose, R.S., Zhang, H.M., 2015. Possible artifacts of data biases in the recent global surface warming hiatus. Science 348, 1469–1472.

Kikuyama, S., Suzuki, T., Sasaki, J., Achiari, H., Soendjoyo, S.A., HIga, H., Wiyono, A., 2017. A study on coastal erosion and deposition processes in Subang, Indonesia. https://doi.org/10.1142/9789813233812_0046.

Kudale, M.D., 2010. Impact of Port Development on the Coastline and the Need for Protection.

Kumar, V.S., 2006. Variations of wave directional spread parameters along the Indian Coast. Appl. Ocean Res. 28 (2), 93–102.

Kumar, A.B., Ravinesh, R., 2011. Will shoreline armouring support marine biodiversity? Curr. Sci. 100 (10), 1463.

Kriauciuniene, J., Zilinskas, G., Pupienis, D., Jarmalavicius, D., Gailiusis, B., 2013. Impact of Sventoji port jetties on coastal dynamics of the Baltic Sea. J. Environ. Eng. Landsc. Manag. 21 (2), 114–122.

Kummu, M., Moel, H.D., Salvucci, G., Viviroli, D., Ward, P.J., Varis, O., 2016. Over the hills and further away from coast: Global geospatial patterns of human and environment over the 20th–21st centuries. Environ. Res. Lett. 11, 034010.

Lageweg, W.V.D., Slangen, A., 2017. Predicting dynamic Costal Delta change in response to sea-level rise. J. Mar. Sci. Eng. 5 (2), 24. https://doi.org/10.3390/jmse5020024.

Leatherman, S.P., Zhang, K., Douglas, B.C., 2000. Sea level rise shown to drive coastal erosion. Eos Trans. Am. Geophys. Union 81 (6), 55–57.

Li, X., Liu, J.P., Saito, Y., Nhuyen, V.P., 2017. Recent evolution of the Mekong Delta and the impacts of dams. Earth Sci. Rev. 175, 1–17.

Lin, L., Pussella, P., 2017. Assessment of vulnerability for coastal erosion with GIS and AHP techniques case study: Southern coastline of Sri Lanka. Nat. Resour. Model. https://doi.org/10.1111/nrm.12146.

Lopez, G., Alonso, R., Mosquera, R.L., Teixeira, L., 2010. Coastal erosion in Uruguay. In: Coasts, Marine Structures and Breakwaters: Adapting to change—Proceedings of the 9th International Conference. https://doi.org/10.1680/cmsb.41301.0042.

Mallik, T.K., Vasudevan, V., Verghese, P.A., Machado, T., 1987. The black sand placer deposits of Kerala beach, Southwest India. Mar. Geol. 77 (1–2), 129–150.

Martinez, C., Contreras-Lopez, M., Winckler, P., Hidalgo, H., Godoy, E., Agredano, R., 2018. Coastal erosion in central Chile: A new hazard? Ocean Coast. Manag. 156, 141–155.

Masalu, D.C.P., Shalli, M.S., Kitula, R.A., 2010. Customs and Taboos: The Role of Indegenious Knowledge in the Management of Fish Stocks and Coral Reefs in Tanzania. University of Dar Es Salaam. Instititute of Marine Science, Melbourne, Australia.

Meehl, G.A., Washington, W.M., Collins, W.D., Julie, M.A., Hu, A., Lawrence, E.B., et al., 2005. How much more global warming and sea level rise. Science 307, 1769–1772.

Meher-Homji, V.M., 2007. Sea-walls—A necessary evil. Curr. Sci. 92, 878.

Moore, L.J., Benumof, B.T., Griggs, G.B., 1999. Coastal erosion hazards in Santa Cruz and San Diego counties, California, coastal Erosion mapping and management. J. Coast. Res. 28, 121–139 M. Crowell and S.P. Leatherman, (Eds.), SI.

Morelock, J., Barreto, M., 2003. An update on coastal erosion in Puerto Rico. Shore Beach 71 (1), 7–11.

Mörner, N.A., Matlack-Klein, P., 2017. New records of sea level changes in the Fiji Islands. J. Coast. Res. submitted for publication.

Morton, R.A., McKenna, K.K., 1999. Analysis and projection of erosion hazard areas in Brazoria and Galveston counties, Texas. J. Coast. Res. 106–120.

Muzaffar, M., Inam, A., Hashmi, M., Zia, I., 2017. Impact of reduction in upstream fresh water and sediment discharge in Indus deltaic region. J. Biodivers. Environ. Sci. 10, 208–216.

Nageswara Rao, K., Sadakata, N., 2006. Coastal zone management along the Yumigahama Peninsula, Japan—An appraisal. Ann. Yamaguchi Geogr. Assoc. 35, 21–29.

Nageswara Rao, K., Vardhan, D.A., Subraelu, P., 2007. Coastal topography and tsunami impact: GIS/GPS based mobile mapping of the coastal sectors affected by 2004 tsunami in Krishna-Godavari delta region. Eastern Geographer 13 (1), 67–74.

Nageswara Rao, K., Subreelu, P., Rajawat, A.S., Ajay, 2008a. Beach Erosion in Visakhapatnam: Causes and remedies. Eastern Geographer 16 (1), 1–6.

Nageswara Rao, K., Subraelu, P., Venkateswara Rao, T., Hema Malini, B., Ratheesh, R., Bhattacharya, S., Rajawat, A.S., Ajai, 2008b. Sea-level rise and coastal vulnerability: an assessment of Andhra Pradesh coast, India. J. Coast. Conserv. 12, 195–207.

Nageswara Rao, K., Subraelu, P., NagaKumar, K.C.V., Demudu, G., HemaMalini, B., Rajawat, A.S., Ajai, 2010. Impacts of sediment retention by dams on delta shoreline recession: Evidences from the Krishna and Godavari deltas, India. Earth Surf. Process. Landf. 35, 817–827.

Nageswarao Rao, K., Subraelu, P., NagaKumar, K.C.V., Demudu, G., HemaMalini, B., Ratheesh, R., Rajawat, A.S., Ajai, 2011. Climate change and sea-level rise: Impact on agriculture along Andhra Pradesh coast—a Geomatics analysis. J. Indian Soc. Remote Sens. 39 (3), 415–422.

Nageswara Rao, K., Saito, Y., Nagakumar, K.Ch.V., Demudu, G., Basavaiah, N., Rajawat, A.S., Tokanai, F., Kazuhiro, K., Nakashima, R., 2012. Holocene environmental changes of the Godavari Delta, east coast of India, inferred from sediment core analyses and AMS ^{14}C dating. Geomorphology, pp. 175–176, 163–175.

Nageswara Rao, K., Saito, Y., NagaKumar, K.C.V., Demudu, G., Rajawat, A.S., Kubo, S., Li, Z., 2015. Palaeogeography and evolution of the Godavari delta, east coast of India during the Holocene: an example of wave-dominated and fan-delta settings. Palaeogeogr. Palaeoclimatol. Palaeoecol. 444, 213–233.

Neelamani, S., 2017. Coastal erosion and accretion in Kuwait–problems and management strategies. Ocean Coast. Manage.

Neumann, B., Vafeidis, A.T., Zimmermann, J., Nicholls, R.J., 2015. Future coastal population growth and exposure to sea-level rise and coastal flooding – A global assessment. PLoS ONE 10 (3), e0118571.

Noujas, V., Thomas, K.V., 2015. Erosion hotspots along southwest coast of India. Aquat. Procedia 4, 548–555.

NSIDC (National Snow and Ice Data Center), 2007. Arctic sea ice news Fall. p. 2007. http://nsidc.org/arcticseaicenews/2007/10/589/, Accessed 22 October 2017.

Palalane, J., Larson, M., Hanson, H., 2015. Modeling Dune Erosion Overwash and Breaching at Macaneta Spit, Mozambique.The Proceedings of the Coastal Sediments.

Pandian, P.K., Ramesh, S., Ramana Murthy, M.V., Ramachandran, S., Thayumanavan, S., 2004. Shoreline changes and near shore processes along Ennore coast, east coast of South India. J. Coast. Res. 20, 828–845.

Petrakis, S., Karditsa, A., Alexandrakis, G., Monioudi, I., Andreadis, O., 2014. Coastal erosion: causes and examples from Greece. In: Coastal Landscapes, Mining Activities & Preservation of Cultural Heritage, Milos Island..

Pietron, J., 2017. Sediment transport from source to sink in the Lake Baikal basin: Impacts of hydroclimatic change and mining. https://su.diva-portal.org/smash/get/diva2:1129557/FULLTEXT01.pdf.

Pinto, C.A., 2016. Coastal erosion and sediment management in Portugal. In: CEDA Iberian Conference-Dredging for Sustainable Port Development, Portuguese Engineers Association, Lisbon, Portugal.

Pozzato, L., Cathalot, C., Berrached, C., Toussaint, F., Stetten, E., Caprais, J.C., Pastor, L., Olu, K., Rabouille, C., 2017. Early Diagenesis in the Congo Deep-Sea Fan Sediments Dominated by Massive Terrigenous Deposits: Part I—Oxygen Consumption and Organic Carbon Mineralizationusing a Micro-Electrode Approach. Deep-Sea Research (II), Tropical Studies in Oceanography.

Rajawat, A.S., Chauhan, H.B., Ratheesh, R., Rode, S., Bhanderi, R.J., Mahapatra, M., Mohit Kumar Yadav, R., Abraham, S.P., Singh, S.S., Keshri, K.N., Ajai, P., 2015. Assessment of coastal erosion along the Indian coast on 1:25,000 scale using satellite data of 1989-1991 and 2004-2006 time frames. Curr. Sci. 109 (2), 347–353.

Ramkumar, M., Gandhi, M.S., 2000. Beach rocks in the modern Krishna delta. J. Geol. Asson. Res. Centre 8, 22–34.

Ramkumar, M., Pattabhi Ramayya, M., Gandhi, M.S., 2000. Beach rock exposures at wave cut terraces of modern Godavari delta: Their genesis, diagenesis and indications on coastal submergence and sealevel rise. Ind. J. Mar. Sci. 29, 219–223.

Ramkumar, M., 2003. Progradation of the Godavari delta: A fact or empirical artifice? Insights from coastal landforms. J. Geol. Soc. Ind. 62, 290–304.

Ramkumar, M., 2015. Spatio-Temporal Analysis of Magnetic Mineral Content as a Tool to Understand Morphodynamics and Evolutionary History of the Modern Godavari Delta: Implications on Deltaic Coastal Zone Environmental Management. In: Ramkumar, M., Kumaraswamy, K., Mohanraj, R. (Eds.), Environmental Management of River Basin Ecosystems. Springer-Verlag, Heidelberg, pp. 177–199 DOI: 10.1007/978-3-319-13425-3-10.

Ramkumar, M., Menier, D., Manoj, M.J., Santosh, M., 2016. Geological, geophysical and inherited tectonic imprints on the climate and contrasting coastal geomorphology of the Indian peninsula. Gondwana Res. 36, 52–80.

Rignot, E., Bamber, J.L., Vander Broeke, M.R., et al., 2008. Recent Antarctic ice mass loss from radar interferometry and regional climate modeling. Nat. Geosci. 1, 106–110.

Sachin Pavithran, A.P., Menon, N.R., Sankaranarayanan, K.C., 2014. An analysis of various Coastal Issues in Kerala. Int. J. Sci. Res. Educ. 2 (1), 1993–2001.

Saito, Y., Chainanee, N., Jarupongsakul, T., Syvitski, J.P.M., 2007. Shrinking megadeltas in Asia: Sea-level rise and sediment reduction impacts from case study of the Chao Phraya delta, vol. 2. Inprint, pp. 3–7.

Saito, Y., 2008. Coastal characteristics and changes in coastal features. In: Mimura, N. (Ed.), Asia-Pacific Coasts and Their Management: States of Environment. Springer, pp. 65–78 (Chapter 3.1).

Sane, M., Yamagishi, H., 2004. Coastal Erosion in Dakar, Western Senegal. J. Jpn. Soc. Eng. Geol. 44 (6), 360–366.

Sasgen, I., van den Broeke, M., Bamber, J.L., Rignot, E., Sorensen, L.S., Wouters, B., Martinec, Z., Velicogna, I., Simonsen, S.B., 2012. Timing and origin of recent regional ice mass loss in Greenland. Earth Planet. Sci. Lett. 333, 293–303.

Shareef, N.M., 2007. Disappearing beaches of Kerala. Curr. Sci. 92, 157–158.

Stanley, D.J., Warne, A.G., 1998. Nile delta in its destruction phase. J. Coast. Res. 14, 794–825.

Stroeve, J., Holland, M.M., Meier, W., et al., 2007. Arctic Sea ice decline: faster than forecast. Geophys. Res. Lett. 34, L09501.

Stroeve, J.C., Kattsov, V., Barrett, A., Serreze, M., Pavlova, T., Holland, M., Meier, W.N., 2012a. Trends in Arctic Sea ice extent from CMIP5, CMIP3 and observations. Geophys. Res. Lett. 39, L16502.

Stroeve, J.C., Serreze, M.C., Holland, M.M., Kay, J.E., Malanik, J., Barrett, A.P., 2012b. The Arctic's rapidly shrinking sea ice cover: a research synthesis. Clim. Chang. 110 (3–4), 1005–1027.

Sundar, V., Sannasiraj, S.A., Murali, K., Sundaravadivelu, R., 2007. Runup and inundation along the Indian peninsula, including the Andaman Islands, due to great Indian Ocean tsunami. J. Waterw. Port Coast. Ocean Eng. 133 (6), 401–413.

Syvitski, J.P.M., Peckham, S.D., Hilberman, R., Mulder, T., 2003. Predicting terrestrial flux of sediment to the global ocean: a planetary perspective. Sediment. Geol. 162, 5–24.

Syvitski, J.P.M., Vorosmarty, C.J., Kettner, A.J., Green, P., 2005. Impact of humans on the flux of terrestrial sediment to the global coastal ocean. Science 308, 376–380.

Syvitski, J.P.M., Saito, Y., 2007. Morphodynamics of deltas under the influence of humans. Glob. Planet. Chang. 57, 261–282.

Syvitski, J.P.M., Kettner, A., 2011. Sediment flux and the Anthropocene. Phil. Trans. R. Soc. A 369, 957–975.

Tedstone, A.J., Bamber, J.L., Cook, J.M., Williamson, C.J., Fettwels, X., Hodson, A.J., Tranter, M., 2017. Dark ice dynamics of the south-West Greenland ice sheet. Cryosphere 11, 2491–2506.

Tessler, Z., Vorosmarty, C.J., Overeem, I., Syvitski, J.P.M., 2017. A model of water and sediment balance as determinants of relative sea level rise in contemporary and future deltas. Geomorphology. DOI: 10/1016/j.geomorph.2017.09.040.

Thornton, E.B., Sallenger, A., Sesto, J.C., Egley, L., McGee, T., Parsons, R., 2006. Sand mining impacts on long-term dune erosion in southern Monterey Bay. Mar. Geol. 229, 45–58. https://doi.org/10.1016/j.margeo.2006.02.005.

Tien, P.H., Thanh, T.D., Long, B.H., Van Cu, N., 2003. Coastal Erosion and Sedimentation in Vietnam. In: Collection of Works on Marine Environment and Resources. Science and Technics Publishing House, Hanoi, pp. 20–33.

Unnikrishnan, A.S., Rup Kumar, K., Fernandes, S.E., Michael, G.S., Patwardhan, S.K., 2006. Sea level changes along the Indian coast: Observations and projections. Curr. Sci. 90, 362–368.

Unnikrishnan, A.S., Nidheesh, A.G., Lengaigne, M., 2015. Sea-level-rise trends off the Indian coasts during the last two decades. Curr. Sci. 108, 966–971.

Vongvisessomjai, S., Polsi, R., Manotham, C., Srisaengthong, D., Charulukkana, S., 1996. Coastal erosion in the Gulf of Thailand. In: Sea-Level Rise and Coastal Subsidence. Springer, Dordrecht, pp. 131–150.

Walling, D.E., 2006. Human impact on land-ocean sediment transfer by the world's rivers. Geomorphology 79, 192–216.

Watson, R.L., 2003. Severe beach erosion at surfside, Texas caused by engineering modifications to the coast and rivers. http://texascoastgeology.com/papers/surfside.pdf/> Accessed 22 October 2017.

WHOI (Woods Hole Oceanographic Institution), 2000. Shoreline Change and the Importance of Coastal Erosion. https://web.whoi.edu/seagrant/wp-content/uploads/sites/24/2015/01/Shoreline-Change-and-the-Importance-of-Co.pdf/> Accessed 22 October 2017.

Wigley, T.M.L., 1983. The pre-industrial carbon dioxide level. Clim. Change 5, 315–320.

Woodroffe, C.D., Nicholls, R.J., Saito, Y., Chen, Z., Goodbred, S.L., 2006. Landscape Variability and Response of Asian Megadeltas to Environmental Change. In: Harvey, N. (Ed.), Global Change and Integrated Coastal Management. In: SpringerVol. 10. Dordrecht, The Netherlands, pp. 277–314.

Wang, H., We, X., Bi, N., Li, S., Yuan, P., Wang, A., Syvitski, J.P.M., Saito, Y., Yang, Z., Liu, S., 2017. Impacts of the dam-oriented water-sediment regulation scheme on the lower reaches and delta of the Yellow River (Huanghe): a review. Glob. Planet. Chang. 157, 93–113.

Wu, S., Yarnal, B., Fisher, A., 2002. Vulnerability of coastal communities to sea-level rise: A case study of Cape May County, New Jersey, USA. Clim. Res. 22, 255–270.

Wu, X., Bi, N., Xu, J., Nittrouer, J.A., Yang, Z., Saito, Y., Wang, H., 2017. Stepwise morphological evolution of the active Yellow River (Huanghe) delta lobe (1976–2013): Dominant roles of riverine discharge and sediment grain size. Geomorphology 292, 115–127.

Zhang, K., Douglas, B.C., Leatherman, S.P., 2004. Global warming and coastal erosion. Climate Change 64, 41–58.

Further Reading

Abessolo Ondoa, G., Almar, R., Kestenare, E., Bahini, A., Houngue, G.H., Jouanno, J., Anthony, E.J., 2016. Potential of video cameras in assessing event and seasonal coastline behaviour: Grand Popo, Benin (Gulf of Guinea). J. Coast. Res. 75 (sp1), 442–446.

Caton, B., Harvey, N., 2015. Coastal Management in Australia. University of Adelaide Press, p. 342.

Coleman, J.M., Huh, O.K., 2004. Major World Deltas: A Perspective from Space. http://cires1.colorado.edu/science/groups/wessman/projects/wdn/papers/colemanManuscript.pdf/ (Accessed 22 October 2017).

Guiqiu, X.D.W.W.W., Fulin, C.J.L., 1993. Coastal Erosion in China. Acta Geograph. Sin. 5, 009.

IPCC, n.d. Summary for Policymakers. In: Soloman, S.D., Manning, Q.M., Chen, Z., Miller, H.L. (Eds.), Climate Change 2007—The Physical Science Basis (Contribution of Working Group I to the Fourth Assessment Report of the Intergovernmental Panel on Climate Change), Cambridge University Press, Cambridge, 1–18.

Kale, V.S., 2002. Fluvial geomorphology of Indian rivers: An overview. Prog. Phys. Geogr. 26, 400–433.

Kumar, V.S., Singh, J., Pednekar, P., Gowthaman, R., 2011. Waves in the nearshore waters of northern Arabian Sea during the summer monsoon. Ocean Eng. 38 (2–3), 382–388.

Lopes, R.P., Oliveira, L.C., Figueiredo, A.M.G., Kinoshita, A., Baffa, O., Buchmann, F.S.C., 2010. ESR dating of pleistocene mammal teeth and its implications for the biostratigraphy and geological evolution of the coastal plain, Rio Grande do Sul, southern Brazil. Quat. Int. 212 (2), 213–222. https://doi.org/10.1016/j.quaint.2009.09.018.

Mahapatra, M., Ramakrishnan, R., Rajawat, A.S., 2015. Coastal vulnerability assessment using analytical hierarchical process for South Gujarat coast, India. Nat. Hazards 76 (1), 139–159.

McNinch, J.E., 2004. Geologic control in the nearshore: Shore-oblique sandbars and shoreline erosional hotspots, mid-Atlantic bight, USA. Mar. Geol. 211 (1–2), 121–141.

Monioudi, I.N., Karditsa, A., Chatzipavlis, A., Alexandrakis, G., Andreadis, O.P., Velegrakis, A.F., Hasiotis, T., 2016. Assessment of vulnerability of the eastern Cretan beaches (Greece) to sea level rise. Reg. Environ. Chang. 16 (7), 1951–1962.

Nair, R.R., 1974. Holocene Sea levels on the western continental shelf of India. Proc. Indian Acad. Sci. 79, 197–203.

Nhuan, M.T., Van Ngoi, C., Nghi, T., Tien, D.M., van Weering, T.C., Van den Bergh, G.D., 2007. Sediment distribution and transport at the nearshore zone of the Red River delta, northern Vietnam. J. Asian Earth Sci. 29 (4), 558–565.

Pavithran, S., Menon, N.R., Sankaranarayanan, K.C., 2014. An analysis of various coastal issues in Kerala. Int. J. Sci. Res. Educ. 2 (10), 1993–2001.

Rajendra, K., Balaji, R., Mukul, P., 2017. Review of Indian Research on Innovative Breakwaters.

Selvam, V., 2003. Environmental classification of mangrove wetlands of India. Curr. Sci. 84, 757–765.

Silva, R., Martínez, M.L., Hesp, P.A., Catalan, P., Osorio, A.F., Martell, R., Cienguegos, R., 2014. Present and future challenges of coastal erosion in Latin America. J. Coast. Res. 71 (sp1), 1–16.

Vethamony, P., Sudeesh, K., Rupali, S.P., Babu, M.T., Jayakumar, S., Saran, A.K., Basu, S.K., Kumar, R., Sarkar, A., 2006. Wave modelling for the North Indian Ocean using MSMR analysed winds. Int. J. Remote Sens. 27 (18), 3767–3780.

CHAPTER 8

Influences of Inherited Structures, and Longshore Hydrodynamics Over the Spatio-Temporal Coastal Dynamics Along the Gâvres-Penthîevre, South Brittany, France

Soazig Pian*, David Menier*, Hervé Régnauld*,†, Mu. Ramkumar‡, Mouncef Sedrati*, Manoj Joseph Mathew§

*Géosciences Océan Laboratory UMR CNRS 6538, University of South Brittany, Vannes Cedex, France, †LETG-Rennes Laboratory (Littoral, Environment, Remote Sensing Laboratory), 2 UMR CNRS 6554, University of Rennes, Rennes Cedex, France, ‡Department of Geology, Periyar University, Salem, India, §Institute of Oceanography and Environment, Universiti Malaysia Terengganu, Kuala Nerus, Malaysia

1 Introduction

Coastal systems function over a variety of temporal and spatial scales, owing to the diversified interactions between topography, sedimentary, and hydrodynamic processes that lead to changes in coastal landforms, ranging from the instantaneous to the geological scale (Cowell and Thom, 1994). Documentation of these temporal and spatial scales enhances our understanding of sediment transport trends and the evolution of coastal landscapes (Schwarzer et al., 2003). Numerous studies have been carried out to assess the shorter time scales' morphodynamic behavior of beaches and their topographic responses to changing sedimentation, and wave and tidal current conditions (Levoy et al., 1998; Ramkumar, 2000, 2003; Wijnberg and Kroon, 2002; Masselink et al., 2006; Ramkumar, 2015; Ramkumar et al., 2015). Over longer periods of time and larger spatial scales, coastal changes are assessed within the scope of the sediment cell concept (Stapor and May, 1983; Bray et al., 1995). A coastal segment is an area where sediment are moved updrift to downdrift by alongshore currents, leading to beach erosion and accretion (Carter, 1999). These studies also include analysis of coastline variations measured using remote sensing and aerial photography data (Ruggiero et al., 2003; Ferreira et al., 2006). Factors driving coastal changes over a regional scale refer to a wide range of data, which can be addressed through the use of GIS (Robin and Gourmelon, 2005). Although various authors have discussed the benefits of such a tool for improving the understanding of coastal system behavior (Robin and Gourmelon, 2005), only a few studies have taken advantage of its new analytical applications. For example, Anfuso et al. (2007) used both GIS facilities and field data to assess the medium- and short-term coastal morphological evolution of Moroccan beaches located between Ceuta and Cabo, and Backstrom et al. (2009) and Dawson and Smithers (2010) used GIS to assess topographic evolution. In most cases, however, the use of GIS in coastal studies is restricted to coastal management and risk analyses (Brown, 2006; Rodriguez et al., 2009). The study described in this chapter focuses on evolution of the Gâvres-Penthièvre beach-dune system, located in South Brittany; Northwestern France (Fig. 1). Over a regional scale, the South Brittany coast displays specific morphological characteristics, including the roughness of the coast and the presence of nearshore and offshore bedrocks (Pian, 2010; Menier et al., 2010; Dubois et al., 2011). The Gâvres-Penthièvre sand dune extends over >50 km between the Gâvres and Penthièvre headlands. It represents the only sandy coastline at South Brittany mainly composed of beaches backed by sand dunes and orientated oblique to the prevailing wind and wave directions. In spite of these peculiar regional morphological features, except the works of Pinot (1974) and Vanney (1977), no systematic studies have ever been made. Using both field data and a set of spatial data supporting spatial analyses, the study aims at identifying the main direction of sediment transport schemes driving coastal changes at different time and space scales with an objective to evaluate the beach evolution, for which suitable management programs can be planned.

2 Regional Setting

The study area is located in South Brittany and extends for about 50 km SW between the Gâvres headland in the north and the Penthièvre isthmus toward the south. It consists of sandy beaches backed by sand dunes. The coastline is interrupted by the Etel Ria (Menier et al., 2006; Estournès et al., 2012) (Fig. 2). On the Gâvres headland (Fig. 2), there are cliffs that are approximately 3 M high that consist of weathered periglacial deposits (Horrenberger, 1972). On the western part of this headland, a climbing dune has developed covering the top of the cliffs. The foot of the cliff is fronted by pocket beaches, partly fed by the material eroded from the cliffs. At Gâvres, the beach is backed by a sea wall. Over the entire beach-dune system, sediment grain size is heterogeneous, ranging from fine to coarse sand and pebble (Estournès et al., 2012). The beach faces are wave-dominated, with coarser material deposited on the upper part of the beaches.

Offshore outcrops and a rocky barrier running parallel to the coastline act as a buffer and reduce the energy of the waves (Vanney, 1977) reaching the coast. Based on previous analyses of bathymetric data, Pinot (1974) identified a slope-break at around 17 m water depth. More recent grain-size and bathymetric analyses confirmed the prevalence of closure depth at about 15–20 m off the Gavres Peninsula and the Penthièvre Isthmus (Estournès et al., 2012). Migniot (1989) analyzed sediment movements near the shore and proposed that around 6000 m^3/yr of

Influences of Inherited Structures, and Longshore Hydrodynamics

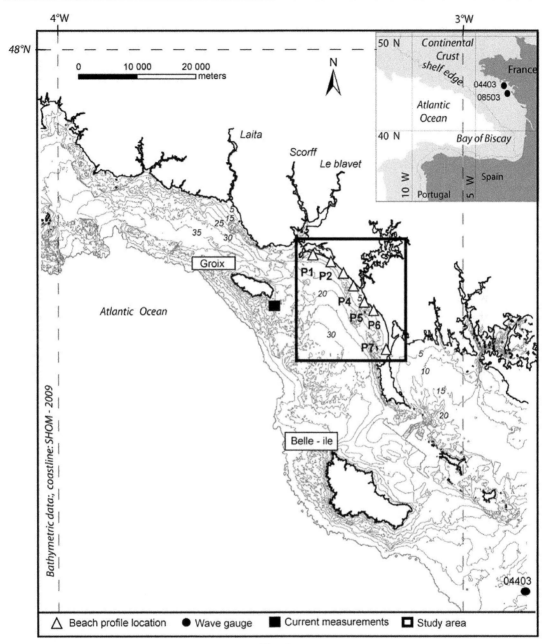

Fig. 1
Location map of the study area.

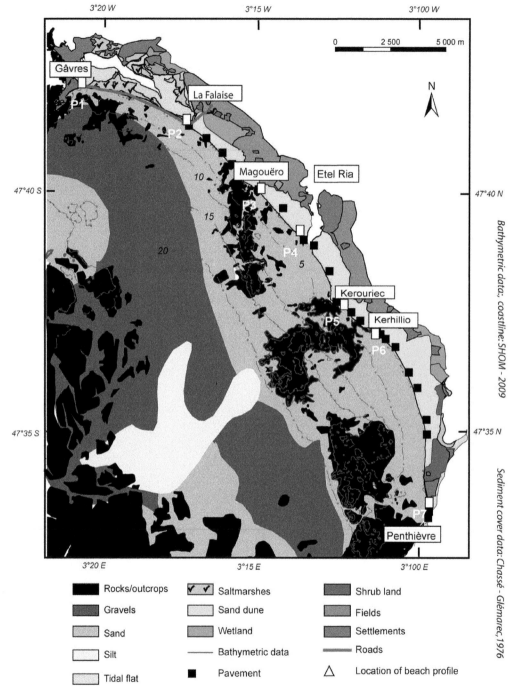

Fig. 2
Geomorphological setting of the Gâvres Penthièvre beach-dune system.

sediments are transported from Gâvres to the Etel Ria by a south-easterly longshore drift confirming the results of previous studies (Pinot, 1974; Bos and Quélennec, 1988).

Typical of the southern coast of Brittany, the studied area is subjected to westerly and southerly winds (Fig. 1). Wind records at Belle-Ile, about 30 km offshore from the studied sites, indicate that the strongest winds blow from the south and the west (Pirazzoli et al., 2004). These winds are generated by a high-pressure center, located south of Morocco and Spain, combined with a low-pressure area close to the English Channel and British Isles. Prevailing waves are characterized by a height ranging from 0.5 to 2.5 m with a period of 5–9 s. They mainly propagate from the northwest and west (Fig. 1). The most energetic waves, with a significant height of >9 m, show an annual periodicity and generally reach the coast from the west and southwest (Tessier, 2006).

3 Material and Methods

Currents, aerial photographs, and spatial data were used to assess coastal changes on a wider spatial scale and longer time scales using GIS spatial analysis. Beach profiles, and wind and wave records were used to analyze coastal changes over a shorter time scale.

3.1 Waves and Currents

Wave data were obtained from two offshore buoys, named l'Iled'Yeu Nord (8503) and Le plateau du Four (04403), which are owned by CETMEF (Centre d'Etudes Techniques Maritimes Et Fluviales), the French national office for shipping and river studies. Wave data are incorporated within the CANDHIS (Centre d'Archivage National de Données de Houle In-Situ) database managed by the CETMEF. The wave characteristics used here include significant wave height (H_s) and its associated period (T_s) recorded from March 2008 to March 2009, except for February 2009, due to a temporary interruption in buoy records. Wind data were provided by the METEO France database and cover the whole period (March 2008–March 2009) of beach surveying. The data were recorded at Le Talut station, located on the south coast of Belle Isle.

Current data were derived from the Mars S4 hydro-numerical model, set up by the design office SAFEGE (2008) (multidisciplinary engineering subsidiary of SUEZ Environment), which aims at modeling the main direction of coastal currents according to tide, wave, and wind data. The model was built over a grid of 100 m cell size, extending between the coordinates: 47°49′13 N, 47°19′22 N, 3°7′17 W, and 3°29′31 W, including offshore, shallow, and coastal waters. Tidal current direction and intensity were modeled using a 2-D depth approach and assessed with current data obtained from an ADCP (Acoustic Doppler Current Profiler) deployed at a water depth of 12.5 m. Based on the current data, sediment transport simulations were carried out using the Inglis and Lacey method. Correlations between modeled and measured data were tested for three different periods characterized by different tidal stages and wind conditions. Results show that current intensity is correctly described by the model, but current directions vary according to the wind direction. Hence, many simulations were undertaken using the 3-D depth approach for different wave and wind directions and intensities. They suggested that refraction of prevailing waves induced the coastal waves and currents, and orientated the longshore with wave convergence around the Gâvres and Magouëro headlands. Residual current direction in shallow waters is strongly controlled by wind direction. The South Brittany Coast is exposed to prevailing winds blowing from the west, and the coastline forms a vast curb that is southwest orientated, suggesting that coastal currents in shallow waters are frequently orientated toward the southeast. Such assertion agrees with previous empirical observations made along the coastline over a regional scale (Bos and Quélennec, 1988; Migniot, 1989).

3.2 Analyses of Beach Changes Over a Short Time Scale

To analyze seasonal beach changes, a beach monitoring program was carried out from March 2008 to May 2009. Seven beaches were surveyed at intervals of 2 months. Twenty-seven beach profiles were leveled from the toe of the dune to the water level line using a TRIMBLE electronic theodolite. For each site, the theodolite was placed on a fixed feature, and the location was carefully recorded using D-GPS data, marked with visual indicators, and referenced to a benchmark of the French National Geodetic Service (I.G.N. 69). The location of each beach profile was defined according to coastline orientation and nearshore morphology in order to monitor foreshores likely to record significant changes in regard to sediment transport schemes.

The coastline orientation, distribution of nearshore bedrocks, and beach access were considered for the distribution of beach profile sampling along the coastline. After Dail et al. (2000), Masselink and Pattiaratchi (2001), Sedrati and Anthony (2008), or Frihy et al. (2008) the temporal variations of offshore significant wave height (H_s) and period (T_s) (Fig. 3) were used to characterize hydrodynamic conditions prevalent during each survey period and interpret the morphodynamic changes. The dates of each beach survey are reported on the graphs describing the wave and wind regimes (Figs. 3 and 4), and categorized into six survey periods (SP1-SP6).

The first survey period (SP1), which ended in March 2008, was characterized by extremely rough weather conditions associated with the most severe storm of the period under study (Cariolet et al., 2010), with significant wave height (H_s) higher than 2.5 m (Fig. 3) and strong southwesterly winds (Fig. 3) exceeding speeds of $12\,\text{m}\,\text{s}^{-1}$. This storm occurred during spring high tide, leading to a storm surge of >3.80 m according to PREVIMER (FRENCH PRE-OPERATIONAL COASTAL OCEAN FORECASTING) wave modeling data (2008). No similar storm surge was recorded during the period of beach survey. The second survey period (SP2) extended from March 27, 2008, to June 23, 2008. It was characterized by two important subperiods of about a month and a half each. The first subperiod ended in May 2008, and was

Influences of Inherited Structures, and Longshore Hydrodynamics 187

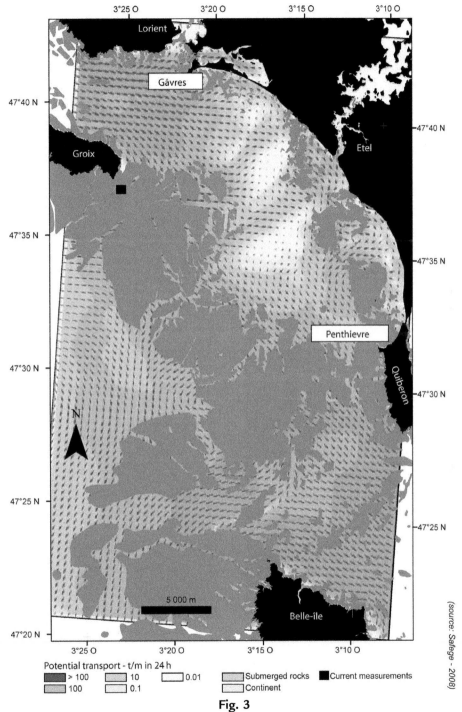

Fig. 3
Variations in wave characteristics between February 2008 and May 2009.

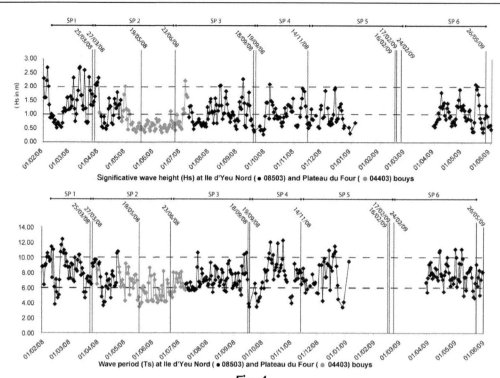

Fig. 4
Variations in wind characteristics between February 2008 and May 2009.

associated with rough weather conditions: H_s was comprised between 1 and 2 m, wind speed between 10 and 15 m s^{-1}, with winds flowing mainly from the southwest. The second subperiod ended in July 2008, and was characterized by fair weather conditions with easterly winds with speeds that did not exceed 10 m s^{-1}, and modal wave conditions with a mean H_s of 0.50 m. The third period (SP3) started in June 2008 and ended on September 19, 2008. It was characterized by agitated conditions with H_s between 1 and 2 m during most of the survey period. Three storms with H_s exceeding 2 m occurred in July, August, and September 2008. During the storm events, winds flew from the northwest and the southwest. At the end of July, a period of more modal wave conditions occurred with H_s <1 m, and accompanied easterly and northeasterly winds. The fourth survey period (SP4) extended from the September 19, 2008 to the November 14, 2008. It is characterized by two storms (H_s > 2 m and T_s > 10 s) occurring at the beginning of October 2008 and at the end of November 2008. These storms were associated with strong northwesterly and southwesterly winds, respectively. Agitated wave conditions occurred a few days after and before these storms. Between these two agitated subperiods, fair weather wave conditions prevailed. The fifth survey period (SP5) extended from November 14, 2008 to

February 24, 2009. For this survey period, wave data are not available. In early December, the wave conditions were characterized by the occurrence of two storms ($H_s > 2$ m and T_s comprised between 8 and 11 s) during 1 and 2 days, respectively, dominated by strong southeasterly winds. These storms were followed by several days of fair weather conditions. Then, the last 2 weeks of December 2008 were characterized by fair weather conditions, with easterly winds with a mean speed of $<10\,\mathrm{m\,s^{-1}}$. The end of the period, in January 2009 and early February 2009, was characterized by more severe wind conditions. Wind speeds exceeded $20\,\mathrm{m\,s^{-1}}$ several times during this period. On the February 9, southwesterly winds exceeded a speed of $25\,\mathrm{m\,s^{-1}}$. These meteorological data suggest a prevalence of rough sea-surface conditions, associated with strong waves coming from the west. The sixth and last period of survey (SP6) ended on May 26, 2009, and only concerns the Penthièvre West Beach. In April 2009 and in May 2009, two storms occurred with $H_s > 2$ m and $T_s = 10$ s. The later storm continued for 3 days. The days following these storms were characterized by rough conditions. Winds were northwesterly and southwesterly. Between each of these severe episodes, a few days of fair weather conditions occurred.

3.3 Analysis of Coastline Variation

Coastal changes occurring on the spatial scale of the sediment cell were assessed through the use of spatial and aerial photographic data. Aerial photographs were used to estimate coastline variations over different time intervals, from 1952 to 2004, at a regional scale (Fletcher et al., 2003).

Spatial data were digitized both from existing maps and from the 2004 orthophotograph produced by the IGN (French National Geographic Institute). The orthophotographs were projected in the Lambert II extended geographic coordinate system, and then integrated within a GIS database (Pian and Menier, 2011). Offshore and nearshore bathymetric data were digitized from the bathymetric map sheets 7139L to 7144L published by the SHOM (Service Hydrographique and Océanographique de la Marine) and projected in the Lambert II extended geographic coordinate system. The scale of these maps ranges from 1:15,000 to 1:2000. Offshore and nearshore sediment cover was digitalized from published maps (Pinot, 1974; Chasse and Glémarec, 1976) and from the SHOM maps. Included within this layer is the information on the presence of submarine outcrops that, in turn, are likely to influence the wave refraction process in terms of wave convergence or divergence. Coastal morphology and orientation were digitized from the 2004 orthophotograph on two different polygon layers. Four data attributes were created to describe coastline morphology: beach, sand-dune, cliff, and salt marsh. Coastline orientation was digitized at a scale of 1:2000. At this scale, five orientations were recognized along the coast: southeast, south, southwest, west, and northwest. One of these orientations was assigned to each polygon as an attribute.

As the spatial migration of the coastline is an indicator of coastal change in the long- and medium-term, positions of the coastline were plotted on aerial photographs dating from 1952 to 2004. First, aerial photographs were scanned with a spatial resolution of 1200 and 1000 dpi. Then, each photograph was geo-rectified, using ArcGis9.2 software. For each photograph, between 12 and 15 ground control points were plotted from the orthophotograph taken in 2004. Then a second-degree transformation was applied. For each geo-rectified photograph, the RMS error is around 4 and the cell sizes varied from 0.68 to 0.96 m (Table 1). After geo-rectification, all photographs covering each site for a given year were assembled into a mosaic using Envi3.0 software. The vegetation covering the toe of the dune and the cliff top line were plotted onto each document (Kroon et al., 2008) as a polyline by means of the ArcGis Editor tool.

Then, the polylines representing the shoreline position for each site at two different dates were merged into a new layer using the ArcGis toolbox function, and then converted into a polygon layer. An attribute describing the direction of shoreline migration (accretion vs. erosion) was assigned to each polygon. This polygon layer was intersected with the polygon layer describing coastline morphology (Fig. 5). Thus, a polygon layer describing the migration of coastline features was obtained for each site and time interval. By measuring the polygon areas, surface area eroded or accreted for each site and each time interval was thus estimated. Before measuring the polygon areas, the margins of error linked to geo-rectification and digitalization processes were extracted from the polygons (Table 1).

3.4 Identification of Sediment Cell Boundaries

To explain both coastline advances and retreats, the position of the coastline within the sediment cell was assessed using GIS and spatial analysis techniques. Fig. 5 summarizes the operations carried out using these GIS layers to relate coastline variations with position within the sediment cells (Pian and Menier, 2011).

Bathymetry, sediment cover, coastline morphology, and orientation data were used to identify discontinuities in these transport patterns that are likely to represent the boundaries of coastal subcells or cells (Carter, 1986; Battiau-Queney et al., 2003). The GIS layers were overlaid onto the 2004 orthophotograph in order to identify the boundaries of the sedimentary cells. Then, the coastline was segmented into three sites according to position of each straight with respect to

Table 1 Error margins associated with georectification and digitization

Air Photograph	RMS Error (m)	Cell Size (m)	Margin of Error (m)
1952	3.60	0.77	4.37
1985	4.42	0.96	5.38
1999		0.50	0.50
2004		0.50	0.50

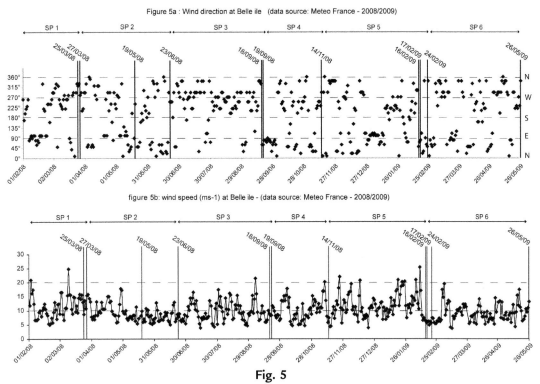

Fig. 5
Schematic diagram showing the relationships between GIS layers used to analyze coastline variations in relation to coastline position within sediment cells.

the boundaries of the sediment cells: (i) the source attribute refers to the sectors where the coastline evolution is mainly driven by coastal erosion processes, (ii) the sink attribute characterizes zones where sediments accumulate as a consequence of coastal currents, and (iii) transport areas are characterized by either temporal and spatial alternating sequences of erosion and accretion in time and space or minor coastal changes, both related to sediment input and output rhythms. Initially, based on current data, approximate coastal cell boundaries were defined: the northern part of the sand dune system is assimilated to a source area, the central part to a transport area, and the southern part to a sink area. This first delimitation was then refined by discontinuities likely to interfere with longshore sediment transport that allowed subdivision of the sediment cells into different units. These discontinuities include the Etel Ria, bathymetric variations controlling coastal currents convergence and divergence, and nearshore bedrocks and slight variations of the coastline at Kerhillio. Recognizing these sediment cell delimitations and discontinuities, each segment of the coastline was digitized either as source, transport, or sink site as a polygon layer. Using the identify function, this last polygon layer was intersected with each of the polygon layers describing the evolution of the coastline for a given time interval. The identity function allowed subdivision of a polygon layer according to the polygon boundaries stored in another polygon layer, and kept all the attribute values associated with both polygon layers. With this procedure, an attribute value was assigned to each polygon in relation to coastline variations, using it to describe the position of the coastline within the sediment cell. By the same method, attributes describing the orientation of the coastline were also associated with each polygon layer referring to coastline variations. These operations led to the construction of a spatial database describing the behavior of the coastline at different time intervals in relation to its morphology, orientation, and location within the sedimentary cell.

4 Results

4.1 Sediment Transport

In the most common fair weather conditions characterized by mean annual wave conditions and moderate westerly winds ($5 \, \text{m s}^{-1}$), sediment transport was found to be broadly oriented alongshore, in a southeasterly direction, except around Kerhillio Beach, where cross-shore sediment transport was locally dominant (Fig. 6). At Penthièvre, the longshore drift was northwesterly. Under storm conditions characterized by southwest waves with a period of 7 s and a height of 6 m, an alongshore coastal current orientated southeast was observed between the Gâvres headland and the Etel Ria. Between Etel Ria and Kerhillio Beach, offshore sediment transports occur. Under storm conditions characterized by westerly winds and waves, coastal currents in shallow waters were orientated southeast along the coastline.

4.2 Beach Profile Evolution

Beach profile variations were analyzed in relation to wave and wind regime changes (Figs. 3 and 4). Fig. 7 shows the observed variations in beach profiles (Pian et al., 2014). At Gâvres, the beach is backed by a sea wall. These profiles display the most pronounced vertical variations recorded along the beaches of the sand dune system. The vertical range of the profile envelope is about 2.5 m, and is located at the upper shoreface, between mean sea level (MSL) and the mean high water spring (MHWS) lying within the surf zone. Erosion prevailed between March 2008 and February 2009 (Fig. 7).

At La Falaise Beach, the profile is SW-NE oriented. The vertical range of the beach profile was about 1 m. The most marked beach profile variations were recorded on the lower shoreface, between the MSL and MLWS (mean low water spring). On the upper part of the profile, beach cusps and transverse bars were identified, especially during May 2008. Erosion prevailed between May 2008 and February 2009 (Fig. 7).

At Magouëro Beach, the profile was oriented SW-NE and embedded within a dense zone of emerged outcrops of bedrock (Fig. 2). The vertical range of the profile envelope was about 2 m. Except for the profile leveled in May 2008, beach profile variations were more pronounced

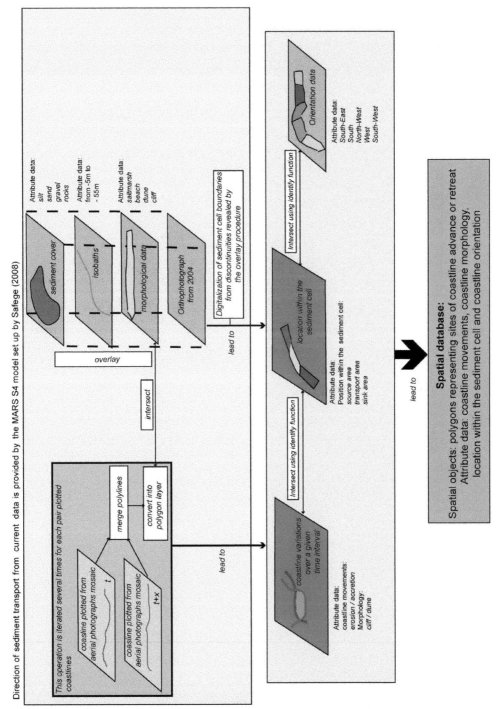

Fig. 6
Simulation of sediment transport by wave and tidal currents under both fair and storm weather conditions using the MARS S4 model.

194   Chapter 8

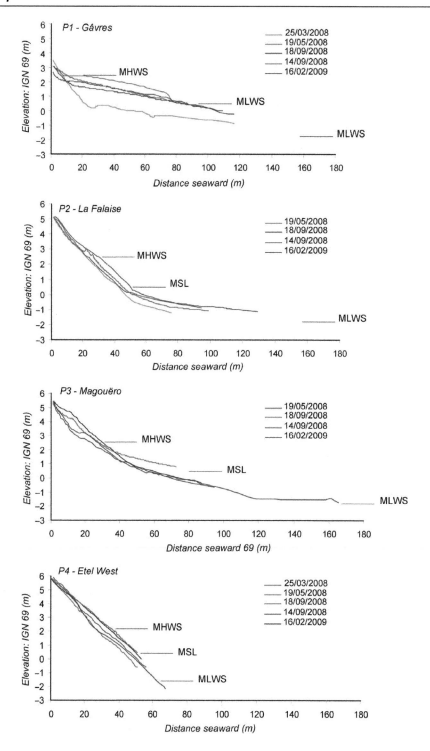

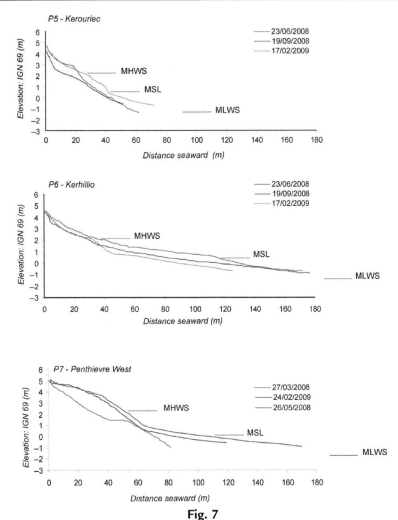

Fig. 7
Topographic profiles of the studied beaches and their variations between February 2008 and May 2009.

on the upper shoreface, partly within the surf zone. Accretion prevailed between May 2008 and February 2008 (Fig. 7). The Etel West profile was SE-NW oriented. Located west of the mouth of the Etel Ria, the profile displayed a much more reflective shape, with the surf zone at a lower elevation than other profiles. The vertical envelope of profile variation was about 1 m. Between March 2008 and February 2009, the profile variations displayed a probable cyclic evolution (Fig. 7). At Kerouriec, the profile was oriented SW-NE and located in front of a dense zone of offshore emerged rocks (Fig.7). The vertical range of the profile envelope was about 1 m. These variations were restricted within the surf zone, between the MHWS and the LHWL. Accretion prevailed between June 2008 and February 2009 (Fig. 7).

At Kerhillio, the profile was also oriented SW-NE, but outcrops of emerged rocks were less abundant. The maximum vertical range of variation was about 1 m, with erosion prevailing between June 2008 and February 2009 (Fig. 7). At Penthièvre West, the profile was E-W oriented. Bedrock and emerged rocks were found to be extending offshore. The beach is backed by a small sea wall, <1 m in height. The beach profile was located on the upper shoreface lying within the surf zone, and showed a vertical range of >2 m. Accretion prevailed between March 2008 and June 2009 (Fig. 7).

4.3 Temporal Coastline Variations

Between 1952 and 2004, the beach system experienced both coastline retreat and advance (Pian and Menier, 2011; Pian et al., 2014). In terms of surface area, 8% of the analyzed coastline experienced dune retreat, and 91.8% dune advance (Table 2). This evolution is highly variable, and the rate of shoreline migration was not linear. The beach-dune system was characterized by a severe erosion of frontal dune during 1952–84 and 1999–2004, when the total eroded area exceeded 51% and 89%, respectively (Table 2). On the contrary, during 1984–99, the beach-dune system underwent a period of dune recovery, with >92% of the area in accretion (Table 2). In addition, during the four periods under consideration, several areas, including the northern part of the sand dune system, around Gâvres, Etel Ria, and Kerhillio beaches (Fig. 8) underwent frontal dune retreat. On the contrary, throughout the time interval under study, the coastline advanced in the southern part of the sand dune system. The most variable evolution was related to the occurrences of pockets of erosion in the southern part of the beach-dune system, which were more numerous during 1952–84, and even more abundant during 1999–2004.

5 Discussion

5.1 Coastal Dynamics as Revealed by Profiling

The profiles were found to be leveled at the end of the SP1, 5–7 days after the storm occurred. At Gâvres, the profile was well fed, especially on the upper shoreface, and it can be interpreted as a consequence of onshore sediment transport feeding the upper beach.

Table 2 Migration of the coastline in m^2 between 1952 and 2004, for each time interval

	Erosion (m^2)	Accretion (m^2)	Balance (m^2)	% Erosion (m^2)	% Accretion (m^2)
1952–84	330,455	315,959	−14,496	51.21%	48.79%
1984–99	2606	1,159,705	1,157,099	0.28%	92.72%
1999–2004	184,833	20,888	−163,945	89.86%	10.14%
1952–2004	**80,384**	**933,688**	**853,484**	**8.04%**	**91.86%**

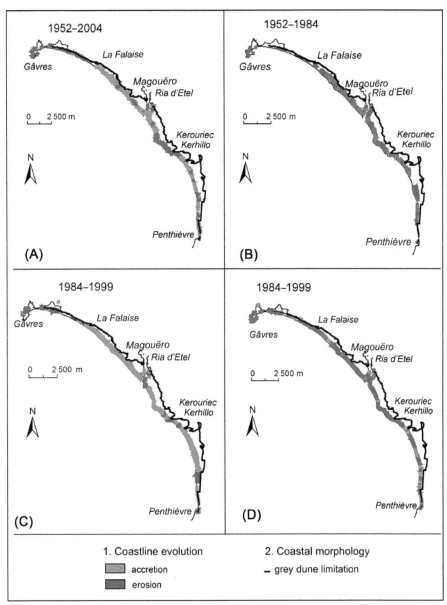

Fig. 8
Spatial distribution of coastline advance and retreat between 1952 and 2004, for each time interval.

This inference of onshore sediment transport is supported by evidence of overwash processes occurring during the storm, with sand accumulating behind the sea wall on the car parks. At the Etel Ria and Penthièvre West, the effect of the storm was likely to have eroded the beach because the profiles leveled during March were the lowest among all the profiles on both of these beaches. During the SP2, the beach profile was eroded at Gâvres, but it remained more nourished than the profiles leveled subsequently. This could have resulted from the redistribution along the profile of sand accumulated during the storm of the preceding survey period in March 2008. The profile leveled at La Falaise was located in the middle of the vertical range of profile envelopes, and was characterized by the presence of cusps on the upper shoreface. At Magouëro beach, the profile was fed and a berm formed on the upper shoreface as a consequence of sand accumulation under fair weather conditions with low-amplitude waves. At Etel West, the profile variations indicated accretion as the consequence of beach profile recovery. At Kerouriec, the profile was marked by the presence of a berm resulting from sand accumulation on the upper shoreface during fair weather conditions. At Kerhillio, the same phenomenon was recorded with lesser amplitude. During the SP3, the profile became lower at Gâvres, especially between the MHWS and MSL, owing to the upper shoreface erosion during rough wave conditions. On the contrary, at La Falaise Beach, the SW-NE-oriented beach profile experienced accretion in the surf zone, with beach cusps occurring between the MHWS and MSL. This suggested first a longshore transport direction toward the southwest between these beaches, and then differential responses of beach profiles according to their location, but under similar wave conditions within the sediment cell. At Magouëo, the lower shoreface was severely eroded over a vertical range of 1 m. The sub-aerial beach was also eroded, with a migration of the berm seaward. At Etel beach, the profile was also eroded, especially in the surf zone. Similar characteristics were observed also on Kerouriec and Kerhillio beaches. Such changes can be linked with the occurrence of rough wave conditions and the effects of both offshore and longshore transport.

The profiles leveled at La Falaise and Magouëo were lowered and displayed the effect of the last storm that occurred in November before the beach survey of SP4. At Gâvres, the upper part of the beach in front of the sea wall was eroded, but the rest of the profile recorded accretion associated with onshore transport similar to that recorded during the storms of March 10, 2008, but with lesser amplitude. At Etel West, the profile evolution during this period was almost stable, except around the MHWS, where slight accretion was recorded. The beach profile leveled in January 2009 displayed two types of responses during SP5. At Gâvres, La Falaise, and Kerhillio, morphological changes were dominated by erosion resulting from longshore sediment transport in the surf zone mainly driven by longshore drift currents toward the southeast. By contrast, at Magouëro, Etel West, Kerouriec, and Penthièvre West, morphological changes were dominated by accretion. Except at Etel West, these beaches were located in front of emerged rocks (Fig. 2) suggesting that these beaches trap sediment transported to the southeast under the action of longshore currents. Morphological changes recorded by the profile leveled at Penthièvre West indicated slight accretion during SP6.

5.2 Dynamics of Sediment Transport and Control of Beach Morphology

In a sandy coastal system, coastal cells provide a useful framework for analyzing the occurrence of erosion and accretion processes that control coastline variations at a regional scale (Komar, 1996). Relationships between onshore/shoreface sediment supply and frontal dune evolution have been well documented (Aagaard et al., 2004; Saye et al., 2005; Anthony et al., 2006). As a result, significant advance or retreat of the frontal dune can be interpreted in terms of predominant accretion or erosion processes favoring sand accumulation on the upper part of the beach and, in turn, the sand dune system. GIS analysis carried out to identify the sediment cell boundaries and the zones of active accretion and erosion within the cell revealed that, on a regional pattern, the coastline faces southwest coastal currents that moved sediments alongshore, from the north to the southeast. The Gâvres Beach, at the northern end of the beach-dune system, was recognized to be a source area, recording frontal dune retreat over the entire studied period (Fig. 8). Over intermediate studied time intervals, the evolution of the front dune is also characterized by retreat.

Dune foot vegetation smooths short-term events and provides indirect data about sand accumulation in the long or medium term (Anfuso et al., 2007). The retreat of frontal-dune vegetation, in association with a reduction of foredune surface area, reflects the occurrence of erosive processes and a negative sediment budget (Ojeda Zujar et al., 2003). The central part of the sand dune system is characterized by spatial and temporal alternations of coastline retreat and advance (Fig. 8) and behave as a transport area due to sediment input and output variability (Fig. 9). Beaches located in the south of the sand dune system are associated with frontal dune advance, although local retreats have occurred, especially since 1999 (Fig. 8). These dune advances principally occurring in sectors defined as a sink area (Fig. 9) were fed by the longshore drift. Thus, on a regional scale, the spatial distribution of areas of frontal dune retreat and advance matches with the sediment cell boundaries. On a finer scale, the sediment transport was disturbed by coastal features forming discontinuities, thus dividing the transport area into different sub-cells between which sediment passes. These discontinuities are mainly due to the variations in coastline orientation and/or the presence of submarine rocks and outcrops (Fig. 9), and include breaks in the coastline, such as the Etel Ria and Kerhillio Beach. The Kerhillio Beach was the source for the Penthièvre Beach between 1952 and 2004, and exhibited frontal dune retreat over the entire study period (Fig. 8). By contrast, the Magouëro and Kerouriec beaches received sufficient sediment supply over most of this period to favor frontal dune advance. In light of these observations, the transport area has been divided into three different sub-units extending from Gâvres to Magouëro, from Etel to Kerouriec, and from Kerhillio to Penthièvre West.

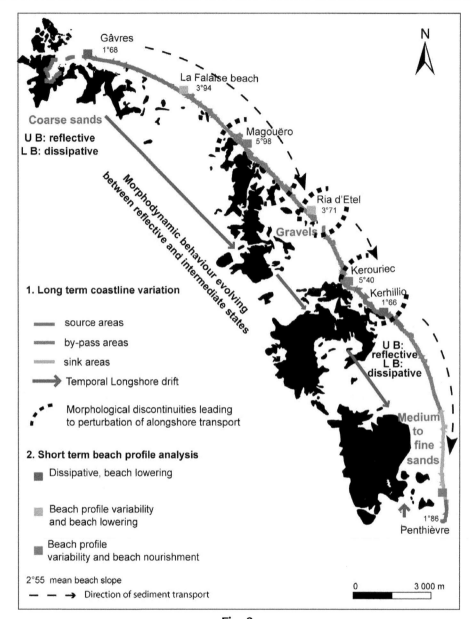

Fig. 9
Geomorphological behavior of the sand dune system over different temporal and spatial scales.

Over a long-term time and regional space scales, the front dune evolution along the Gâvres-Penthièvre beaches was driven by a southeast oriented longshore transport. The roughness of the coastline, as well as the location of nearshore bedrocks, interfere with this sediment transport and favor the relative stability of erosional or accretionary areas, which in turn matches with the sediment cell delimitations.

Beach profile variations over the short term also suggest the occurrence of longshore transport directed toward the southeast, both updrift and downdrift of the Etel Ria (Fig. 9). When onshore sediment transport dominates during storm surge conditions, sand accumulation occurs over the upper shoreface at Gâvres, but most of the time sediment is transported downdrift. According to the time interval between the two beach surveys, as well as the rate of sediment transport processes, sediments accumulate either on the beaches of La Falaise or Magouëo. Southeast of the Etel Ria, profiles leveled at Kerhillio and Penthièvre west result from erosion and accretion processes, respectively, which suggests sediment transportation toward the southeast. The morphological changes recorded were also controlled by seasonal variations of wave and wind conditions (Dubois, 1988; Hills et al., 2004). During survey period 2, when fair weather and modal wave conditions prevailed, accretion dominated in the upper shoreface as a consequence of onshore sediment transport processes. At La Falaise, Magouëro, and Kerouriec, onshore sediment transport led to the formation of a berm. Moreover, at Magouëro, Kerouriec, and Penthièvre west, beach accretion was favored by the presence of emerged outcrops of bedrock. The topographic relief of the emerged outcrops exercised control by acting as a sediment trap, and through reducing the bypass. Furthermore, the emerged outcrops can generate an atypical hydrodynamic circulation (gyre currents), which may interrupt the longshore transport and promote cross-shore transfer of sediments (Dubois et al., 2011). The dynamics of the Etel west beach were different, and were influenced by tidal currents associated with the river mouth. However, grain-size analyses carried out by Estournès et al. (2012) indicate sediment bypassing from the north to the southeast. Morphological changes recorded along the beaches of the sand dune system are thus partly explained by SE-directed longshore transport combined with seasonal wave and wind variations that alternately favor onshore and offshore transport. In addition, beach profile variations were prominent at Gâvres and Penthièvre west, which are both backed by sea walls. Thus, the higher amplitude of beach profile variations could be related to the presence of sea walls (Basco et al., 1997).

The morphological features of the coastline influenced the sediment transport over both the short and long term (Pian et al., 2014). They divide the sediment cell into different units and interrupt sediment transport over a short time scale, leading to temporary sediment accumulation, especially during storm events when higher amounts of sediments are transported. The responses of the beach to seasonal changes in wave regime are also coherent with their location within the sediment cell. In turn, sediment cell boundaries are controlled by the presence of emerged outcrops (Storlazzi and Field, 2000). Beach profiles located in source areas are eroded between February 2008 and June 2009. Inversely, beach profiles located in sinks experienced accretion. In transport areas, beach profiles experienced either erosion or accretion due to sediment output and input variability, suggesting a strong relationship between short-term profile behavior and long-term beach sediment budget evolution, and the important role played by the rocky barriers and outcrops on the wave regime attenuation, atypical current generation, and sediment transport regulation (Pian et al., 2014). The rocky outcrops limit offshore sand movement under moderate to lower high energy conditions, enhancing and concentrating the longshore sediment transport into a cross-shore corridor limited between the upper-shoreface and the rocky barrier. However, during high energy conditions, the offshore sediment can bypass this barrier within strong onshore-directed current, and may enrich the sedimentary stock in the sediment cell.

6 Conclusions

Analyses of spatial and temporal field and remotely sensed data and integration with oceanographic, geomorphologic, and sediment data through GIS has permitted the interpretation of prevalence of stable sediment transport direction on a long term and short term, and recognition of the impact of the coastline roughness on its evolution. Over a long period of time and at a large spatial scale, the coastline evolution is likely to evolve in relation to SE-directed longshore drift. The longshore redistribution of sediments from the north to the southeast leads to coastline retreat in the north, and sand accumulation downdrift to the southeast. Over the different time intervals studied, coastline retreat has occurred at least at two similar locations corresponding to downdrift areas. Coastline evolution displays no linear trend, and two of the survey periods were associated with severe frontal dune retreats. In addition, coastline retreat may have increased during the most recent survey period. Between 1952 and 2004, coastline evolution was also characterized by the occurrence of local pockets of erosion in downdrift areas. These pockets were more abundant between 1999 and 2004. On the short term and at a smaller scale, a dominant proportion of sediment transport was orientated southeast, although beach erosion was also related to seasonal variations in wind and wave conditions. This study shows that coastal morphodynamics recognized by aerial photographs on a long period do not necessarily reflect the short-term volumetric changes displayed by beach profile evolution.

Acknowledgments

The authors thank Michael Carpenter for his linguistic corrections, Pr Brian Greenwood for his comments improving the manuscript, and Alexandre Dubois for his assistance on the field.

References

Aagaard, T., Davidson-Arnott, R., Greenwood, B., Nielsen, J., 2004. Sediment supply from shoreface to dunes: linking sediment transport measurements and long term morphological evolution. Geomorphology 60, 205–224.

Anfuso, G., Dominguez, L., Gracia, E.J., 2007. Short and medium-term evolution of a coastal sector in Cadiz, SW Spain. Catena 70, 229–242.

Anthony, E.J., Vanhee, S., Ruz, M., 2006. Short-term beach-dune budgets on the North Sea coast of France: sand supply from shoreface to dunes, and the role of wind and fetch. Geomorphology 81, 316–329.

Backstrom, J.T., Jackson, D.W.T., Cooper, J.A.G., 2009. Shoreface morphodynamics of a high-energy, steep and geologically constrained shoreline segment in Northern Ireland. Mar. Geol. 257 (1–4), 94–106.

Basco, D.R., Bellomo, D.A., Hazelton, J.M., Jones, B.N., 1997. The influence of seawalls on subaerial beach volumes with receding shorelines. Coast. Eng. 30, 203–233.

Battiau-Queney, Y., Billet, J.F., Chaverot, S., Lanoy-Ratel, P., 2003. Recent shoreline mobility and geomorphologic evolution of macrotidal sandy beaches in the north of France. Mar. Geol. 194, 31–45.

Bos, P., Quélennec, R.E., 1988. Etude de l'évolution du littoral nord ouest du Morbihan entre Guidel et la Trinité sur Mer - Eléments pour une politique de protection côtière. Bureau de recherches géologiques et minières - Service Géologique Régional de Bretagne.

Bray, J.M., Carter, D.J., Hooke, J.M., 1995. Littoral cell definition and budgets for central southern England. J. Coast. Res. 11 (2), 381–400.

Brown, I., 2006. Modelling future landscape change on coastal floodplains using a rule-based GIS. Environ. Model. Softw. 21, 1479–1490.

Cariolet, J.M., Costa, S., Caspar, R., Ardhuin, F., Magne, R., Goasguen, G., 2010. Aspects météo marins de la tempête du 10mars 2008 en Atlantique et en Manche. Norois 215 (2), 11–31.

Carter, R.W.G., 1999. Coastal Environments, An Introduction to the Physical, Ecological and Cultural Systems of Coastlines, San Diego. Academic Press, New York.

Carter, R.W.G., 1986. The morphodynamics of beach-ridge formation: Magilligan, Northern Ireland. Mar. Geol. 73, 191–214.

Chasse, C., Glémarec, M., 1976. Atlas du littoral français: Atlas des fonds meubles du plateau continental du Golfe de Gascogne - Cartes Biosédimentaires. Laboratoire d' Océanographie Biologique - Université de Bretagne Occidentale.

Cowell, P.J., Thom, B.G., 1994. Morphodynamics of Coastal Evolution. In: Carter, R.W.G. (Ed.), Coastal Evolution: Late Quaternary Shoreline Morphodynamics. Cambridge University Press, Cambridge, UK.

Dail, H.J., Merrifield, M.A., Bevis, M., 2000. Steep beach morphology changes due to energetic wave forcing. Mar. Geol. 162, 443–458.

Dawson, J.L., Smithers, S.C., 2010. Shoreline and beach volume change between 1967 and 2007 at Raine Island, great barrier reef, Australia. Glob. Planet. Chang. 72 (3), 141–154.

Dubois, R.N., 1988. Seasonal changes in beach topography and beach volume in Delaware. Mar. Geol. 8, 79–96.

Dubois, D., Sedrati, M., Menier, D., 2011. Morphologic response of four pocket beaches to high energy conditions: including the Xynthia storm (South Brittany, France). J. Coast. Res. SI 64, 1845–1849.

Estournès, G., Menier, D., Guillocheau, F., Le Roy, P., Paquet, F., Goubert, E., 2012. Neogene to Holocene sedimentary infill of a Neogene incised valley on a low sediment supply inner shelf: Bay of Etel, southern Brittany, Bay of Biscay. Marine Geol., 329–331, 75–92.

Ferreira, O., Garcia, T., Matias, A., Taborda, R., Dias, J.A., 2006. An integrated method for the determination of set-back lines for coastal erosion hazards on sandy shores. Cont. Shelf Res. 26, 1030–1044.

Fletcher, C., Rooney, J., Barbee, M., Lian Siang, C., Richmond, B., 2003. Mapping shoreline change using digital orthophotogrammetry on Maui, Hawai. J. Coast. Res. 38, 106–124.

Frihy, O.E., Hassan, M.S., Deabes, E.A., Badr, A.E.M., 2008. Seasonal wave changes and the morphodynamic response of the beach-inner shelf of Abu Qir Bay, Mediterranean coast, Egypt. Mar. Geol. 247, 145–158.

Hills, H.W., Kelley, J.T., Belknap, D.F., Dickson, S.M., 2004. The effects of storms and storm-generated currents on sand beaches in Southern Maine, USA. Mar. Geol. 210, 149–168.

Horrenberger, J., 1972. Carte Géologique au 1/50 000—Lorient 383. Institut Géographique National.

Komar, P.D., 1996. The Budget of Littoral Sediments: Concepts and Applications. In: Shore and Beaches.Vol. 3, pp. 18–26.

Kroon, A., Larson, M., Möller, I., Yokoki, H., Rozynski, G., Cox, J., Larroude, P., 2008. Statistical analysis of coastal morphological data sets over seasonal to decadal time scales. Coast. Eng. 55 (7–8), 581–600.

Levoy, F., Anthony, E.J., Barusseau, J.P., Howa, H., Tessier, B., 1998. Morphodynamique d'une plage macrotidale à barres. C. R. Acad. Sci.—Serie IIA—Earth Planet. Sci. 327 (12), 811–818.

Masselink, G., Kroon, A., Davidson-Arnott, R.G.D., 2006. Morphodynamics of intertidal bars in wave-dominated coastal settings—a review. Geomorphology 73 (1–2), 33–49.

Masselink, G., Pattiaratchi, C.B., 2001. Seasonal changes in beach morphology along the sheltered coastline of Perth, Western Australia. Mar. Geol. 172, 243–263.

Migniot, C., 1989. Etudesédimentologique de l'épi de Plouhinec à l'entrée de la rivière d'Etel: aménagement possible de l'embouchure. Unpublished rapport of the Direction Départementale de l'Equipement, p. 32.

Menier, D., Reynaud, J., Proust, J., Guillocheau, F., Guennoc, P., Bonnet, S., Tessier, B., Goubert, E., 2006. Basement control on shaping and infilling of valleys incised at the southern coast of Brittany, France. Soc. Sediment. Geol. 85, 37–55.

Menier, D., Tessier, B., Proust, J.N., Baltzer, A., Sorrel, P., Traini, C., 2010. The Holocene transgression as recorded by incised-valley infilling in a rocky coast context with low sediment supply (southern Brittany, western France). Bull. Soc. Géol. France 181 (2), 115–128.

Ojeda Zujar, J., Borgniet, L., Pérez Romero, A.M., Loder, J., 2003. Monitoring coastal morphological changes using topographical methods, softcopy photogrammetry and GIS, Huelva (Andalucia, Spain). In: Green, D.R. (Ed.), Coastal and Marine Geo Information Systems: Applying the Technology to the Environment. Dordrecht. pp. 137–152.

Pian, S., 2010. Analyse multiscalaire et multifactorielle de l'évolution et du comportement géomorphologique des systèmes côtiers sud bretons. (PhD thesis). 2 Université de Rennes, p. 377.

Pian, S., Menier, D., Sedrati, M., 2014. Analysis of Morphodynamic Beach States along the South Brittany Coast. Géomorphologie: Relief, Processus, Environnement, 3/2014, pp. 261–274.

Pian, S., Menier, D., 2011. The use of a geodatabase to carry out a multivariate analysis of coastline variations at various time and space scales. J. Coast. Res. 64, 1722–1726. SI.

Pinot, J.P., 1974. Le précontinent breton entre Penmarc'h, Belle-ile et l'escarpement continental. (PhD thesis). Université de Paris VII, p. 256.

Pirazzoli, P.A., Regnauld, H., Lemasson, L., 2004. Changes in storminess and surges in western France during the last century. Mar. Geol. 210, 307–323.

Ramkumar, M., 2000. Recent changes in the Kakinada spit, Godavari delta. J. Geol. Soc. Ind. 55, 183–188.

Ramkumar, M., 2003. Progradation of the Godavari delta: a fact or empirical artifice? Insights from coastal landforms. J. Geol. Soc.Ind. 62, 290–304.

Ramkumar, M., 2015. Spatio-temporal analysis of magnetic mineral content as a tool to understand morphodynamics and evolutionary history of the Modern Godavari Delta: implications on deltaic coastal zone environmental management. In: Ramkumar, M., Kumaraswamy, K., Mohanraj, R. (Eds.), Environmental Management of River Basin Ecosystems. Springer-Verlag, Heidelberg, pp. 177–199. https://doi.org/10.1007/978-3-319-13425-3-10.

Ramkumar, M., Kumaraswamy, K., Mohanraj, R., 2015. Land use dynamics and environmental management of River Basins with emphasis on deltaic ecosystems: need for integrated study based development and nourishment programs and institutionalizing the management strategies. In: Ramkumar, M., Kumaraswamy, K., Mohanraj, R. (Eds.), Environmental Management of River Basin Ecosystems. Springer-Verlag, Heidelberg, pp. 1–20. https://doi.org/10.1007/978-3-319-13425-3-1.

Robin, M., Gourmelon, F., 2005. La télédétection et les S.I.G. dans les espaces côtiers: éléments de synthèse à travers le parcours de François Cuq. Norois 196 (3), 11–21.

Rodriguez, I., Montoya, I., Sanchez, M.J., Carreno, F., 2009. Geographic information systems applied to integrated coastal zone management. Geomorphology 107, 100–105.

Ruggiero, P., Kaminsky, G.M., Gelfenbaum, G., 2003. J. Coast. Res. 38, 57–82.

SAFEGE, 2008. Modélisation hydrodynamique et sédimentologique du secteur de l'Ile de Groix – Belle Ile, unpublishedscientific rapport, p. 120.

Saye, S.E., van der Wal, D., Pye, K., Blott, S.J., 2005. Beach-dune morphological relationships and erosion/accretion: an investigation at five sites in England and Wales using LIDAR data. Geomorphology, 128–155.

Schwarzer, K., Diesing, M., Larson, M., Niedermeyer, R.O., Schumacher, W., Furmanczyk, K., 2003. Coastline evolution at different time scales—examples from the Pomeranian Bight, southern Baltic Sea. Mar. Geol. 194, 79–101.

Sedrati, M., Anthony, E.J., 2008. Sediments dynamics and morphological change on the upper beach of a multi-barred macrotidal foreshore, and implications for mesoscale shoreline retreat: Wissant Bay, Northern France. Z. Geomorphol. 52 (3), 91–106.

Stapor, F.W., May, J., 1983. The cellular nature of littoral drift along the Northeast Florida Coast. Mar. Geol. 51, 217–237.

Storlazzi, C.D., Field, M.E., 2000. Sediment distribution and transport along a rocky, embayed coast: Monterey Peninsula and Carmel Bay, California. Mar. Geol. 170, 289–316.

Tessier, C., 2006. Caractérisation et dynamique des turbidités en zone côtière: l'exemple de la région marine Bretagne sud. (PhD thesis). Université de Bordeaux, p. 428.

Vanney, J.R., 1977. Géomorphologie de la marge continentale sud-armoricaine. S.E.D.E.S, Paris.

Wijnberg, K.M., Kroon, A., 2002. Barred beaches. Geomorphology 48 (1–3), 103–120.

CHAPTER 9

Temporal Trends of Breaker Waves and Beach Morphodynamics Along the Central Tamil Nadu Coast, India

V. Joevivek[*,†], N. Chandrasekar[†,‡]

[*]Akshaya college of Engineering and Technology, Coimbatore, India, [†]Centre for GeoTechnology, Manonmaniam Sundaranar University, Tirunelveli, India, [‡]Francis Xavier Engineering college, Tirunelveli, India

1 Introduction

Morphological changes in the beach are generally the outcome of interactions involving waves, tides, currents, sediment supply, river discharge, and longshore sediment transport. Among these parameters, beaches display spatio-temporal changes at different scales in relation to the prevailing wave conditions. However, identification of the breaker wave type and resultant morphodynamics of the beach are yet incomplete. Galvin (1968) attempted a study on breaker wave type classification through laboratory simulations of synthetic wave forms with known physical parameters. The study suggests that four types of breaker wave types, namely spilling, plunging, collapsing, and surging, prevail in the surf zone. Among these, collapsing waves are intermediate breaker types between plunging and surging breakers. Followed by this, Battjes (1974) classified three types of wave groups (excluding collapsing) in relation to the Iribarren dimensionless number (ξ). These classifications are based on uniform waves in laboratory conditions, and similar breaker wave types were recorded in field investigations as well (Komar, 1998; Masselink and Hughes, 2003; Saravanan et al., 2011; Merlotto et al., 2013; Masselink and Gehrels, 2014; Joevivek and Chandrasekar, 2017).

Morphodynamic behavior of the beach with respect to the foreshore slope, sediment discharge, and wave climate were also the focus of related research (Guza and Inman, 1975; Short, 1978; Short, 1979; Short and Hesp, 1982; Wright et al., 1982a, b; Nielsen, 1982; Wright and Short, 1984; Wright et al., 1985; Van Rijn, 1998; Stive et al., 2002; Benedet et al., 2004; Pereira et al., 2014; and many others). Wright and Short (1984) proposed a generalized beach morphodynamic state model for classifying a beach system based on particle settling velocity and wave propagation. Previous works on beach and wave dynamics suggested that the beach reworking process is influenced by the foreshore slope and grain size distribution (Bagnold, 1940; Bascom, 1951, 1953; Rector, 1954; Dubois, 1972; Van Hijum, 1974; Wright and Short, 1984; Larson, 1991; Kaiser and Frihy, 2009; Joevivek and Chandrasekar, 2016; Joevivek et al., 2017; Joevivek et al., 2018). Short (2012) demonstrated a common relationship between the morphodynamic state of the beach system and grain size distribution. It revealed that a beach with surging breakers and a reflective morphodynamic state encompasses medium grain size at foreshore region, and a beach with plunging breakers and an intermediate morphodynamic state promote deposition of medium to fine grain sizes at the foreshore region, and a beach with spilling breakers and a dissipative morphodynamic state entrains fine grain size at the foreshore region.

Recent studies have explored the tidal response on different wave types, temporal changes in intensity, and duration of the wave processes operating on the cross-shore profile and behavior of morphological patterns in a sandbar system (Ruessink and Terwindt, 2000; Van Enckevort and Ruessink, 2003a, b; Ranasinghe et al., 2004; Quartel et al., 2007; Ruessink et al., 2007; Van der Zanden et al., 2017). Wave dominated beaches produce three types of intertidal bar systems, namely slip-face bars, low amplitude ridges, and sand waves (Masselink et al., 2006). In a microtidal regime, the intertidal and subtidal bar morphodynamics depend on the tidal water level variations, rather than wave height variability. Converse responses were recorded in a macrotidal beach.

Despite these advancements in our understanding, there exists a knowledge gap regarding the trends in breaker wave type and beach morphodynamic state. This study contributes to the understanding of the temporal variation of beach and wave dynamics according to the breaker wave type and morphodynamics of the beach system on a temporal scale of months to seasons.

2 Study Area

The eastern coastal plain of India is drained by many major rivers that have built large deltas abutting the Bay of Bengal (Ramkumar et al., 2016). The shoreline extends from Kanyakumari on the south to the West Bengal in the northeast. The waves in the Bay of Bengal are highly active in the end phase of SW monsoon, and the entire NE monsoon period. During these periods, low pressures are formed in the Bay of Bengal, and intensify due to the favorable warm temperatures of the sea to become cyclones. The cyclones then travel east and cross the shores of the eastern states of India. Thus, the entire east coast of India is prone to cyclones during both of the monsoon seasons (Ramkumar et al., 2016).

The eastern coastal plain of Tamil Nadu contains sedimentary formations ranging in age from Permian to recently deposited over Precambrian crystalline basement rocks. In the central Tamil Nadu coast, Cauvery is the major river system that flows from west to east, and debauches into the Bay of Bengal. The coastal tract of central Tamil Nadu consists of upland plain (denudation), a flood plain (fluvial), a deltaic plain, and a coastal plain (marine) (Mohan et al., 2000).

The present study area is located on the Nagapattinam and Karaikkal coasts, covering 51 km of the coastal stretch between Thirukadaiyur in the north and Velankanni in the south (Fig. 1). The beaches extend up to 45–180 m, and tides are semidiurnal. Beaches have experienced an average high tide level of 0.68 m and a low tide level of 0.28 m based on the chart datum (chart no. 3007, scale 1:35000, year 2010) published by NHO, Dehradun. The winds approach the coast north–easterly during the period between November and February, south-easterly during the period between March and April and south-westerly in the rest of the period.

The relief of the Nagapattinam coast is interspersed by estuaries, lagoons, and creeks. Beach ridges present along the coast lie almost parallel to the present shoreline. The beaches are straight and gentle in slope. The width of the beach ridges vary from 45 to 150 meters. Clay, clay black, and alluviam sandy clay are the major litho units present in the coastal region (Joevivek and Chandrasekar, 2014). The period from April to July is usually hot, with maximum temperatures of 37°C. The coldest months are December and January, with temperatures dropping to a minimum of 22°C. The average rainfall rate on the Nagapattinam coast is 1252 mm, and on the Karaikkal coast it is 1315 mm. Humidity in these regions ranges between 52% and 86%.

3 Data and Methods

3.1 Field Data

Ten profile sites with an approximate interval of 5 km were fixed. Field investigations were carried out on a monthly and seasonal basis during the period between January 2011 and January 2013 along ten profile sites fixed at approximately 5 km intervals in the region between Thirukadaiyur and Velankanni, on the central Tamil Nadu coast in southern India. Care was taken to select measurement spots away from river confluence and anthropogenic activities. The parameters, namely, breaking wave height (m), wave period (s), foreshore slope (β) (in degree), and particle settling velocity (m/s) were measured. Breaking wave height (H_b) was measured by using a millimeter (mm) scale calibrated staff. The staff fixed at low-tide region; height of breaker waves measured by line of sight to wave crest and horizon (Bascom, 1964). Significant wave height was estimated from the top one third of the breaker waves. The wave period was obtained by observing the time taken by waves for passing two successive crests at a fixed point. Predominant high and low wave periods were estimated from the observation of 300 successive waves. The foreshore slope (θ) of each profile site was determined from the following procedure: (a) measure the elevation difference between the berm crest (Be) and low tide (Te) by using a calibrated staff; (b) measure the length (L) between

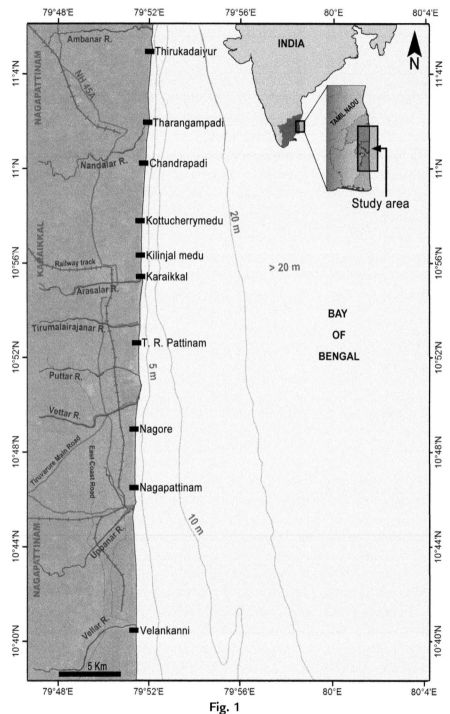

Fig. 1
Location map of the study area with river and bathymetry features.

the berm crest and low tide by using tape; (c) the foreshore slope (θ) was calculated by the inverse tangent of the ratio between the elevation difference (Be − Te) and length (L).

Sediment samples were collected at the intertidal region by an aluminium grabber with a single open edge. Samples were collected from unmodified beaches, away from the engineering structures and nourishment beaches. They were then packed and labeled. In the laboratory, after being soaked in water, the samples were agitated by a mechanical stirrer to disaggregate them and to remove the clay fractions. The samples were processed by H_2O_2 and dil. HCL for removing organic and inorganic contents, and calcareout shell fragments. About 100 g of treated samples were removed by using the coning and quartering method. These samples were sieved in a Ro-Top machine with quarter phi interval mesh grids ranging from +40 to +230 ASTM units. The weight percentage of each sieve fraction was tabulated, and grain size characteristics were determined by the method of moments (Folk and Ward, 1957; Blott and Pye, 2001).

3.2 Breaker Wave Type and Beach Morphodynamic State Model

Temporal changes in the breaker wave type and beach morphodynamic state model were estimated from the field dataset. The key parameters for calculating the breaker wave type are breaking wave height, deep water wavelength, and foreshore slope. A semiempirical formula proposed by Battjes (1974) was used.

$$\xi = \frac{\tan\beta}{(H_b/L_0)^{0.5}} \quad (1)$$

Where, ξ is the breaker wave type, H_b is the breaker wave height, L_0 is the deep water wavelength, and β is the foreshore slope (in degrees). The deep water wavelength (L_0) was estimated from the mathematical formula (CETN, 1985).

$$L_0 = \frac{g \times T^2}{2 \times \pi} \quad (2)$$

Where L_0 is the deep water wavelength, T is the wave period, and g is the gravitational acceleration. According to Battjes (1974), ξ value less than 0.4 is classified as spilling breakers, ξ value between 0.4 and 2 is classified as plunging breakers, and ξ value greater than 2 is classified as surging breakers.

Similarly, the morphodynamic state of the beach system was calculated based on the model proposed by Wright and Short (1984).

$$\Omega = \frac{H_b}{w_s T} \quad (3)$$

Where, w_s is the particle settling velocity and T is the wave period. Particle settling velocity is calculated from the generalized formula proposed by Zhiyao et al. (2008).

$$W_s = \frac{v}{d}d_*^3\left[38.1 + \left(0.93 \times d_*^{\frac{12}{7}}\right)\right]^{-\frac{7}{8}}$$

Where W_s is the particle settling velocity, v is the fluid kinematic viscosity, d is the particle diameter and d_* is the dimensionless particle diameter. According to the morphodynamic state model (Wright and Short, 1984), Ω value less than 1 is reflective of the beach morphodynamic state, Ω value between 1 and 6 is the intermediate beach morphodynamic state, and Ω value greater than 6 is a dissipative beach morphodynamic state.

4 Results

4.1 Wave Climate

The data obtained for the period between January 2011 and January 2013 on a monthly and seasonal basis were used to determine the temporal changes in wave climate. The wave period (T) for each profile site during the study period is shown in Table 1. The wave period has consistently varied between 7 and 13 s, depending on the monsoon. The data presented in the table also show that propagation of the wave cycle is shorter during northeast monsoons and longer in southwest monsoons. The longer wave period indicates less undertow current action on wave propagation. Hence, turbulence became negligible at the foreshore region, and as an outcome, the suspended sediments get settled. Due to this, sediments are deposited at the tide and berm region, creating a steep slope between the berm crest and low tide region. The shorter wave period during northeast monsoons establishes two different conditions: the first one is that the waves create strong turbulence to erode sediments at tide – in the berm region and transported to the nearshore; the second one is the collision between successive and backwash waves, leading to the settling of sediment at the nearshore region, resulting in the formation of sandbars.

The variations in breaking wave height (H_b) observed in the study region show marked differences between monsoons (Table 2). During northeast monsoons, the waves attain a higher breaker height due to the actions of low pressure depressions of the northeast Indian Ocean. A maximum wave height of 1.18 m was observed during the Thane cyclone. The successive waves tend to hit on well-developed sandbars at the nearshore, leading to higher breaking waves. The breaker wave height was less (0.41–0.7 m) during southwest monsoons due to the absence of sandbars in the nearshore region. Hence, the nearshore region seems to be flat, and it leads to shallow waveforms.

4.2 Foreshore Slope

The foreshore slope of beach is the angle formed at the intersection of the plane of the foreshore and nearshore region. Visual observation in field studies revealed that a steep foreshore slope is associated with accretion and flat slope is associated with erosion. The study area has

Table 1 Monthly and seasonal variation in wave period (T) along the study area

Month-Wise (Jan-2011 to Jan-2012)/Season-Wise (Jan-2012 to Jan-2013)	Station										
	TKR	TGP	CNP	KCM	KJL	KRL	TRP	NGR	NGP	VKN
January 2011	8	7	8	8	7	8	7	8	8	8
February 2011	9	8	9	9	8	8	8	8	9	9
March 2011	9	9	8	9	9	9	8	9	9	8
April 2011	9	9	10	9	9	10	9	9	10	9
May 2011	11	10	11	10	10	11	11	9	10	10
June 2011	10	9	10	11	11	9	10	9	9	11
July 2011	11	12	11	12	11	10	11	10	11	11
August 2011	11	11	12	11	12	12	12	11	10	12
September 2011	11	12	12	13	10	13	11	11	12	11
October 2011	10	11	10	11	9	11	10	10	11	10
November 2011	9	9	8	9	8	9	9	8	9	9
December 2011	8	8	8	9	7	8	8	9	8	9
January 2012	8	7	7	8	8	8	7	7	8	8
April 2012	10	8	9	8	8	9	8	9	9	9
July 2012	12	11	11	12	13	12	11	13	12	12
October 2012	9	10	8	9	10	9	9	9	9	10
January 2013	8	7	7	8	8	8	7	8	7	7

TKR, Thirukadaiyur; *TGP*, Tharangampadi; *CNP*, Chandrapadi; *KCM*, Kottucherrymedu; *KJL*, Kilinjal medu; *KRL*, Karaikkal; *TRP*, T. R. Pattinam; *NGR*, Nagore; *NGP*, Nagapattinam; *VKN*, Velankanni.

Table 2 Monthly and seasonal variation in breaker wave height (H_b) along the study area

Month-Wise (Jan-2011 to Jan-2012)/Season-Wise (Jan-2012 to Jan-2013)	Station									
	TKR	TGP	CNP	KCM	KJL	KRL	TRP	NGR	NGP	VKN
January 2011	0.88	0.82	0.74	0.94	0.88	0.83	0.92	0.92	0.98	0.7
February 2011	0.84	0.8	0.74	0.92	0.94	0.78	0.83	0.84	0.81	0.74
March 2011	0.72	0.74	0.72	0.84	0.79	0.72	0.78	0.66	0.78	0.74
April 2011	0.68	0.72	0.94	0.74	0.78	0.68	0.75	0.74	0.86	0.67
May 2011	0.76	0.68	0.84	0.74	0.72	0.66	0.71	0.78	0.76	0.71
June 2011	0.62	0.51	0.63	0.61	0.52	0.65	0.48	0.58	0.54	0.63
July 2011	0.58	0.54	0.57	0.55	0.53	0.66	0.46	0.68	0.59	0.6
August 2011	0.5	0.62	0.44	0.49	0.46	0.64	0.43	0.47	0.63	0.48
September 2011	0.48	0.67	0.49	0.46	0.42	0.68	0.38	0.41	0.53	0.46
October 2011	0.56	0.64	0.66	0.76	0.66	0.65	0.61	0.74	0.61	0.62
November 2011	0.87	0.73	0.82	0.91	0.87	0.73	0.77	0.73	0.91	0.72
December 2011	0.92	1.02	0.98	0.98	1.13	1.06	1.06	1.02	1.18	0.92
January 2012	0.93	0.88	0.82	0.89	0.84	0.93	0.96	0.94	0.91	0.87
April 2012	0.73	0.76	0.73	0.85	0.84	0.82	0.73	0.79	0.75	0.76
July 2012	0.51	0.47	0.5	0.42	0.46	0.68	0.39	0.61	0.52	0.53
October 2012	0.66	0.54	0.66	0.76	0.55	0.71	0.68	0.64	0.61	0.62
January 2013	0.96	0.94	0.96	0.98	1.02	0.89	1.02	0.96	0.98	0.92

TKR, Thirukadaiyur; *TGP*, Tharangampadi; *CNP*, Chandrapadi; *KCM*, Kottucherrymedu; *KJL*, Kilinjal medu; *KRL*, Karaikkal; *TRP*, T. R. Pattinam; *NGR*, Nagore; *NGP*, Nagapattinam; *VKN*, Velankanni.

experienced a maximized steep slope during SW monsoons, and a gentle slope during the northeast monsoon and fair weather periods (Table 3). A less prevalent under-tow current during the SW monsoons promoted deposition of suspended sediments at the foreshore. Thus, the backwash eroded lesser sediments, leading to the deposition. As a result, beaches developed a steep foreshore slope during the SW monsoons. During the monsoonal changes from SW to NE, these steep slopes were eroded by high energy waves that also transported the sediment to the nearshore and offshore regions. This process created numerous sandbars in the surf zone, and was further stabilized by the fair weather period, and thus an equilibrium condition was established on the coast (Fig. 4).

4.3 Particle Settling Velocity

Particle settling velocity at the foreshore region is widely used for understanding the variations in the hydraulic regime, and interactions between sediment and fluid in the surf zone. Table 4 illustrates the particle settling velocity observed during the study period. The particle settling velocity is highly irregular over time because of the non-stationary turbulence conditions in the surf zone, particularly as the wave breaks. It is also found that the particle settling velocity tends to decrease with increasing wave energy, and thus, higher turbulence forces promote relatively larger amounts of suspended sediments. Backwash carries these suspended sediments from the foreshore, and deposits them in the nearshore and offshore regions. In contrast, during the SW monsoons, high particle settling velocities were recorded in the foreshore region due to less wave turbulence. The feeble turbulence force deposited the suspended sediments around the tide and berm regions, promoting accretion of the beach.

4.4 Grain Size Distribution

Grain size characteristics reveal the influence of their sources, and the influence of hydrodynamics that rework and redistribute the sediments during the transport process (Liu and Zarillo, 1990). The weight percentages of grain size distribution at the foreshore region varied with seasonal changes. The mean reflects the overall average size of the sediment as influenced by the source of the supply and environment of the deposition. It is the function of (i) the total amount of sediment availability, (ii) the amount of energy imparted to the sediments, and (iii) the nature of the transporting agent. The energy of the transporting agent includes the degree of turbulence and the role played by currents and waves (Saravanan et al., 2013). The studied samples are medium- to fine-grain sized (Table 5).

Sediment sorting helps researchers understand the uniformity of the particle size distribution. Sediment sorting exposes the energy of the depositional environment, and the presence or absence of coarse- and fine-grained fractions (McKinney and Friedman, 1970). The standard deviation values range from 0.36 to 0.75ϕ, which fall under well sorted to moderately sorted (Blott and Pye, 2001). The sediments were well sorted during the SW monsoons,

Table 3 Monthly and seasonal variation in foreshore slope (degree) along the study area

Monthly (Jan-2011 to Jan-2012)/ Season (Jan-2012 to Jan-2013)	Station									
	TKR	TGP	CNP	KCM	KJL	KRL	TRP	NGR	NGP	VKN
January 2011	0.8	0.6	0.5	0.4	0.6	0.7	0.6	0.9	0.9	1
February 2011	1.96	1	2.77	2.82	2.87	2.08	2.4	2.26	1.8	2.43
March 2011	2.91	2.18	3.26	3.15	3.14	2.87	2.58	2.55	2.14	2.85
April 2011	3.58	3.35	3.51	3.4	3.08	3.24	3.12	3.18	3.24	3.42
May 2011	3.05	2.9	2.94	3.49	3.47	3.34	2.9	3.37	3.11	2.98
June 2011	3.39	3.24	3.4	3.1	3.58	3.45	3.16	3.58	3.62	3.23
July 2011	6.39	6.24	7.4	6.1	7.13	6.94	7.96	7.58	7.06	7.23
August 2011	7.3	6.74	5.19	7.69	7.18	8.37	8.63	8.78	9.08	8.93
September 2011	9.78	10.49	10.44	9.45	9.31	10.51	11.51	10.82	10.79	9.64
October 2011	8.35	8.82	8.02	7.04	9.17	8.65	9.73	7.65	7.45	9.55
November 2011	1.3	2.47	1.23	1.84	1.19	1.23	1.17	1.23	1.44	1.31
December 2011	0.5	0.4	0.3	0.4	0.5	0.6	0.5	0.4	0.7	0.3
January 2012	0.8	0.6	0.5	0.4	0.6	0.7	0.6	0.9	0.9	1
April 2012	2.62	2.35	2.51	3.4	3.08	3.24	3.12	3.18	3.24	3.42
July 2012	7.35	7.82	7.02	6.04	8.17	7.65	8.73	6.65	6.45	8.55
October 2012	8.83	7.72	7.53	8.35	7.94	8.88	8.55	7.62	7.59	8.21
January 2013	0.8	0.6	0.5	0.4	0.6	0.7	0.6	0.9	0.9	1

TKR, Thirukadaiyur; TGP, Tharangampadi; CNP, Chandrapadi; KCM, Kottucherrymedu; KJL, Kilinjal medu; KRL, Karaikkal; TRP, T. R. Pattinam; NGR, Nagore; NGP, Nagapattinam; VKN, Velankanni.

Table 4 Monthly and seasonal variation in particle settling velocity (m/s) along the study area

Month-Wise (Jan-2011 to Jan-2012)/Season-Wise (Jan-2012 to Jan-2013)	Station									
	TKR	TGP	CNP	KCM	KJL	KRL	TRP	NGR	NGP	VKN
January 2011	0.032	0.027	0.025	0.035	0.025	0.031	0.026	0.034	0.030	0.034
February 2011	0.024	0.023	0.021	0.023	0.026	0.022	0.025	0.020	0.025	0.025
March 2011	0.018	0.021	0.018	0.019	0.022	0.019	0.024	0.018	0.024	0.023
April 2011	0.015	0.027	0.029	0.021	0.029	0.036	0.047	0.019	0.019	0.037
May 2011	0.021	0.031	0.022	0.019	0.030	0.020	0.020	0.035	0.023	0.030
June 2011	0.014	0.017	0.015	0.009	0.012	0.032	0.010	0.020	0.012	0.017
July 2011	0.042	0.043	0.057	0.037	0.045	0.100	0.081	0.081	0.063	0.106
August 2011	0.057	0.111	0.073	0.052	0.055	0.140	0.199	0.112	0.104	0.105
September 2011	0.164	0.198	0.163	0.162	0.093	0.177	0.216	0.138	0.113	1.175
October 2011	0.047	0.064	0.044	0.076	0.055	0.063	0.041	0.074	0.034	0.034
November 2011	0.037	0.028	0.030	0.032	0.023	0.018	0.021	0.025	0.028	0.027
December 2011	0.034	0.032	0.022	0.044	0.035	0.050	0.035	0.032	0.049	0.050
January 2012	0.046	0.044	0.034	0.057	0.048	0.063	0.048	0.044	0.061	0.063
April 2012	0.016	0.028	0.030	0.022	0.030	0.037	0.048	0.020	0.020	0.038
July 2012	0.041	0.042	0.056	0.036	0.044	0.099	0.080	0.080	0.062	0.105
October 2012	0.046	0.063	0.043	0.075	0.054	0.062	0.040	0.073	0.033	0.033
January 2013	0.031	0.026	0.024	0.034	0.024	0.030	0.025	0.033	0.029	0.033

TKR, Thirukadaiyur; *TGP*, Tharangampadi; *CNP*, Chandrapadi; *KCM*, Kottucherrymedu; *KJL*, Kilinjal medu; *KRL*, Karaikkal; *TRP*, T. R. Pattinam; *NGR*, Nagore; *NGP*, Nagapattinam; *VKN*, Velankanni.

Table 5 Monthly and seasonal changes in grain size distribution along the study area

Period	Station										
	TKR	TGP	CNP	KCM	KJL	KRL	TRP	NGR	NGP	VKN
	Mean (φ)										
January 11	2.02	2.03	2.03	2.03	2.03	2.04	2.04	2.06	2.06	2.08
February 11	1.92	1.92	1.91	1.92	1.92	1.93	1.92	1.93	1.96	1.97
March 11	1.92	1.92	1.92	1.91	1.92	1.93	1.94	1.95	1.95	1.98
April 11	1.88	1.89	1.88	1.88	1.89	1.9	1.9	1.92	1.92	1.94
May 11	1.84	1.84	1.83	1.84	1.83	1.85	1.85	1.86	1.87	1.9
June 11	1.79	1.79	1.8	1.79	1.8	1.81	1.81	1.81	1.83	1.85
July 11	1.76	1.77	1.76	1.76	1.76	1.77	1.78	1.79	1.8	1.82
August 11	1.73	1.74	1.73	1.74	1.75	1.77	1.82	1.82	1.82	1.81
September 11	1.75	1.76	1.75	1.75	1.75	1.77	1.79	1.8	1.8	1.82
October 11	1.94	1.95	1.93	1.94	1.94	1.95	1.94	1.96	1.97	2
November 11	2.09	2.1	2.1	2.09	2.09	2.11	2.11	2.11	2.13	2.16
December 11	2.34	2.35	2.35	2.35	2.35	2.36	2.36	2.38	2.37	2.4
January 12	2.31	2.31	2.31	2.3	2.31	2.32	2.32	2.33	2.34	2.37
April 12	1.96	1.95	1.95	1.96	1.96	1.97	1.97	1.98	1.99	2.01
July 12	1.81	1.8	1.8	1.8	1.81	1.82	1.81	1.83	1.84	1.86
October 12	1.93	1.93	1.93	1.94	1.94	1.95	1.95	1.96	1.98	1.99
January 13	2	2	2	2	2.01	2.02	2.02	2.03	2.03	2.07
	Sorting (φ)										
January 11	0.66	0.67	0.67	0.67	0.66	0.66	0.65	0.66	0.66	0.67
February 11	0.61	0.6	0.6	0.6	0.6	0.6	0.59	0.6	0.61	0.62
March 11	0.6	0.6	0.6	0.59	0.6	0.6	0.6	0.6	0.6	0.62
April 11	0.56	0.57	0.56	0.56	0.56	0.57	0.55	0.57	0.57	0.59
May 11	0.52	0.51	0.5	0.51	0.5	0.51	0.5	0.51	0.52	0.54
June 11	0.46	0.46	0.46	0.46	0.45	0.46	0.45	0.46	0.47	0.5
July 11	0.43	0.44	0.42	0.41	0.41	0.43	0.42	0.43	0.45	0.47

	TKR	TGP	CNP	KCM	KJL	KRL	TRP	NGR	NGP	VKN
August 11	0.36	0.39	0.36	0.37	0.39	0.41	0.48	0.47	0.47	0.45
September 11	0.4	0.42	0.39	0.39	0.39	0.4	0.42	0.43	0.43	0.46
October 11	0.59	0.6	0.58	0.58	0.59	0.59	0.57	0.59	0.59	0.6
November 11	0.68	0.68	0.67	0.67	0.67	0.67	0.67	0.67	0.67	0.68
December 11	0.75	0.75	0.75	0.74	0.74	0.74	0.73	0.73	0.73	0.73
January 12	0.74	0.74	0.74	0.74	0.74	0.73	0.73	0.73	0.73	0.73
April 12	0.62	0.62	0.61	0.62	0.61	0.62	0.61	0.61	0.62	0.63
July 12	0.49	0.48	0.47	0.48	0.49	0.48	0.47	0.49	0.49	0.51
October 12	0.6	0.6	0.59	0.6	0.6	0.6	0.59	0.6	0.6	0.61
January 13	0.65	0.64	0.64	0.64	0.64	0.64	0.64	0.64	0.64	0.66

TKR, Thirukadaiyur; *TGP*, Tharangampadi; *CNP*, Chandrapadi; *KCM*, Kottucherrymedu; *KJL*, Kilinjal medu; *KRL*, Karaikkal; *TRP*, T. R. Pattinam; *NGR*, Nagore; *NGP*, Nagapattinam; *VKN*, Velankanni.

moderately well sorted during the NE monsoons, implying relatively higher and lesser energy conditions, respectively, during these monsoon seasons.

4.5 Trends in Breaker Wave Type and Beach Morphodynamic State

The results of the breaker wave type and beach morphodynamic state obtained from each station is summarized in Table 6 and 7. From Figs. 2 and 3 and Tables 6 and 7, it is found that beaches experienced a dissipative morphodynamic state and spilling breakers during the northeast (NE) monsoons. The wave climate during this period exhibited more than 1 m of breaking wave height, and less than 8 s of wave period. Beaches with an intermediate state and plunging breakers during the fair weather period exhibited a breaking wave height with a range between 0.6 and 1 m, and a wave period ranging between 8 and 10 s. Beaches with a reflective state and surging breakers during the southwest monsoon revealed that the breaking wave height was less than 0.6 m, and the wave period within a cycle was more than 10 s (Fig. 4). These results imply a cyclic process of beach and wave dynamics. From the graphical results, it is understood that a near equilibrium state prevailed in the coast.

5 Discussion

Wind-generated waves are the prime sources of energy driving sediment discharge, beach morphology, and breaker waves. However, surf zones express complex processes due to the non-uniform turbulence conditions. Four types of fluid motion at the surf zone have been considered: (a) a sediment-agitated wave group due to the broken and un-broken waves, (b) standing edge waves, (c) longshore currents and rip currents by wave energy dissipation, and (d) nonwave generated currents, including tidal currents and currents generated by local wind (Wright and Short, 1984). All of these were observed during the present study. The relation between the foreshore slope and nearshore bar was detailed by Bascom (1953). He suggested that the process of seasonal wave action is to shift the sand materials from the foreshore to the nearshore, and vice versa. This physical phenomenon is strongly correlated with the temporal variation in breaker wave type and beach morphodynamic state. High energy waves that prevailed during the NE monsoons exhibited spilling breakers and a dissipative beach state that were associated with a gentle foreshore slope and nearshore bars (Fig. 4). It was also revealed that excess wave energy eroded the materials from the foreshore and deposited them at offshore (Joevivek and Chandrasekar, 2014). In contrast, the low energy waves during the southwest monsoons lost their turbulence force at the foreshore, resulting in deposition of the suspended sediments, which ultimately led to a steep slope at the foreshore. Absence of a sandbar in the nearshore region (Fig. 4) during this season supports this inference. The beach with an intermediate state exhibited both reflective and dissipative conditions (Fig. 4).

Table 6 Monthly and seasonal variation in breaker wave type along the study area

Month-Wise (Jan-2011 to Jan-2012)/Season-Wise (Jan-2012 to Jan-2013)	Station									
	TKR	TGP	CNP	KCM	KJL	KRL	TRP	NGR	NGP	VKN
January 2011	0.093	0.072	0.063	0.045	0.070	0.084	0.055	0.082	0.079	0.104
February 2011	0.233	0.122	0.422	0.385	0.388	0.308	0.345	0.269	0.218	0.308
March 2011	0.599	0.498	0.671	0.675	0.694	0.664	0.510	0.616	0.423	0.578
April 2011	0.853	0.776	0.632	0.777	0.685	0.772	0.629	0.646	0.686	0.821
May 2011	0.916	0.921	0.910	1.152	1.072	1.077	0.901	0.916	0.857	0.926
June 2011	1.128	1.189	1.122	0.953	1.301	1.121	1.294	1.232	1.291	1.155
July 2011	2.021	2.231	2.364	1.981	2.147	1.872	3.091	2.420	2.216	2.251
August 2011	2.490	2.063	2.053	2.651	2.785	2.528	3.470	3.378	3.019	3.684
September 2011	3.420	3.392	3.947	3.373	3.161	2.811	4.540	4.476	3.925	3.755
October 2011	1.470	1.212	1.084	0.885	1.490	1.179	1.372	0.976	1.255	1.602
November 2011	0.152	0.315	0.178	0.252	0.167	0.188	0.174	0.157	0.198	0.202
December 2011	0.057	0.035	0.033	0.026	0.041	0.038	0.053	0.035	0.042	0.020
January 2012	0.053	0.054	0.045	0.026	0.051	0.046	0.054	0.060	0.077	0.067
April 2012	0.602	0.471	0.449	0.644	0.660	0.703	0.638	0.547	0.735	0.771
July 2012	2.708	2.753	2.394	2.448	3.174	2.280	3.380	2.052	2.155	2.838
October 2012	2.912	2.043	1.948	2.206	1.799	2.029	2.640	1.592	2.080	2.225
January 2013	0.089	0.054	0.045	0.044	0.065	0.065	0.065	0.100	0.079	0.091

TKR, Thirukadaiyur; *TGP*, Tharangampadi; *CNP*, Chandrapadi; *KCM*, Kottucherrymedu; *KJL*, Kilinjal medu; *KRL*, Karaikkal; *TRP*, T. R. Pattinam; *NGR*, Nagore; *NGP*, Nagapattinam; *VKN*, Velankanni.

Table 7 Monthly and seasonal variation in beach morphodynamic state along the study area

Month-Wise (Jan-2011 to Jan-2012)/Season-Wise (Jan-2012 to Jan-2013)	Station									
	TKR	TGP	CNP	KCM	KJL	KRL	TRP	NGR	NGP	VKN
January 2011	5.500	6.189	6.041	5.438	7.071	5.430	9.020	6.779	8.167	5.213
February 2011	7.000	7.000	6.000	6.708	6.000	6.000	5.587	8.282	6.480	6.000
March 2011	5.000	4.000	5.000	5.000	4.000	4.138	4.000	4.000	4.000	4.000
April 2011	5.000	3.000	4.000	4.000	3.000	2.125	2.000	5.000	5.000	2.000
May 2011	3.000	1.838	3.000	3.000	2.000	2.694	3.000	2.000	3.000	2.000
June 2011	3.577	2.488	3.566	5.951	3.714	1.688	3.789	2.417	3.682	2.930
July 2011	1.269	1.043	0.910	1.340	1.171	0.660	0.473	0.703	0.855	0.516
August 2011	0.800	0.510	0.500	0.857	0.700	0.416	0.120	0.350	0.460	0.180
September 2011	0.120	0.080	0.250	0.258	0.450	0.118	0.160	0.090	0.340	0.028
October 2011	2.000	2.000	3.000	2.000	2.000	2.053	3.000	2.000	3.000	3.000
November 2011	4.661	5.214	4.598	4.707	6.214	6.704	6.039	5.959	5.491	4.454
December 2011	5.458	8.000	9.000	7.350	8.000	7.000	6.000	8.000	8.000	6.075
January 2012	7.027	5.376	7.043	5.992	5.461	5.204	4.934	7.244	4.189	5.097
April 2012	5.034	3.434	3.433	4.929	3.123	2.492	1.906	5.788	4.144	2.209
July 2012	1.048	1.014	0.812	0.964	0.866	0.657	0.443	0.697	0.766	0.460
October 2012	0.876	0.873	1.337	0.830	1.217	1.337	1.148	1.057	1.558	1.570
January 2013	6.194	9.216	10.213	5.838	8.540	7.524	8.327	5.831	8.448	7.061

TKR, Thirukadaiyur; *TGP*, Tharangampadi; *CNP*, Chandrapadi; *KCM*, Kottucherrymedu; *KJL*, Kilinjal medu; *KRL*, Karaikkal; *TRP*, T. R. Pattinam; *NGR*, Nagore; *NGP*, Nagapattinam; *VKN*, Velankanni.

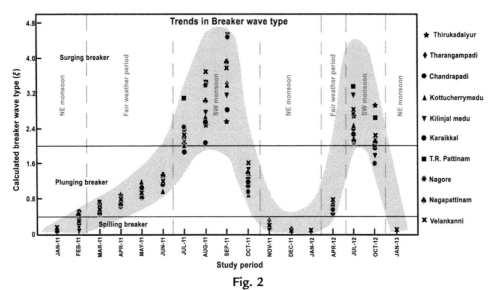

Fig. 2
Temporal trend of breaker wave type present in the study area.

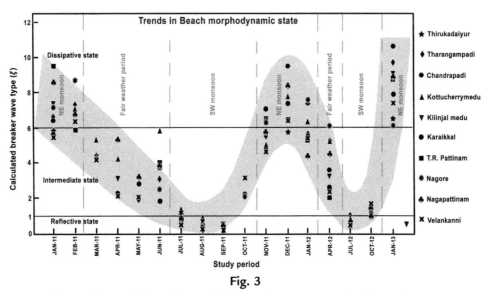

Fig. 3
Temporal trend of beach morphodynamic state present in the study area.

Short (2012) provided a spatial relationship between grain size distribution and the beach morphodynamic state. In this regard, present observations of medium grain size distribution in beaches with surging breakers and a reflective morphodynamic state, medium-fine grain size distribution in beaches with plunging breakers, and an intermediate morphodynamic state and fine grain size distribution in beaches with spilling breakers and dissipative morphodynamics are in conformity with the propositions of Short (2012). In addition, the

Fig. 4
(Continued)

Fig. 4
(A and B) Field photographs show beach and wave response with respect to the different monsoon conditions. The empirical results of breaker wave type and beach morphodynamic state are apparently correlated with the in-situ beach condition.

variations in the grain size depended on the wave conditions and slope variations in the foreshore region. For example, surging waves were associated with beaches of surging breakers and a reflective morphodynamic state. During this process, coarser material was left behind, and medium sands were deposited at the foreshore slope during SW monsoons. These trends in wave condition and beach morphology were severely modified by the Thane cyclone during the NE monsoons. The Thane event caused a major shift of the beach morphodynamic state and collapsing breakers due to the high wave energy. It took a few months of reworking to return to the normal, pre-cyclone state.

Cherian (2003) inferred that steep and gentle foreshore slopes occur due to the cyclic erosion and accretion caused by the NE and SW monsoonal effects. Holly et al. (2000) studied the beach and wave dynamics of Waimea Bay, Oahu, Hawaii, and found that the foreshore faced formation and removal of foreshore dunes, and the sand was redistributed on the beach, but not offshore. Another study was carried out on the Sendai coast in Japan by Hoang Cong et al. (2014) that portrayed that the sand was deposited near the tip of the breakwater at the foreshore region, implying deposition by down drift. The sand redistributed in the nearshore formed a shallow zone due to the seasonal action. Larson et al. (2003) and Adams et al. (2016) reported the prevalence of bar-berm sediment transition and the absence of cyclonic impacts on long-term trends of sediment size distribution. Conversely, Oak Island in North Carolina has experienced high morphodynamic changes due to frequent hurricane action. Due to this, seaward movement of sediment during the winter, and landward during the summer were reported by Baldwin (2008). Together, from these observations of previous works and present study, it can be construed that the morphodynamics of the beach system and breaker wave types are influenced by the seasonal wave action that in turn is reflected in beach morphology, sediment deposition/erosion, and grain size.

6 Conclusions

Wind, waves, monsoonal action, and water levels during storms are the driving forces that create temporal changes in beach topography. The trends of breaker wave type and beach morphodynamic state show that beaches are exposed to cyclic occasions of erosion and accretion in response to the NE/SW monsoon and non-monsoon seasons. Physical processes of waves and currents influence sand bars near the surf zone, where no headlands and terrace occur along the coast. Also, grain size distribution is apparently correlated with beach and wave dynamics on the coast.

Acknowledgments

The authors acknowledge Mr. Vincent Jayaraj, Mrs. Xavier Leema Rose and Mrs. Suganya Jenifer for their support. The corresponding author acknowledges The Management of Akshaya college of Engineering and Technology,

Kinathukadavu, Coimbatore, for permission to utilize the lab facility and also the Indian Academy of Sciences, Bangalore, for the IASc–INSA–NASI summer research fellowship (SRF. No. ENGT69–2010).

References

Adams, P.N., Keough, K.M., Olabarrieta, M., 2016. Beach Morphodynamics influenced by an ebb-tidal delta on the north Florida Atlantic coast. Earth Surf. Process. Landf. 41, 936–950. https://doi.org/10.1002/esp.3877.

Bagnold, R.A., 1940. Beach formation by waves: some model experiments in a wave tank. J. Inst. Civ. Eng. 15, 27–52. https://doi.org/10.1680/ijoti.1940.14279.

JH Baldwin., 2008. Variability in Beach Topography and Forcing Along Oak Island, North Carolina. Postgraduate Thesis, University of North Carolina Wilmington, 80 p.

Bascom, W.N., 1951. The relationship between sand size and beach-face slope. Trans. Am. Geophys. Union 32, 866. https://doi.org/10.1029/TR032i006p00866.

Bascom, W.N., 1953. Characteristics of natural beaches. In: Proceedings of the 4th Coastal Engineering Conference. ASCE, pp. 163–180.

Bascom, W., 1964. Waves and Beaches: The Dynamics of the Ocean Surface. Doubleday and Company, Garden City, New York, 268 p.

Battjes, J.A., 1974. Surf similarity. Coast. Eng. Proc. 14, 466–480. https://doi.org/10.9753/icce.v14.%p.

Benedet, L., Finkl, C.W., Campbell, T., Klein, A., 2004. Predicting the effect of beach nourishment and cross-shore sediment variation on beach morphodynamic assessment. Coast. Eng. 51, 839–861. https://doi.org/10.1016/j.coastaleng.2004.07.012.

Blott, S.J., Pye, K., 2001. GRADISTAT: a grain size distribution and statistics package for the analysis of unconsolidated sediments. Earth Surf. Process. Landf. 26, 1237–1248. https://doi.org/10.1002/esp.261.

CETN, 1985. Direct Methods for Calculating Wavelength. Coastal Engineering Technical Note, CETN-1-17, revised 6/851-4. http://chl.erdc.usace.army.mil/library/publications/chetn/pdf/cetn-i-17.pdf (Accessed 4 July 2016.).

A. Cherian, 2003. Sedimentological Studies in the Beaches Between Valinokkam and Tuticorin, Tamil Nadu. Unpublished Ph.D. Thesis, Tamil University, Tamil Nadu, India, 189 p.

Dubois, R.N., 1972. Inverse relation between foreshore slope and mean grain size as a function of the heavy mineral content. Geol. Soc. Am. Bull. 83, 871. https://doi.org/10.1130/0016-7606(1972)83[871:IRBFSA]2.0.CO;2.

Folk, R.L., Ward, W.C., 1957. Brazos River bar: a study in the significance of grain size parameters. J. Sediment Petrol. 27, 3–27.

Galvin, Jr., C.J., 1968. Breaker type classification on three laboratory beaches. J. Geophys. Res. 73 (12), 3651–3659. https://doi.org/10.1029/JB073i012p03651.

Guza, R.T., Inman, D.L., 1975. Edge waves and beach cusps. J. Geophys. Res. 80, 2997–3012. https://doi.org/10.1029/JC080i021p02997.

Hoang Cong, V.O., Yuta, M., Hitoshi, T., 2014. Analysis of shoreline behavior on Sendai coast before and after the 2011 Tsunami. Coast. Eng. Proc. 1 (34), 82.

Holly, J.D., Mark, A.M., Mike, B., 2000. Steep beach morphology changes due to energetic wave forcing. Mar. Geol. 162 (2–4), 443–458. https://doi.org/10.1016/S0025-3227(99)00072-9.

Joevivek, V., Chandrasekar, N., 2014. Seasonal impact on beach morphology and the status of heavy mineral deposition—central Tamil Nadu coast, India. J. Earth Syst. Sci. 123 (1), 135–149. https://doi.org/10.1007/s12040-013-0388-6.

Joevivek, V., Chandrasekar, N., 2016. ONWET: a simple integrated tool for beach morphology and wave dynamics analysis. Mar. Georesour. Geotechnol. 34 (6), 581–593. https://doi.org/10.1080/1064119X.2015.1040904.

Joevivek, V., Chandrasekar, N., 2017. Data on nearshore wave process and surficial beach deposits, central Tamil Nadu coast, India. Data Brief 13, 306–311. https://doi.org/10.1016/j.dib.2017.05.052.

Joevivek, V., Chandrasekar, N., Saravanan, S., Anandakumar, H., Thanushkodi, K., Suguna, N., Jaya, J., 2017. Spatial and temporal correlation between beach and wave processes: implications for bar–berm sediment transition. Front. Earth Sci. https://doi.org/10.1007/s11707-017-0655-y.

Joevivek, V., Chandrasekar, N., Shree Purniema, K., 2018. Influence of porosity in quantitative analysis of heavy mineral placer deposits. Oceanogr. Fish. Open Access J. 6 (3), 555689. https://doi.org/10.19080/OFOAJ.2018.06.555689.

Kaiser, M.F.M., Frihy, O.E., 2009. Validity of the equilibrium beach profiles: Nile Delta Coastal Zone, Egypt. Geomorphology 107, 25–31. https://doi.org/10.1016/j.geomorph.2006.09.025.

Komar, P.D., 1998. Beach Processes and Sedimentation. Prentice Hall. 544 p.

Larson, M., 1991. Equilibrium profile of a beach with varying grain size. In: Proceedings Coastal Sediments '91. ASCE, New York, pp. 905–919.

Larson, M., Capobianco, M., Jansen, H., Rozynski, G., Southgate, H., Stive, M., Wijnberg, K., Hulscher, S., 2003. Analysis and modeling of field data on coastal morphological evolution over yearly and decadal time scales. Part 1: background and linear techniques. J. Coast. Res. 19 (4), 760–775.

Liu, J.T., Zarillo, G.A., 1990. Shoreface dynamics: evidence from bathymetry and surficial sediments. Mar. Geol. 94, 37–53. https://doi.org/10.1016/0025-3227(90)90102-P.

Masselink, G., Gehrels, W.R., 2014. Coastal Environments and Global Change. American Geophysical Union, John Wiley & Sons. 448 p.

Masselink, G., Hughes, M.G., 2003. Introduction to Coastal Processes and Geomorphology. Oxford University Press, New York. 354 p.

Masselink, G., Kroon, A., Davidson-Arnott, R.G.D., 2006. Morphodynamics of intertidal bars in wave-dominated coastal settings—a review. Geomorphology 73, 33–49. https://doi.org/10.1016/j.geomorph.2005.06.007.

McKinney, T.F., Friedman, G.M., 1970. Continental shelf sediments off Long Island, New York. J. Sediment. Petrol. 40, 213–248.

Merlotto, A., Bértola, G.R., Piccolo, M.C., 2013. Seasonal morphodynamic classification of beaches in Necochea municipality, Buenos Aires Province, Argentina. Cienc. Mar. 39, 331–347.

Mohan, P.M., Shepherd, K., Suresh Gandhi, M., Rajamanickam, G.V., 2000. Evolution of quaternary sediments along the coast between Vedaranyam and Rameshwaram, Tamil Nadu. J. Geol. Soc. India 56 (3), 271–283.

Nielsen, P., 1982. Explicit formulae for practical wave calculations. Coast. Eng. 6, 389–398. https://doi.org/10.1016/0378-3839(82)90008-4.

Pereira, L.C.C., Pinto, K.S.T., Vila-Concejo, A., 2014. Morphodynamic variations of a macrotidal beach (Atalaia) on the Brazilian Amazon Coast. J. Coast. Res. 70, 681–686. https://doi.org/10.2112/SI70-115.1.

Quartel, S., Ruessink, B.G., Kroon, A., 2007. Daily to seasonal cross-shore behaviour of quasi-persistent intertidal beach morphology. Earth Surf. Process. Landf. 32, 1293–1307. https://doi.org/10.1002/esp.1477.

Ramkumar, M., Menier, D., Manoj, M.J., Santosh, M., 2016. Geological, geophysical and inherited tectonic imprints on the climate and contrasting coastal geomorphology of the Indian peninsula. Gondwana Res. 36, 52–80.

Ranasinghe, R., Symonds, G., Black, K., Holman, R., 2004. Morphodynamics of intermediate beaches: a video imaging and numerical modelling study. Coast. Eng. 51, 629–655. https://doi.org/10.1016/j.coastaleng.2004.07.018.

Rector, R.L., 1954. Laboratory Study of the Equilibrium Profiles of Beaches. U.S. Army Corps of Engineers, Beach Erosion Board, Technical Memo, No. 41.

Ruessink, B., Terwindt, J.H., 2000. The behaviour of nearshore bars on the time scale of years: a conceptual model. Mar. Geol. 163, 289–302. https://doi.org/10.1016/S0025-3227(99)00094-8.

Ruessink, B.G., Coco, G., Ranasinghe, R., Turner, I.L., 2007. Coupled and noncoupledbehavior of three-dimensional morphological patterns in a double sandbar system. J. Geophys. Res. 112C07002. https://doi.org/10.1029/2006JC003799.

Saravanan, S., Chandrasekar, N., Mujabar, P.S., Hentry, C., 2011. An overview of beach morphodynamic classification along the beaches between Ovari and Kanyakumari, Southern Tamil Nadu Coast, India. Phys. Oceanogr. 21, 129–141. https://doi.org/10.1007/s11110-011-9110-x.

Saravanan, S., Chandrasekar, N., Joevivek, V., 2013. Temporal and spatial variation in the sediment volume along the beaches between Ovari and Kanyakumari (SE INDIA). Int. J. Sediment Res. 28 (3), 384–395. https://doi.org/10.1016/S1001-6279(13)60048-7.

Short, A.D., 1978. Wave power and beach-stages: a global model. Coast. Eng. Proc. 16, 1145–1162. https://doi.org/10.9753/icce.v16.%25p.

Short, A.D., 1979. Three dimensional beach-stage model. J. Geol. 87, 553–571.

Short, A.D., 2012. Coastal processes and beaches. Nat. Educ. Knowl. 3 (10), 15.

Short, A.D., Hesp, P.A., 1982. Wave, beach and dune interactions in southeastern Australia. Mar. Geol. 48, 259–284. https://doi.org/10.1016/0025-3227(82)90100-1.

Stive, M.J., Aarninkhof, S.G., Hamm, L., Hanson, H., Larson, M., Wijnberg, K.M., Nicholls, R.J., Capobianco, M., 2002. Variability of shore and shoreline evolution. Coast. Eng. 47, 211–235. https://doi.org/10.1016/S0378-3839(02)00126-6.

Van der Zanden, J., Van der A, D.A., Hurther, D., Aceres, I.C., O'Donoghue, T., Hulscher, S.J.M.H., Ribberink, J.S., 2017. Bedload and suspended load contributions to breaker bar morphodynamics. Coast. Eng. 129, 74–92.

Van Enckevort, I.M., Ruessink, B., 2003a. Video observations of nearshore bar behaviour. Part 1: alongshore uniform variability. Cont. Shelf Res. 23, 501–512. https://doi.org/10.1016/S0278-4343(02)00234-0.

Van Enckevort, I.M., Ruessink, B., 2003b. Video observations of nearshore bar behaviour. Part 2: alongshore non-uniform variability. Cont. Shelf Res. 23, 513–532. https://doi.org/10.1016/S0278-4343(02)00235-2.

Van Hijum, E., 1974. Equilibrium profiles of coarse material under wave attack. In: 14th International Conference on Coastal Engineering. ASCE, pp. 939–957. https://doi.org/10.1061/9780872621138.057.

Van Rijn, L.C., 1998. Principles of Coastal Morphology. Aqua Publications, The Netherlands. 730 p.

Wright, L., Short, A., 1984. Morphodynamic variability of surf zones and beaches: a synthesis. Mar. Geol. 56, 93–118. https://doi.org/10.1016/0025-3227(84)90008-2.

Wright, L., Guza, R., Short, A., 1982a. Dynamics of a high-energy dissipative surf zone. Mar. Geol. 45, 41–62. https://doi.org/10.1016/0025-3227(82)90179-7.

Wright, L., Nielsen, P., Short, A., Green, M., 1982b. Morphodynamics of a macrotidal beach. Mar. Geol. 50, 97–127. https://doi.org/10.1016/0025-3227(82)90063-9.

Wright, L., Short, A., Green, M., 1985. Short-term changes in the morphodynamic states of beaches and surf zones: an empirical predictive model. Mar. Geol. 62, 339–364. https://doi.org/10.1016/0025-3227(85)90123-9.

Zhiyao, S., Tingting, W., Fumin, X., Ruijie, L., 2008. A simple formula for predicting settling velocity of sediment particles. Water Sci. Eng. 1, 37–43.

Further Reading

Dail, H.J., Merrifield, M.A., Bevis, M., 2000. Steep beach morphology changes due to energetic wave forcing. Mar. Geol. 162, 443–458.

PART 3

Coastal Hydrogeology

CHAPTER 10

Assessing Coastal Aquifer to Seawater Intrusion: Application of the GALDIT Method to the Cuddalore Aquifer, India

Subbarayan Saravanan*, Parthasarathy K.S.S.[†], S. Sivaranjani[‡]

*Department of Civil Engineering, NIT Trichy, Tiruchirappalli, India, [†]Department of Applied Mechanics and Hydraulics, NIT Surathkal, Mangalore, India, [‡]Department of Geography, Bharathidasan University, Tiruchirappalli, India

1 Introduction

Coastal zones are undergoing a tremendous change due to the development and utilization of coast in the recent decades. Thus, the trend in socio-economic and environmental change along the coast is expected to continue in the near future (Neumann et al., 2015). Most of the important cities in the world are located along the coastal zones amounting to 40% of the world's population. The demand for water in coastal areas is increasing due to ever increasing pressures from urbanization, population growth, industrial development, agriculture, and tourism (Kummu et al., 2016). Seawater intrusion (SWI) is an inherent threat to coastal aquifers due to depletion and pollution of surface water in many coastal areas of the world. Groundwater usage should be managed appropriately in coastal aquifers to maintain a freshwater barrier to control seawater intrusion. Unregulated water abstraction and high pumping rates in coastal aquifers often cause seawater movement toward the freshwater zone (Jayasekera et al., 2011). Superimposed over this are the attenuation effects of climate change and associated sea level rise over coastal hydrologic systems, leading to rapid deterioration of groundwater quality.

Albinet and Margat (1970) defined vulnerability as a degree of protection in the hydrological settings to tolerate against the ingress of pollutants to the aquifer. They defined groundwater vulnerability (Lobo-Ferreira and Cabral, 1991) to seawater intrusion as "the sensitivity of groundwater quality to an imposed groundwater pumping, or sea level rise, or both in the coastal belt, which is determined by the intrinsic characteristics of the aquifer" (Trabelsi et al., 2016). The prevalence and degree of contamination of groundwater by SWI has been studied through (a) radioactive isotopes in order to explain the increase of salinity (Yechieli et al., 2006), (b) variations of salinity and cation and anion concentrations (Polemio et al., 2006;

Pulido-Leboeuf, 2004), and (c) geophysical and geochemical approaches to obtain a more comprehensive picture of this phenomenon (Di Sipio et al., 2006; Melloul and Goldenberg, 1997; Sodde and Barrocu, 2006). Geographic information system (GIS) based SWI models are also used more widely for their ease of applicability and simplicity.

Satellite remote sensing and geographic information systems help integrate geospatial data with hydrological models. Recently, several GIS -based vulunarability models have been developed. The most common are: DRASTIC (Aller et al., 1987), GOD (Foster, 1987), the aquifer vulnerability index (AVI) (Stempvoort et al., 1993), SINTACS (Civita, 1994), the EPIK model (Doerfliger et al., 1999), and GALDIT (Chachadi et al., 2002). Vulnerability to seawater intrusion was also quantified for various stress situations, including sea-level rise (Werner and Simmons, 2009; Marfai and King, 2008; Kouz et al., 2017), change in recharge due to climate change, change in seaward discharge (El-Raey et al., 1999; Lobo-Ferreira and Chachadi, 2005; Werner et al., 2012), and excessive groundwater extraction (Gorgij and Moghaddam, 2016). To determine vulnerability of groundwater to SWI, the GALDIT model is most widely used. This weight-driven approach was developed by Chachadi and Lobo-Ferreira (2001) for assessing the vulnerability of coastal aquifers to saltwater intrusion using hydro-geological parameters and geochemistry prospecting techniques.

Sundaram et al. (2008) modeled SWI using GALDIT and GIS. This study also demonstrated methods for identification of favorable zones for artificial recharge and quantitative assessment of artificial recharge to nullify the effects of the landward intrusion of seawater in coastal aquifers. Mahesha et al. (2012) studied SWI in coastal aquifers between the Gurpur and Pavanje rivers in Karnataka, India. This study attempted to assess the impact of sea level rise on the groundwater quality for an anticipated climate change. The aquifer was studied through pumping tests, electrical resistivity tests, and groundwater quality analyses. Sophiya and Syed (2013) studied vulnerability to seawater intrusion along the coastal Ramanathapuram district of southeastern India using the GALDIT index model. The key contributions of this work include a comprehensive assessment of seawater intrusion in the study area, over a significantly large period of 10 years. In addition, the impact of global mean sea level rise, as a consequence of a progressively warming climate, has been assessed in terms of the spatial heterogeneity of seawater intrusion in the study area for the year 2050. Gorgij and Moghaddam (2016) developed a GAPDIT method ("L" being replaced by "P") and applied it to a coastal aquifer in a hyper-saline lake in the vicinity of Urmia, Iran. An analytic hierarchy process (AHP) was employed to enhance the prediction of vulnerability distribution and pumping, and the rate has been proposed in the study. Moghaddam et al. (2015) investigated the impact of rising seawater levels, as well as SWI the resulted from overdrafting in a coastal aquifer in northern Iran. In this study, the GALDIT model was applied to obtain a vulnerability zoning distribution map for a coastal aquifer. Finally, the output map of the GALDIT model was compared with the DRASTIC index by a few chosen quality parameters. DRASTIC and GALDIT indices were applied to determine the groundwater vulnerability to contamination from anthropogenic activities, and seawater intrusion in Kapas Island, northeastern Malaysia (Kura ct al., 2014).

The study described in this chapter applied a GALDIT vulnerability index in an alluvial coastal aquifer in the Cuddalore district of Tamil Nadu State, India. A strip of land measuring 54 km length and 17.5 km width was chosen for the SWI analysis. This area consists of a cluster of industries in which seawater intrusion potential is very pronounced. It would be useful to evolve a proper management strategy to protect the coastal aquifer from the adverse effects of overexploitation.

2 Regional Geological Setting of the Study Area

The Cuddalore district is located between 11°08′5″N-11°52′26.6″N and 78°51′39.77″E-79° 48′50.4″E in Tamil Nadu, South India. It has a geographical area of about 3,677.81 km^2 with a population of 2,285,395. The coastal strip chosen for the present study located in the Cuddalore District covers a land of area about 931 km^2 (Fig. 1). It is a fast growing industrial city with a burgeoning population. The study area is bounded by the basement crystalline rocks (Archaean age)/Cretaceous rocks in the west, and in the east by the Bay of Bengal, whereas, the northern and southern boundaries are marked by the Gadilam and Vellar rivers, respectively. A major part of the study area is underlain by Cuddalore sandstone of the late Miocene age in the north, while recent alluvium deposited by the Vellar river and Manimukta Nadi overlies these older deposits. The northern boundary is defined by an up dip along the Gadilam River, while the southern boundary is defined by the Vellar River. The Archaeans marks the western boundary, while the Bay of Bengal Sea constitutes the eastern boundary of the Cuddalore coastal aquifer system. The largest known deposit of fossil fuel (lignite) occurs within the Cuddalore coastal aquifer system (CGWB, 2015). These confined aquifers are interconnected by discontinuous layers of clay beds of the present at variable depths within aquifer sands. Recent alluvium and tertiary Cuddalore sandstone constitute the principal and potential aquifer system the study area.

The central part of the study area is marked by the depositional regime of many rivers manifested by typical fluvial features such as levees, channel bars, paleo channels, back swamps, and vast flood plains. The coastal landforms include strand-lines, raised beaches, sand dunes, mangrove swamps, and tidal flats with predominantly sandy beaches on the northern side, and mangrove swamps to the south. Black, red, ferruginous and arenaceous soil types cover the study area (CGWB, 2009). Alluvial soils are found as patches along the stream and river courses. The climate of the study area is hot and humid throughout the year. The estimated average daily maximum temperature ranges between 31°C and 39°C, and the minimum ranges between 21°C and 27°C. The average annual rainfall is 1110–1400 mm, with the majority of rains occurring from October to December.

2.1 Hydrogeology of the Study Area

In many parts of the aquifer, the groundwater level is close to the ground surface, and water intake areas are located along the coastline, where the groundwater level may often fall below the sea level. The pumping test data of 52 selected wells (after 12 h of constant pumping rates) were adopted from the Central Ground Water Board (CGWB, 2015). The data show typical

236 Chapter 10

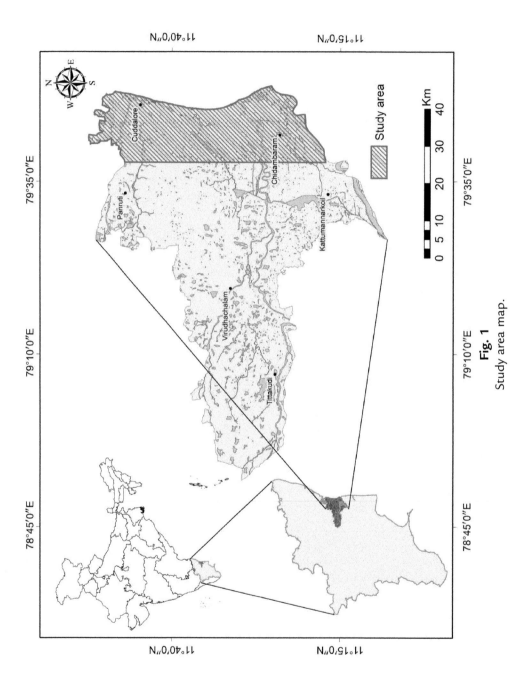

Fig. 1
Study area map.

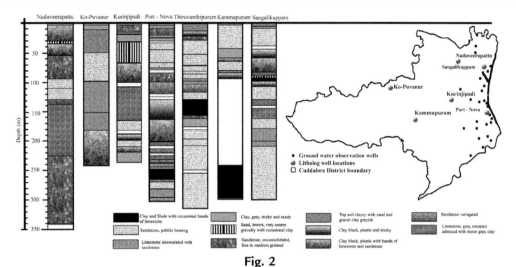

Fig. 2
Lithology data, Bore log and groundwater sampling locations.

values for alluvial aquifer layers composed of sand and gravel. Transmissivity, T, ranges between 1.15×10^{-3} and 9.971×10^{-2} m²/s (showing an average of 2.395×10^{-2} m²/s), hydraulic conductivity ranges between 2.49×10^{-5} and 2.22×10^{-5} m/s, with an average value of 4.67×10^{-4} m/s, and storativity ranges between 1.3×10^{-4} and 2.9×10^{-2} with an average value of 5.06×10^{-3} (CGWB, 2015). The major water-bearing formations of this area constitute weathered and fractured charnockite, sandstone, limestone, and alluvium. The area has a water level of 1.5–17.54 mbgl, and 0.04–7.46 mbgl during pre-monsoon season, and post-monsoon season, respectively. The geological formations are broadly classified into fissured and fractured formations and porous formations. A detailed hydrogeology of the bore log well and observation wells is shown in Fig. 2.

3 Methodology

The GALDIT method incorporates a weight range and importance rating that defines the relative importance of each factor under varied hydrogeological settings. The factors that control the saltwater intrusion are indicated in the acronym, GALDIT (Lobo-Ferreira and Chachadi, 2005), which stands for *G*roundwater occurrence (aquifer type; unconfined, confined and leaky confined), *A*quifer hydraulic conductivity, Height of groundwater *L*evel above sea level, *D*istance from the shore (distance inland perpendicular from shoreline), *I*mpact of existing status of saltwater intrusion in the area, and *T*hickness of the aquifer that is being mapped. The system contains three significant components: weights, ranges, and importance ratings (Table 1) (after Chachadi, 2005). The basic assumption made in the analysis is that the bottom of the aquifer lies below the mean sea level.

Table 1 Weights rank and range of GALDIT parameters

Sl No.	Indicator	Weight	Indicator Variable Class & Ranging	Importance Ratings
1	Groundwater occurrence (G)	1	Unconsolidated	7.5
			Semi-unconsolidated	5
2	Aquifer hydraulic conductivity (m/day) (A)	3	63.42	10
			22.18	7.5
			9.27	5
			2.78	2.5
3	Depth of groundwater level (m) (L)	4	<1	10
			1–1.5	7.5
			1.5–2	5
			>2	2.5
4	Distance from coastline (m) (D)	4	<500	10
			500–750	7.5
			750–1000	5
			>1000	2.5
5	Impact of seawater intrusion (I)	1	2.03–3.59	10
			1.46–2.03	7.5
			0.95–1.46	5
			0.3–0.95	2.5
6	Thickness of the aquifer (m) (T)	2	20–25	10
			10–20	10
			7.5–10	7.5
			5–7.5	5
			<5	2.5

Each factor's weight depicts the relative importance to the process of saltwater intrusion on a scale of 4 (most significant) to 1 (least significant). The factor's weight is decided based on the results from the studies related to saltwater intrusion from peers in the field (Chachadi and Lobo Ferreira, 2001). Considering the relative importance of each factor, its rank is determined. The importance rating ranges between 2.5, 5, 7.5, and 10. The higher the value of importance ratings, the more vulnerable is the aquifer to saltwater intrusion. Table 1 shows the three significant parts of factors (i.e., weights, ranges, and importance ratings).

3.1 Computation of the GALDIT Index

The local index of vulnerability was computed through multiplication of the value attributed to each parameter (rating) by its relative weight, and adding up all six products, following the expression (Lobo-Ferreira and Chachadi, 2005; Chachadi and Lobo-Ferreira, 2005) (Eq. 1).

$$GALDIT = \frac{1 \times G + 3 \times A + 4 \times L + 4 \times D + 1 \times I + 2 \times T}{15} \quad (1)$$

Where the numerical coefficients are the weights (already set up) and the $G, A, L, D, I,$ and T are the ratings of factors involved in the GALDIT vulnerability index (Eq. 2).

$$GALDIT = \sum_{i=1}^{6} \left[\frac{W_i * R_i}{15}\right] \quad (2)$$

Where W_i is the weight of the *i*th indicator and R_i is the importance rating of the *i*th indicator.

4 Results

Characterization of the aquifer seems to be affected by saltwater intrusion at depths ranging from 15 to 55 m from the ground surface. The wells are located at an elevation almost equal to the mean sea level, or lower, and are within 800 m from the sea or estuary.

4.1 Groundwater Occurrence (G)

The aquifer in the study area is divided into two types, namely unconsolidated and consolidated aquifers. Unconsolidated aquifers formed by deltaic depositions are present in the coastal areas. Good storage potential is offered by these aquifers under high rainfall and recharge conditions. They cover about 60% of the study area. Another type of aquifer present in the study area is a semi-consolidated aquifer that consists of sand interbedded with silt, clay, and minor amounts of carbonate rocks, possessing moderate to high aquifer conductivity (Fig. 3). The porosity of these regions is intergranular, and they are overlain in fluvial and deltaic regions. They cover 25% of the study area, and occupy the western part.

4.2 Aquifer Hydraulic Conductivity (A)

Hydraulic conductivity of the material can be defined as the ability of the fluid to pass through the pores and fractured rocks. The conductivity depends on the type of the soils that are found in the region. Thus, the hydraulic conductivity near the coastal region is relatively high (63.42 m/day) due to the highly permeable soils of the regions. The southeastern part of the study area has the hydraulic conductivity of about 22.18 m/day. The hydraulic conductivity decreases from 9.27 (coast) to 2.78 m/day (inland) (Fig. 4).

4.3 Depth to Water Table (L)

The groundwater level plays a crucial role in maintaining the hydraulic pressure along the coast to resist saltwater. When the height of water above the sea level is at its maximum level, the vulnerability ranking will be at its minimum level, and vice versa. Water table elevation less than 1 from the mean sea level is considered as the highest importance rating of 10, indicating the region's high vulnerability to saltwater intrusion. Water table elevation greater than 2 m

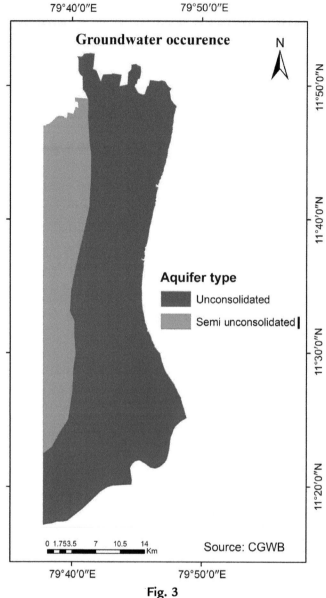

Fig. 3
Groundwater occurrence.

from the mean sea level is considered to be the least vulnerable, with a rating of 2.5. The lowest water table elevation may be considered for estimation as a conservative approach during a year. The height above the groundwater is obtained from the CGWB data for the pre-monsoon season for the year 2006 (Fig. 5). The data indicate the maximum height of the water level in the post-monsoon season (December to February). The regions where the Gadilam and Vellar River meets the Bay of Bengal consist of very low groundwater level depth with the maximum amount of vulnerability to sea water intrusion. This scenario occurs to a greater extent during the pre-monsoon seasons.

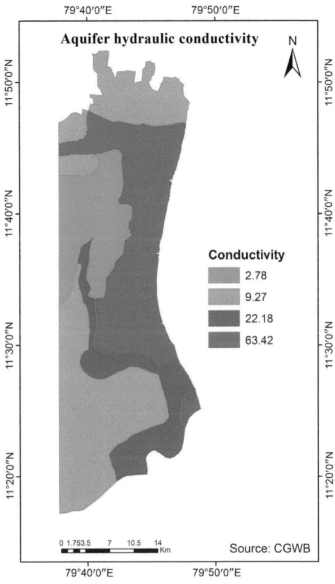

Fig. 4
Aquifer hydraulic conductivity.

4.4 Distance From the Shore/River (D)

The magnitude to which the salt water intrusion may take place depends on the distance from the coast. Accordingly, the coastline buffer is taken with the range of maximum 1000 m to minimum of 500 m. The region >1000 m away from the coast is considered as far, and thus vulnerability to SWI may be comparatively less, whereas the region within <500 m limit has

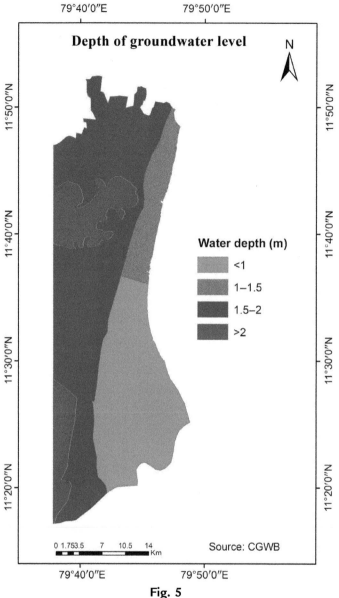

Fig. 5
Depth of groundwater level.

maximum vulnerability. Accordingly the ratings given are: 10 for closest to the coast, and 2.5 for regions away from the coast (Fig. 6). The two estuaries that bring in saline water far inland are also given consideration, and accordingly, appropriate ratings are given to the regions adjoining these estuarine tracts.

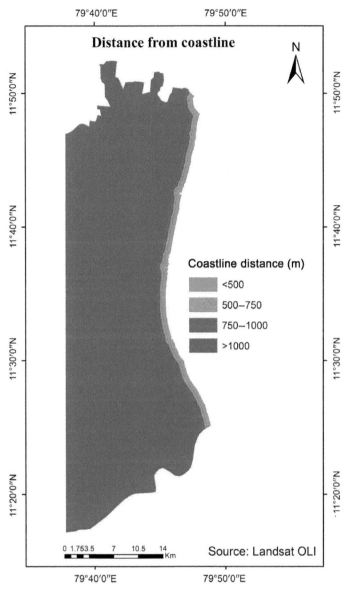

Fig. 6
Distance from the Shore/River.

4.5 Impact of Existing Status of Saltwater Intrusion (I)

A balanced hydraulic gradient will be established between the fresh water and salt water, if the region remains more stable and unstressed. Thus, the extraction of groundwater resources in the coastal areas may be considered to promote SWI into the land. A common measure to identify the incursion of salt water into the groundwater is the ratio of Cl to that of the HCO_3+CO_3 (Revelle, 1941). The formula for calculating sea water intrusion is shown in Eq. (3).

$$I = \frac{Cl^-}{[CO_3^{2-}+HCO_3^-]} \qquad (3)$$

From the study, it is found that the sea water intrusion is greater at the regions of confluences of rivers (Fig. 7).

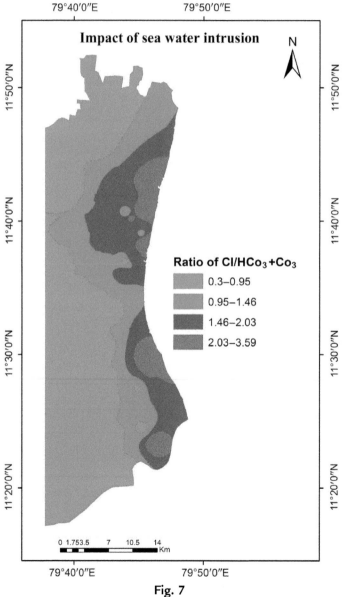

Fig. 7
Impact of existing status of saltwater intrusion.

4.6 Thickness of the Aquifer (T)

The thickness of the aquifer is directly proportional to salt water intrusion potential in the region. An aquifer thickness of 20–25 m is observed near the coast where the vulnerability ranking is the highest. Whereas the interior portion has less than 5 m of aquifer thickness, and the lowest vulnerability ranking of 2.5 (Fig. 8).

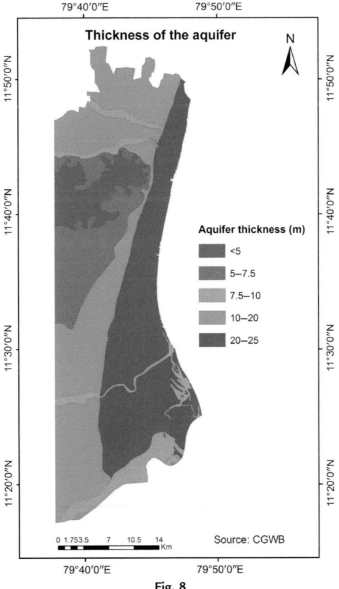

Fig. 8
Thickness of the aquifer.

5 Discussion

The GALDIT indices were calculated for the study area in the month of December, 2006, by using linear rank and weights methods. The ranking of the parameters ranges from 2.5 to 10 in a 10-point scale with 2.5 being the least importance and 10 being the most (Table 2). The high importance shows greater vulnerability to the salt water intrusion. From the overlay analysis and reclassification, the study area is classified in to four categories of vulnerability i.e., very low, low, moderate and high vulnerable to saltwater intrusion.

From Fig. 9, it is perceivable that 34.46% of the study area (inclusive of Thondamanatham, Kothandaramapuram, Anukkambattu, and Allapakkam) covered by Cuddalore sandstone formations and fertile agriculture land falls under moderate vulnerability to SWI. The area on the northern side, namely, Thiruvandipuram, Pudhur, and Valudambattu, are less vulnerable to SWI. They cover, at the western and northern part of the study area, about 33.54 %. In the western and northern part of the study area, nearly 22.03% of the area is considered to have very low vulnerability to SWI. The high, elevated highland with the laterite soils region is attributed to very low GALDIT values.

The regions such as Samiyaar Pettai, Pudukkuppam, and Cuddalore old town are found to be highly vulnerable to SWI due to the presence of industries and developmental activities that substantially increase vulnerability, as shown in Fig. 9. The ratio of Cl/HCO_3 of the groundwater also evidenced the study area to have been worst affected due to SWI. The hydraulic conductivity is very high in this region due to high permeability of soil, except for the old harbor and Gadilam River mouth areas. These places fall into the high vulnerability category due to the tidal water influx into the estuarine tracts of the river. Overall, 9.97% of the areas studied are highly vulnerable to SWI. Furthermore, the highly industrialized and populated nature of these regions suggests an overdraft of groundwater from the aquifers that could have attenuated SWI. Similar inferences of intense SWI in regions of industrialization and urbanization are also reported by Steyl and Dennis (2010) and Trabelsi et al. (2016) based on application of the GALDIT model.

Table 2 Vulnerability Classes of GALDIT index

Index Range	Vulnerability Classes
≥7.5	Very high vulnerability
5–7.5	High vulnerability
5–2.5	Moderate vulnerability
<2	Low vulnerability

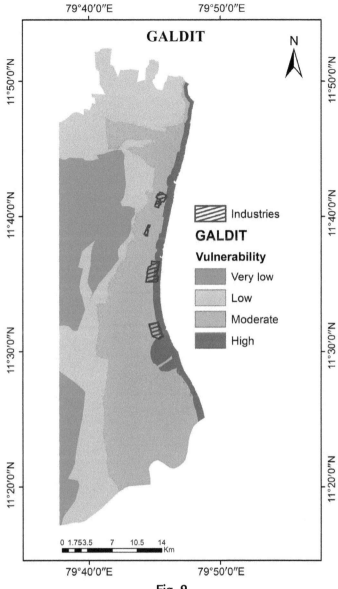

Fig. 9
GALDIT index of Cuddalore.

6 Conclusions

- The present study area is in one of the regions where industrial development, rapid urbanization, and high population growth is being witnessed, along with agriculture and tourism acquiring immense importance. Attendant with the intensive agricultural,

industrial, and domestic usage is the heavy dependence and drafting of groundwater. Natural and anthropogenic factors have contributed toward SWI, as demonstrated explicitly in the GALDIT model output.
- The study identified 9.97% of the study area to be highly vulnerable to SWI, whereas 34.46% the area is moderately vulnerable. Only a third of the study area shows low vulnerability. This is alarming and needs to be addressed.
- The study had demonstrated that the novel method of combining remote sensing, GIS, and the GALDIT model can delineate areas of higher vulnerability better than a simple geochemical assessment of groundwater against SWI.

References

Albinet, M., Margat, J., 1970. Cartographie de la vulnerabilit'e ala pollution des nappes d'eau souterraine (Contamination vulnerability mapping of groundwater). Bulletin de la Bureau de Recherches G'eologiques et Mini'eres 2nd serves 3, 13–22.

Aller, L., Bennett, T., Lehr, J., Petty, R., Hackett, G., 1987. DRASTIC: a standardized system for evaluating groundwater pollution potential using hydrogeologic settings. U.S. Environmental Protection Agency Report 600/2-87/035, 622.

CGWB (Central Ground Water Board), 2009. Ministry of Water Resources. Aquifer Systems of Tamilnadu and Puducherry 2009–10, p. 19.

CGWB, 2015. Pilot Project Report on Aquifer Mapping in Lower Vellar Watershed, Cuddalore District, Tamil Nadu. CGWB, South Eastern Coastal Region, Chennai.

Chachadi, A.G., Lobo Ferreira, J.P., Noronha, L., Choudri, B.S., 2002. Assessing the impact of sea-level rise on salt water intrusion in coastal aquifers using GALDIT model. Coastin 7, 27–32.

Chachadi, A.G., 2005. Seawater intrusion mapping using modified Galdit indicator model—case study in Goa. Jalvigyan Sameek. 20, 29–45.

Chachadi, A.G., Lobo Ferreira, J.P., 2001. Seawater intrusion vulnerability mapping of aquifers using the GALDIT method. Coastin 4, 7–9.

Chachadi, A.G., Lobo-Ferreira, J.P., 2001. Sea water intrusion vulnerability mapping of aquifers using GALDIT method.Process workshop on modelling in hydrogeology, Anna University, Chennaipp. 143–156.

Chachadi, A.G., Lobo-Ferreira, J.P., 2005. Assessing aquifer vulnerability to sea-water intrusion using GALDIT method: part 2—GALDIT indicators description.Fourth Inter-Celtic Colloquium on Hydrogeology and Management of Water Resources, Portugal, 11–14 July.

Civita, M., 1994. Le carte della vulnerabilità degli acquiferi allìnquinamento: Teoria e pratica [Contamination Vulnerability Mapping of the Aquifer: Theory and Practice]. Quaderni di Tecniche di Protezione Ambientale, Pitagora.

Di Sipio, E., Galgaro, A., Zuppi, G.M., 2006. New geophysical knowledge of groundwater systems in Venice estuarine environment. Estuar. Coast. Shelf Sci. 66 (1–2), 6–12.

Doerfliger, N., Jeannin, P., Zwahlen, F., 1999. Water vulnerability assessment in karst environments: a new method of defining protection areas using a multiattribute approach and GIS tools (EPIK method). Environ. Geol. 39 (2), 165–176.

El-Raey, M., Frihy, O., Nasr, S.M., Dewidar, K.H., 1999. Vulnerability Assessment of Sea Level Rise Over Port Said Governorate. Egypt. J. Environ. Monitor. Assess. 56, 113–128.

Foster, S., 1987. Fundamental concept in aquifer vulnerability pollution risk and protection strategy. In: Proceedings of the International Conference on the Vulnerability of soil and groundwater to pollution Nordwijk, The Netherlands, April 1987.

Gorgij, A.D., Moghaddam, A.A., 2016. Vulnerability assessment of saltwater intrusion using simplified GAPDIT method: a case study of Azarshahr Plain Aquifer, East Azerbaijan, Iran, Arab. J. Geosci. 9, 106.

Jayasekera, D.L., Kaluarachchi, J.J., Villholth, K.G., 2011. Groundwater stress and vulnerability in rural coastal aquifers under competing demands: a case study from Sri Lanka. Environ. Monit. Assess. https://doi.org/10.1007/s10661-010-1563-8.

Kouz, T., CherkaouiDekkaki, H., Mansour, S., HassaniZerrouk, M., Mourabit, T., 2017. Application of GALDIT index to assess the intrinsic vulnerability of coastal aquifer to seawater intrusion case of the Ghiss-Nekor Aquifer (North East of Morocco). Environ. Earth Sci. 169–177. https://doi.org/10.1007/978-3-319-69356-9_20.

Kummu, M., De Moel, H., Salvucci, G., Viviroli, D., Ward, P.J., Varis, O., 2016. Over the hills and further away from coast: global geospatial patterns of human and environment over the 20th–21st centuries. Environ. Res. Lett. 11 (3), 034010.

Kura, Y., Joffre, O., Laplante, B., Sengvilaykham, B., 2014. Redistribution of water use and benefits among hydropower affected communities in Lao PDR. Water Resour. Rural Dev. 4, 67–84.

Lobo-Ferreira, J.P., Cabral, M., 1991. Proposal for an operational definition of vulnerability for the European community's atlas of groundwater resources. In: Meeting of the European Institute for Water (Groundwater Work Group, Brussels, February 1991).

Lobo-Ferreira, J.P., Chachadi, A.G., 2005. Assessing aquifer vulnerability to saltwater intrusion using GALDIT method: part 1—application to the portuguese aquifer to Monte Gordo. In: Proceedings of 4th Interseltic Colloquium on Hydrology and Management of Water Resources, Portugal, pp. 1–12.

Mahesha, A., Vyshali, Lathashri, U., Ramesh, H., 2012. Parameter estimation and vulnerability assessment of coastal unconfined aquifer to saltwater intrusion. J. Hydrol. Eng. 17, 933–943.

Marfai, M.A., King, L., 2008. Potential vulnerability implications of coastal inundation due to sea-level rise for the coastal zone of Semarang city, Indonesia. Environ. Geol. 54, 1235–1245.

Melloul, A.J., Goldenberg, L.C., 1997. Monitoring of seawater intrusion in coastal aquifers: basics and local concerns. J. Environ. Manage. 1997. https://doi.org/10.1006/jema.1997.0136.

Moghaddam, H.K., Jafari, F., Javadi, S., 2015. Evaluation vulnerability of coastal aquifer via GALDIT model and comparison with DRASTIC index using quality parameters. Hydrol. Sci. J. https://doi.org/10.1080/02626667.2015.1080827.

Neumann, B., Vafeidis, A.T., Zimmermann, J., Nicholls, R.J., 2015. Future coastal population growth and exposure to sea-level rise and coastal flooding—a global assessment. PLoS ONE 10 (3), e0118571. https://doi.org/10.1371/journal.pone.0118571.

Polemio, M., Limoni, P.P., Mitolo, D., Virga, R., 2006. Il degrado qualitativo delle acque sotterranee pugliesi. Giorn. Geol. Appl. 3, 25–31.

Pulido-Leboeuf, P., 2004. Seawater intrusion and associated processes in a small complex aquifer (Castell de Ferro, Spain). Appl. Geochem. 19, 17–27. https://doi.org/10.1016/j.apgeochem.2004.02.004.

Revelle, R., 1941. Criteria for recognition of sea water in groundwaters. Trans. Am. Geophys. Union 22, 593–597.

Sodde, M., Barrocu, G., 2006. Seawater intrusion and arsenic contamination in the alluvial plain of Quirra and Flumini Pisale rivers, South-Eastern Sardinia, 2006. Atti del 1st SWIM-SWICA Conference—Cagliari.

Sophiya, M.S., Syed, T.H., 2013. Assessment of vulnerability to seawater intrusion and potential remediation measures for coastal aquifers: a case study from eastern India. Environ. Earth Sci. 70 (3), 1197–1209.

Stempvoort, D.V., Ewert, L., Wassenaar, L., 1993. Aquifers vulnerability index: a GIS-compatible method for groundwater vulnerability mapping. Can. Water Resour. J. 18, 25–37.

Steyl, G., Dennis, I., 2010. Review of coastal-area aquifers in Africa. Hydrogeol. J. 18, 217–225.

Sundaram, L.K.V., Dinesh, V., Ravikumar, G., 2008. Vulnerability assessment of seawater intrusion and effect of artificial recharge in Pondicherry coastal region using GIS. Ind. J. Sci. Technol. 1 (7), 1–7.

Trabelsi, N., Triki, I., Hentati, I., Zairi, M., 2016. Aquifer vulnerability and seawater intrusion risk using GALDIT, GQISWI and GIS: case of a coastal aquifer in Tunisia. Environ. Earth Sci. 75, 669.

Werner, A.D., Simmons, C.T., 2009. Impact of sea-level rise on sea water intrusion in coastal aquifers. Ground Water 47 (2), 197–204.

Werner, A.D., Ward, J.D., Morgan, L.K., Simmons, C.T., Robinson, N.I., Teubner, M.D., 2012. Vulnerability indicators of seawater intrusion. Ground Water 50, 48–58.

Yechieli, Y., Abelson, M., Bein, A., Crouvi, O., Shtivelman, V., 2006. Sinkhole "swarms" along the Dead Sea coast: Reflection of disturbance of lake and adjacent groundwater systems. Geol. Soc. Am. Bull. 118, 1075–1087.

Further Reading

Najib, S., Grozavu, A., Mehd, K., Breaban, I.G., Guessir, H., Boutayeb, K., 2012. Application of the method GALDIT for the cartography of groundwaters vulnerability: aquifer of Chaouia coast (Morocco). Sci. Annals Alexandru Ioan Cuza Univ. Iasi 63 (3), 77.

Recinos, N., Kallioras, A., Pliakas, F., Schuth, C., et al., 2014a. Application of GALDIT index to assess the intrinsic vulnerability to seawater intrusion of coastal granular aquifers. Environ. Earth Sci. 59 (72), 1866–6299.

Recinos, N., Kallioras, A., Pliakas, F., Schuth, C., 2014b. Application of GALDIT index to assess the intrinsic vulnerability to sea water intrusion of coastal granular aquifers. Environ. Earth Sci. 73, 1017–1032.

Reddy, A., Saibaba, B., Sudarshan, G., 2012. Hydrogeo chemical characterization of contaminated groundwater in Patancheru industrial area, southern India. Environ. Monit. Assess. 184, 3557–3576.

Sherif, M.M., Hamza, K.I., 2001. Mitigation of seawater intrusion by pumping brackish water. Transp. Porous Media 43 (1), 29–44.

Singh, K.P., Malik, A., Singh, V.K., Mohan, D., Sinha, S., 2005. Chemometric analysis of groundwater quality data of alluvial aquifer of Gangetic plain, North India. Anal. Chim. Acta 550 (1–2), 82–91.

Steinich, B., Escolero, O., Marin, L.E., 1998. Salt-water intrusion and nitrate contamination in the valley of Hermosillo and EI Sahuaral coast aquifers, Sonora, Mexico. Hydrogeol. J. 6, 518–526.

Tomaszkiewicz, M., AbouNajm, M., El-Fadel, M., 2014. Development of a groundwater quality index for seawater intrusion in coastal aquifers. Environ. Model Softw. 57, 13–26.

CHAPTER 11

An Assessment of the Administrative-Legal, Physical-Natural, and Socio-Economic Subsystems of the Bay of Saint Brieuc, France: Implications for Effective Coastal Zone Management

Nerea Del Estal[*,†], Manoj Joseph Mathew[*,‡], David Menier[*], Erwan Gensac[*], Hugo Delanoe[*], Romain Le Gall[*], Thomas Joostens[*], Angélique Rouille[*], Béatrice Guillement[*], Gaëlle Lemaitre[*]

[*]Géosciences Océan Laboratory UMR CNRS 6538, University of South Brittany, Vannes cedex, France, [†]Mott MacDonald Limited, Mott MacDonald House, Croydon, United Kingdom, [‡]Institute of Oceanography and Environment, Universiti Malaysia Terengganu, Kuala Terengganu, Malaysia

1 Introduction

The coastal region is of great value from the point of view of sustenance to coastal population and ecosystem services to a variety of stakeholders that need nourishment. The littoral zone is of vital importance because it houses much of the trade and anthropogenic activities, and a large percentage of economic activities depend on them. The littoral zone houses numerous ecosystems and habitats that deteriorated or have been eliminated because of infrastructural development in these areas. Coastal areas are experiencing a population growth (demographic concentration) wherein, according to UNEP (2006), half of the world's population is living not more than 60 km away from the sea. The oceanic domain changes continually on an upward trend because of the intensification of population, and the behavior patterns of the population (Barragán Muñoz, 2010). Since the 1980s, integrated coastal zone management (ICZM) has become an objective set by the governing bodies of countries with deteriorating coasts, and it is gaining momentum all over the world. This constitutes a substantial tool to maintain the sustainability of economic development among the frameworks of natural processes and environmental quality (Isla and Lasta, 2006). In this context, integrated coastal zone management is characterized as an intrinsic tool for defining the coastal area as a combination of physical, socio-economic, and political interdependence within dynamic coastal systems (Sorensen, 1997; Pickaver et al., 2004).

The coast of the Bay of Saint Brieuc, located in the north of France, can be divided into three distinct but complementary and interdependent subsystems to elucidate the complex dynamics that occur along this bay (Fig. 1). Currently, there is an explicit inadequacy of specific projects of ICZM for the Bay of Saint Brieuc. Here we demonstrate various coastal hazards that exist in the Bay of Saint Brieuc, and propose measures to revamp and improve integrated coastal management strategies for this area.

2 The Subsystems

2.1 Administrative and Legal Subsystem

During the 1970s in France, more than half of the area near the seashore was under property development, and around 20% could be classified as overdeveloped (López, 2008). This situation led the French government and the DATAR (Délégation à l'aménagement du territoire et à l'action régionale) in 1975 to work on the law (*Loi n°75-602 du 10 juillet 1975 portant création du conservatoire de l'espace littoral et des rivages lacustres*) that aims to regulate coastal management. As a result, an institution was created with the objectives of monitoring and guaranteeing land use. Unfortunately, this law did not stop urban development, and was improved by another law in 1986 (*Loi n°86-2 du 3 janvier 1986 relative à l'aménagement, la protection et la mise envaleur du littoral*). This is the only document that has managed to slow down or contain coastal urbanization, and constitutes the basis on which coastal uses in France are currently conducted.

At the end of the 1990s, the European Union for Coastal Conservation (EUCC) conducted a study for the development of Integrated Coastal Zone Management (ICZM) in Europe Member States (EUCC, 1999) based on the Rio Earth Summit of 1992 that aimed at comprehensive management of littoral zones. The goal of the Integrated Coastal Zone Management program was to develop tools and regulatory structures to ensure and restore the balance between human activities and natural dynamics, avoiding overexploitation. Finally, in 2014, France consolidated the ICZM and proposed numerous projects involving public contribution. Specifically, for the north of France, including the Bay of Saint Brieuc, a project entitled *"La Charte des espacescôtiers Bretons,"* intended to elaborate a plan for future development and conservation. Other than this, municipal regulations and ordinances of the Bay of Saint Brieuc are closely linked to certain urban services, such as waste and pollution management, water supply and sanitation, and so forth.

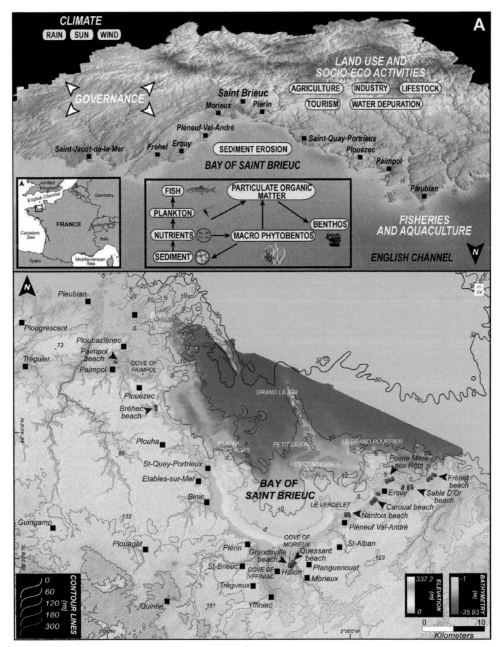

Fig. 1
(A) The three distinct, complementary and interdependent subsystems of the Bay of Saint Brieuc.
(B) Map showing the bathymetric and topographic features of the Bay of Saint Brieuc. Also shown are the sampling points for granulometric analysis conducted in this study along 10 sandy beaches.

2.1.1 Protected natural areas

The Bay of Saint Brieuc has a natural environment with areas of high ecological and scenic value in the Natural Reserve of the Bay of Saint Brieuc (Fig. 2). This area was declared a protected area in 1998. In addition to this natural reserve, along the coastal area of the Bay of Saint Brieuc, there are two Special Conservation Zones/Sites of Community Interest (SCZ/SCI) belonging to the Natura 2000 Network. There are also five Special Protection Areas for Birds (SPAB), two natural areas of ecological interest for fauna and flora (NAEIFF), as well as a coastal conservatory (Fig. 2).

2.1.2 The natural reserve of the Bay of Saint Brieuc

This natural reserve is in the Bay of Saint Brieuc, on the north side of Brittany. It is located within the cove of *Yffiniac* and the cove of *Morieux*, which extends for about ∼2900 ha of sandy foreshore. At the top of the foreshore, there are salt meadows of ∼110 ha that provide a link with the shore of the *Yffiniac* cove. The surface area of the reserve is ∼1140 ha of the 3000 ha of the intertidal zone at the bottom of the Bay of Saint Brieuc. Almost all of the reserve is located in

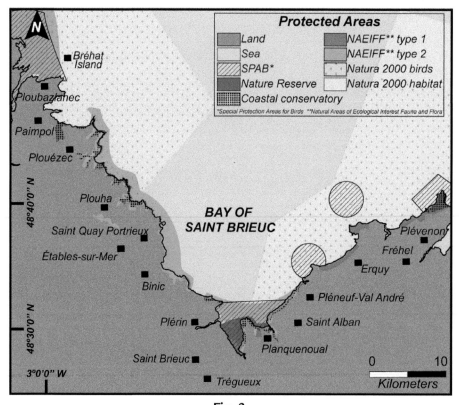

Fig. 2
Map showing the different protected areas and natural reserve of the Bay of Saint Brieuc.

the maritime public domain, in the communes of *Langreux*, *Yffiniac,* and *Hillion*, and in the municipalities of *Morieux* and *Saint Brieuc*. The terrestrial part of the reserve is in the commune of *Hillion*, where the dunes of Bon Abri are located, which constitute around ∼7 ha.

Proclamation of an area as a natural reserve could be defined as a legal protection tool on land and on the maritime public domain. Natural reserves are intended for strict preservation of fragile, rare, or endangered natural environments of high ecological and scientific value. These are classified to be natural reserves by decree, after public inquiry, if the natural environment is of importance or should be removed from any artificial interventions that are likely to degrade them. Its mission is to protect, conserving the spaces and the species present in the territory that are in danger of extinction or are threatened.

2.1.3 Coastal conservatory

One of the most important instruments of protection that France takes into account, and particularly in the study area, is the coastal conservatory. It was amended in 1975, and finally ratified in 1986. According to López (2008), the concept of ecology was included in the French law through the law of 1975. The coastal conservatory aims to carry out a policy of land use that leads to the protection of coastal areas with an aim to conserve and maintain ecological balance of natural spaces.

2.2 Physical and Natural Subsystem

2.2.1 General geography and geomorphology

On the northwest part of France, in the region of Brittany, at the southwest limit of the Breton Gulf, opens the deep Bay of Saint Brieuc (Fig. 1). The area consists of two almost linear coasts, forming a right angle from Saint Brieuc. The eastern coast with southwest/northeast orientation is hilly with some rocky headlands that do not exceed 60 m (e.g., Cap d'*Erquy* and Cap *Fréhel*). The cliffs are cut by rivers or by depressions that coincide with the major directions of basement fractures (Beigbeder, 1964). The western coast with northwest/southeast orientation is composed of cliffs that are among the highest on the Breton coast, reaching altitudes of 100 m at *Plohua*'s peak. These cliffs are made of hard rocks and are brittle (Fig. 3). Erosion slowly flushes the rocky shelf forward from the cliff (Rue, 1988). The faulted nature of the geological structures individualizes resistant islands, separated by depressed areas occupied by watercourses that reach the sea via small sandy loops.

On the coast, the foreshore domain extends mainly to the two great coves of *Yffiniac* and *Morieux*, the western strikes of *Rosaires* and *Binic* to *Saint-Quay-Portieux*, and the eastern strikes of *Pléneuf-Val-André* to *Erquy* and *Sables-d'Or-les-Pins*. The rest of the area is comprised of multiple small beaches or coves limited by rocky outcrops (Fig. 3).

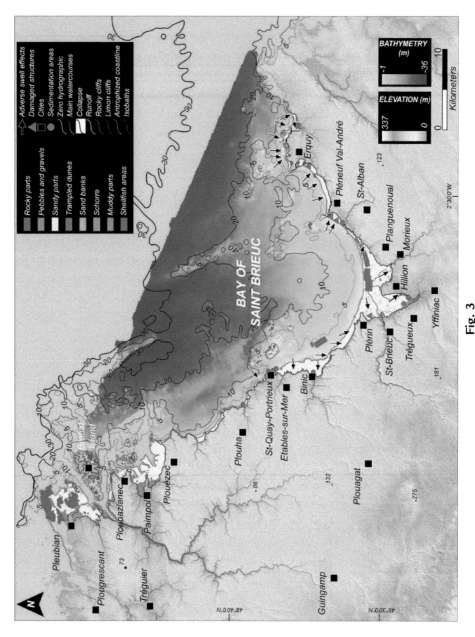

Fig. 3
Salient coastal geomorphic features of the Bay of Saint Brieuc. *From Bonnot-Courtois, C., Bousquet-Bressolier, C., 1998. Géomorphologie et vulnérabilité des rivages de la baie de Saint Brieuc. Norois 45(179), 495–506.*

The bay occupies an area of ~800 km² up to the 30 m isobaths, which is located more than three miles from the bottom of the *Yffiniac* cove (Fig. 1). It shows a very low slope of ~0.1%. Multiple reliefs of up to ~20 m height characterize the underwater morphology. These are either high rocky shoals, such as those of the *Roches de Saint-Quay*, *Plateau des Jaunes*, *Grand Léjon,* and *Grand* Pourier, or sandy banks, such as the one at *La Horaine*. These reliefs, mostly elongated in a northwest/southeast direction, are aligned parallel to the left shore of the bay, and extend along the principal headlands of the eastern shore, and delimit depressions or basins of variable sizes. The principal one being the "central depression," open to the northwest. A relatively smaller depression aligned northeast/southwest, perpendicular to the west coast, distinctly individualizes northwest of the *Roches de Saint-Quay* and off the tip of *Minard* by the alignment of reliefs of a few meters in height.

2.2.2 Geology

The shape and morphology of the Bay of Saint Brieuc is the result of the geological history of the region. The rectilinear course of the western and eastern coast is controlled by old fractures (Fig. 4) that have been reactivated many times. The Bay of Saint-Brieuc is placed in the context of the Armorican Massif "North Breton Cadomian" area. The geological history of this mountain range can be explained by the superposition of two orogenesis: The Cadian orogeny (~750 to ~520 Ma) and the Hercynian (or Variscan) orogeny, dated ~360 to ~300 Ma. This orogenesis is responsible for the origin of the faults, including two main shears cutting the territory from west to east: the north Armorican shear (NAS) and the south Armorican shear (SAS).

The Bay of Saint Brieuc is mainly constructed of rocks dating from the Cadomian orogeny (Fig. 4), and later subjected to the tectonic activity of the NAS. During the Cadian orogeny, the northern Breton domain underwent subduction (L'Homer et al., 1999) resulting in crustal shortening, especially along the NNE-SSW compressional faults (Brun, et al., 2001) This led to the formation of a rear arc basin that was partially metamorphosed thereafter, at high temperature and pressure. The end of the Cadomian activity is marked by the formation of layers of Brieurian shale and the emplacement of granites. During the Hercynian orogeny, these rocks of the northern Breton domain underwent slight deformation due to the reactivation of Cadomian collisional faults. The Triassic and Jurassic sediments encountered in other sedimentary basins (Paris Basin or Aquitaine) are therefore absent in this region. Only a few dolerite veins have survived, and these allow us to conclude a period of extension in correlation with the opening of the Atlantic Ocean. During the Cenozoic (66 Ma), rocks emplaced during the Cadian and Hercynian orogeny underwent alteration. During the Lower Eocene (hot and humid period), thick layers of laterites were formed from the alteration of the substratum. At the same time, grabens were set up by NW-SE brittle faults. Sediments (clay, sand, and limestone) subsequently filled these collapsed basins during the Oligocene and Miocene (Guillocheau et al., 1998). The west coast of the bay follows the same direction as the faults. The beginning of

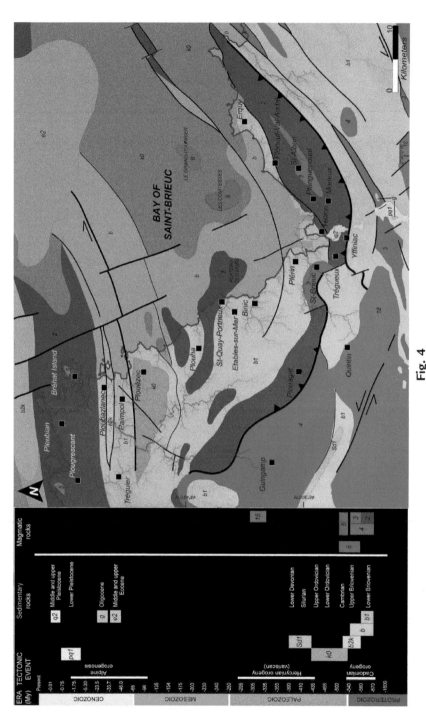

Fig. 4
General geology and structural lineaments of the Bay of Saint Brieuc.

the Quaternary marks the advent of an ice age, which resulted in the lowering of the sea level and the closure of the English Channel. This climatic condition led to the emplacement of glacial deposits, such as loess, that are observable in some areas of the Bay (Bigot and Monnier, 1987). Until the current period, there was an increase in temperature (interglacial period) that caused sea levels to rise and flood the valleys that were created during the previous ice age, and this process formed rias that are still observable today in Brittany.

2.2.3 Hydrodynamics

The tide is one of the physical phenomenon that determines the Bay of Saint Brieuc. The tidal range is semidiurnal (M2), with a tide period of 12 h and 25 min (Augris et al., 1996), leading to two tidal cycles per day. The tidal range varies between the two sides of the Bay, to the west it is 4.1 m, while at the eastern end it reaches 12.8 m (Schroëtter et al., 2015). The topography of the bottom of the Bay of Saint Brieuc, and in general the shape of the coast, has a great influence on the currents (Fig. 5). Thus, the recorded maximum is in front of the caps (1.5 m/s in front of *Cap Fréhel*), while the bottom of the bay is an area of weak tidal currents, where the lowest tidal currents are, in particular in Erquy, Dahouët and Pordic (less than 0.2 m/s) (Garreau, 1993). The area where the bay is located is globally affected by strong semidiurnal tidal currents. The resulting wave, propagating on a west-east axis to cross the English Channel, causes an increase

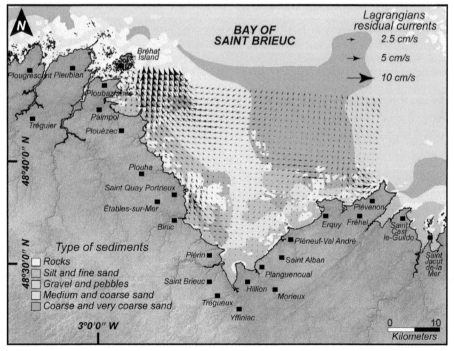

Fig. 5
Map of the seafloor sedimentology and prevailing tidal currents.

in the effect of the tide because of the orientation of the Normandy-Breton Gulf, where the bay of Saint Brieuc is located. The strong agitation of the north-west does not affect the west coast. This region seems to be protected also from the north-east and north swells by the rocks of *Saint-Quay* and the *Hors* Plateau. It is devoid of large sedimentary transits. In contrast, currents of tide are strengthened in the channel of Saint-Quay, especially through the new port immediately to the south that has relatively calm conditions.

Although the middle of the Bay is relatively stable because it is well protected from swells (NW) by the *Pointe du Roselier*, it remains sensitive to the action of tidal currents that contribute to the accretion of sand at the areas of *Yffiniac* and *Morieux*. Their tidal flats have steep slopes and are exposed during half of the tidal cycle. Channels that cut them wander and migrate, in particular that of the *Gouessant* River, which has significantly migrated to the west. The main factor of erosion is not the action of the sea, but the saturation of rainwater in the Quaternary deposits that leads to solifluction phenomena. The concentrated rainwater drains into the bottoms of valleys and destabilizes the cliffs. These phenomena are particularly visible on the coastline of the *Hillion* peninsula, where heavy rains in 1987-1988 caused significant damage. The fragility of these areas sometimes doubles as an anthropogenic degradation, which affects more particularly the dune environments. The east coast, due to the nature of the shoreline and its orientation in relation to swell domination, could evoke significant damage by wave attack or the littoral drift, and this is reflected in the beach of *ValAndré*. Besides, the promontory of *Erquy* is subjected to very strong agitation, and the coastline recedes at a rate of $\sim$1 m/year. In addition, the solifluction destabilizes the silt cliffs.

2.2.4 Climate

The Bay of Saint Brieuc belongs to the Breton region and has a strong oceanic influence, with a mild climate (average temperature of $\sim$11°C), with low thermal amplitudes and few extreme events. Because of the genesis of nitrogen fluxes, the controlling factor of coastal eutrophication, soil mineralization occurs with little limitation throughout the year. Rainfall variations influence the leaching of nitrogen, and once mineralized, they are very important in this zone that straddles the western and eastern sides of Brittany. Rainfall in the upstream parts of *Gouët* in the west is recorded to be $\sim$1040 mm/year, as compared with $\lesssim$640 mm/year on the eastern coastal fringe (Fig. 6).

The prevailing winds are mainly from the west, and secondarily from the east-northeast. The seasonal distribution of the winds is such that the frequency of the strong westerly winds is distributed during the entire year, and follows the order: winter, autumn, spring, summer. For the eastern sector, the seasons are arranged differently, that is, winter, spring, autumn, summer. Wind gusts (speeds >25 m/s or $\sim$90 km/h) from the western sector occur mainly in winter and autumn, while those in the eastern sector occur in winter and spring. Due to the configuration of the bay, meridian (north-south) winds are strengthened at the expense of west and east winds.

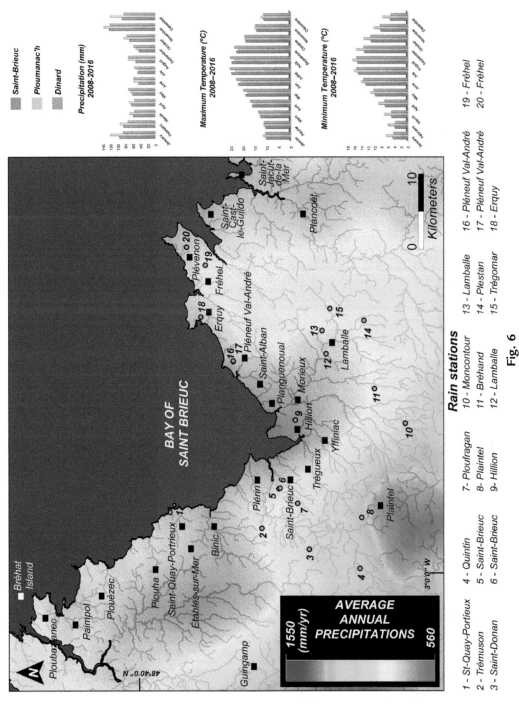

Fig. 6

Mean annual precipitation, and maximum and minimum temperatures. Also shown are rain gauge stations in and around the Bay.

Wind circulation brings changes, sometimes significantly, locally at sea and rivers, but does not significantly change the dynamics induced by the tide.

2.2.5 Sedimentology

Marine region

The oceanographic campaigns conducted in 1960 by a research group from the Laboratoire de Géomorphologie de Dinard, defined a majority of seabed sediments of the Bay of Saint Brieuc into five categories (Fig. 5).

- Mud: particle size <40 μm. These fine sediments are very poorly represented in the bay, with the highest percentages of fraction <40 μm constituting 30%. Two richer areas stand out: one in the west, close to the coast in front of Plouha, the other in the center of the bay, to the southeast of the Plateau de Hors, lying in a north-south direction between isobaths 10 and 20 m.
- Silt and fine sand: particle size between 40 and 200 μm. These extend near the coast in the southern part of the bay. On the western side, fine sands are distributed from Minard Point to the tip of the Bec de Vir, and show a coarsening trend toward Roches de Saint Quay. To the southeast of Saint Quay, the silt and fine sand cover the entire bottom of the bay, from the Yffiniac and Morieux coves to the Erquy point.
- Medium and coarse sand: particle size between 200 μm to 1 mm. This class is largely dominant in the Bay. It succeeds fine sands offshore, and in some areas near rocky shoals, medium and coarse sands are exclusive. In addition, a broad band of medium and coarse sands extend in a northwesterly/southeast direction, between the point of Minard and the Comtesses, as well as east of Petit-Léjon and north of the plateau of Great Pourrier.
- Coarse and very coarse sand: particle size between 1 and 5 mm. The coarse sands occupy all the northern parts of the bay, as well as around the rocky areas. The coarse sands are located near the coast between the tip of Plouézec and the tip of Minard, and north of the Rocks of Saint Quay. A vast zone essentially characterized by these sediment size classes extends east of Grand-Léjon. To the east, along the coast of Cape Fréhel, coarse sands become the majority, even below the 20 m isobath.
- Gravel and pebbles: particle size >5 mm. These very coarse sediments are observed only in the eastern part of the bay, in Cape Frehel. In the northern part of the bay, the gravel shows a gradual transition to the west, with very coarse sands up to the Grand Pourrier and the Comtesses. To the west, the gravel is completely absent, apart from a few sporadic parts to the north of Gran Léjon and the channels of Saint Quay and Erquy.

Coastal region

Along the coast, the Bay of Saint Brieuc is mainly composed of rocky, fine-to-coarse sand and few pocket beaches. We analyzed ten sandy beaches along the Bay of Saint Brieuc (Fig. 1) to determine granulometric distribution of sediment along the coastline. Three transects were chosen based on the lateral extent of the beach, and three sediment samples were collected at the upper, middle, and lower parts of the intertidal zone along each transect (Fig. 1). In a few cases, we sampled along one transect in the middle of the intertidal zone due to the smaller surface area of the beach. Grain size analysis was carried out to construct graphical plots of grain size distribution. After visual estimation, the sizes of sediment particles were measured using a set of sieves according to the Folk and Ward classification (Folk and Ward, 1957). The samples were analysed using a vibrating sieve machine at the Laboratoire Géoscience Océan at Université Bretagne Sud.

Our results indicated that in Bréhec, Nantois, Caroual, Pointe de la Mare aux Rêts, and Sable D'Orbeaches, much of the sediment is fine sand with a very minor percentage of medium sand (Figs. 7–11). Grandeville beach shows coarse to fine sand in the upper part of the profile (Fig. 12), and in the middle and lower parts of the beach, a trend of fine sand is observed. Quessant beach consists of fine sand (Fig. 13) along the analyzed transects. The Erquy beach is mainly composed of medium sand with varying percentages of fine and coarse sand along the transect (Fig. 14). In the upper part of the intertidal zone of the Fréhel beach, coarse to medium sands are observed, with a small percentage of fine sand, and in the middle part, a bulk of the sediment was composed of medium to fine sand with coarse sand, and in the lower part, the composition shows the occurrence of fine to medium sand (Fig. 15).

2.3 Socio-Economic Subsystem

Between 1962 and 2009, the Saint Brieuc area gained nearly 68,000 inhabitants (Fig. 16), which corresponds to an overall population growth of around 33%. This population increase was very largely fuelled by the population growth of "briochine" agglomeration (*Saint Brieuc* commune). From 1975 to 1999, the share of population of the agglomeration within the municipality of Saint Brieuc was stable at about 59%. The 2009 census shows a decrease of ~2% to represent 57% of the population, and 20% in the department of *Côtesd'Armor*. From 1990 to 2009, the population of Saint Brieuc increased by 30,320 inhabitants, which accounted an increase of 14.78%. These population trends are comparable to that of Brittany (+14.68%) (INSEE, 2009).

Beyond urban centers of the coastal municipalities being densified, it is the extension of urbanization that poses problems. The conditions of urbanization of the villages and coastal hamlets also come with the risk of creating continuous urbanized fringes on the sea front and a habitat on the shoreline. Conflicts of land use emerge between inhabitants of coastal towns out of concern about their quality of life and production.

Brittany's coastal area is currently used for many activities (residential, productive, or recreational) (Fig. 17). The concentration of activities on the same space has consequences in terms of artificialization of the soil, difficulty in accessing certain sites, and traffic congestion.

264 Chapter 11

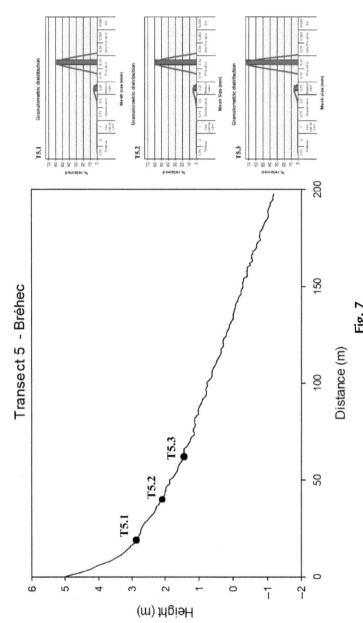

Fig. 7

Graphical cross-section of the Bréhec beach. Also shown are the sampling points and granulometric distribution in the upper, middle, and lower parts of the transect.

Implications for Effective Coastal Zone Management 265

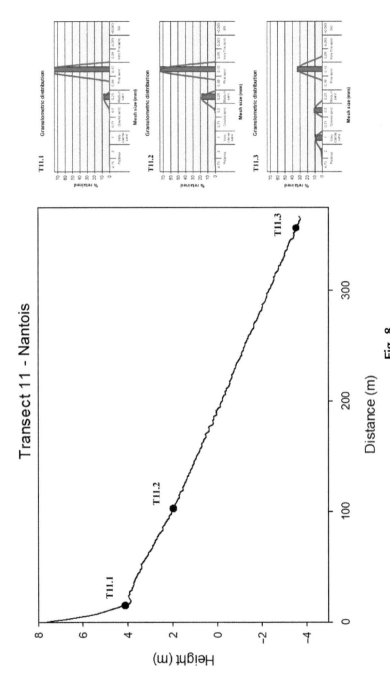

Fig. 8

Graphical cross-section of the Nantois beach. Also shown are the sampling points and granulometric distribution in the upper, middle, and lower parts of the transect.

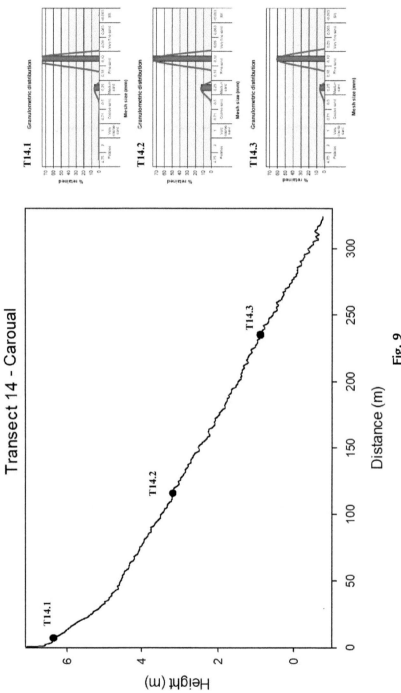

Fig. 9

Graphical cross-section of the Caroual beach. Also shown are the sampling points and granulometric distribution in the upper, middle, and lower parts of the transect.

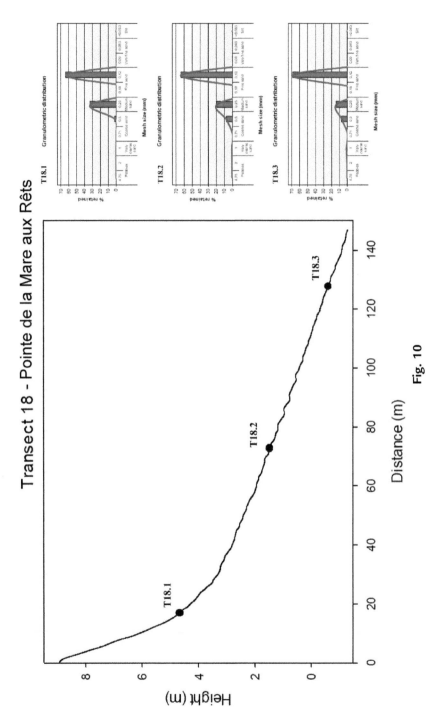

Fig. 10

Graphical cross-section of the Pointe de la Mare aux Rêts beach. Also shown are the sampling points and granulometric distribution in the upper, middle, and lower parts of the transect.

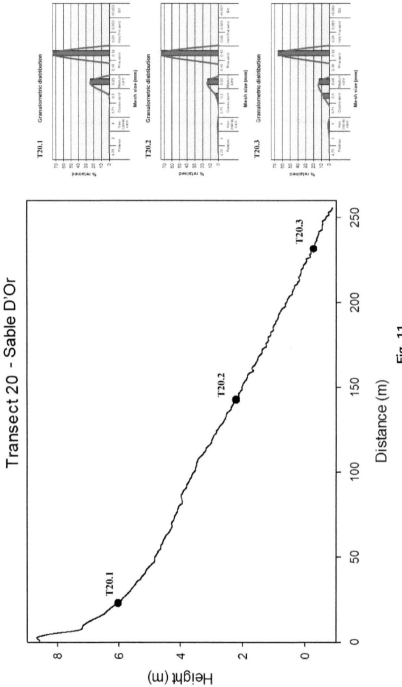

Fig. 11
Graphical cross-section of the Sable D'Or beach. Also shown are the sampling points and granulometric distribution in the upper, middle, and lower parts of the transect.

Implications for Effective Coastal Zone Management 269

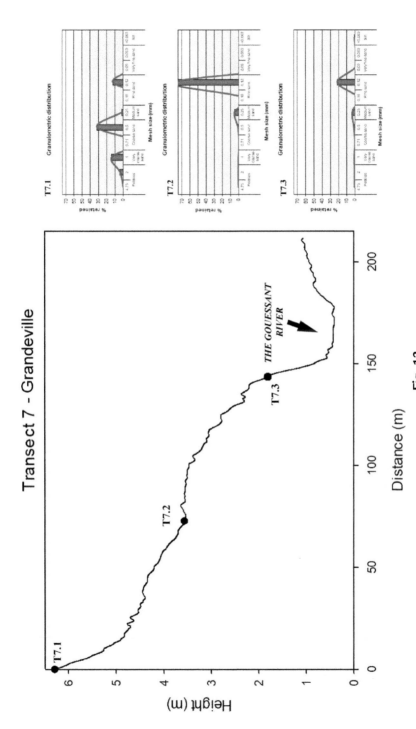

Fig. 12

Graphical cross-section of the Grandeville beach. Also shown are the sampling points and granulometric distribution in the upper, middle, and lower parts of the transect.

270 Chapter 11

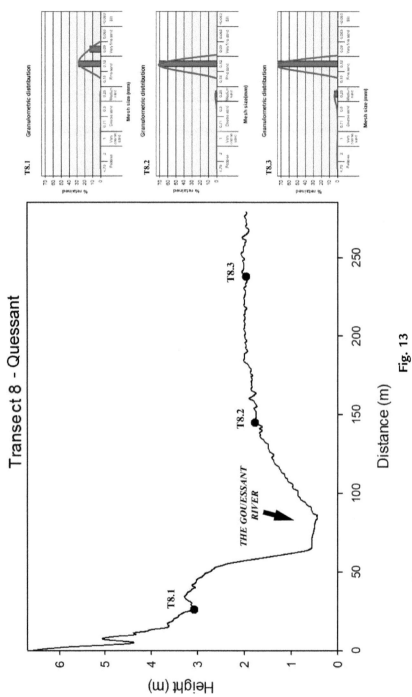

Fig. 13

Graphical cross-section of the Quessant beach. Also shown are the sampling points and granulometric distribution in the upper, middle, and lower parts of the transect.

Implications for Effective Coastal Zone Management 271

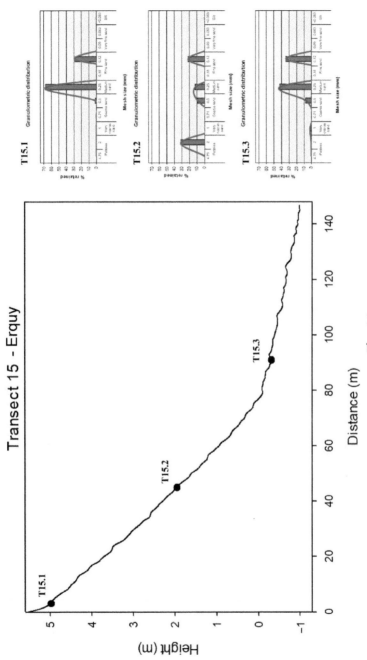

Fig. 14

Graphical cross-section of the Erquy beach. Also shown are the sampling points and granulometric distribution in the upper, middle, and lower parts of the transect.

272 Chapter 11

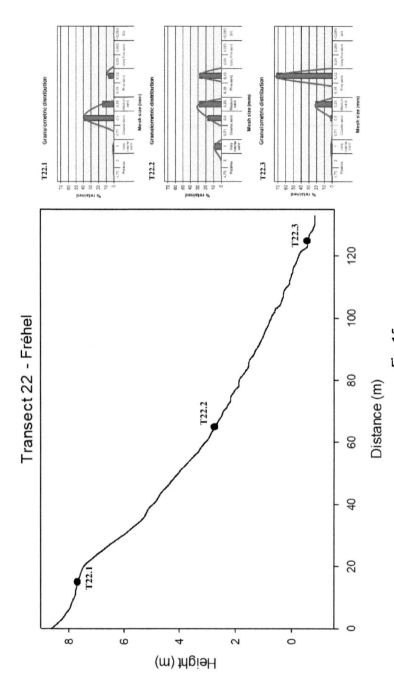

Fig. 15

Graphical cross-section of the Fréhel beach. Also shown are the sampling points and granulometric distribution in the upper, middle, and lower parts of the transect.

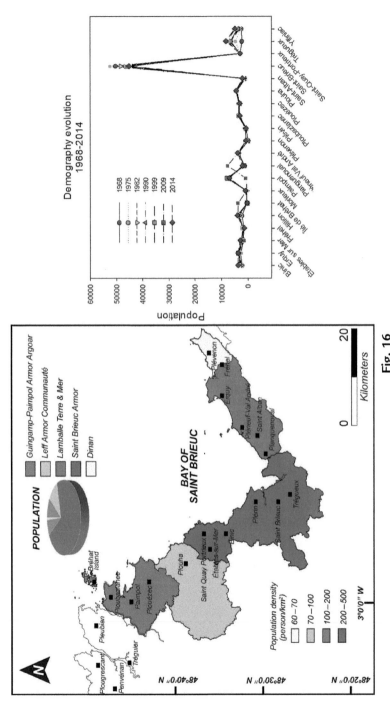

Fig. 16
The population distribution, density, and evolution of the demography in the Bay of Saint Brieuc.

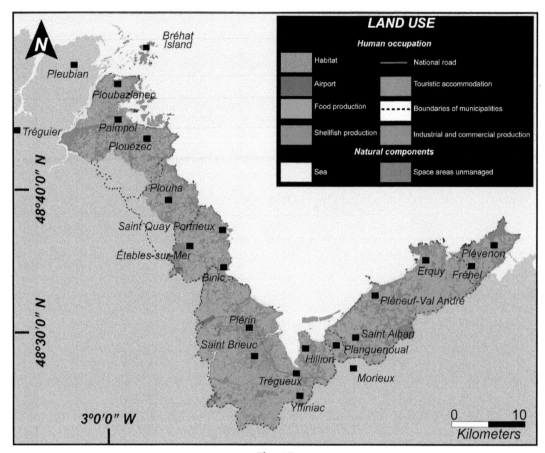

Fig. 17
Map showing land use in the Bay of Saint Brieuc.

Indeed, it presents a great diversity of activities: shellfish farming (mussel farming), professional fishing, tourism, commercial shipping activity, yachting and boating, fishing on foot, marine renewable energies, and so forth. Amalgamation of these varied activities within common territory sometimes creates socio-economic conflicts.

The Bay of Saint Brieuc is an area conducive to the breeding and natural development of resources. Mussel farming in *L'Anse de Morieux* accounts for around 10% of the national production, and oyster farmers in *Paimpol* are the largest producers in North Brittany, and scallop fishery is the main activity in this zone. The area of the Bay of Saint Brieuc is strongly oriented toward agriculture. Farmland occupies more than 60% of the territory's surface, and the main agricultural productions are milk, pig, poultry, beef, cereals, and vegetables. However, agriculture has been declining for several decades due to a sharp reduction in workforce. As it stands, the "Breton" agricultural model based on a production mass does not create jobs anymore (Pays de Saint Brieuc, 2008).

Between a dense urban area (*Saint-Brieuc, Langueux, Trégueux,* and *Yffiniac*) and agricultural sectors practicing the intensive breeding of pork and poultry, the Bay of Saint Brieuc is subjected to pollutants of domestic, agricultural, and industrial origins. The sludge is carried by rivers to the coast, and it accumulates in this region. The poor quality of the water is a major environmental problem in the bay of Saint Brieuc. The development of human activities at the seaside (urbanization, nautical activities, tourist visits, etc.) has a negative impact on the coastal environment and on the foreshore of the bay. Trampling, caused by overfrequenting of capes and various other parts of the bay, leads to the degradation of natural environments and soil erosion.

3 Conclusions

The coastal zone supports diverse activities and interactions between ecological and socio-economic factors, and holds great importance in human development. The interactions among diverse stakeholders and land uses, especially when enforced on a "mutually exclusive" basis, often result in unsustainable practices, that in turn are detrimental to the sustenance of a healthy coastal zone. In order to obviate these, and to establish a comprehensive management framework, the following generalized suggestions are made.

- Planning and regulating the establishment of industrial corridors with adequate provisions to minimize and dispose of hazardous wastes.
- Identifying and regulating coastal defense systems in resonance land use and environmental parameters.
- Reducing the urbanization effects that contribute to degradation of coastal dunes and coastal erosion (Schnack, 2004).
- Integration of coastal zone management, which is a dynamic and continuous process, in which decisions are made toward the sustainable use, development, and conservation of the littoral zone and its resources. In this regard, strategic planning by individuals and organizations with a common goal for nourishment of the entire bay and adjoining regions should be formulated and implemented.

References

Augris, C., Hamon, D., et al., 1996. Atlas thématique de l'environnement marin en baie de Saint-Brieuc (Côtes-d'Armor). Ifremer. https://archimer.ifremer.fr/doc/00031/14246/.

Barragán Muñoz, J.M., 2010. Manejo Costero Integrado y Política Pública en Iberoamérica: Un diagnóstico. Necesidad de Cambio. Red IBERMAR (CYTED), Cádiz. 380 p.

Beigbeder, Y., 1964. Contribution à l'étude géomorphologique et sédimentologique de la partie orientale de la baie de Saint Brieuc. Thèse Doct. Spéc. Géographie. Univ. Rennes, 342 p.

Bigot, B., Monnier, J.L., 1987. Stratigraphie et sédimentologie des lœss récents du Nord de la Bretagne. Données nouvelles d'après l'étude des coupes de Sables-D'or-les-Pins et de Port-Lazo (Côtes-du-Nord, France). In: Bull. AFEQ, vol 24, n° 1, 1987 pp. 27-36. DOI: https://doi.org/10.3406/quate.1987.1829.

Brun, J.P., Guennoc, P., Truffert, C., Vairon, J., The ARMOR Working Group of the GeoFrance3D Program, 2001. Cadomian tectonics in northern Brittany: a contribution of 3D crustal scale modelling. In: Ledru, P. (Ed.), The Cadomian Crust of Brittany (France): 3D Imagery From Multisource Data (GéoFrance 3D).In: Tectonophysics, vol. 331. pp. 229–246.

EUCC, 1999. Progress of ICZM Development in European Countries: A Pilot Study.

Folk, R.L., Ward, W.C., 1957. Brazon River bar: a study in the significance of grain size parameters. J. Sediment. Res. 27 (1), 3–26.

Garreau, P., 1993. Conditions hydrodynamiques sur la côte Nord-Bretagne. Rapport IFREMER-DEL/93-12.

Guillocheau, F., Bonnet, S., Bourquin, S., Dabard, M.P., Outin, J.M., Thomas, E., 1998. Mise en évidence d'un réseau de paléo vallées ennoyées (paléo rias) dans le Massif Armoricain: une nouvelle interprétation des sables pliocènes armoricains. C.R. Acad. Sci 327, 237–243.

INSEE, 2009. Bretagne. Analyse économique du département des Côtes-d'Armor, Juillet 2008.

Isla, F.I., Lasta, C., 2006. Isla, F.I., Lasta, C. (Eds.), Manual de manejo costero para la provincia de Buenos Aires. 1st ed. Universidad Nacional de Mar de Plata, Mar de Plata. 280 p.

L'Homer, A., Courbouleix, S., Beurrier, M., Bonnot-Courtois, C., Caline, B., Ehrhold, A., Lautridou, J.P., Le Rhun, J., Siméon, Y., Thomas, Y., Villey, M., 1999. Notice explicative, Carte géol. France (1/500000), feuille Baie du Mont-Saint-Michel, 208.

López, E., 2008. Le Conservatoire du littoral: pourquoi et pourquoi faire? Edit. Conservatoire de l'espace littoral et des rivages lacustres, 15 p.

Pays de Saint Brieuc, 2008. Schéma de cohérence territoriale du Pays de Saint Brieuc. Diagnostics du territoire. Syndicat Mixte du Pays de Saint Brieuc, 260 p.

Pickaver, A.H., Gilbert, C., Breton, F., 2004. An indicator set to measure the progress in the implementation of integrated coastal zone management in Europe. Ocean Coast. Manag. 47, 449–462.

Rue, O., 1988. Sédimentologie et morphogenèse des rivages et des fonds de la baie de Saint Brieuc. Thèse Dr. Université. Univ. Paris XI, Orsay, 254 p.

Schnack, E.J., 2004. Gestión Integrada de la Zona Costera. Comisión de Investigaciones Científicas de la Provincia de Buenos Aires, Facultad de Ciencias Naturales y Museo, Universidad Nacional de La Plata.Primer Taller de Manejo Costero Integrado "Hacia un Plan de Costas Bonaerense". San Clemente, Partido de La Costa, Provincia de Buenos Aires 91 p.

Schroëtter J.M., Blaise, E., Debert, V., et al., 2015. Atlas des aléas littoraux (Érosion et Submersion marine) des départements d'Ille-et-Vilaine, des Côtes-d'Armor et du Finistère: Phase 1, Rapport final, BRGM/RP- 65212-FR, 1282, 861 ill., 19 annexes, 1 CD.

Sorensen, J., 1997. National and international efforts at integrated coastal management: definitions, achievements, and lessons. Coast. Manag. 25 (1), 3–41.

UNEP, 2006. Marine and Coastal Ecosystems and Human Well-Being: A Synthesis Report Based on the Findings of the Millennium Ecosystem Assessment.

PART 4
Coastal Sediment Geochemistry

CHAPTER 12

Geochemical Characterization of Beach Sediments of Miri, NW Borneo, SE Asia: Implications on Provenance, Weathering Intensity, and Assessment of Coastal Environmental Status

R. Nagarajan*, A. Anandkumar*, S.M. Hussain[†], M.P. Jonathan[‡],
Mu. Ramkumar[§], S. Eswaramoorthi*, A. Saptoro[¶], Han Bing Chua[¶]

*Department of Applied Geology, Curtin University, Miri, Malaysia, [†]Department of Geology, University of Madras, Guindy Campus, Chennai, India, [‡]Centro Interdisciplinario de Investigaciones y Estudios sobre Medio Ambiente y Desarrollo (CIIEMAD), Instituto Politécnico Nacional (IPN), Mexico City, Mexico, [§]Department of Geology, Periyar University, Salem, India, [¶]Department of Chemical Engineering, Curtin University, Miri, Malaysia

1 Introduction

The coastal zone is the interface between the land and the sea (Christian and Mazzilli, 2007). It is the area where the freshwater and seawater regimes (i.e., the estuarine region, the coastal beaches, and the shelf) are influenced by each other, making it a transitional, yet complex environment (Woodroffe, 2003). Most of the world rivers discharge their freshwater and sediments into the coastal region (Milliman and Katherine, 2011; Whitefield et al., 2015; Déry et al., 2016). This discharge drives the global flux of the elements to the ocean, and exerts control over the composition of the seawater in the coastal zone. Some of the elements, namely, C, Fe, Mn, N, P, and Si, have significant biogeochemical significance, and are essential for the understanding of global climate change (Carey et al., 2002; Harrison et al., 2010; Dürr et al., 2011; Wang et al., 2011; Amon et al., 2012; Moyer et al., 2012; Spencer et al., 2013).

The developmental activities in the river basins (such as land use/land cover changes due to deforestation, mining, agricultural activities with the application of pesticides and fertilizers, etc.) have a profound impact in the coastal area (Ferguson et al., 2011; Gupta et al., 2012). The delivery of dissolved and particulate loads through these rivers generates gradients in the physicochemical composition of the water column in the coastal zone, and paves the way for several biogeochemical interactions (Liu et al., 2010; Aufdenkampe et al., 2011). An understanding of these processes is vital for the estimation of global elemental flux from the continents to the ocean.

The rich biodiversity of the coastal zone supports a highly complex food-web that accompanies a large quantity of the population, including humans. A significant amount of the world's population is living in the coastal zone (37% of people within 100 km of the coastal zone; Cohen and Small (1998); and 23% of people near the coastal zone; elevation <100 m and closer than 100 km from the coast (Small and Nicholls, 2003)), and by 2015, the population in coastal areas is expected to rise 70% within 200 km of the coastline (Hinrichsen, 1996). Together, the dense human population (Lichter et al., 2010; Neumann et al., 2015) and attendant interference with the natural processes induce high pressure on coastal ecosystems and natural resources (Patterson and Hardy, 2008). Several recent studies have demonstrated the deterioration of the coastal zone due to natural calamities, rising sea levels, and human influences (Balica et al., 2012; Jongman et al., 2012), reduction in sediment supply (Ramkumar, 2003; Wang et al., 2011) and accelerated coastal erosion (Ramkumar, 2000), global climate change associated with rising sea levels (Sallenger Jr et al., 2012), coastal erosion and submergence (Hanebuth et al., 2013; Luo et al., 2015), degradation of mangrove forests (Anthony and Gratiot, 2012; Lovelock et al., 2015), shipping, exploration for oil and gas, fishing, and so forth (Wang et al., 2013; Vikas and Dwarakish, 2015). These studies have emphasized the urgency to protect coastal environments (Spalding et al., 2014).

About 40% of the global coastlines (Bird, 2000; Carranza-Edwards et al., 2009) are sandy-gravelly due to riverine and onland sediment transport and deposition. Although most of the world's beaches consist of quartz-rich sand, their textural and geochemical characteristics are controlled by many factors, namely, waves, wind, longshore currents, climate, relief, and source composition (Folk, 1974; Kasper-Zubillaga and Carranza Edwards, 2005; Armstrong-Altrin et al., 2012, 2014, 2015, 2018; Tonyes et al., 2017; Papadopoulos et al., 2015; Papadopoulos, 2018). The winnowing and sorting actions of coastal processes concentrate heavy minerals (i.e., garnets, ilmenites, rutiles, zircons, etc.) of economic and strategic importance, and as a result, several studies have been conducted to attempt to understand their conditions of occurrence based on textural, mineralogical, and geochemical characteristics (Acharya et al., 2015; Ravindra Kumar and Sreejith, 2010; Papadopoulos et al., 2016). Beach sediments were also studied for an understanding of depositional environments (Kasper-Zubillaga and Zolezzi-Ruiz, 2007; Hernández-Hinojosa et al., 2018), provenance (Le Pera and Critelli, 1997; Dubrulle et al., 2007; Carranza-Edwards et al., 2009; Armstrong-Altrin et al., 2012, 2014, 2015; Kasper-Zubillaga et al., 2013), tectonic settings (Kasper-Zubillaga et al., 1999; Armstrong-Altrin et al., 2014, 2015, 2018; Ramachandran et al., 2016), and to assess their environmental status (Yalcin and Ilhan, 2008; Nagarajan et al., 2013; Yalcin et al., 2016).

The accumulations of heavy metals in the coastal environments "above" the natural levels of concentrations are often ascribed to anthropogenic activities. In order to define the "natural level," baseline studies are being carried out around the world. The baseline studies are aimed at evaluating the actual surface concentration of an element, independent of its origin against the "background" that varies due to the nature of the area (such as basic geology and genesis of overburden), and diffuse anthropogenic pollution (Salminen and Gregorauskien, 2000). The concept of a baseline is used to differentiate contamination arising from point sources, while background is used to define a range of values within which an element concentration is expected to prevail. In environmental geochemistry, the baseline studies are more relevant, as they are useful to identify the pollution level. The geochemical baseline studies commenced during 1988 by the International Geological Correlation Program (IGCP) Project 259-International Geochemical Mapping (Darnley, 1990, 1995, 1997). Following this, survey studies have been conducted at the global, regional, and local scales (e.g., Xuejing et al., 1997; Plant et al., 2001; Johnson et al., 2005; Garrett et al., 2008). Baseline studies using sediments and organisms were reported for different coastal regions of the world, namely, King George Island of Antarctica (Ahn et al., 1996; Dos Santos et al., 2006), the United States (Hanson et al., 1993; Daskalakis and O'connor, 1995), the Gulf of Oman (de Mora et al., 2004), New Zealand (Abrahim and Parker, 2008), the Mediterranean (Renzi et al., 2015), the Kanyakumari coast of India (Peter et al., 2017), Sundarban mangroves (Kumar et al., 2016), the Guangdong coast of China (Zhao et al., 2016), Egypt (Okbah et al., 2014), the Persian Gulf (Pourang et al., 2005), Ennore Creek (Jayaprakash et al., 2005, 2014), and the coral reef sedimentary environment in Lakshadweep (Gopinath et al., 2010).

The accumulated heavy metals can be remobilized under suitable environmental conditions, and the solubility of the elements directly influences their bioavailability (Xian, 1987). Because bioavailability is directly related to bioaccumulation and toxicity, the study of living organisms often provides information vital to understanding the impact of trace metal contamination. Along these lines, studies on the taxonomy, ecological distribution, and biogeography of shallow water benthic foraminifera provide quick and cost-effective methods of assessing the impact of pollution and other environmental changes on shallow marine biota (Fursenko, 1978; Saraswat et al., 2004; Le Cadre and Debenay, 2006; Frontalini et al., 2009). Morphological abnormalities in benthic foraminifera may also be due to either natural or anthropogenic impacts. Sometimes, the morphological abnormalities of benthic foraminifera tests are considered pollution indicators, because the proportion of deformed tests increases significantly in the polluted areas, including sediments contaminated by heavy metals (Sharifi et al., 1991; Yanko et al., 1998; Samir, 2000; Vilela et al., 2004; Burone et al., 2006; Le Cadre and Debenay, 2006; Mancin and Darling, 2015).

In Malaysia, studies were reported on trace metals in the stream sediments of the Bernam River (Kadhum et al., 2016), Sungai Puloh Estuary (Udechukwu et al., 2015), the coastal marine sediments of Sabah (Ashraf et al., 2017), Bidong Island off the east coast of Peninsular

Malaysia (Ong et al., 2015), the west coast of Peninsular Malaysia (Buhari and Ismail, 2016; Redzwan et al., 2014), the Strait of Malacca (Looi et al., 2015), Sungai Puloh in Peninsular Malaysia (Abubakar et al., 2018), Kelantan Estuary (Wang et al., 2017), and Marudu Bay (Soon et al., 2016). Trace element levels in biological tissues were also reported from the Miri Estuary of Sarawak (Billah et al., 2017), the Miri coastal area (Anandkumar et al., 2017), Baram River Port Klang in Selangor (Sany et al., 2013), the Tanjung Pelepas harbor (Yusoff et al., 2015), the Langat River in Selangor (Haris et al., 2017), and its estuary (Mokhtar et al., 2015), and recently, Prabakaran et al. (2017) have reviewed the application and status of biomonitoring in Malaysia.

The Sarawak state encompasses a 1035-km-long coastline accounting for about a quarter of the 4809-km coastline of all of Malaysia. The Miri coast of the Sarawak consists of sandy and rocky beaches with scenic geological structures, such as rock cliffs, arches, caves, wave cut platforms, stocks, beach rocks, and so forth, which makes them frequently visited tourist beaches. Three major rivers (Baram, Miri, and Sibuti) that originate inland of Sarawak and debauch in the South China Sea deliver vast amounts of sediments to the coastal area of the South China Sea. Among these three rivers, the sediment influx from the Baram River is greater than the combined quantities of other two (Collins et al., 2017). Other than these, research on the source and nature of these sediments is scarce (Nagarajan et al., 2013; Ismail, 1993; Sim et al., 2017; Prabakaran, 2017).The land use/land cover changes of the coastal area have been assessed recently for the Miri coastal region (Anandkumar et al., 2018). Sand spit formation in front of the Baram River has been reported by Nagarajan et al. (2015a). Dodge-Wan and Nagarajan (2019) (Chapter 3, in this volume) have documented the typology and mechanism of coastal erosion in the clastic rocks of the same study region (Tanjung Lobang from the NE part, and Beraya, Tusan, Peliau, Bungai beaches from the SW part of the study area).

Comparison between the source and sink relationship for NW Borneo sediments are still unclear, except for the overall expectation that the sediments of NW Borneo are recycled from the Rajang Group of sediments (van Hattum et al., 2006, 2013; Nagarajan et al., 2014a, 2015b, 2017a, b). Also, the influences of riverine influx on the coastal sediment quality are not well documented. Although there have been a few studies on the trace element levels in the coastal areas of Malaysia, an integrated approach that addresses the nature of the source, the possible sinks, and the environmental and biological impact of the trace elements, is lacking. Moreover, comparatively few studies have been conducted on the east Malaysian coastal area (i.e., Nagarajan et al., 2013), other than what has been documented in Peninsular Malaysia. Herein we attempt to provide baseline data on present environmental conditions, which have been assessed and recorded. This dataset can be a baseline with which future changes can be compared, and accordingly, remedial measures can be implemented.

Thus, this study mainly reports the total trace metal concentrations of various beach sediments, geochemical characterizations, and their constraints on weathering, sorting, recycling, and provenance, and finally, environmental quality of the beach sands have been analyzed using selected trace elements and indices. In addition, identification and characterization of foraminifera and ostracoda in the sands and their response to the present environmental conditions are also taken into consideration.

2 Regional Setting

The regional geology of the Sarawak state is divided into three major zones, namely, Kuching, Sibu, and Miri. The Miri zone consists of sedimentary rocks deposited under shallow marine, estuarine, and fluvial systems during Neogene. The major Neogene sedimentary formations are (a) Sibuti, (b) Lambir, (c) Tukau, and (d) Miri (Fig. 2). The Sibuti Formation is mainly exposed along the Bungai Beach, and the inland area of the SW part of the study area. The lithologies are mudstone and shale, with interbeds of limestone/marl and sandstone, particularly toward the Lambir Formation. This formation is underlain by the Suai Formation, with a fault contact, and overlain by the Lambir Formation along a conformable and abrupt boundary (Hutchison, 2005). The Sibuti Formation is considered a member of the Setap Shale Formation. However, it is distinguished from the Setap Shale Formation based on high fossil content, predominant marl lenses, and thin limestone beds (Peng et al., 2004). The age of the Sibuti Formation is assigned as Early to Late Middle Miocene (Liechti et al., 1960; Simmons et al., 1999; Vermeij and Raven, 2009; Simon et al., 2014). The clastic rocks of the Sibuti Formation are classified into quartz arenites, litharenites, sublitharenites, arkoses, subarkoses, and wackes, in which litharenites and sublitharenites are dominant. Mineralogically these rocks consist of quartz, mica, clay minerals (illite, kaolinite, and chlorite), zeolites with little feldspar and heavy minerals (zircon, ilmenite, tourmaline, rutile, and cassiterite) (Nagarajan et al., 2015b) and pyrite rich concretions. The Lambir Formation is exposed along the beaches, and consists of mainly massive sandstone, sandstone inter-bedded with shales, and some limestone/marl (Hutchison, 2005; Nagarajan et al., 2017a), and are classified into shale, wacke, arkoses, litharenites, Fe-sands, and quartz arenites (Nagarajan et al., 2017a). Mineralogically the sandstones of the Lambir Formation consist of quartz, illite/muscovite, and a minor amount of plagioclase; whereas mudstones are dominated by quartz, illite/muscovite, amorphous phase, chlorite, plagioclase, and calcite. The limestone consists of calcite, ankerite, quartz, chlorite, illite/muscovite, and a trace amount of aragonite (Nagarajan et al., 2017a). The Lambir Formation is underlain by the Sibuti or Setap Shale Formations, and is overlain by the Tukau Formation. The Tukau Formation is exposed in the western part of Miri city, and along the beaches, which extends inland. The basal part of the Tukau Formation conformably overlies the Lambir Formation near Sungai Liku in the eastern Lambir Hill area, and is conformably overlain by the Liang Formation near Miri. Considered to be from the Upper Miocene to Lower Pliocene periods (Wilford, 1961), the Tukau Formation is located on a coastal plain, based on the absence of foraminifers (except some brackish water forms), and the presence of lignite layers and amber balls (Hutchison, 2005). The lithology consists of sandstones, shales, and alternate interbedded shales and sandstones with minor conglomerates. These are classified into wacke, arkoses, litharenites, and quartz arenites (the lower part of the Tukau Formation) (Nagarajan et al., 2014a). Occurrences of shales are reported, in addition to these sandstone types from the upper part of the Tukau Formation (Nagarajan et al., 2017b). Mineralogically, these are dominated by quartz, clay minerals (illite is more dominant than kaolinite), feldspars, and some heavy minerals such as tourmaline, zircon, rutile/anatase, chromian spinels, ilmenite, and pyrite (Nagarajan et al., 2017b). Pyrite concretions are common, which are enriched by Fe, S, Ni, Mo, As, and Nb, compared with parent rocks. Mineralogically these concretions are made up of pyrite, quartz, arsenopyrite, cattierite, goethite, and berlinite (Nagarajan et al., 2014b). The Miri Formation is geologically a very complex NE-SW elongated anticline, bound to the east by a steep NW dipping fault (Shell Hill Fault). Further east, and underneath the Shell Hill Fault, three NW dipping thrust faults (Canada Hill Thrust, Inner and Outer Kawang Thrusts) are reported. The rocks, exposed in and around Miri city, belong to the Miri Formation (Middle Miocene), and are the uplifted part of the subsurface, oil-bearing sedimentary strata of the Miri oilfield (Jia and Rahman, 2009).The Miri Formation is divided into (1) the Upper Miri Formation, which is more arenaceous; (2) the Lower Formation, which consists of repeated and irregular sandstone-shale alternations, with the sandstone beds passing gradually into clayey sandstone and sandy or silty shale. Seepage of crude oil from the fault zones is common during the rainy season. The geochemistry and mineralogy of the clastic rocks of the Miri Formation have not been reported on yet.

The Miri River Basin consists of Neogene sedimentary rocks, while the Baram River Basin consists predominantly of meta-sedimentary to sedimentary rocks aged from the Paleogene to Recent eras. The upstream region is primarily covered by Oligocene, Miocene, and Eocene sediments; whereas the downstream region consists of Quaternary river fluvial and coastal alluvium (Caline and Huong, 1992). The alluvium in the lower river section includes thick peat soils (Anderson, 1964). There are very small minor acid/basic intrusive rocks in the SW part of the Baram River Basin (Usun Apau area). Deep water turbidites are common in the eastern part of the basin and near Long Loma. The Miocene to Pliocene sediments are mostly recycled materials (van Hattum et al., 2013; Nagarajan et al., 2014a, 2015b, 2017a, b). Recycling is a common phenomenon in this region. Structural trend lines are oriented in a NE-SW direction. Limestones belonging to the Melinau Limestone Formation are common in the Baram River Basin, and minor exposures are common near the Long Loma area.

Miri is the second largest city in the Sarawak state of Malaysia, and is located in the coastal area of NW Borneo between the latitudes N4°14″ and N4°36″, and longitudes E113°50″ and E114°08″. It is a fast growing city based on industries and tourism, and thus known as "Resort City." Miri has 300,000 inhabitants with diverse ethnicities, in addition to the multicultural tourists that may match that of the local population in numbers. Miri and its coastal region contribute to the Malaysian economy by providing oil and gas resources, and being a northern gateway of the Sarawak State of Malaysia. The beaches of Miri are dominated by sands with open stands of *Casuarina equisetifolia* (Fig. 1C and E), coarse grasses, and

Fig. 1
(A–H) Field photographs of the study area. (A) Peat and log accumulation in the Kualabaram beach; (B) thick pile of peat accumulation within sand deposits; (C) pheneromic view of Lutong beach; (D) Tanjung Lobang beach; (E) Kpg. Beraya beach; (F) Tusan beach with rock cliff; (G) Bedrock exposures in the Peliau beach; and (H) Folded sandstones in the Bungai beach.

shallow swamps running parallel to the coast in most places. Many tourist beaches and other landmarks are located along the 75-km stretch from Kaulabaram to Bungai Beach, such as the Baram River mouth/Estuary (Fig. 1A and B); the fish landing center, Lutong Beach (Fig. 1C); Piasau Beach (the Boat Club); the Miri River Mouth/Estuary; Park Everly Beach; Luak Bay; Tanjung Lobang Beach (Fig. 1D); Esplanade Beach; Hawaii beach; Kpg. Beraya Beach (Fig. 1E); Tusan Beach (Fig. 1F); Peliau Beach (Fig. 1G), and Bungai Beach (Fig. 1H). Among the Miri beaches, Tusan Beach is famed for seasonal blue luminescence waters (Chong, 2016) by the algae "Dinaflagellates." The study area is classified into the northeast part (Baram River mouth to the Miri River Estuary) and the southwest part (Miri River Estuary to Bungai Beach) (Fig. 2). The NE part of the study area consists of river mouths (Baram River Estuary and Miri Estuary), fish landing centers, Lutong Beach, Piasau Beach, and other places of

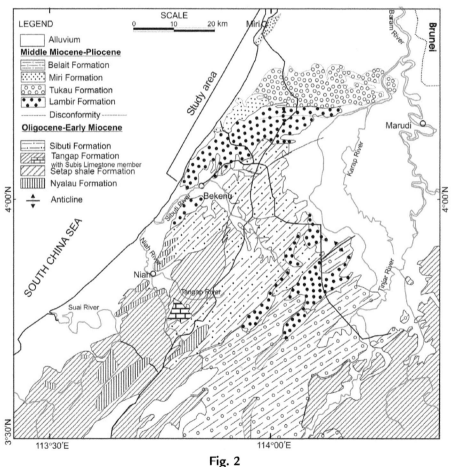

Fig. 2
Geology map of the northern Sarawak showing the study area. *After Liechti, P.R., Roe, F.W., Haile, N.S., 1960. The geology of Sarawak, Brunei and the western part of North Borneo. British Borneo Geol. Surv. Bull. 3, 1–360.*

industrial activities, fishing, the oil and gas (offshore) industries, shipyard industries, plywood industries, agriculture, and so forth. The SW part consists of Park-Everley, Tanjung Lobang, Esplanade, Hawaii, Kpg. Beraya, Tusan, and Peliau and Bungai Beaches, which support tourism and agriculture. The man-made activities observed are usage of anti-corrosive/fouling paints, dyes in the fishing vessel hulls and nets; which contain zinc oxide (Al-Fori et al., 2014). In addition, small industries such as ceramics, fiber boats, and concrete curing industries are located here. Shrimp hatcheries are also common in this area. The beaches in this area consist of sandy shores (i.e., Kualabaram, Fish Landing Center, Lutong, Miri River Estuary from NE part; Luak Bay, Hawaii and Beraya from SW), rock shores (Peliau and Tusan from the SW part) and mixed rocky and sandy shores (Tanjung Lobang from the NE part and Bungai from the SW part) (Fig. 1A–H).

The study area mainly consists of the Quaternary-Recent alluvium and Neogene sedimentary rocks, such as sandstone, siltstone, mudstone, shale (arenaceous and carbonates), and limestone/marl beds. The Quaternary terrace deposits are exposed along the northern Sarawak coastline, and made an angular unconformity over the Neogene sedimentary formations (Kessler and Jong, 2014). The alluvial coastal plain can be observed along the shoreline of the Miri coast, and the alluvium formation can be seen on the banks of the rivers in the study area. The shoreline areas near the Baram and Miri rivers are characterized by peat soils, mangroves, nipahs, swamp forests, and tidal inundation. As a result of geological conditions and the tropical monsoon climate in central Borneo, erosion rates are among the highest in the world since the Eocene era (Sandal, 1996; Straub et al., 2000). The Miri coast consists of well-developed sandy beaches, resulting from a strong SW longshore drift and relatively high offshore wave action (see Anandkumar et al., 2018 for more information).

3 Materials and Methods

The study described in this chapter was carried out to document textural, major, trace, and rare earth element geochemical, mineralogical studies and microfossil contents of beach sediment samples of a coastal region located between the confluences of the Baram River and Sibuti River. Of a total of 59 seasonal samples (December 2013; $n=28$; June 2014; $n=31$) collected, 31 samples were collected (Fig. 3) during June 2014 to study textural, mineralogical, and geochemical characterization. In addition, 48 samples were collected for paleontological studies. Prior to the sampling, sample containers were cleaned with diluted HNO_3 (20%) and rinsed with deionized water.

The samples were washed with ultrapure water to remove the salt content and dried in an oven at 60°C. After drying, all the roots, gravels, and rock fragments were removed, and we ensured that all the samples could pass through 2mm sieve. A selected portion of the samples, after using the cone and quartering method, was ground to <63μm using agate mortar and pestle. Bulk geochemical analyses (major, trace, and rare earth elements) were carried out at

288 Chapter 12

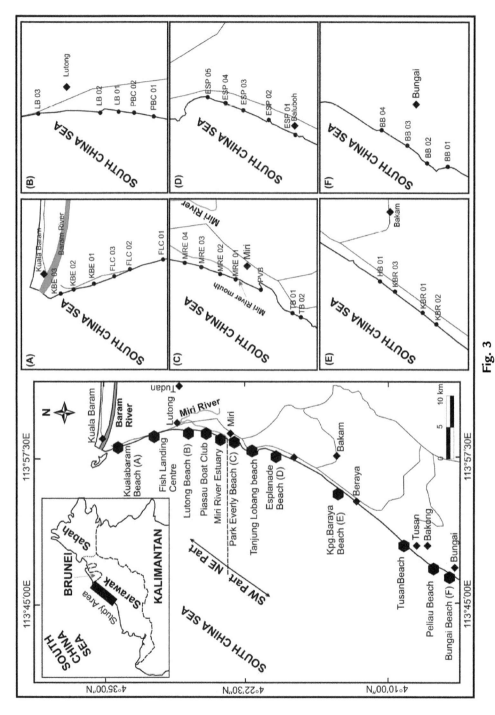

Fig. 3
Map of the study shows sample locations and from the two parts of the study area.

Actlabs in Canada, as per the method stated in Ramkumar et al. (2018). Certified reference materials, such as NIST694, DNC-1, W-2a (for major and trace elements), NCSDC70009, OREAS100a, 101a, and JR-1 (for REE) were used to calibrate the trace element concentrations during the ICP-MS analyses. The major and trace element concentrations were normalized against average upper-continental crustal values (UCC; McLennan, 2001). Chondrite values of McDonough and Sun (1995) were used for normalizing rare earth elements (REE). Eu anomaly (Eu/Eu*) was calculated using the formula proposed by McLennan (1989). In order to construct different weathering and provenance diagrams, the major oxides' values were calculated on an LOI-free (anhydrous) basis.

The grain size analysis was carried out using a mechanical shaker at half ϕ intervals. Due to lack of differences between the intervals, the results were grouped into three fractions as coarse sand (2–0.5 mm), medium + fine sand (500–125 μm) and very find sand+mud (<125 μm) and were reported in percentages.

Approximately 50 g of each sample were taken, and the foraminiferal tests and ostracod carapaces were handpicked from all the samples using a 0.00 winds or sable haired brush adopting standard micropaleontological techniques. The handpicked faunal specimens from each sample were transferred to 24-chambered micropaleontological slides and mounted over a thin layer of tragacanth gum according to the family, genus, and species, wherever possible. The different genera and species were identified; type specimens of each species were selected and transferred to round punch microfaunal slides with cover slips. The widely utilized foraminiferal classification proposed by Loeblich and Tappan (1987) and ostracods classification proposed by Hartman and Puri (1974) were followed. The taxonomy and morphology of the individual species were studied under a stereo-binocular microscope at the University of Madras, India. The selected fossils from the sediments were stored for future study, and are available at the Department of Geology, University of Madras.

Statistical multivariate analyses were carried out using SPSS Statistics 17 software and MATLAB version 8.2.0.701 (R2013b), following the protocols detailed in Ramkumar (1999, 2001). Recorded parameters with concentrations below detection limits (BDL) were excluded from the data analyses. Factor analysis was performed for these data, and the resultant principal components/factor scores in the latent space were generated from the original datasets. The first three principal components representing the majority of the total variance were then plotted to graphically analyze the levels of contributions and correlations among the parameters. Only the correlation coefficient values $r \geq 0.5$ with significant levels $P \leq .05$ were considered for discussion.

3.1 Calculation of Environmental Indices

The environmental indices, namely, the Contamination Factor (CF), the Geoaccumulation Index ($Igeo$), and the Pollution Load Index (PLI) were calculated. Average Upper Continental Crustal (UCC; McLennan, 2001) values were used as background values for all the calculations. The contamination factor was calculated based on Pekey et al. (2004), using the ratio between the metal content in the sediment at a sample location and the background values. The Geoaccumulation Index (*Igeo*) was calculated using the equation proposed by Muller (1969). $Igeo = \log_2(Cn/1.5Bn)$; where, Cn is the metal (n) concentration recorded in the sediments of the study area, Bn is the background value of the corresponding metal (n), and factor 1.5 is the background matrix correction due to lithogenic effects (Salomons and Forstner, 1984). The *Igeo* classification proposed by Muller (1969) was used to determine the level of contamination. The seven classes in this classification are: uncontaminated ($Igeo \leq 0$; Class 0), uncontaminated to moderately contaminated (*Igeo* 0–1; Class 1), moderately contaminated (*Igeo* 1–2; Class 2), moderately to highly contaminated (*Igeo* 2–3; Class 3), highly contaminated (*Igeo* 3–4; Class 4), highly to very highly contaminated (*Igeo* 4–5; Class 5), and very highly contaminated ($Igeo \geq 5$; Class 6) (Muller, 1969). The PLI was calculated by using formulae proposed by Tomlinson et al. (1980) based on the contamination factor values of individual elements.

$$\text{PLI} = (Cf_1 \times Cf_2 \times Cf_3 \times \cdots \times Cf_n)^{1/n}$$

Where, Cf is the contamination factor and "n" is the number of metals. This empirical index provides a simple, comparative means for assessing the level of trace/heavy metal pollution.

4 Results

The study area has been divided into two sectors/parts based on geographic features (Fig. 3) as NE and SW parts. Geochemical (major oxides, trace and rare earth elements (REEs)) results of samples collected from these respective sectors are also presented separately (NE: $n=15$; SW: $n=16$) in Table 1. Confluence points of two major rivers (Baram River and Miri River) are in the NE part; whereas the SW part mostly consists of small rivulets and tidal channels that have confluences in the South China Sea.

4.1 Grain Size

The sediment characteristics of NE and SW parts show distinct changes between them, perhaps as a result of differences in terms of physiographic, fluvial, and oceanographic conditions. The NE part consists of medium to fine sand fractions and very fine sand to mud fractions, in particular, very fine sand and mud fractions are recorded higher near the confluences of the Baram and Miri Rivers. A decrease of medium to fine sand fractions was observed toward the SW up to Esplanade Beach, followed by an increase toward Park Everly Beach in the SW part. The very fine sand and mud ($\leq 125\,\mu m$ fraction) shows a clear negative trend against the medium to fine sand fraction (Fig. 4). The coarse sand (2–0.5 mm) is recorded at 0% to 5.86% (avg. 0.75 wt%) in the SW, part and 0 to 0.25 wt% (avg. 0.04 wt%) in the NE part, respectively. Medium to fine sand (500–125 μm) is the major constituent in both parts of the study area.

Table 1 Statistical summary of the bulk geochemical analysis of the beach sediments from the Miri coastal region

Parameters	SW Part (n = 16)			NE part (n = 15)		
	Min	Max	Avg.	Min	Max	Avg.
C Sand	0	5.86	0.75	0	0.25	0.04
M-F Sand	50.96	94.67	78.74	20.68	83.9	65.21
VF Sand-Mud	2.43	48.8	20.51	15.85	79.22	34.75
SiO_2	94.96	98.7	96.57	93.35	97.27	95.78
Al_2O_3	0.51	1.85	1.05	1.05	3.19	1.74
$Fe_2O_3(T)$	0.3	0.85	0.55	0.41	1.02	0.60
MnO	0.003	0.011	0.01	0.004	0.013	0.01
MgO	0.07	0.27	0.13	0.13	0.36	0.18
CaO	0.05	0.76	0.16	0.08	0.17	0.11
Na_2O	0.03	0.19	0.10	0.11	0.3	0.19
K_2O	0.09	0.36	0.20	0.18	0.58	0.33
TiO_2	0.046	0.153	0.10	0.125	0.533	0.20
P_2O_5	0.01	0.02	0.01	0.01	0.02	0.02
LOI	0.45	1.3	0.67	0.51	1.31	0.79
Total	98.46	100.9	99.51	98.7	100.8	99.93
Sc	1	2	1.09	1	3	1.73
V	9	15	11.56	10	21	15.00
Ba	22	62	39.00	40	87	56.73
Sr	6	54	17.63	16	28	20.47
Y	3	6	4.44	4	19	6.53
Zr	35	287	79.94	55	1677	259.60
Cr	20	30	23.33	20	150	45.00
Co	1	2	1.29	1	3	2.07
Cu	130	260	217.50	110	490	226.67
Zn	60	120	95.63	50	240	104.00
Ga	1	2	1.60	1	4	2.13
Ge	1	1	1.00	1	2	1.10
As	5	9	7.33	BDL	BDL	–
Rb	2	14	6.94	7	23	12.27
Nb	BDL	BDL	–	1	9	2.69
Ag	BDL	BDL	–	0.5	3.4	1.56
Sn	1	5	2.63	1	17	3.93
Sb	BDL	BDL	–	0.5	0.5	0.50
Cs	0.6	0.6	0.60	0.5	1.1	0.62
La	4.1	7.3	5.66	6	30	9.77
Ce	7.4	14	10.63	11.2	60.9	18.59
Pr	0.85	1.54	1.18	1.23	6.67	2.02
Nd	3.2	5.3	4.27	3.9	24.7	7.43
Sm	0.6	1.1	0.84	0.8	4.5	1.38
Eu	0.11	0.22	0.16	0.17	0.58	0.25
Gd	0.5	0.9	0.70	0.7	3.3	1.06
Tb	0.1	0.1	0.10	0.1	0.5	0.15
Dy	0.5	0.9	0.66	0.6	2.9	1.05
Ho	0.1	0.2	0.14	0.1	0.6	0.23

(Continued)

Table 1 Statistical summary of the bulk geochemical analysis of the beach sediments from the Miri coastal region—Cont'd

Parameters	SW Part (n=16)			NE part (n=15)		
	Min	Max	Avg.	Min	Max	Avg.
Er	0.3	0.6	0.43	0.4	2.3	0.73
Tm	0.05	0.09	0.07	0.07	0.39	0.12
Yb	0.3	0.6	0.46	0.5	3	0.85
Lu	0.04	0.11	0.07	0.07	0.56	0.14
Hf	0.8	6.2	1.88	1.4	45.4	6.39
Ta	0.1	0.4	0.19	0.2	0.9	0.31
W	2	2	2.00	1	2	1.67
Pb	5	11	7.00	5	19	8.29
Th	1.2	2.1	1.69	1.6	8.4	2.73
U	0.4	0.8	0.54	0.5	3.3	0.98

BDL, below detection limit.

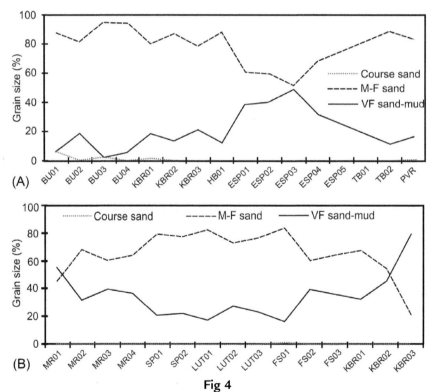

Fig 4
Grain size distribution map of the beach sediments of Miri coastal region SW part (A) and NE part (B) of the study area.

The very fine sand and mud content (<125 μm) ranges from 2.43 to 48.8 wt% (avg. 20.5 wt%) in the SW part, and 15.85 to 79.22 wt% in the NE part. The average fine sand recorded a higher proportion in the SW part, and very fine sand+mud fraction recorded higher in the NE part (Fig. 4).

4.2 Major Oxides

The SiO_2 concentrations are similar between the SW and NE parts of the study area (94.96–98.70 wt% and 93.35–97.27 wt% respectively). Al_2O_3 content is recorded higher in the NE part (1.05–3.19 wt%) than the SW part (0.51–1.85 wt%). Relatively lower contents of Fe_2O_3 (avg. 0.55 wt%), MgO (0.13 wt%), Na_2O (0.10 wt%), K_2O (0.20 wt%), and TiO_2 (0.10 wt%) are recorded in the SW part than in the NE part (avg. Fe_2O_3=0.60 wt%, MgO=0.18 wt%, Na_2O=0.19 wt%, K_2O=0.33 wt%, and TiO_2=0.20 wt%). However, MnO, P_2O_5 and LOI contents are relatively similar, or show only minor variations between the SW and NE parts.

4.3 Trace Elements

Trace elements such as Ni, In, Tl, and Bi are recorded below detection limits (BDL) for all the studied samples. The elements W, Cs, Ge, Cr, Ag, and Nb are also recorded below detection limits for many samples, especially from the samples from the SW part of the study area. Other than these, the average concentrations of other trace elements are higher in the NE part than the SW part, except As and W, which are recorded higher in the SW part. In particular, Cu and Zn are recorded higher (226.67 and 104.00 ppm respectively) in the NE part, and in the SW part, their average concentrations are 217.50 and 95.63 ppm, respectively. The high field strength elements, Zr and Hf, show enrichment (1677 and 45.4 ppm respectively) in a sample (FS 01) from the NE part, and the average of these elements are recorded as 79.94 and 259.60 ppm for Zr; and 1.88 and 6.39 ppm for Hf in the SW and NE parts, respectively.

4.4 UCC Normalization

The major oxides and trace elements were normalized against UCC, and the normalized values are presented in Fig. 5. Except SiO_2, other major oxides are depleted compared with UCC. The trace elements Cu and Zn are enriched in all the studied samples, while As, Zr, Hf, and Sn are enriched in selected samples. Samples from the SW part are depleted in Al_2O_3, Na_2O, TiO_2, Ba, Sr, and Rb; whereas MnO, MgO, CaO, Zr, Sn Hf, and Ta show fluctuating values between the samples. SiO_2, Cu, and Zn are uniformly enriched, compared with the UCC in all the samples.

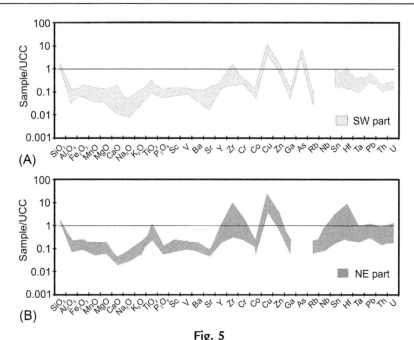

Fig. 5

UCC normalized multielements plot shows the trends for the beach sediments of Miri coastal region SW part (A) and NE part (B).

4.5 Rare Earth Elements

The total REE ($\sum$REE) is recorded higher in the NE part of the study area (26.0–140.9 ppm; avg. 43.8 ppm), with very high content of $\sum$REE (Sample FS01; 140.9 ppm) in one sample; whereas the range is low in the SW part (18.5–32.8 ppm; avg. 25.3 ppm). The chondrite normalized REE plot shows (Fig. 6) a fractionated LREE pattern (La/Yb$_{CN}$=4.87–11.21 and 6.67–9.96 for SW and NE parts, respectively), flat HREE (Gd/Yb$_{CN}$=1.08–1.62 and 0.81–1.21, respectively, in the SW and NE parts), and a negative Eu anomaly with similar variations regardless of NE or SW parts (0.40–0.79; avg. 0.64 and 0.46–0.78; avg.0.68 for the SW and NE parts, respectively). Expect one sample from NE, all the samples show lesser contents of $\sum$REE than PAAS and UCC.

4.6 Systematic Paleontology

A systematic paleontological study was conducted from the NE part of the study area, where two major rivers confluence with the South China Sea, to understand the effect of pollution on the biota (foraminifera and ostracoda species), and to use foraminifera and ostracoda as bioindicators. In this study, 14 ostracod taxa belonging to 11 genera, 7 families, and 2 suborders were identified (Table 2) and 21 benthic foraminifera species belonging to 13 genera, 10 families, 8 superfamilies, and 3 suborders were identified (Table 3). The foraminifera and ostracoda populations were observed as 2252 and 205 specimens, respectively.

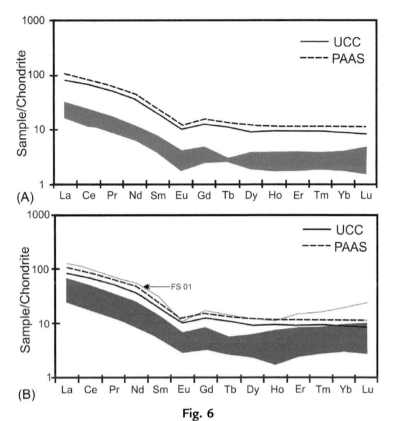

Fig. 6
Chondrite normalized REE plot for the beach sediments of Miri coastal region (A) SW Part and (B) NE part.

Table 2 Taxonomic chart of ostracoda of the study area

Order	Suborder	Super Family	Family	Genus	Species
Podocopida	Platycopa		Cytherellidae	Cytherelloidea	Cytherelloidea sp.
			Cytheridae	Hemicytheridea	Hemicytheridea sp.
	Podocopa	Cytheracea		Actinocythereis	Actinocythereis scutigera
					Actinocythereis sp.
			Trachyleberididae	Stigmatocythere	Stigmatocythere indica
				Alocopocythere	Alocopocythere reticulate indoaustralica
				Keijella	Keijella karwarensis
				Lankacythere	Lankacythereis sp.
			Brachycytheridae	Pterygocythereis	Pterygocythereis sp.
			Cytherettidae	Neocytheretta	Neocytheretta snellii
					Neocytheretta murilineata
			Loxoconchidae	Loxoconcha	Loxoconcha mandviensis
					Loxoconcha gruendeli
		Cypridacea	Candonidae	Phlyctenophora	Phlyctenophora orientalis

Table 3 Taxonomic chart of foraminifera of the study area

Order	Sub Order	Super Family	Family	Genus	Species
Foraminiferida	Miliolina	Miliolacea	Spiroloculinidae	Spiroloculina	Spiroloculina communis
					Spiroloculina orbis
					Spiroloculina costifera
					Spiroloculina sp
			Hauerinidae	Quinqueloculina	Quinqueloculina agglutinans
					Quinqueloculina costata
				Triloculina	Triloculina terquemiana
				Miliolinella	Miliolinella circularis
					Miliolinella Hauerina
Foraminiferida	Rotalina	Rotaliacea	Elphidae	Elphidium	Elphidium hispidulum
			Rotalidae	Ammonia	Ammonia beccari
				Asterorotalia	Asterorotalia trispinosa
		Nummulitoidea		Pseudorotalia	Pseudorotalia schroeteriana
		Planorbulinacea	Nummulitidae	Operculina	Operculina ammonoides
			Cibicididae	Cibicides	Cibicides lobatulus
					Cibicides refulgens
		Nonionacea	Nonionoidae	Nonionoides	Nonionoides elongatum
					Nonionoides boueanum
		Asterigerinacea	Amphisteginidae	Amphistegina	Amphistegina lessonii
					Amphistegina radiate
		Discorbacea	Glabratellidae	Glabratella	Glabratella australensis

5 Discussion

5.1 Weathering and Mobility of Elements

The intensity of the source area can be calculated by using alkali cations and residual Al_2O_3. An increase in weathering intensity results in a decrease of alkali cations, and enrichment of residual Al_2O_3 in the soil profile (Nesbitt and Young, 1982; Selvaraj and Chen, 2006;

Armstrong-Altrin et al., 2014, 2015; Nagarajan et al., 2014a, b, 2015b, 2017a, b). The chemical indices, such as the Chemical Index of Alteration (CIA; Nesbitt and Young, 1982), Chemical Index of Weathering (CIW; Harnois, 1988), and the Weathering Index of Parker (WIP; Parker, 1970) are commonly used to determine the weathering intensity using molecular proportions of mobile alkali and alkaline Earth metal concentrations after correcting Ca in apatite. For this study, no correction was required for Ca in carbonates because all the studied samples lacked carbonates; however, Ca content in these samples may be influenced at a minor level by the fossil content in them. In those cases, the method proposed by Bock et al. (1998) was adopted. The CIA values for SW part ranged from 64 to 78, and for the NE part 63 to78, with average values of 68 (NE part) and 67 (SW part), indicating moderate weathering intensity. This is further supported by CIW values of 72–84 and 71–82 for the SW and NE parts. The weathering intensity is also calculated based on single element mobility by comparing mobile elements with the nonmobile elements, which have similar magmatic compatibility (Gaillardet et al., 1999). However, the nonmobile elements were replaced with Al (Garzanti et al., 2009) to overcome the anomalous concentration of heavy minerals by hydrodynamic processes (that can show enrichment of nonmobile elements in the sediments; i.e., Ti, Th, Nd, and Sm (Gaillardet et al., 1999), which are preferentially hosted in ultradense minerals such as monazite, allanite, titanite, ilmenite, rutile, etc. (Garzanti et al., 2013a)). The αAl values calculated for both SW and NE parts based on Garzanti et al. (2013a) show the bulk-sediment mobility sequence as follows.

$$\propto_{Na}^{Al} > \propto_{Ca}^{Al} > \propto_{Sr}^{Al} > \propto_{Mg}^{Al} > \propto_{K}^{Al} > \propto_{Rb}^{Al} > \propto_{Ba}^{Al} > \propto_{Cs}^{Al} \text{ [SW Part]}$$
$$\propto_{Ca}^{Al} > \propto_{Na}^{Al} > \propto_{Sr}^{Al} > \propto_{Mg}^{Al} > \propto_{K}^{Al} > \propto_{Ba}^{Al} > \propto_{Rb}^{Al} > \propto_{Cs}^{Al} \text{ [NE Part]}$$

The bulk mobility sequence of the individual elements is slightly different between the SW and NE parts. The slightly high values for Na, Ca, Sr, and Mg indicate relatively higher mobility than the larger cations, such as K, Rb, Ba, and Cs. However, the αAl values are relatively lower for the studied sands due to the dilution effect by quartz, although recycling and true elements such as Mg, K, Rb, and Cs are mainly associated with phyllosilicates. This inference is also supported by low concentrations of alkali and alkaline-earth metals and a high content of SiO_2 against UCC (Fig. 5). Depletion of these metals is not necessarily the imprint of weathering, because quartzose polycyclic detritus also displays similar characters (Garzanti et al., 2006). The samples from the NE part show αAl values higher than the SW part, due to the very fine sand and mud dominated sediments in the NE part than the SW part, where samples are medium to fine sand dominated. The A-CN-K plot (Fig. 7) plot was also utilized to ascertain the weathering trends inferred from CIA and CIW, and it shows that the studied samples are plotted along a granodiorite weathering trend up to the average shale, confirming the interpretation of moderate weathering based on other weathering indices.

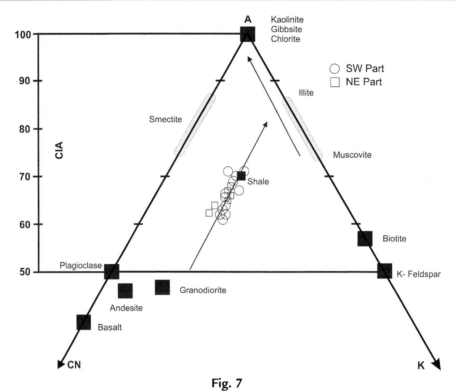

Fig. 7

A-CN-K plot showing the weathering trend for the beach sediments from the Miri coastal region. *After Nesbitt, H.W., Young, G.M., 1984. Prediction of some weathering trends of plutonic and volcanic rocks based on thermodynamic and kinetic consideration. Geochim. Cosmochim. Acta 48(7), 1523–1534.*

5.2 Sorting and Recycling

Sedimentary sorting during transportation and deposition leads to enrichment of selected primary (i.e., quartz, feldspar) and heavy minerals (i.e., zircon, rutile, etc.) in the coarse fraction of the sediments; whereas fine suspended sediments are enriched with the secondary and light minerals (Wu et al., 2013). The sediment recycling and sorting increases the zircon grain in the sediments that enhance the Zr/Sc ratio of the ensued sediments. This compositional variation emanating from sediment recycling and sorting can be tracked through positive trend of the Th/Sc ratio, with Zr/Sc in the bi-blot of Zr/Sc versus Th/Sc (McLennan et al., 1993). Accordingly, the beach sediments fall near the upper end of trend 1, and many samples have followed trend 2 in the plot (Fig. 8), indicating the addition of zircons due to recycling, and thus zircon enrichment. The Th/Sc ratio ranges from 0.90 to 2.10, and from 0.85 to 4.20. The corresponding samples fall along the compositional variation line, suggestive of differences in provenances of the studied sediments. The Zr/Sc ratios are higher and scattered along trend 2, indicative of significant reworking, and consistent with zircon enrichment (McLennan et al., 1993). However, La/Th versus Hf plot (Fig. 9), in which samples mostly fall

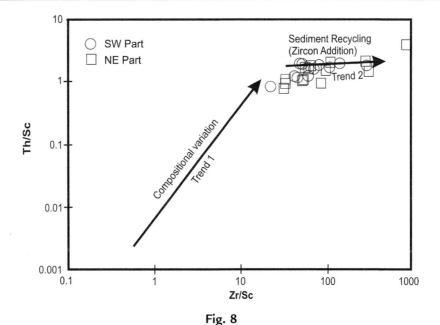

Fig. 8

Th/Sc vs. Zr/Sc biplot shows the effect of sediment recycling in the beach sediments of Miri region.

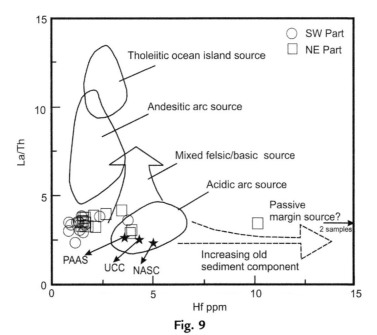

Fig. 9

Hf vs. La/Th bi-blot shows the provenance characters for the sediments from the Beaches of Miri coastal region. *After Floyd, P.A., Leveridge, B.E., 1987. Tectonic environment of the Devonian Gramscatho basin, south Cornwall: framework mode and geochemical evidence from turbiditic sandstones. J. Geol. Soc. 144(4), 531–542.*

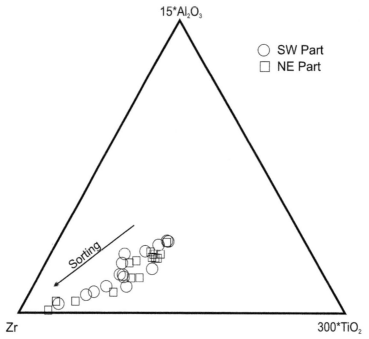

Fig. 10

Al-Zr-Ti plot shows the sorting trend toward Zr apex due to sedimentary sorting effect. *After Garcia, D., Fonteilles, M., Moutte, J., 1994. Sedimentary fractionation between Al, Ti, and Zr and genesis of strongly peraluminous granites. J. Geol. 102, 411–422.*

near PAAS and UCC. In addition, three samples from the NE part show enrichment of Hf (two of them are plotted in this plot due to high Hf content). Together, these may be indicative of the recycling and sorting effects that selectively favored enrichment of zircon against rutile or anatase. This assumption gains support from the triplot Al-Zr-Ti (Fig. 10). Garzanti et al. (2013b) used a CIA/WIP ratio to differentiate the weathering effect from recycling. The WIP values decrease linearly when compared with weathering indices such as CIA and αAl values. In this case, the recycled sediments should show higher CIA/WIP ratios with lower WIP values (and can approach zero for recycled sand) by the addition of quartz to the sediment. The WIP ratios of this study ranged between 2–6 and 3–8 for the SW and NE parts, respectively, which confirms the quartz enrichment (quartz rich lithic fragments; i.e., metasandstones) by a recycling effect. In the CIA versus WIP plot (Fig. 11), the sands from the Miri beaches deviated from the weathering trend and clustered together with very low WIP values, particularly in the SW part compared with the NE part.

5.3 Provenance

The provenance of the sediments was interpreted based on the geochemical signatures of the beach sediments from the Miri coastal region. The major oxides-based provenance discrimination diagram (Roser and Korsch, 1988) was used, in which all the samples are plotted

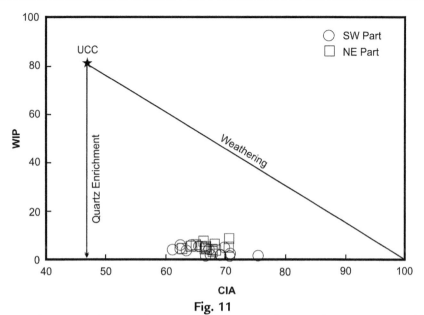

Fig. 11

CIA/WIP plot shows enrichment of quartz in the beach sediments due to recycling effect. *After Garzanti, E., Padoan, M., Andò, S., Resentini, A., Vezzoli, G., Lustrino, M., 2013a. Weathering at the equator: petrology and geochemistry of East African Rift sands. J. Geol. 121(6), 547–580. doi: 10.1086/673259.*

as a cluster in the quartzose sedimentary provenance (Fig. 12), indicating a recycled passive margin source. This is further supported by the WIP values, CIA/WIP ratios, and WIP versus CIA plot (Fig. 11). In the Hf versus La/Th plot (Fig. 9, Floyd and Leveridge, 1987), most of the samples are plotted in the mixed felsic and mafic source, and few samples are in the acidic arc field, indicating felsic to intermediate source rocks. Only a few other samples ($n=3$ from NE part) were plotted away from the rest of the samples by the enrichment of zircon due to recycling. Although all the samples showed a recycled nature, only three samples were highly affected by the sorting effect, which resulted in enrichment of heavy minerals (i.e., zircon). Except those samples, all the samples show felsic to intermediate source rock characteristics. Similarly, Sc versus Th/Sc plot (Fig. 13) also suggested felsic to intermediate characteristics of the source rocks. In this plot, the samples are plotted between UCC and granite, with the majority of the samples falling near granite source rocks. The geochemical ratios, such as La/Sc, La/Co, Th/Sc, Th/Co, Cr/Th, $(La/Lu)_{CN}$, and Eu/Eu* of the studied sediments were compared with the sediments derived from felsic and mafic sources, and UCC and PAAS values (Table 4). From this table, it is clear that the Miri Beach sediments were dominantly derived from a felsic dominated source region. The Zr/Hf ratios are related to the zircon minerals, and the values recorded were comparable to UCC and PAAS. However, a few samples showed a higher Zr/Hf ratio, which may be related to selective dissolution of the heavier isovalent element due to long exposure in the alkaline seawater

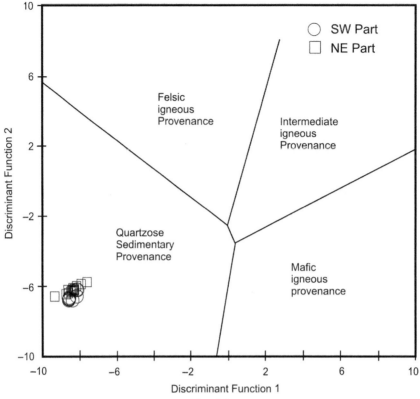

Fig. 12

Discriminant function diagram for the provenance signatures of the beach sediments from Miri coastal region sandstones (Roser and Korsch, 1988). The discriminant functions are: Discriminant Function $1 = (-1.773 \times TiO_2) + (0.607 \times Al_2O_3) + (0.760 \times Fe_2O_3) + (-1.500 \times MgO) + (0.616 \times CaO) + (0.509 \times Na_2O) + (-1.224 \times K_2O) + (-9.090)$; Discriminant Function $2 = (0.445 \times TiO_2) + (0.070 \times Al_2O_3) + (-0.250 \times Fe_2O_3) + (-1.142 \times MgO) + (0.438 \times CaO) + (1.475 \times Na_2O) + (-1.426 \times K_2O) + (-6.861)$.

(El-Kammar et al., 2011). The recorded Zr/Hf ratios are identical to the zircons derived from most of the crustal rocks (37; Brooks, 1970), and particularly granitic rocks (~39; Erlank et al., 1978; 38.5; Wang et al., 2010).

5.4 Statistical Analysis

The correlation analysis and factor analysis with varimax rotation were carried out to elucidate the geochemical processes controlling the chemistry of beach sands of the Miri coastal area. The parameters that exhibited the values below detection limit were eliminated. Similarly, parameters that do not show any variation between the samples were also eliminated from the analysis.

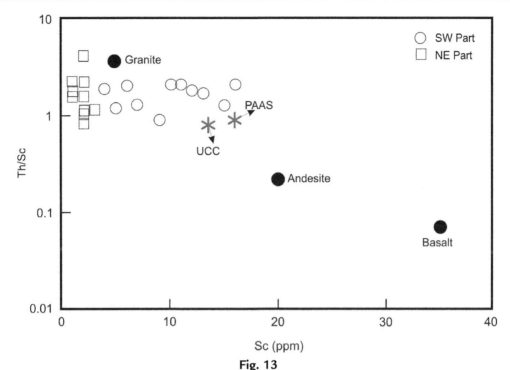

Fig. 13
Th/Sc versus Sc bivariate plot for beach sediments from the Miri coastal region. PAAS and UCC (Taylor and McLennan, 1985; McLennan, 2001, respectively), and granite, andesite, and basalt (Condie, 1993).

5.4.1 Correlation coefficient

The main purpose of applying statistical techniques is to bring out the inherent associations of the elements in a dataset, thereby understanding possible mineralogical compositions of the sediments, the geochemical processes that the sediments have undergone, and how the environmental factors exerted control over sediment composition, including grain size. Worldwide beach sediments are known to have been dominated by sand (Armstrong-Altrin et al., 2012; Pereira et al., 2016). There are several recent studies on the trace element content of beach sediments (Armstrong-Altrin et al., 2015; Pit et al., 2017), demonstrating that beach sediments are indeed a highly complex matrix. In this regard, an attempt was made with the use of Pearson's correlation analysis to understand the elemental associations. The correlation coefficient values ($r > 0.5$) were considered for the discussion.

SW part (Miri River Mouth to Bungai Beach)

The SiO_2 shows weak to moderate negative correlation with many elements (Al_2O_3, MgO, K_2O, TiO_2, V, Sr, Rb, LREEs, Yb, Th and <125 μm fraction, which indicates that the silica is controlled by quartz) (e.g., Osman, 1996; Rahman and Suzuki, 2007; Tiju et al., 2018), and

Table 4 Range of elemental ratios of beach sediments from the Miri Coastal region, compared with range of sediments derived from felsic and mafic rocks, Upper Continental Crust, and Post-Archaean Australian Shale

Element Ratio	Range of Elemental ratio in the Sediments of Miri Beaches[a]		Range of Sediments From[b]		UCC[c]	PAAS[c]
	SW Part[1]	NE Part[2]	Felsic Source	Mafic Source		
La/Sc	3.05–7.30 (5.72; n=11)	3.00–15.00 (5.93; n=15)	2.50–16.3	0.43–0.86	2.21	2.40
La/Co	3.05–7.30 (4.92; n=11)	3.00–10.00 (4.75; n=15)	1.80–13.8	0.14–0.38	1.76	1.66
Th/Sc	0.90–2.10 (1.67; n=11)	0.85–4.20 (1.63; n=15)	0.84–20.5	0.05–0.22	0.79	0.90
Th/Co	0.90–2.00 (1.45; n=14)	0.85–2.80 (1.31; n=15)	0.67–19.4	0.04–1.40	0.63	0.63
Cr/Th	10.53–14.29 (12.72; n=3)	8.70–17.86 (11.71; n=8)	4.00–15.0	25–500	7.76	7.53
(La/Lu)$_{CN}$	4.46–11.94 (8.30; n=16)	5.56–10.15 (8.21; n=15)	3.00–27.0	1.10–7.00	9.73	9.22
Eu/Eu*	0.40–0.79 (0.64; n=16)	0.46–0.78 (0.68; n=15)	0.40–0.94	0.71–0.95	0.65	0.65
Zr/Hf	31.82–51.25 (43.26)	35.77–60.20 (42.43)	–	–	33	42

[a](1, 2), This Study.
[b]Cullers (1994, 2000); Cullers and Podkovyrov (2000); Cullers et al. (1988).
[c]Taylor and McLennan (1985) and McLennan (2001).
–, Not determined.

quartz is not associated with any particular fraction of the sediments. However, a weak positive correlation exists between SiO_2 and medium-fine sands ($r=0.49$). Similarly, Al_2O_3 shows strong positive correlation with Na_2O ($r=0.97$; $P\leq.01$), K_2O (0.98; $P\leq.01$), Ba (0.96; $P\leq.01$), Zn (0.72; $P\leq.01$), Rb (0.98; $P\leq.01$), and <125 μm fraction (0.85; $P\leq.01$), and a weak to moderate correlation with TiO_2, Sr, Co, Eu, and Th, which indicates their association with clay minerals and phyllosilicates (e.g., Hofer et al., 2013), and also their strong association with the very fine sand-mud fraction ($r=0.85$; $P\leq.01$). The significant negative correlation of CIA with Al_2O_3 (−0.62; $P=.06$), Na_2O (−0.77; $P\leq.01$), K_2O (−0.74; $P\leq.01$), and Ba (−0.80; $P\leq.01$) reveal that weathering processes are responsible for the mobilization of these elements from the source rock (Meybeck, 1987; Amiotte Suchet et al., 2003). The $\sum$REEs are mainly controlled by heavy minerals (Ti-silicates, rutile, zircon etc.). This is evident by their positive correlation with TiO_2 (0.90; $P\leq.01$), Th (0.79; $P\leq.01$), U (0.76; $P\leq.01$), Zr (0.59; $P\leq.01$), and Hf (0.59; $P\leq.01$). However, HREE shows moderate positive correlation with Fe_2O_3 (0.64; $P\leq.05$), and MnO (0.64; $P\leq.05$), indicative of adsorption of HREEs over Fe-Mn oxy-hydroxides (Koppi et al., 1996; Mutakyahwa et al., 2003; Pokrovsky et al., 2006). REEs occur as carbonate complexes in seawater (Ohta and Kawabe, 2000). In addition, REEs in seawater form cationic and anionic carbonate complexes, which are used by two hydrogenetic constituents; hydrous δ-MnO_2 and Fe-Oxyhydroxides (Ohta and Kawabe, 2000). The δ-MnO_2 fractionates LREE complexes, whereas Fe-oxyhydroxides fractionate HREE complexes. While the SW part showed Fe and Mn complexation behavior with HREEs, the coarse sand does not represent any role in the SW part. However, <125 µm fraction showed control over selected elements associated with the phyllosilicates (i.e., Al_2O_3, Na_2O, K_2O, Ba, Co, Rb and Eu: $r \geq 0.68$; $P \leq .05$). Cu and Zn show different characteristics in the SW part compared with the NE part in terms of a weak positive correlation ($r = 0.57$). However, Cu was not controlled by clay minerals, grain size, and Fe-Mn oxy-hydroxides indicative of characteristic features related to the source rocks, and Zn was weakly ($r = 0.51$) associated with <125 µm fraction.

NE part (Baram River Mouth to Miri River Mouth)

The SiO_2 content in the sands showed moderate negative correlation with Al_2O_3 (−0.75; $P \leq .01$), Fe_2O_3 (−0.63; $P \leq .01$), Na_2O (−0.64; $P \leq .01$), K_2O (−0.70; $P \leq .01$), Sc (−0.60; $P \leq .01$), Ba (−0.65; $P \leq .01$), Sr (−0.75; $P \leq .01$), Ga (−0.68; $P \leq .01$), Rb (−0.68; $P \leq .01$), and Cs (−0.81; $P \leq .01$). Representation of quartz as the exclusive form of Si content can be inferred from this correlation (Osman, 1996; Rahman and Suzuki, 2007; Babu, 2017; Tiju et al., 2018). In addition, association of quartz with medium-fine sand fractions ($r = 0.76$; $P \leq .01$), and significant quartz dilution in the beach sediments (e.g., Malick et al., 2012) are also inferred. Aluminum showed moderate to strong positive correlation with Fe_2O_3 (0.74; $P \leq .01$), Na_2O (0.90; $P \leq .01$), K_2O (0.99; $P \leq .01$), Sc (0.65; $P \leq .01$), Ba (0.97; $P \leq .01$), Sr (0.90; $P \leq .01$), Ga (0.90; $P \leq .01$), Rb (0.99; $P \leq .01$), and Cs (0.72; $P \leq .01$), indicative of the grain size (<125 µm fraction of the sediments; $r = 0.76$; $P \leq .01$) and mineralogical (clay, phyllosilicates) control over these elemental abundances (e.g., Hofer et al., 2013). Fe_2O_3 and MnO showed moderate positive correlations (0.61; $P \leq .01$), indicating their association as Fe-Mn oxides and oxyhydroxides, and scavenged V, Co, Eu, and Th (Koppi et al., 1996; Mutakyahwa et al., 2003; Pokrovsky et al., 2006). However, most of the trace elements are mainly controlled by Mn-oxides (e.g., Koppi et al., 1996; Mutakyahwa et al., 2003) and detrital heavy minerals (i.e., zircon, rutile, allanite etc.). The positive correlations of MnO with MgO, TiO_2, Y, Zr, Co, Cu, Zn, Sn, Hf, Ta, Pb, Th, U, and REEs ($r > 0.67$; $P \leq .01$) and their association with coarse sand fractions indicates that this fraction is the result of occurrences of lithic fragments and heavy minerals. The Fe-Mn oxides generally accumulate in fine grained sediments and control the REE abundance (Mao et al., 2014), but in this case, the REEs are associated with sand fraction, which indicates that the heavy minerals and lithic fragments control the REE abundance more than the clay minerals. This observation is further supported by the moderate to strong positive correlations of REEs with trace elements, namely, Y ($r = 0.93$–0.99), Zr ($r = 0.93$–0.99), Co ($r = 0.51$–0.71), Cu ($r = 0.74$–0.83), Zn ($r = 0.80$–0.88), Sn ($r = 0.76$–0.83), Hf ($r = 0.94$–0.99), Ta ($r = 0.91$–0.98), Pb ($r = 0.71$–0.82), Th ($r = 0.94$–0.99), and U ($r = 0.96$–1.00). The strong positive correlations of Y with REEs suggest that the lanthanides are mainly controlled by Y bearing minerals (i.e., xenotime)

(Lopez et al., 2005). The zirconium showed moderate to strong positive correlations with REEs, Hf, Th, U, Pb, Cu, and Zn, suggestive of its association almost exclusively with the heavy mineral zircon (e.g., Gu et al., 2013) and also the occurrence of this mineral with sand fraction ($r \geq 0.80$; $P \leq .01$). However, the moderate positive correlations of Cu ($r=0.70$; $P \leq .01$), Zn ($r=0.72$; $P \leq .01$), V ($r=0.61$; $P \leq .01$) and Pb ($r=0.62$; $P \leq .01$) with coarse sand fractions may indicate that they are mainly controlled by recycling of sedimentary lithic fragments (shale and meta-shale) derived from the source area.

5.4.2 Factor analysis

With the correlation analysis, albeit it being a powerful tool to explore the relationship among various constituents in the sediments, it becomes difficult to understand the complexities involved in the distribution of elements in the sediments in accordance with the grain size distribution. The factor analysis reduces the dimensions of the data, and highlights the latent variables. The Principal Component Analysis (PCA) was run to extract important components, and factor loadings were obtained after a Varimax rotation. The data belonging to the sediments collected from the SW and NE parts are separately treated so that the sedimentary processes in these two areas could be distinguished. The Principal Components are presented in Tables 5 and 6 and Figs. 14 and 15 for SW and NE parts, respectively.

SW part

For the SW part, the factor model brought out seven factors, explaining 94% of total variances (Table 5). Factor 1 explains 25% of the total variance, and is represented by the positive loadings of TiO_2, Zr, Hf, U, LREEs, and $\sum REE$. While confirming the observations made based on correlation analysis, this factor unequivocally establishes the association of these elements exclusively in heavy mineral phases, such as Ti-silicates, rutile, and zircon (e.g., Lopez et al., 2005). However, these heavy minerals are not controlled by the particular size fraction of the sediments. The strong positive loadings of LREEs with TiO_2, rather than Zr and Hf, indicates that the LREEs are mainly controlled by titanites and they have a relatively higher REE content (Gromet and Silver, 1983; Lopez et al., 2005) up to 5000 ppm (Basir and Balakrishnan, 1999). Factor 2 explains 21% of the total variance with positive loadings of Y, Th, Ta, and HREEs, and negative loading by Cu. The Cu concentrations are controlled by the terrigenous lithic fragments, such as meta-sandstone and meta-mudstone, and the HREEs are mainly controlled by Y and Th bearing minerals such as monazite, xenotime, and allanite (e.g., McLennan, 1989; Preston et al., 1998; Lopez et al., 2005).The preferential loading of LREE on the first factor and HREE on the second factor suggest that the distributions of LREE and HREE in the beach sediments are being controlled by two distinct factors, and there is no influence from grain size distribution. Among various processes responsible for REEs fractionation, a few processes, namely, the enrichment of HREEs in the riverine sediments, preferential removal of LREEs in low-salinity regions, and adsorption and coprecipitation

Table 5 Rotated Matrix components for SW part of the study area (for the discussion, only the factor loadings with absolute value >0.5 are considered and are indicated in "bold font")

Parameters	Rotated Component Matrix						
	Factor 1	Factor 2	Factor 3	Factor 4	Factor 5	Factor 6	Factor 7
C Sand	−0.455	0.084	−0.204	−0.035	0.321	**0.733**	0.033
M-F Sand	−0.155	0.048	**−0.803**	−0.264	−0.352	0.175	0.122
VF Sand-Mud	0.201	−0.056	**0.791**	0.256	0.298	−0.253	−0.121
SiO_2	−0.413	−0.320	−0.337	−0.355	−0.103	0.367	0.320
Al_2O_3	0.219	0.129	**0.947**	0.151	0.018	−0.028	−0.090
Fe_2O_3	0.141	0.495	−0.282	**0.687**	0.367	0.015	−0.056
MnO	0.137	0.463	−0.201	**0.795**	0.156	0.136	−0.142
MgO	0.329	0.090	0.189	**0.886**	0.057	−0.061	0.019
CaO	0.137	0.141	0.145	**0.939**	−0.163	0.019	0.032
Na_2O	0.205	−0.039	**0.953**	0.117	−0.085	−0.048	0.037
K_2O	0.252	−0.020	**0.951**	0.123	0.005	−0.063	−0.002
TiO_2	**0.850**	0.020	0.358	0.182	−0.026	0.015	0.045
LOI	0.226	0.187	0.391	**0.855**	0.043	−0.061	−0.027
V	0.474	0.167	−0.015	0.599	0.480	−0.311	0.100
Ba	0.248	−0.068	**0.947**	0.106	−0.055	−0.044	0.084
Sr	0.209	0.104	0.369	**0.862**	−0.176	−0.009	0.078
Y	0.236	**0.842**	0.029	0.059	−0.040	−0.172	0.085
Zr	**0.655**	0.257	−0.472	0.020	−0.068	0.047	**0.508**
Co	0.008	0.182	**0.554**	**0.706**	0.141	−0.142	0.152
Cu	−0.147	**−0.639**	**0.530**	−0.390	0.150	0.188	−0.168
Zn	−0.182	−0.015	**0.855**	−0.167	−0.139	0.253	−0.030
Rb	0.120	0.256	**0.936**	0.127	−0.060	0.009	−0.118
Sn	−0.234	−0.078	0.049	−0.086	**−0.878**	−0.195	0.123
La	**0.935**	0.224	0.190	0.105	0.083	−0.098	−0.006
Ce	**0.938**	0.162	0.198	0.142	0.111	−0.098	0.077
Pr	**0.903**	0.345	0.084	0.166	0.110	−0.103	−0.020
Nd	**0.913**	0.276	0.155	0.151	0.104	−0.096	−0.067
Sm	**0.782**	**0.507**	0.081	0.077	0.182	−0.034	−0.049
Eu	**0.731**	0.237	0.467	0.352	−0.043	0.089	0.072
Gd	0.369	**0.821**	0.052	0.341	0.034	0.089	−0.125
Dy	**0.545**	**0.728**	−0.086	−0.009	0.314	−0.014	0.010
Ho	0.213	**0.823**	0.141	0.280	0.016	0.020	0.132
Er	0.349	**0.889**	−0.001	0.052	0.007	0.228	−0.003
Tm	0.040	**0.810**	0.041	0.161	0.045	−0.372	0.375
Yb	0.408	**0.816**	0.074	0.274	0.117	0.136	−0.061
Lu	0.454	**0.816**	−0.084	0.201	−0.047	0.171	0.064
Hf	**0.624**	0.322	−0.437	0.127	−0.164	0.100	0.484
Ta	−0.277	**0.751**	−0.253	−0.143	−0.104	−0.246	−0.239
Pb	−0.144	0.020	−0.267	**0.562**	**−0.638**	0.025	−0.184
Th	**0.535**	**0.633**	0.240	0.364	0.048	−0.083	−0.218
U	**0.630**	0.477	−0.027	0.450	−0.044	0.196	0.345
CIA	−0.127	0.472	**−0.724**	0.022	0.069	0.066	−0.377
LREE	**0.935**	0.221	0.184	0.136	0.105	−0.095	0.023
HREE	0.456	**0.849**	0.044	0.193	0.105	0.106	−0.044
$\sum$REE	**0.913**	0.306	0.174	0.149	0.109	−0.074	0.016
Variance %	24.66	21.36	20.88	15.52	5.14	3.40	3.06
Eigen value	11.09	9.61	9.40	6.98	2.31	1.53	1.38

Table 6 Rotated Matrix components for NE part of the study area

Parameters	Rotated Component Matrix			
	Factor 1	Factor 2	Factor 3	Factor 4
C Sand	**0.865**	0.012	−0.148	0.022
M-F Sand	0.092	**−0.761**	0.497	−0.285
VF Sand-Mud	−0.095	**0.760**	−0.496	0.285
SiO_2	0.027	**−0.775**	0.196	−0.470
Al_2O_3	−0.199	**0.976**	−0.044	−0.047
Fe_2O_3	0.481	**0.861**	0.004	0.015
MnO	**0.917**	0.167	0.263	0.047
MgO	**0.932**	0.274	0.138	0.040
CaO	−0.127	0.472	**0.758**	0.041
Na_2O	0.005	**0.930**	0.279	−0.004
K_2O	−0.254	**0.956**	0.000	−0.087
TiO_2	**0.977**	0.020	−0.138	0.020
LOI	0.031	**0.970**	0.041	−0.023
Sc	0.319	**0.731**	−0.010	−0.008
V	**0.740**	0.580	−0.070	−0.128
Ba	−0.232	**0.943**	0.068	−0.120
Sr	0.100	**0.956**	0.131	0.026
Y	**0.959**	−0.027	0.068	−0.079
Zr	**0.974**	−0.108	0.093	0.060
Co	0.628	0.540	0.010	−0.301
Cu	**0.831**	−0.007	0.344	−0.257
Zn	**0.873**	0.000	0.313	−0.229
Ga	0.045	**0.922**	0.055	−0.039
Rb	−0.180	**0.961**	−0.073	−0.134
Sn	**0.818**	−0.038	0.116	0.392
Cs	0.146	**0.751**	−0.423	0.295
La	**0.998**	−0.006	0.006	0.013
Ce	**0.997**	−0.016	0.019	0.031
Pr	**0.996**	−0.010	0.024	0.041
Nd	**0.996**	0.002	0.010	0.047
Sm	**0.992**	−0.017	0.054	0.065
Eu	**0.987**	0.122	−0.001	−0.016
Gd	**0.994**	0.003	0.021	0.058
Tb	**0.965**	−0.028	−0.008	0.170
Dy	**0.994**	0.020	−0.011	0.051
Ho	**0.953**	−0.076	−0.112	−0.042
Er	**0.991**	−0.028	−0.018	−0.018
Tm	**0.995**	−0.030	−0.007	−0.004
Yb	**0.997**	−0.015	0.020	0.012
Lu	**0.996**	−0.037	0.053	0.021
Hf	**0.986**	−0.116	0.078	0.046
Ta	**0.971**	−0.068	−0.104	−0.112
Pb	0.771	0.048	0.114	**0.585**
Th	**0.994**	0.089	0.013	0.014
U	**0.995**	0.012	0.015	0.000

Table 6 Rotated Matrix components for NE part of the study area—Cont'd

Parameters	Rotated Component Matrix			
	Factor 1	Factor 2	Factor 3	Factor 4
CIA	−0.229	0.351	**−0.868**	−0.068
LREE	**0.997**	−0.009	0.015	0.031
HREE	**0.998**	−0.011	0.001	0.029
$\sum$REE	**0.998**	−0.010	0.014	0.031
Variance %	61.57	24.47	5.26	2.70
Eigen value	30.17	11.99	2.58	1.32

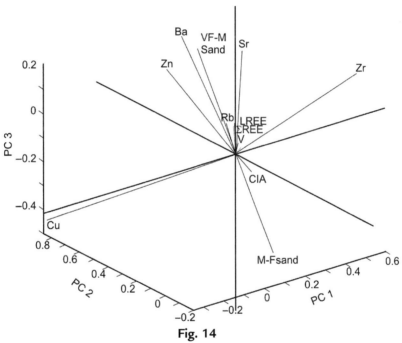

Fig. 14
Triplot of the first three Principal components for SW data.

reactions in the estuarine region (Elderfield et al., 1990; Munksgaard et al., 2003) are considered to be important and widely prevalent. Selective sediment dispersal could fractionate REEs due to the enrichment of LREE with clays, while, Zr and Hf result in preferential concentration with sand and silt fractions, rather than clay fractions (Klaver and van Weering, 1993). Accordingly, REE fractionation in the coastal sediments of Miri are interpreted, though the exact mechanisms by which such a fractionation has occurred are not known. Factor 3 is explained in about 21% of the total variance, and is positively loaded with Al_2O_3, Na_2O, K_2O, Ba, Co, Cu, Zn, Rb, and <125 μm fraction, which indicates their association with the clay minerals and phyllosilicates (Fig. 14). The medium to fine sand fraction and Chemical Index of

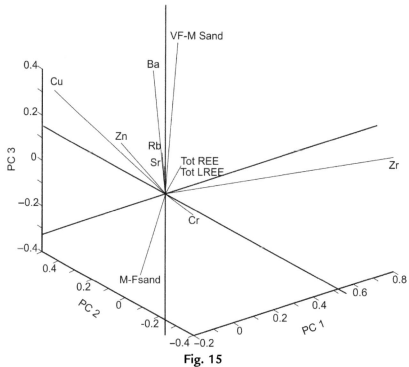

Fig. 15
Triplot of the first three principal components for NE data.

Alteration (CIA) are negatively loaded in this factor (Fig. 13). It seems that with an increase in medium to fine sand content, the CIA also increases. However, the Cu is loaded with both 2nd and 3rd factors, suggestive of their association with clay and lithic fragments. In the 3D plot of the first three factors, (Fig. 14), medium to fine sand and <125 μm fraction are the major controlling factors, loaded independently, which indicates the control exercised by grain size over selected elements. Factor 4 is represented by the positive loadings of Fe_2O_3, MnO, MgO, CaO, LOI, V, Sr, Pb, and Co, and explains 16% of the total variance. This factor indicates that these elements are scavenged by Fe-Mn oxy-hydroxides. The coated carbonates (Young and Harvey, 1992; Gu et al., 1995) present as Fe-Mn bearing calcareous clay minerals, and adsorbed V, Pb, and Co on their surfaces, which can also be inferred from this factor. Factor 5 explains 5% of the total variance represented by the negative loadings of Sn and Pb, which indicates their association with chalcophile minerals such as pyrite, chalcopyrite, arseno-pyrite, and so forth. Pyrites are common in the source area of the Baram River Basin (Oligocene deepwater sediments) and Miri River Basin (concretions in the Neogene sedimentary rocks; Nagarajan et al., 2014b; Table 7). They might have been transported by the rivers to the South China Sea and redistributed by littoral currents onto the beach. The 6th factor is represented by coarse sand, which explains 3.4% of the total variance. The SiO_2 loading is not strong, and there is no significant correlation between coarse sand and silica. Therefore, this factor can be thought of

Table 7 Comparison of selected elements of the present study with the source area/adjacent geological formations and river sediments

Area of Study	Fe %	Mn %	Cu	Pb	Co	Ni	Zn	Cr	As	V	Reference
SW part of the Miri coastal area	0.21–0.59 (0.38)	0.002–0.009 (0.005)	130–260 (217.5)	5–11 (7.0)	1–2 (1.3)	BDL	60–120 (95.6)	20–30 (23.3)	5–9 (7.3)	9–15 (11.6)	Present study
NE part of the Miri Coastal area	0.29–0.71 (0.42)	0.003–0.01 (0.005)	110–490 (226.7)	5–19 (8.3)	1–3 (2.1)	BDL	50–240 (104)	20–150 (45.0)	BDL	10–21 (15.00)	Present study
Miri River (n=27)	0.18–4.27 (2.20)	0.002–0.042 (0.012)	20–270 (82.6)	6–20 (11.6)	1–12 (6.0)	20–30 (26.4)	40–220 (102.4)	30–130 (84.6)	5–20 (11.1)	5–120 (64.9)	Nagarajan (per. comm.)
Lower Baram River	1.30–3.62 (4.04)	0.008 to 0.08 (0.03)	30–410 (142.8)	11–123 (28.5)	5–18 (9.8)	20–40 (27.8)	60–250 (123.1)	50–90 (64.7)	5–10 (6.5)	42–102 (83.9)	Prabakaran (2017)
Upper Tukau Formation (n=14)	0.75	0.007	42.9	16.9	5.29	18.14	36.3	172	–	62.36	Nagarajan et al. (2014a, b)
Lower Tukau Formation (n=50)	1.39	0.007	41.32	16.34	6.94	32.14	60.56	74.19	8.03	70.22	Nagarajan et al. (2017b)
Lambir Formation (n=30)	2.12	0.023	77	18.23	7.97	27.86	73.46	54.07	13.71	76.63	Nagarajan et al. (2017a)
Sibuti Formation	3.08	0.038	65.3	47.6	13.4	65.4	117.3	95.0	–	104.9	Nagarajan et al. (2015b)
Concretions from Tukau (n=30)	24.5	0.072	–	60.7		660	–	–	–		Nagarajan et al. (2014b)

BDL, below detection limit.

as representing the separation of coarse sand from the medium to fine sand, and <125 μm fractions by the wave and tidal energy. Zr and Hf are weakly positively loaded in the last factor, which can be related to zircon population derived from different sources, in addition to their association with factor 1 (polycyclic zircons vs fresh young populations).

NE part

The Principal Components plot is presented in Fig. 15 and Table 6. Four factors were extracted, and account for 94% variance in the data pertaining to the NE part. Factor 1 explains 62% of the total variance, and exhibits positive loadings of MnO, MgO, TiO_2, V, Y, Zr, Co, Cu, Zn, Sn, Hf, Ta, Pb, Th, U REEs, and coarse sand. The association of these elements in this factor can be explained by the presence of heavy minerals (rutile and zircons mainly controls TiO_2, Zr, Hf, Th, U, and REEs) and lithic fragments (metasandstones and metamudstone/shale from upstream of the Baram River). In addition, grain size-dependent associations (e.g., Lopez et al., 2005) of MnO, MgO, Cu, Zn, V, Ta, and so forth, are established. This factor implies the sorting of heavy minerals in the coarse sand fraction of the sediments in response to the wave action. Grouping of Zr and Cu in this factor, but taking into account that their plot is in to two different directions (Fig. 15), could suggest association of Zr with heavy mineral occurrence, while lithic fragments are associated with Cu. Occurrences of both types of grains in the beach might be due to buoyancy, wave reworking, and a relatively lower rate of sediment influx than the rate of reworking and redistribution under wave action. Factor 2 explains 24% of the total variance with positive loadings of Al_2O_3, Fe_2O_3, Na_2O, K_2O, LOI, Sc, Ba, Sr, Ga, Rb, Cs, and <125 μm fractions, which implies their association with clay mineral and phyllosilicates. The medium - fine sand and SiO_2 are negatively loaded, indicative of quartz as the dominant constituent of the medium - fine sand sediment fraction. Factor 3 explains 5% of the total variance represented by the positive loadings of CaO and negative loading of CIA. Such an observation points to the fact that whenever CIA is more, the CaO content of the sediments is poised to go down, and vice versa. The CIA takes into consideration only the CaO content of silicate minerals (not from carbonate), and the average CaO content of the sediments in the NE part is less (0.11%) than the SW part (0.16%). Thus, it can be concluded that the silicate minerals in the NE part of the study area are Ca-poor, perhaps due to the presence of microfossils and shell fragments in the NE part. However, the presence of foraminifera towards the Baram River mouth is negligible or rare due to rapid sedimentation. The weak positive loading of medium-fine sand fractions in this factor supports the association of foraminifera in the medium-fine sand fraction of the sediments. The αAl mobility index for Ca is high in the NE part, suggestive of minimal occurrences of carbonate fragments, and tests of foraminifera and ostracoda fossils are common in the NE part. Factor 4 explains 3% of the total variance represented by weak positive loading of Pb, indicating anthropogenic input. The possible sources of this element in the NE part may be due to intense vessel navigation and harbor activities near the mouths of the Baram River and Miri River.

5.5 Environmental Significance

The environmental status of the tourist beaches of the Miri coastal region was assessed based on environmental indices, such as contamination factors, the geoaccumulation index, and the pollution load index. The observations and inferences are detailed herein.

5.5.1 Contamination factor

According to Pekey et al. (2004), the contamination factor can be classified into four groups as low ($CF<1$), moderate ($1 \leq CF \leq 3$), considerable ($3 \leq CF \leq 6$), and very high ($CF>6$). The average values of the elements in the studied beach samples are in the order of Cu>As>Zn>Sn>Pb>Cr>Ta=U>Th>V>Co=Mn>Ba>Sr>Fe for the SW part, and Cu>Zn>Sn>Cr>Pb>U>Ta>Th>V>Co>Ba>Mn>Sr>Fe>As for the NE part. Overall, sediments from the NE part show higher CF values for many elements, except Fe, Mn (recorded same), and As (BDL in all the samples from NE part). Based on the CF values, the sediments of the SW part are highly contaminated with Cu, considerably contaminated by As, and moderately contaminated by Zn; whereas the sediments from the NE part are very highly contaminated by Cu and moderately contaminated by Zn (Fig. 16).

5.5.2 Geoaccumulation index

The $Igeo$ values in the SW part are recorded >0 for the elements Cu, Zn, and As, and for Cr, Cu, Zn, and As in the NE part of the study area (Fig. 17). However, the average positive $Igeo$ values are recorded only for Cu and As, in both parts of the study area. Indeed, As was recorded in the

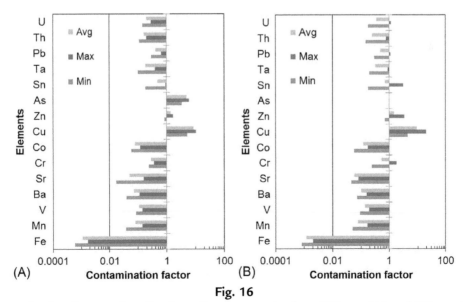

Fig. 16
Contamination Factor values for the sediments samples from SW part (A) and NE part (B).

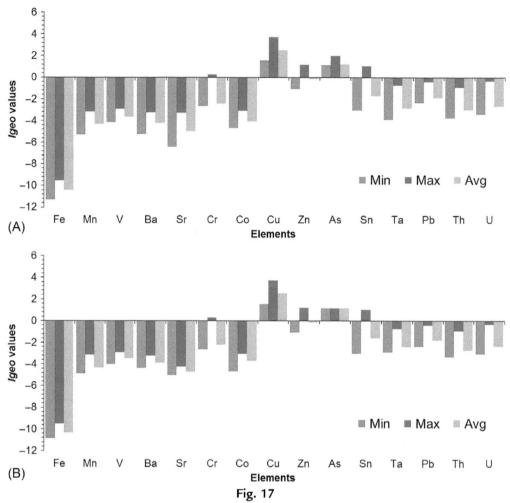

Fig. 17
Range of *Igeo* values for the beach sediments from SW part (A) and NE part (B).

range of *Igeo* Class—2 (avg. 1.25; 1.15 for SW and NE part respectively) for all the samples, while Cu has recorded higher values (*Igeo* = 2.51 and 2.50 for SW and NE part respectively) throughout the study area. All these may imply a natural source, rather than an anthropogenic source, and could be related to the sedimentary to metasedimentary rock provenance.

5.5.3 Pollution load index

Overall pollution load, its variation, and severity of the pollution were calculated using the PLI (Tomlinson et al., 1980). The calculated PLI values range from 0.11 to 0.26 (avg. 0.16) for the SW part, and 0.13 to 0.45 (avg. 0.20) for the NE part, respectively. All the samples show PLI values <1, which implies no contamination. Nevertheless, Cu (most of the locations),

As, and Zn (selected locations) show higher contamination factor and *Igeo* values. These observations are further confirmed by ERL-ERM values of Cu.

5.5.4 Biological effect of total trace metals

The elevated concentrations of heavy metals in the aquatic ecosystems may cause chronic effects on the living organisms (Tellez and Merchant, 2015). Selected trace elemental concentrations were compared with Effects-Range Low (ERL; low probability of effects on biota the elements' concentration below this level) and Effects-Range Median (ERM>50%; above this level elements will have high probability of effect on biota) values proposed by Long et al. (1995), based on multivariate analyses of the relationships between the metal concentration in sediments versus the benthic community's health in various environments, to evaluate the biological effect linked with the degree of contamination. Similarly, trace elements' concentration exceedance above the ERM values may induce ecological concern. The estimated adverse effect of selected metals in the beach sands of NW Borneo is illustrated in Fig. 18. Accordingly, Cu in all the samples from the SW part ranges between ERL and ERM values, and is linked with possible effects on aquatic organism. In addition, 6.7% of As also falls between ERL and ERM values. In the NE part, 81.25% of Cu, 6.25% each of Cr and Zn are recorded between ERL and ERM values. In addition, 18.75% of Cu was recorded above the ERM values, which are related to the higher probability of effect on biota. Overall, the studied beach sediments are contaminated with Cu, and their concentration can affect the biota on the coastal environment. In particular, one location from the NE part is highly contaminated by Cr, Cu, Zn, Sn, Pb, and Mn, perhaps related to both natural and anthropogenic influences.

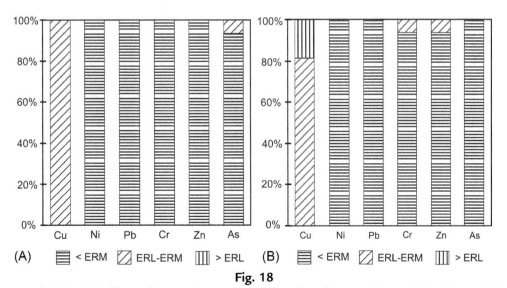

Fig. 18
Level of Biological effects of trace elements in the beach sediments SW part (A); NE part (B).

5.5.5 Comparison of chemistry between costal sediments and source rock/sediment

The geochemistry of coastal sediments from Miri beaches was compared with the chemistry of adjacent geological formations (i.e., Lambir Formation, Tukau Formation, and Sibuti Formation) and river sediments (the Lower Baram River and the Miri River) (Table 7). From this table, it is clear that the chemistry is dominantly controlled by source area geology. Selected elements are enriched by the anthropogenic input to the rivers, which in turn drain to the coastal regions of Miri. In particular, Cu and Zn are enriched in both parts of the study area, and are comparable to the Baram and Miri Rivers, which indicates that riverine sediment influx exercises dominant control over the beach sediment geochemistry. Other elements, such as Cr, V, Pb, Mn, and Fe are found to be comparable to the source area rock composition. However, the concretions from the Tukau and Sibuti Formations may be the source for many chalcophile elements, and need to be studied in detail for ascertaining this inference.

5.5.6 Paleontology and environmental impact through bio-monitoring

Asterorotalia trispinosa, followed by *Psuedorotalia schroeteriana*, are dominant and abundant in the study area. Most of the forms of Ostracoda species are found to be either moderately calcified, pitted, or highly ornate forms, suggestive of either adoptative characteristics to higher energy, sandy substrate, or impacts of turbulent/agitated tidal environmental conditions. Among all the specimens studied, 54 specimens were found to have complete carapaces, while the rest (151 specimens) have open valves, and 18 of them have shown signs of predation. The open valves are recorded higher than the distribution of closed carapaces. Predation can be stated as an interaction between two organisms that results in negative effects on the growth and survival of one of the populations (Odum, 1971). Known predators include coelenterates and many gastropods (muricids and naticids). Muricids drill oblique holes in a deep-water environment, whereas naticids drill vertical holes in a shallow marine environment. There are active predators of many pelecypods, gastropods, scaphopods, ostracods, and barnacles. The following species, namely, *Hemicytheridea paiki*, *Neomonoceratina iniqua*, *Neocytheretta snellii*, and *Actinocytheries scutigera* exhibit predation. Out of these, *Hemicytheridea paiki* shows double predation and *Neomonoceratina iniqua* shows multiple predation.

Foraminifera and ostracoda fossils provide excellent baseline information to understand short-term variability in meiobenthic populations. Such information is essential to evaluate the potential impacts of short-term environmental disturbances from sand mining, pollution, and nutrient influx, or high-frequency climatic variability on coastal and continental shelf ecosystems. The foraminifera population was observed to be more ($n=2252$) than the Ostracod population ($n=205$). The distribution of C/V reveals that the open valves outnumbered closed carapaces. From the preceding observations, it may be concluded that the slow rate of sedimentation prevails on Miri Beach. In order to assess the occurrence and intensity of pollution, morphological deformities of foraminiferal tests and the variation in the color of the ostracod carapace were attempted (Fig. 19A–F). Few of the foraminifera species show some

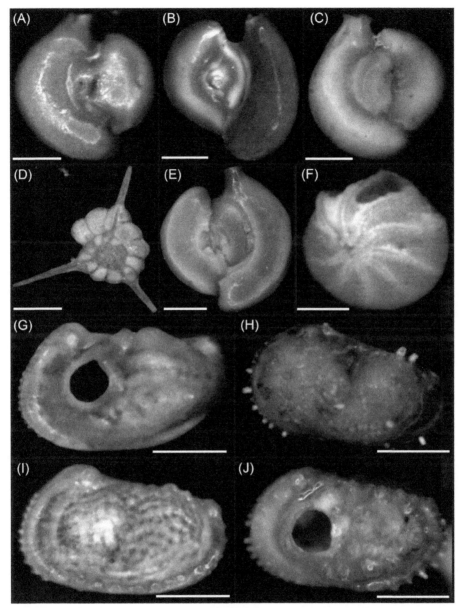

Fig. 19
Selected foraminifera and ostracoda species from Miri Coastal region.(A) *Spiroloculina orbis*, (B) *Spiroloculina orbis (Chamber deformation due to pollution)*, (C) *Quinqueloculina costata*, (D) *Asterorotalia trispinosa*, (E) *Spiroloculina sp.*, (F) *Elphidium hisphidulum*, (G) *Neocytheretta snellii (Predated)*, (H) *Pterygocythereis sp.*, (I) *Neocytheretta murilineata*, and (J) *Actinocytheries scutigera* (predated) (Scale bar=100 μm for A–F; 200 μm for G–J).

abnormalities due to morphological deformation, widening of apertures, and so forth (Fig. 19A–C). The morphological deformation in the selected foraminifer test can be related to enrichment of copper in the sediments (i.e., Le Cadre and Debenay, 2006; Fabrizio and Rodolfo, 2012). The impact of pollution is noticed in the foraminiferal tests through the widening of the aperture, a sudden increase of the chambers, and grey, brown, and pyritized coloration of ostracod carapaces (Fig. 19). However, compared with the total population of foraminifera and ostracoda species, the impact of pollution on the living organisms are less significant. The presence of pale yellow, white, and light colored specimens of foraminifera and ostracod indicate that the sediments are deposited under normal oxygenated environmental conditions (Honnappa and Venkatachalapathy, 1978; Hussain et al., 2004). The species, namely *Atjehella semiplicata*, *Neocytheretta snelii* (Fig. 19G), and *Actinocytheries scutigera* (Fig. 19J), exhibit predation ($n=15$) near the Miri River confluence, which might be due to gastropods.

6 Implications on Local and Regional Scale Coastal Zone Management Strategies

Abundant literature is available on the geochemical assessment of beaches with an objective of creating environmental baseline data and coastal zone management strategies. However, this study is the first comprehensive database focused on the Miri Coast. This baseline study is vindicated considering the proposed dam construction in the upland areas of the Baram River. The effect of upland freshwater and sediment retention by man-made reservoirs on the coastal processes, and their environmental impact can be solicited with this baseline data. As dams in the upper reaches of the river systems are known to drastically affect the riverine sediment influx and alter the coastal equilibrium (Ramkumar, 2000, 2003), and given the interpreted fragility of the studied beaches in terms of paucity of sediments, recycling, and sorting effects; due projections and appropriate remedial measures are necessary. Occurrences of lithic fragments, together with heavy minerals and spatial variability of sediment characteristics, as documented through this study, warrant site-specific measures rather than a one size fits all approach for coastal nourishment. Further, the longshore transport of the sediments delineates the coastal area into two sections—one dominated by the prevalence of very fine sand—mud sediments (NE part), and the other by the prevalence of medium to fine sediments (SW part). One possible explanation for such differentiation may arise from the consideration that the Baram River plume can influence long-shore sediment transport (Nagarajan et al., 2015a). The Baram River plume is constrained to flow parallel to the coast due to the growth of a sand bar in front of its mouth. A reduction in freshwater flow will drastically affect the flow and mixing regime, and alter the geochemical processes along the coastal zone. And, all such processes may have unknown ramifications on the environmental settings and ecological balance of this area.

Damming and expected reductions in freshwater input to the coastal zone will decrease groundwater recharge, and expose more peat lands in the coastal region, which may lead to drying peat lands, and have a higher potential to raise greenhouse gas emissions (Dommain et al., 2011; Cole et al., 2015; Pärn et al., 2018). The proposed damming could potentially aggravate the situation, with negative impacts on not only the coastal ecosystem, but also on the global phenomenon of greenhouse gas emissions. In order to obviate this, systematic scientific studies on determining "ecological flow" required to maintain the optimum equilibrium conditions, considering the coastal flux of fresh water, sediment, and nutrient budgets are necessary. Once such a database on the ecological flow conditions is established, and the equilibrium conditions are known, the effects of damming on coastal environments and the ecological equilibrium conditions can be managed effectively. The coastal area of Miri is currently experiencing conditions similar to the estuary—low salinity coastal waters due to the influx of fresh water from the Baram River. Thus, many of the geochemical processes that are akin to the estuarine region are observed along the coast. By defining the baseline values of trace elements in the coastal zone, this study provides a robust platform for future studies. This study was able to identify and delineate two different areas - the southwest and northeast parts, which show significantly different geochemical signatures. The mobility of the elements in these two segments has shown differences.

7 Conclusions

- The coastal region of Miri in NW Borneo shows significant variations in the sediment characteristics, that is, grain size, sand and mud content, and associated geochemical composition. The geochemical compositions of the sediments are mainly influenced by the provenance of the source area, followed by the major rivers in the study area (i.e., the Baram River and the Miri River). Particularly, the geochemistry of the NE sediments is mainly controlled by the Baram River, while SW sediments are controlled by many factors in which influence of the Baram River via longshore currents is also one of the factors.
- The sediments of the study area show moderate weathering, and are strongly influenced by the recycling effect, and thus enriched with quartz and heavy minerals, which are reflected in their geochemical signatures.
- The sediments are recycled from the Rajang Group of sedimentary to meta-sedimentary rocks, and exhibit felsic to intermediate characteristics. The sediments of the Miri coastal regions show affinity with rivers in the vicinity and the source area (i.e., incompatible elements). Chalcophile elements (i.e., Cu and Zn) are enriched in the sediments, and are related to both natural and anthropogenic sources. Their accumulation has been enhanced by the grain size (very fine sand and mud fractions) and adsorption processes.
- The environmental indices clearly demarcated that the sediments of Miri coastal regions are moderate to highly contaminated by Cu and Zn. However, local enrichment of trace metal (e.g., Pb) was noticed in selected locations.

- Among the studied elements, only Cu has exceeded the ERM values. The Cr, Zn, and As depict a trend between the ERL and ERM values, which suggests that there is a possibility of impact on biota by these elements. The impact of trace metals on biota was noticed through deformity of foraminifera species. However, the impact on biota is less significant compared with the overall population of the studied species.

Acknowledgments

This research was conducted with the Curtin Sarawak Research Institute Academic Grant (CSRI2011/Grants/01) and Curtin University Research Performance fund and Curtin Malaysia Research incentive fund awarded to RN, for which RN and AA are grateful. We thank the Editors for the invitation to contribute to this book. We thank the reviewers for their critical comments and suggestions that helped improve the quality of the presentation.

References

Abrahim, G.M.S., Parker, R.J., 2008. Assessment of heavy metal enrichment factors and the degree of contamination in marine sediments from Tamaki Estuary, Auckland, New Zealand. Environ. Monit. Assess. 136 (1–3), 227–238.

Abubakar, U.S., Zulkifli, S.Z., Ismail, A., 2018. Heavy metals bioavailability and pollution indices evaluation in the mangrove surface sediment of Sungai Puloh, Malaysia. Environ. Earth Sci. 77 (6), 225.

Acharya, B.C., Nayak, B.K., Das, S.K., 2015. Mineralogy and mineral chemistry of placer deposit around Jhatipodar, Odisha. J. Geol. Soc. India 86, 137–147. https://doi.org/10.1007/s12594-015-0293-5.

Ahn, I.Y., Lee, S.H., Kim, K.T., Shim, J.H., Kim, D.Y., 1996. Baseline heavy metal concentrations in the antarctic clam, Laternula elliptica in Maxwell Bay, King George Island, Antarctica. Mar. Pollut. Bull. 32 (8–9), 592–598.

Al-Fori, M., Dobretsov, S., Myint, M.T.Z., Dutta, J., 2014. Antifouling properties of zinc oxide nanorod coatings. Biofouling 30 (7), 871–882. https://doi.org/10.1080/08927014.2014.942297.

Amiotte Suchet, P., Probst, J.L., Ludwig, W., 2003. Worldwide distribution of continental rock lithology: implications for the atmospheric/soil CO_2 uptake by continental weathering and alkalinity river transport to the oceans. Glob. Biogeochem. Cycles 17 (2), 1038.

Amon, R.M.W., Rinehart, A.J., Duan, S., Louchouarn, P., Prokushkin, A., Guggenberger, G., Bauch, D., Stedmon, C., Raymond, P.A., Holmes, R.M., McClelland, J.W., 2012. Dissolved organic matter sources in large Arctic rivers. Geochim. Cosmochim. Acta 94, 217–237.

Anandkumar, A., Nagarajan, R., Prabakaran, K., Rajaram, R., 2017. Trace metal dynamics and risk assessment in the commercially important marine shrimp species collected from the Miri coast, Sarawak, East Malaysia. Reg. Stud. Mar. Sci. 16, 79–88. https://doi.org/10.1016/j.rsma.2017.08.007.

Anandkumar, A., Vijith, H., Nagarajan, R., Jonathan, M.P., 2018. Evaluation of decadal shoreline changes in the coastal region of Miri, Sarawak, Malaysia. In: Krishnamurthy, R.R., Jonathan, M.P., Srinivasalu, S., Glaeser, B. (Eds.), Coastal Management: Global Challenges and Innovations. Elsevier. https://doi.org/10.1016/B978-0-12-810473-6.00008-X.

Anderson, J.A.R., 1964. The structure and development of the peat swamps of Sarawak and Brunei. J. Trop. Geogr. 18, 7–16.

Anthony, E.J., Gratiot, N., 2012. Coastal engineering and large-scale mangrove destruction in Guyana, South America: averting an environmental catastrophe in the making. Ecol. Eng. 47, 268–273.

Armstrong-Altrin, J.S., Lee, Y.I., Kasper-Zubillaga, J.J., Carranza-Edwards, A., Garcia, D., Eby, G.N., Balaram, V., Cruz-Ortiz, N.L., 2012. Geochemistry of beach sands along the western Gulf of Mexico, Mexico: implication for provenance. Chem. Erde Geochem. 72 (4), 345–362.

Armstrong-Altrin, J.S., Nagarajan, R., Lee, Y.I., Kasper-Zubillaga, J.J., Córdoba-Saldaña, L.P., 2014. Geochemistry of sands along the San Nicolás and San Carlos beaches, Gulf of California, Mexico: implications for provenance and tectonic setting. Turk. J. Earth Sci. 23, 533–558.

Armstrong-Altrin, J.S., Nagarajan, R., Balaram, V., Natalhy-Pineda, O., 2015. Petrography and geochemistry of sands from the Chachalacas and Veracruz beach area, Western Gulf of Mexico, Mexico: constraints on provenance and tectonic setting. South Am. J. Earth Sci. 64 (1), 199–216. https://doi.org/10.1016/j.jsames.2015.10.012.

Armstrong-Altrin, J.S., Ramos-Vázquez, M.A., Zavala-León, A.C., Montiel-García, P.C., 2018. Provenance discrimination between Atasta and Alvarado beach sands, western Gulf of Mexico, Mexico: constraints from detrital zircon chemistry and U–Pb geochronology. Geol. J. https://doi.org/10.1002/gj.3122.

Ashraf, A., Saion, E., Gharibshahi, E., Kamari, H.M., Yap, C.K., Hamzah, M.S., Elias, M.S., 2017. Distribution of trace elements in core marine sediments of coastal East Malaysia by instrumental neutron activation analysis. Appl. Radiat. Isot. 122, 96–105.

Aufdenkampe, A.K., Mayorga, E., Raymond, P.A., Melack, J.M., Doney, S.C., Alin, S.R., Aalto, R.E., Yoo, K., 2011. Riverine coupling of biogeochemical cycles between land, oceans, and atmosphere. Front. Ecol. Environ. 9 (1), 53–60.

Babu, K., 2017. Geochemical characteristics of sandstones from Cretaceous Garudamangalam area of Ariyalur, Tamil Nadu, India: implications of provenance and tectonic setting. J. Earth Syst. Sci. 126 (3), 45. https://doi.org/10.1007/s12040-017-0821-3.

Balica, S.F., Wright, N.G., van der Meulen, F., 2012. A flood vulnerability index for coastal cities and its use in assessing climate change impacts. Nat. Hazards 64 (1), 73–105.

Basir, S.R., Balakrishnan, S., 1999. Geochemistry of sphene from granodiorites surrounding the Hutti-Maski Schist Belt: significance to rare earth element (REE) modelling. J. Geol. Soc. India 54, 107–119.

Billah, M.M., Kamal, A.H.M., Idris, M.H., Ismail, J., 2017. Mangrove macroalgae as biomonitors of heavy metal contamination in a tropical estuary, Malaysia. Water Air Soil Poll. 228 (9), 347.

Bird, E., 2000. Coastal Geomorphology: An Introduction. John Wiley & Sons, Ltd., Chichester, England 322 p.

Bock, B., McLennan, S.M., Hanson, G.N., 1998. Geochemistry and provenance of the Middle Ordovician Austin Glen Member (Normanskill Formation) and the Taconian Orogeny in New England. Sedimentology 45 (4), 635–655. https://doi.org/10.1046/j.1365-3091.1998.00168.x.

Brooks, C.K., 1970. The concentrations of zirconium and hafnium in some igneous and metamorphic rocks and minerals. Geochim. Cosmochim. Acta 34, 411–416.

Buhari, T.R.I., Ismail, A., 2016. Heavy metals pollution and ecological risk assessment in surface sediments of west coast of Peninsular Malaysia. Int. J. Environ. Sci. Dev. 7 (10), 750.

Burone, L., Venturini, N., Sprechmann, P., Valente, P., Muniz, P., 2006. Foraminiferal responses to polluted sediments in the Montevideo coastal zone, Uruguay. Mar. Pollut. Bull. 52, 61–73.

Caline, B., Huong, J., 1992. New insight into the recent evolution of the Baram Delta from satellite imagery. Geol. Soc. Malaysia Bull. 32, 1–13.

Carey, A.E., Nezat, C.A., Lyons, W.B., Kao, S.J., Hicks, D.M., Owen, J.S., 2002. Trace metal fluxes to the ocean: the importance of high-standing oceanic islands. Geophys. Res. Lett. 29 (23), 2099. https://doi.org/10.1029/2002GL015690.

Carranza-Edwards, A., Kasper-Zubillaga, J.J., Rosales-Hoz, L., Morales-de la Garaza, E.A., Cruz, R.L.S., 2009. Beach sand and composition and provenance in a sector of the southwestern Mexican Pacific. Rev. Mex. Cienc. Geol. 26, 433–447.

Chong, E.V., 2016. Blue Tears and Bioluminescence Phenomenon Back Again in Miri!. Borneo Post Online, 22 November 2016. http://www.theborneopost.com/2016/11/22/blue-tears-bioluminescence-phenomenon-back-again-in-miri/ (Accessed 21 November 2017).

Christian, R.R., Mazzilli, S., 2007. Defining the coast and sentinel ecosystems for coastal observations of global change. Hydrobiologia 577, 55. https://doi.org/10.1007/s10750-006-0417-4.

Cohen, J.E., Small, C., 1998. Recounting coastal population. Issues Sci. Technol. 14, 26.

Cole, L.E., Bhagwat, S.A., Willis, K.J., 2015. Long-term disturbance dynamics and resilience of tropical peat swamp forests. J. Ecol. 103 (1), 16–30.

Collins, D.S., Johnson, H.D., Allison, P.A., Guilpain, P., Damit, A.R., 2017. Coupled 'storm-flood' depositional model: application to the Miocene–Modern Baram Delta Province, north-west Borneo. Sedimentology 64 (5), 1203–1235. https://doi.org/10.1111/sed.12316.

Condie, K.C., 1993. Chemical composition and evolution of the upper continental crust: contrasting results from surface samples and shales. Chem. Geol. 104, 1–37.

Cullers, R.L., 1994. The controls on the major and trace element variation of shales, siltstones and sandstones of Pennsylvanian-Permian age from uplifted continental blocks in Colorado to platform sediment in Kansas, U.S.A. Geochim. Cosmochim. Acta 58, 4955–4972.

Cullers, R.L., 2000. The geochemistry of shales, siltstones and sandstones of Pennsylvanian-Permian age, Colorado, U.S.A.: implications for provenance and metamorphic studies. Lithos 51, 305–327.

Cullers, R.L., Podkovyrov, V.N., 2000. Geochemistry of the Mesoproterozoic Lakhanda shales in southeastern Yakutia, Russia: implications for mineralogical and provenance control, and recycling. Precambrian Res. 104 (1-2), 77–93.

Cullers, R.L., Basu, A., Suttner, L.J., 1988. Geochemical signature of provenance in sand-size material in soils and stream sediments near the Tobacco Root batholith, Montana, U.S.A. Chem. Geol. 70, 335–348.

Darnley, A.G., 1990. International geochemical mapping: a new global project. J. Geochem. Explor. 39 (1–2), 1–13.

Darnley, A.G., 1995. International geochemical mapping—a review. J. Geochem. Explor. 55 (1–3), 5–10.

Darnley, A.G., 1997. A global geochemical reference network: the foundation for geochemical baselines. J. Geochem. Explor. 60 (1), 1–5.

Daskalakis, K.D., O'connor, T.P., 1995. Normalization and elemental sediment contamination in the coastal United States. Environ. Sci. Technol. 29 (2), 470–477.

de Mora, S., Fowler, S.W., Wyse, E., Azemard, S., 2004. Distribution of heavy metals in marine bivalves, fish and coastal sediments in the Gulf and Gulf of Oman. Mar. Pollut. Bull. 49 (5–6), 410–424.

Déry, S.J., Stadnyk, T.A., MacDonald, M.K., Gauli-Sharma, B., 2016. Recent trends and variability in river discharge across northern Canada. Hydrol. Earth Syst. Sci. 20 (12), 4801.

Dodge-Wan, D., Nagarajan, R., 2019. Typology and mechanisms of coastal erosion in siliciclastic rocks of the NW Borneo coastline (Sarawak, Malaysia): a field approach (Chapter 3). In: Ramkumar, M., Arthur James, R., Menier, D., Kumaraswamy, K. (Eds.), Coastal Zone Management: Global Perspectives, Regional Processes, Local Issues. Elsevier. https://doi.org/10.1016/B978-0-12-814350-6.00003-3.

Dommain, R., Couwenberg, J., Joosten, H., 2011. Development and carbon sequestration of tropical peat domes in south-east Asia: links to post-glacial sea-level changes and Holocene climate variability. Quat. Sci. Rev. 30 (7–8), 999–1010.

Dos Santos, I.R., Silva-Filho, E.V., Schaefer, C., Sella, S.M., Silva, C.A., Gomes, V., de ACR Passos, M.J., Van Ngan, P., 2006. Baseline mercury and zinc concentrations in terrestrial and coastal organisms of Admiralty Bay, Antarctica. Environ. Pollut. 140 (2), 304–311.

Dubrulle, C., Lesueur, P., Boust, D., Dugue´, O., Poupinet, N., Lafite, R., 2007. Source discrimination of fine-grained deposits occurring on marine beaches: the Calvados beaches (eastern Bay of the Seine, France). Estuar. Coast. Shelf Sci. 72, 138–154. https://doi.org/10.1016/j.ecss.2006.10.021.

Dürr, H.H., Meybeck, M., Hartmann, J., Laruelle, G.G., Roubeix, V., 2011. Global spatial distribution of natural riverine silica inputs to the coastal zone. Biogeosciences 8 (3), 597.

Elderfield, H., Upstill-Goddard, R., Sholkovitz, E.R., 1990. The rare earth elements in rivers, estuaries, and coastal seas and their significance to the composition of ocean waters. Geochim. Cosmochim. Acta 54 (4), 971–991.

El-Kammar, A.A., Ragab, A.A., Moustafa, M.I., 2011. Geochemistry of economic heavy minerals from Rosetta Black Sand of Egypt. J. King Abdulaziz Univ. Earth Sci. 22 (2), 69–97. https://doi.org/10.4197/Ear.22-2.4.

Erlank, A.J., Smith, H.S., Marchant, J.W., Cardoso, M.P., Ahrens, L.H., 1978. Hafnium. In: Wedepohl, K.H. (Ed.), Handbook of Geochemistry. Springer-Verlag, Berlin, Heidelberg, New York pp. 72C1–72O1.

Fabrizio, F., Rodolfo, C., 2012. Response of benthic foraminiferal assemblages to copper exposure: a pilot mesocosm investigation. J. Environ. Prot. 3 (4), 342–352. https://doi.org/10.4236/jep.2012.34044.

Ferguson, J.W., Healey, M., Dugan, P., Barlow, C., 2011. Potential effects of dams on migratory fish in the Mekong River: lessons from salmon in the Fraser and Columbia Rivers. Environ. Manag. 47 (1), 141–159.

Floyd, P.A., Leveridge, B.E., 1987. Tectonic environment of the Devonian Gramscatho basin, south Cornwall: framework mode and geochemical evidence from turbiditic sandstones. J. Geol. Soc. 144 (4), 531–542.

Folk, R.L., 1974. Petrology of Sedimentary Rocks. Hemphill Publishing Company, Austin, Texas 190 p.

Frontalini, F., Buosi, C., Coccioni, R., Cherchi, A., Da Pelo, S., Bucci, C., 2009. Benthic foraminifera as a bioindicator of the environmental quality: a case study from the lagoon of Santa Gilla (Cagliari, Italy). Mar. Pollut. Bull. 58, 858–877.

Fursenko, A.V., 1978. Introduction to Study of Foraminifera. Nauka, Novosibirsk 242 p (in Russian).

Gaillardet, J., Dupré, B., Louvat, P., Allegre, C.J., 1999. Global silicate weathering and CO_2 consumption rates deduced from the chemistry of large rivers. Chem. Geol. 159 (1), 3–30.

Garrett, R.G., Reimann, C., Smith, D.B., Xie, X., 2008. From geochemical prospecting to international geochemical mapping: a historical overview. Geochem. Explor. Environ. Anal. 8 (3–4), 205–217.

Garzanti, E., Andò, S., Vezzoli, G., Megid, A.A.A., El Kammar, A., 2006. Petrology of Nile River sands (Ethiopia and Sudan): sediment budgets and erosion patterns. Earth Planet. Sci. Lett. 252 (3), 327–341.

Garzanti, E., Andò, S., Vezzoli, G., 2009. Grain-size dependence of sediment composition and environmental bias in provenance studies. Earth Planet. Sci. Lett. 277 (3), 422–432.

Garzanti, E., Padoan, M., Andò, S., Resentini, A., Vezzoli, G., Lustrino, M., 2013a. Weathering at the equator: petrology and geochemistry of East African Rift sands. J. Geol. 121 (6), 547–580. https://doi.org/10.1086/673259.

Garzanti, E.M., Padoan, M., Setti, M., Najman, Y., Peruta, L., Villa, I.M., 2013b. Weathering geochemistry and Sr-Nd fingerprints of equatorial upper Nile and Congo muds. Geochem. Geophys. Geosyst. 14, 292–316. https://doi.org/10.1002/ggge.20060.

Gopinath, A., Nair, S.M., Kumar, N.C., Jayalakshmi, K.V., Pamalal, D., 2010. A baseline study of trace metals in a coral reef sedimentary environment, Lakshadweep Archipelago. Environ. Earth Sci. 59 (6), 1245–1266. https://doi.org/10.1007/s12665-009-0113-6.

Gromet, L.P., Silver, L.T., 1983. Rare earth element distributions among minerals in a granodiorite and their petrogenetic implications. Geochim. Cosmochim. Acta 47, 925–939.

Gu, B., Schmitt, J., Chen, Z., Liang, L., McCarthy, J.F., 1995. Adsorption and desorption of different organic matter fractions on iron oxide. Geochim. Cosmochim. Acta 59 (2), 219–229.

Gu, J., Huang, Z., Fan, H., Jin, Z., Yan, Z., Zhang, J., 2013. Mineralogy, geochemistry, and genesis of lateritic bauxite deposits in the Wuchuan–Zheng'an–Daozhen area, Northern Guizhou Province, China. J. Geochem. Explor. 130, 44–59. https://doi.org/10.1016/j.gexplo.2013.03.003.

Gupta, H., Kao, S.J., Dai, M., 2012. The role of mega dams in reducing sediment fluxes: a case study of large Asian rivers. J. Hydrol. 464, 447–458.

Hanebuth, T.J., Kudrass, H.R., Linstädter, J., Islam, B., Zander, A.M., 2013. Rapid coastal subsidence in the central Ganges-Brahmaputra Delta (Bangladesh) since the 17th century deduced from submerged salt-producing kilns. Geology 41 (9), 987–990.

Hanson, P.J., Evans, D.W., Colby, D.R., Zdanowicz, V.S., 1993. Assessment of elemental contamination in estuarine and coastal environments based on geochemical and statistical modeling of sediments. Mar. Environ. Res. 36 (4), 237–266.

Haris, H., Looi, L.J., Aris, A.Z., Mokhtar, N.F., Ayob, N.A.A., Yusoff, F.M., Salleh, A.B., Praveena, S.M., 2017. Geo-accumulation index and contamination factors of heavy metals (Zn and Pb) in urban river sediment. Environ. Geochem. Health 39 (6), 1259–1271.

Harnois, L., 1988. The CIW index: a new chemical index of weathering. Sediment. Geol. 55 (3–4), 319–322.

Harrison, J.A., Bouwman, A.F., Mayorga, E., Seitzinger, S., 2010. Magnitudes and sources of dissolved inorganic phosphorus inputs to surface fresh waters and the coastal zone: a new global model. Glob. Biogeochem. Cycles. 24(1)GB1003https://doi.org/10.1029/2009GB003590.

Hartman, G., Puri, H.S., 1974. Summary of neontological and paleontological classification of Ostracoda. Mitt. Hamb. Zool. Mus. Inst. 70, 7–73.

Hernández-Hinojosa, V., Montiel-García, P.C., Armstrong-Altrin, J.S., Nagarajan, R., Kasper-Zubillaga, J.J., 2018. Textural and geochemical characteristics of beach sands along the western Gulf of Mexico, Mexico. Carpath. J. Earth Environ. Sci. 13 (1), 161–174.

Hinrichsen, D., 1996. Coasts in crisis. Issues Sci. Technol. 12 (4), 39–47.

Hofer, G., Wagreich, M., Neuhuber, S., 2013. Geochemistry of fine-grained sediments of the upper Cretaceous to Paleogene Gosau Group (Austria, Slovakia): implications for paleoenvironmental and provenance studies. Geosci. Front. 4, 449–468.

Honnappa, Venkatachalapathy, V., 1978. Some aspects of pyritised ostracode shells: a possible tool in petroleum sedimentology from the sediments of Mangalore Harbor area, Karnataka state, west coast of India.Proceedings of VII Indian Colloquium on Micropaleontology and Stratigraphypp. 65–69.

Hussain, S.M., Ravi, G., Mohan, S.P., Rajeshwara Rao, N., 2004. Recent benthic ostracoda from the inner shelf off Chennai, southeast Coast of India—implication on microenvironments environmental micropaleontology. Microbiol. Meiobenthol. 1, 105–121.

Hutchison, C.S., 2005. Geology of North-West Borneo: Sarawak, Brunei and Sabah. Elsevier, Amsterdam.

Ismail, A., 1993. Heavy metal concentrations in sediments off Bintulu, Malaysia. Mar. Pollut. Bull. 26 (12), 706–707. https://doi.org/10.1016/0025-326X(93)90556-Y.

Jayaprakash, M., Srinivasalu, S., Jonathan, M.P., Rammohan, V., 2005. A baseline study of physico-chemical parameters and trace metals in waters of Ennore Creek, Chennai, India. Mar. Pollut. Bull. 50 (5), 583–589.

Jayaprakash, M., Nagarajan, R., Velmurugan, P.M., Giridharan, L., Neetha, V., Urban, B., 2014. Geochemical assessment of sediment quality using multivariate statistical analysis of Ennore Creek, North of Chennai, SE Coast of India. Pertanika J. Sci. Technol. 22 (1), 315–328.

Jia, T.Y., Rahman, A.H.A., 2009. Comparative analysis of facies and reservoir characteristics of Miri Formation (Miri) and Nyalau Formation (Bintulu), Sarawak. Bull. Geol. Soc. Malaysia 55, 39–45.

Johnson, C.C., Breward, N., Ander, E.L., Ault, L., 2005. G-BASE: baseline geochemical mapping of Great Britain and Northern Ireland. Geochem. Explor. Environ. Anal. 5 (4), 347–357.

Jongman, B., Ward, P.J., Aerts, J.C., 2012. Global exposure to river and coastal flooding: long term trends and changes. Glob. Environ. Chang. 22 (4), 823–835.

Kadhum, S.A., Ishak, M.Y., Zulkifli, S.Z., 2016. Evaluation and assessment of baseline metal contamination in surface sediments from the Bernam River, Malaysia. Environ. Sci. Pollut. Res. 23 (7), 6312–6321.

Kasper-Zubillaga, J.J., Carranza Edwards, A., 2005. Grain Size discrimination between sands of desert and coastal dunes from northwest Mexico. Rev. Mex. Cienc. Geol. 22 (3), 383–390.

Kasper-Zubillaga, J.J., Zolezzi-Ruiz, H., 2007. Grain size, mineralogical, and geochemical studies of coastal and inland dune sands from El Vizcaino Desert, Baja California Peninsula, Mexico. Rev. Mex. Cienc. Geol. 24 (3), 423–438.

Kasper-Zubillaga, J.J., Carranza-Edwards, A., Rosales-Hoz, L., 1999. Petrography and geochemistry of holocene sands in the western Gulf of Mexico: implications for provenance and tectonic setting. J. Sediment. Res. 69 (5), 1003–1010. https://doi.org/10.2110/jsr.69.1003.

Kasper-Zubillaga, J.J., Armstrong-Altrin, J.S., Carranza-Edwards, A., Morton-Bermea, O., Santa-Cruz, R.I., 2013. Control in beach and dune sands of the Gulf of Mexico and the role of nearby rivers. Int. J. Geosci. 4, 1157–1174.

Kessler, F.L., Jong, J., 2014. Habitat and C-14 ages of lignitic terrace deposits along the northern Sarawak coastline. Bull. Geol. Soc. Malaysia 60, 27–34.

Klaver, G.T., van Weering, T.C., 1993. Rare earth element fractionation by selective sediment dispersal in surface sediments: the Skagerrak. Mar. Geol. 111 (3–4), 345–359.

Koppi, A.J., Edis, R., Field, D.J., Geering, H.R., Klessa, D.A., Cockayne, D.J.H., 1996. Rare earth element trends and cerium–uranium–manganese associations in weathered rock from Koongarra, northern territory, Australia. Geochim. Cosmochim. Acta 60, 1695–1707.

Kumar, A., Ramanathan, A.L., Prasad, M., Datta, D., Kumar, M., Sappal, S., 2016. Distribution, enrichment, and potential toxicity of trace metals in the surface sediments of Sundarban mangrove ecosystem, Bangladesh: a baseline study before Sundarban oil spill of December, 2014. Environ. Sci. Pollut. Res. 23 (9), 8985–8999.

Le Cadre, V., Debenay, J.P., 2006. Morphological and cytological response of Ammonia (Foraminifera) to copper contamination. Implication for the use of foraminifera as bioindicators of pollution. Environ. Pollut. 143, 304–317.

Le Pera, E., Critelli, S., 1997. Sourceland controls on the composition of beach and fluvial sand of the tyrrhenian Coast of Calabria, Italy: implications for actualistic petrofacies. Sediment. Geol. 110, 81–97.

Lichter, M., Vafeidis, A.T., Nicholls, R.J., Kaiser, G., 2010. Exploring data-related uncertainties in analyses of land area and population in the "Low-Elevation Coastal Zone" (LECZ). J. Coast. Res. 27 (4), 757–768.

Liechti, P.R., Roe, F.W., Haile, N.S., 1960. The geology of Sarawak, Brunei and the western part of North Borneo. British Borneo Geol. Surv. Bull. 3, 1–360.

Liu, K.K., Atkinson, L., Quiñones, R.A., Talaue-McManus, L., 2010. Biogeochemistry of continental margins in a global context. In: Liu, K.K., Atkinson, L., Quiñones, R., Talaue-McManus, L. (Eds.), Carbon and Nutrient Fluxes in Continental Margins: A Global Synthesis. Springer, Berlin, Heidelberg, pp. 3–24.

Loeblich, A.R., Tappan, H., 1987. Foraminiferal Genera and Their Classification. Von Nostrand Reinhold, New York. 970 p.

Long, E., MacDonald, D., Smith, S., Calder, F., 1995. Incidence of adverse biological effects within ranges of chemical concentrations in marine and estuarine sediments. Environ. Manag. 19, 81–97.

Looi, L.J., Aris, A.Z., Yusoff, F.M., Hashim, Z., 2015. Mercury contamination in the estuaries and coastal sediments of the Strait of Malacca. Environ. Monit. Assess. 187 (1), 4099.

Lopez, J.M.G., Bauluz, B., Fernandex-Nieto, C., Oliete, A.Y., 2005. Factors controlling the trace-element distribution in fine-grained rocks: the Albian kaolinite-rich deposits of the Oliete Basin (NE Spain). Chem. Geol. 214, 1–19. https://doi.org/10.1016/j.chemgeo.2004.08.024.

Lovelock, C.E., Cahoon, D.R., Friess, D.A., Guntenspergen, G.R., Krauss, K.W., Reef, R., Rogers, K., Saunders, M.L., Sidik, F., Swales, A., Saintilan, N., 2015. The vulnerability of Indo-Pacific mangrove forests to sea-level rise. Nature 526 (7574), 559–563.

Luo, S., Cai, F., Liu, H., Lei, G., Qi, H., Su, X., 2015. Adaptive measures adopted for risk reduction of coastal erosion in the People's Republic of China. Ocean Coast. Manag. 103, 134–145.

Malick, B.M.L., Young, S.M., Ishiga, H., 2012. Major and trace element geochemistry of beach sands from northern Kyushu, Japan. Geosci. Rept. Shimane Univ. 31, 1–8.

Mancin, N., Darling, K., 2015. Morphological abnormalities of planktonic foraminiferal tests in the SW Pacific ocean over the last 550ky. Mar. Micropaleontol. 120, 1–19. https://doi.org/10.1016/j.marmicro.2015.08.003.

Mao, L.J., Mo, D.W., Yang, J.H., Guo, Y.Y., Lv, H.Y., 2014. Rare earth elementsgeochemistry in surface floodplain sediments from the XiangjiangRiver, middle reach of Changjiang River, China. Quat. Int. 336, 80–88.

McDonough, W.F., Sun, S.S., 1995. The composition of the Earth. Chem. Geol. 120, 223–253.

McLennan, S.M., 1989. Rare earth elements in sedimentary rocks; influence of provenance and sedimentary processes. In: Lipin, B.R., McKay, G.A. (Eds.), Geochemistry and Mineralogy of Rare Earth Elements. Reviews in Mineralogy and Geochemistry, vol. 21, pp. 169–200.

McLennan, S.M., 2001. Relationships between the trace element composition of sedimentary rocks and upper continental crust. Geochem. Geophys. Geosyst. 2 (4), 1021–1024. https://doi.org/10.1029/2000GC000109.

McLennan, S.M., Hemming, D.K., Hanson, G.N., 1993. Geochemical approaches to sedimentation, provenance and tectonics. Geol. Soc. Am. Spec. Pap. 284, 21–40.

Meybeck, M., 1987. Global chemical weathering of surficial rocks estimated from river dissolved loads. Am. J. Sci. 287 (5), 401–428.

Milliman, J.D., Katherine, F., 2011. River Discharge to the Coastal Ocean—A Global Synthesis. Cambridge University Press. https://doi.org/10.1017/CBO9780511781247.

Mokhtar, N.F., Aris, A.Z., Praveena, S.M., 2015. Preliminary study of heavy metal (Zn, Pb, Cr, Ni) contaminations in Langat River Estuary, Selangor. Procedia Environ Sci 30, 285–290.

Moyer, R.P., Grottoli, A.G., Olesik, J.W., 2012. A multiproxy record of terrestrial inputs to the coastal ocean using minor and trace elements (Ba/Ca, Mn/Ca, Y/Ca) and carbon isotopes ($\delta^{13}C$, $\Delta^{14}C$) in a nearshore coral from Puerto Rico. Paleoceanogr. Paleoclimatol. 27(3). https://doi.org/10.1029/2011PA002249.

Muller, G., 1969. Index of geoaccumulation in the sediments of the Rhine River. GeoJournal 2, 108–118.

Munksgaard, N.C., Lim, K., Parry, D.L., 2003. Rare earth elements as provenance indicators in North Australian estuarine and coastal marine sediments. Estuar. Coast. Shelf Sci. 57 (3), 399–409.

Mutakyahwa, M.K.D., Ikingura, J.R., Mruma, A.H., 2003. Geology and geochemistry of bauxite deposits in Lushoto District, Usambara Mountains, Tanzania. J. Afr. Earth Sci. 36, 357–369.

Nagarajan, R., Jonathan, M.P., Roy, P.D., Wai-Hwa, L., Prasanna, M.V., Sarkar, S.K., Navarrete-López, M., 2013. Metal concentrations in sediments from tourist beaches of Miri City, Sarawak, Malaysia (Borneo Island). Mar. Pollut. Bull. 73 (1), 369–373.

Nagarajan, R., Roy, P.D., Jonathan, M.P., Lozano-Santacruz, R., Kessler, F.L., Prasanna, M.V., 2014a. Geochemistry of Neogene sedimentary rocks from Borneo basin, East Malaysia: paleo-weathering, provenance and tectonic setting. Chem. Erde Geochem. 74 (1), 139–146.

Nagarajan, R., Kessler, F.L., Jonathan, M.P., Srinivasalu, S., Roy, P.D., 2014b. Preliminary studies on the mineralogy and geochemistry of concretions from Neogene Clastic Sediments of NW Borneo. In: Conference on Environmental Earth Sciences: Accomplishments Plans and Challenges, Chennai, Tamil Nadu, India, p. 100.

Nagarajan, R., Jonathan, M.P., Roy, P.D., Muthusankar, G., Lakshumanan, C., 2015a. Decadal evaluation of a spit in the Baram River mouth in Eastern Malaysia. Cont. Shelf Res. 105 (15), 18–25. https://doi.org/10.1016/j.csr.2015.06.006.

Nagarajan, R., Armstrong-Altrin, J.S., Kessler, F.L., Hidalgo-Moral, E.L., Dodge-Wan, D., Taib, N.I., 2015b. Provenance and tectonic setting of Miocene siliciclastic sediments, Sibuti Formation, northwestern Borneo. Arab. J. Geosci. 8, 8549–8565.

Nagarajan, R., Armstrong-Altrin, J.S., Kessler, F.L., Jong, J., 2017a. Petrological and geochemical constraints on provenance, paleo-weathering and tectonic setting of clastic sediments from the Neogene Lambir and Sibuti Formations, NW Borneo (Chapter 7). In: Mazumder, R. (Ed.), Sediment Provenance: Influences on Compositional Change From Source to Sink. Elsevier, Amsterdam, Netherlands, pp. 123–153. https://doi.org/10.1016/B978-0-12-803386-9.00007-1.

Nagarajan, R., Roy, P.D., Kessler, F.L., Jong, J., Vivian, D., Jonathan, M.P., 2017b. An integrated study of geochemistry and mineralogy of the Upper Tukau Formation, Borneo Island (East Malaysia): Sediment Provenance, Depositional Setting and Tectonic Implications. J. Asian Earth Sci. 143, 77–97. https://doi.org/10.1016/j.jseaes.2017.04.002.

Nesbitt, H.W., Young, G.M., 1982. Early Proterozoic climates and plate motions inferred from major element chemistry of lutites. Nature 299 (5885), 715–717.

Neumann, B., Vafeidis, A.T., Zimmermann, J., Nicholls, R.J., 2015. Future coastal population growth and exposure to sea-level rise and coastal flooding- AGlobal Assessment. PLoS One 10(6), e0131375. https://doi.org/10.1371/journal.pone.0131375.

Odum, E.P., 1971. Fundamentals of Ecology, third ed. London Toppan Company Limited, W.B. Saunders, Japan, p. 557.

Ohta, A., Kawabe, I., 2000. Rare earth element partitioning between Fe oxyhydroxide precipitates and aqueous NaCl solutions doped with $NaHCO_3$: Determinations of rare earth element complexation constants with carbonate ions. Geochem. J. 34, 439–454.

Okbah, M.A., Nasr, S.M., Soliman, N.F., Khairy, M.A., 2014. Distribution and contamination status of trace metals in the Mediterranean coastal sediments, Egypt. Soil Sediment Contam. Int. J. 23 (6), 656–676.

Ong, M.C., Joseph, B., Shazili, N.A.M., Ghazali, A., Mohamad, M.N., 2015. Heavy metals concentration in surficial sediments of Bidong Island, South China Sea off the East Coast of Peninsular Malaysia. Asian J. Earth Sci. 8 (3), 74–82. https://doi.org/10.3923/ajes.2015.74.82.

Osman, M., 1996. Recent to Quaternary River Nile Sediments: A Sedimentological Characterization on Samples From Aswan to Naga Hammadi, Egypt (Unpublished Doctoral Dissertation). University of Vienna, Vienna.

Papadopoulos, A., 2018. Geochemistry and REE content of beach sands along the Atticocycladic coastal zone, Greece. Geosci. J. 46. https://doi.org/10.1007/s12303-018-0004-5.

Papadopoulos, A., Christofides, G., Pe-Piper, G., Koroneos, A., Papadopoulou, L., 2015. Geochemistry of beach sands from Sithonia Peninsula (Chalkidiki, Northern Greece). Mineral. Petrol. 109 (1), 53–66.

Papadopoulos, A., Koroneos, A., Christofides, G., Papadopoulou, L., 2016. Geochemistry of beach sands from Kavala, Northern Greece. Ital. J. Geosci. 135 (3), 526–539. https://doi.org/10.3301/IJG.2016.01.

Parker, A., 1970. An index of weathering for silicate rocks. Geol. Mag. 107 (6), 501–504.

Pärn, J., Verhoeven, J.T., Butterbach-Bahl, K., Dise, N.B., Ullah, S., Aasa, A., Egorov, S., Espenberg, M., Järveoja, J., Jauhiainen, J., Kasak, K., 2018. Nitrogen-rich organic soils under warm well-drained conditions are global nitrous oxide emission hotspots. Nat. Commun. 9 (1), 1135. https://doi.org/10.1038/s41467-018-03540-1.

Patterson, M., Hardy, D., 2008. Economic drivers of change and their oceanic-coastal ecological impacts. In: Patterson, M., Glavovic, B.C. (Eds.), Ecological Economics of the Oceans and Coasts. Edward Elgar Publishing, pp. 187–209.

Pekey, H., Karakas, D., Ayberk, S., Tolun, L., Bakoglu, M., 2004. Ecological risk assessment using trace elements from surface sediments of Izmit Bay (Northeastern Marmara Sea) Turkey. Mar. Pollut. Bull. 48, 946–953.

Peng, L.C., Leman, M.S., Hassan, K., Nasib, M.B., Karim, R., 2004. Stratigraphic Lexicon of Malaysia. Geological Society of Malaysia, Kuala Lumpur. 162 p.

Pereira, M.F., Albardeiro, L., Gama, C., Chichorro, M., Hofmann, M., Linnemann, U., 2016. Provenance of Holocene beach sand in the Western Iberian margin: the use of the Kolmogorov–Smirnov test for the deciphering of sediment recycling in a modern coastal system. Sedimentology 63 (5), 1149–1167.

Peter, T.S., Chandrasekar, N., Wilson, J.J., Selvakumar, S., Krishnakumar, S., Magesh, N.S., 2017. A baseline record of trace elements concentration along the beach placer mining areas of Kanyakumari coast, South India. Mar. Pollut. Bull. 119 (1), 416–422.

Pit, I.R., Dekker, S.C., Kanters, T.J., Wassen, M.J., Griffioen, J., 2017. Mobilisation of toxic trace elements under various beach nourishments. Environ. Pollut. 231, 1063–1074.

Plant, J., Smith, D., Smith, B., Williams, L., 2001. Environmental geochemistry at the global scale. Appl. Geochem. 16 (11–12), 1291–1308.

Pokrovsky, O.S., Schott, J., Dupre, B., 2006. Trace element fractionation and transport in boreal rivers and soil porewaters of permafrost-dominated basaltic terrain in Central Siberia. Geochim. Cosmochim. Acta 70, 3239–3260.

Pourang, N., Nikouyan, A., Dennis, J.H., 2005. Trace element concentrations in fish, surficial sediments and water from northern part of the Persian Gulf. Environ. Monit. Assess. 109 (1–3), 293–316.

Prabakaran, K., 2017. Environmental Geochemistry of the Lower Baram River, Borneo. Unpublished Doctoral Dissertation. Curtin University, Malaysia, 335 p.

Prabakaran, K., Nagarajan, R., Franco, F.M., Anandkumar, A., 2017. Biomonitoring of Malaysian aquatic environments: a review of status and prospects. Ecohydrol. Hydrobiol. 17 (2), 134–147.

Preston, J., Hartley, A., Hole, M., Buck, S., Bond, J., Mange, M., Still, J., 1998. Integrated whole-rock trace element geochemistry and heavy mineral chemistry studies: aids to the correlation of continental red-bed reservoirs in the Beryl Field: UK North Sea. Pet. Geosci. 4, 7–16.

Rahman, M.J.J., Suzuki, S., 2007. Geochemistry of sandstones from the Miocene Surma group, Bengal Basin, Bangladesh: implications for provenance, tectonic setting and weathering. Geochem. J. 41, 415–428.

Ramachandran, A., Madhavaraju, J., Ramasamy, S., Lee, Y.I.L., Rao, S., Chawngthu, D.L., Velmurugan, K., 2016. Geochemistry of Proterozoic clastic rocks of the Kerur Formation of Kaladgi-Badami Basin, North Karnataka, South India: implications for paleoweathering and provenance. Turk. J. Earth Sci. 25, 126–144. https://doi.org/10.3906/yer-1503-4.

Ramkumar, M., 1999. Multivariate statistical analysis for deducing controls of carbonate deposition and diagenesis: a case study from South Indian Cretaceous sequence. Ind. J. Geochem. 14, 79–95.

Ramkumar, M., 2000. Recent changes in the Kakinada spit, Godavari delta. J. Geol. Soc. India 55, 183–188.

Ramkumar, M., 2001. Sedimentary environments of the modern Godavari delta: characterization and statistical discrimination towards computer assisted environment recognition scheme. J. Geol. Soc. India 57, 49–63.

Ramkumar, M., 2003. Progradation of the Godavari delta: a fact or empirical artifice? Insights from coastal landforms. J. Geol. Soc. India 62, 290–304.

Ramkumar, M., Santosh, M., Nagarajan, R., Li, S.S., Mathew, M., Menier, D., Siddiqui, N., Rai, J., Sharma, A., Farroqui, S., Poppelreiter, M.C., Lai, J., Prasad, V., 2018. Late Middle Miocene volcanism in the Northern Borneo, Southeast Asia: implications for Tectonics, Paleoclimate and stratigraphic marker. Palaeogeogr. Palaeoclimatol. Palaeoecol. 490, 141–162. https://doi.org/10.1016/j.palaeo.2017.10.022.

Ravindra Kumar, G.R., Sreejith, C., 2010. Relationship between heavy mineral placer deposits and hinterland rocks of southern Kerala: a new approach for source-to-sink link from the chemistry of garnets. Ind. J. Mar. Sci. 39 (4), 562–571.

Redzwan, G., Halim, H.A., Alias, S.A., Rahman, M.M., 2014. Assessment of heavy metal contamination at west and east coastal area of Peninsular Malaysia. Malays. J. Sci. 33 (1), 23–31.

Renzi, M., Bigongiari, N., Focardi, S.E., 2015. Baseline levels of trace elements in coastal sediments from the central Mediterranean (Tuscany, Italy). Chem. Ecol. 31 (1), 34–46.

Roser, B.P., Korsch, R.J., 1988. Provenance signatures of sandstone-mudstone suites determined using discrimination function analysis of major-element data. Chem. Geol. 67, 119–139.

Sallenger Jr., A.H., Doran, K.S., Howd, P.A., 2012. Hotspot of accelerated sea-level rise on the Atlantic coast of North America. Nat. Clim. Chang. 2 (12), 884.

Salminen, R., Gregorauskien, V., 2000. Considerations regarding the definition of a geochemical baseline of elements in the surficial materials in areas differing in basic geology. Appl. Geochem. 15 (5), 647–653.

Salomons, W., Forstner, U., 1984. Metals in the Hydrocycle. Springer, New York.

Samir, A.M., 2000. The response of benthic foraminifera and ostracods to various pollution sources. A study from two lagoons in Egypt. J. Foraminifer. Res. 30, 83–98.

Sandal, S.T., 1996. The Geology and Hydrocarbon Resources of Negara Brunei Darussalam, second ed. Syabas Publisher, Brunei.

Sany, S.B.T., Salleh, A., Rezayi, M., Saadati, N., Narimany, L., Tehrani, G.M., 2013. Distribution and contamination of heavy metal in the coastal sediments of Port Klang, Selangor, Malaysia. Water Air Soil Pollut. 224 (4), 1476. https://doi.org/10.1007/s11270-013-1476-6.

Saraswat, R., Kurtarkar, S.R., Mazumder, A., Nigam, R., 2004. Foraminifers as indicators of marine pollution: a culture experiment with *Rosalinaleei*. Mar. Pollut. Bull. 48, 91–96.

Selvaraj, K., Chen, C.T.A., 2006. Moderate chemical weathering of subtropical Taiwan: constraints from solid-phase geochemistry of sediments and sedimentary rocks. J. Geol. 114 (1), 101–116.

Sharifi, A.R., Croudace, I.W., Austin, R.L., 1991. Benthic foraminiferids as pollution indicators in Southampton Water, Southern England. J. Micropalaeontol. 10, 109–113.

Sim, S.F., Rajendran, M., Nyanti, L., Ling, T.Y., Grianag, J., Liew, J.J., 2017. Assessment of trace metals in water and sediment in a tropical river potentially affected by land use activities in Northern Sarawak, Malaysia. Int. J. Environ. Res. 11, 99–110. https://doi.org/10.1007/s41742-017-0011-9.

Simmons, M.D., Bidgood, M.D., Brenac, P., Crevello, P.D., Lambiash, J.J., Morley, C.K., 1999. Microfossil assemblages as proxies for precise palaeoenvironmental determination—an example from Miocene sediments of northwest Borneo. In: Jones, R.W., Simmons, M.D. (Eds.), Biostratigraphy in Production and Development Geology. vol, 152. pp. 219–242 Geological Society Special Publications.

Simon, K., Hakif, M., Hassan, A., Barbeito, M.P.J., 2014. Sedimentology and Stratigraphy of the Miocene Kampung Opak Limestone (Sibuti Formation), Bekenu, Sarawak. Bull. Geol. Soc. Malaysia 60, 45–53.

Small, C., Nicholls, R.J., 2003. A global analysis of human settlement in coastal zones. J. Coast. Res. 19, 584–599.

Soon, T.K., Denil, D.J., Ransangan, J., 2016. Temporal and spatial variability of heavy metals in Marudu Bay, Malaysia. Oceanol. Hydrobiol. Stud. 45 (3), 353–367.

Spalding, M.D., Ruffo, S., Lacambra, C., Meliane, I., Hale, L.Z., Shepard, C.C., Beck, M.W., 2014. The role of ecosystems in coastal protection: adapting to climate change and coastal hazards. Ocean Coast. Manag. 90, 50–57.

Spencer, R.G., Aiken, G.R., Dornblaser, M.M., Butler, K.D., Holmes, R.M., Fiske, G., Mann, P.J., Stubbins, A., 2013. Chromophoric dissolved organic matter export from US rivers. Geophys. Res. Lett. 40 (8), 1575–1579.

Straub, J.R., Among, H.L., Gastaldo, R.A., 2000. Seasonal sediment transport and deposition in the Rajang River Delta, Sarawak, East Malaysia. Sediment. Geol. 133 (3), 249–264.

Taylor, S.R., McLennan, S.M., 1985. The Continental Crust: Its Composition and Evolution. Blackwell Scientific Publications, Oxford.

Tellez, M., Merchant, M., 2015. Biomonitoring heavy metal pollution using an aquatic apex predator, the American alligator, and its parasites. PLoS One, 10(11), e0142522. https://doi.org/10.1371/journal.pone.0142522.

Tiju, I.V., Prakash, T.N., Nagendra, R., Nagarajan, R., 2018. Sediment geochemistry of coastal environments, southern Kerala, India: implication for provenance. Arab. J. Geosci. 11(61). https://doi.org/10.1007/s12517-018-3406-9.

Tomlinson, D.L., Wilson, J.G., Harris, C.R., Jeffrey, D.W., 1980. Problems in the assessment of heavy metal levels in estuaries and the formation of a pollution index. Helgol. Meeresunter 33, 566–575.

Tonyes, S.G., Wasson, R.J., Munksgaard, N.C., Evans, K.G., Brinkman, R., Williams, D.K., 2017. Understanding coastal processes to assist with coastal erosion management in Darwin Harbour, Northern Territory, Australia. IOP Conf. Ser. Earth Environ. Sci. 55012012. https://doi.org/10.1088/1755-1315/55/1/012012.

Udechukwu, B.E., Ismail, A., Zulkifli, S.Z., Omar, H., 2015. Distribution, mobility, and pollution assessment of Cd, Cu, Ni, Pb, Zn, and Fe in intertidal surface sediments of Sg. Puloh mangrove estuary, Malaysia. Environ. Sci. Pollut. Res. 22 (6), 4242–4255.

van Hattum, M.W.A., Hall, R., Pickard, A.L., Nichols, G.J., 2006. Southeast Asian sediments not from Asia: provenance and geochronology of North Borneo sandstones. Geology 34, 589–592.

van Hattum, M.W.A., Hall, R., Pickard, A.L., Nichols, G.J., 2013. Provenance and geochronology of Cenozoic sandstones of northern Borneo. J. Asian Earth Sci. 76 (25), 266–282.

Vermeij, G.J., Raven, J.G.M., 2009. Southeast Asia as the birthplace of unusual traits: the Melongenidae (Gastropoda) of northwest Borneo. Contrib. Zool. 78, 113–127.

Vikas, M., Dwarakish, G.S., 2015. Coastal pollution: a review. Aquatic Proc. 4, 381–388.

Vilela, C., Batista, D.S., Batista-Neto, J.A., Crapeza, M., Mcallister, J.J., 2004. Benthic foraminifera distribution in high polluted sediments from Niteroi Harbor (Guanabara Bay), Rio de Janeiro, Brazil. An. Acad. Bras. Cienc. 76, 161–171.

Wang, X., Griffin, W.L., Chen, J., 2010. Hf contents and Zr/Hf ratios in granitic zircons. Geochem. J. 44 (1), 65–72.

Wang, H., Saito, Y., Zhang, Y., Bi, N., Sun, X., Yang, Z., 2011. Recent changes of sediment flux to the western Pacific Ocean from major rivers in East and Southeast Asia. Earth Sci. Rev. 108 (1-2), 80–100.

Wang, S.L., Xu, X.R., Sun, Y.X., Liu, J.L., Li, H.B., 2013. Heavy metal pollution in coastal areas of South China: a review. Mar. Pollut. Bull. 76 (1-2), 7–15.

Wang, A.J., Bong, C.W., Xu, Y.H., Hassan, M.H.A., Ye, X., Bakar, A.F.A., Li, Y.H., Lai, Z.K., Xu, J., Loh, K.H., 2017. Assessment of heavy metal pollution in surficial sediments from a tropical river-estuary-shelf system: a case study of Kelantan River, Malaysia. Mar. Pollut. Bull. 125 (1-2), 492–500.

Whitefield, J., Winsor, P., McClelland, J., Menemenlis, D., 2015. A new river discharge and river temperature climatology data set for the pan-Arctic region. Ocean Model. 88, 1–15.

Wilford, G.E., 1961. The geology and mineral resources of Brunei and adjacent parts of Sarawak with Descriptions of the Seria and Miri Oilfields. In: British Geological Survey Memoirs Book. vol. 10. Brunei Press Limited, Brunei.

Woodroffe, C.D., 2003. Coasts: Form, Process and Evolution. Cambridge University Press. 623 p.

Wu, W., Zheng, H., Xu, S., Yang, J., Liu, W., 2013. Trace element geochemistry of riverbed and suspended sediments in the upper Yangtze River. J. Geochem. Explor. 124, 67–78.

Xian, X., 1987. Chemical partitioning of cadmium, zinc, lead, and copper in soils near smelters. J. Environ. Sci. Health A 226, 527–541.

Xuejing, X., Xuzhan, M., Tianxiang, R., 1997. Geochemical mapping in China. J. Geochem. Explor. 60 (1), 99–113.

Yalcin, M.G., Ilhan, S., 2008. Multivariate analysis to determine the origin of potentially harmful heavy metals in Beach and dune sediments from Kizkalesisi Coast (Mersin), Turkey. Bull. Environ. Contam. Toxicol. 81, 57–68.

Yalcin, F., Nyamsari, D.G., Paksu, E., Yalcin, M.G., 2016. Statistical assessment of heavy metal distribution and contamination of beach sands of Antalya-Turkey: an approach to the multivariate analysis techniques. Filomat 30 (4), 945–952. https://doi.org/10.2298/FIL1604945Y.

Yanko, V., Ahmad, M., Kaminski, M., 1998. Morphological deformities of benthic foraminiferal tests in response to pollution by heavy metals: implications for pollution monitoring. J. Foraminifer. Res. 28, 177–200.

Young, L.B., Harvey, H.H., 1992. The relative importance of manganese and iron oxides and organic matter in the sorption of trace metals by surficial lake sediments. Geochim. Cosmochim. Acta 56 (3), 1175–1186.

Yusoff, A.H., Zulkifli, S.Z., Ismail, A., Mohamed, C.A.R., 2015. Vertical trend of trace metals deposition in sediment core off Tanjung Pelepas harbour, Malaysia. Procedia Environ Sci 30, 211–216.

Zhao, G., Lu, Q., Ye, S., Yuan, H., Ding, X., Wang, J., 2016. Assessment of heavy metal contamination in surface sediments of the west Guangdong coastal region, China. Mar. Pollut. Bull. 108 (1–2), 268–274.

Further Reading

Garcia, D., Fonteilles, M., Moutte, J., 1994. Sedimentary fractionation between Al, Ti, and Zr and genesis of strongly peraluminous granites. J. Geol. 102, 411–422.

McGann, M., Sloan, D., 1997. Benthic Foraminifers in the Regional Monitoring Program's San Francisco Estuary Samples. Regional Monitoring Program-Annual Report. San Francisco Estuary Institute, Richmond, CA.

Nagarajan, R., 2018. Geochemistry of Miri River Sediments. (per. comm.).

Nesbitt, H.W., Young, G.M., 1984. Prediction of some weathering trends of plutonic and volcanic rocks based on thermodynamic and kinetic consideration. Geochim. Cosmochim. Acta 48 (7), 1523–1534.

Yanko, V., Kronfeld, J., Flexer, A., 1994. Response of benthic foraminifera to various pollution sources; implications for pollution monitoring. J. Foraminifer. Res. 24, 1–17.

CHAPTER 13

Multimarker Pollution Studies Along the East Coast of Southern India

Murugaiah Santhosh Gokul*, Hans-Uwe Dahms[†], Krishnan Muthukumar*, Thanamegam Kaviarasan*, Santhaseelan Henciya*, Sivanandham Vignesh*, Thiyagarajamoorthy Dhinesh Kumar*, Rathinam Arthur James*

*Department of Marine Science, Bharathidasan University, Tiruchirappalli, India, [†]Department of Biomedical Science and Environmental Biology, KMU—Kaohsiung Medical University, Kaohsiung, City, Taiwan, ROC

1 Introduction

Water is essential to life and human health, economic development, food security, poverty reduction, and sustainable ecological functions (UN Water, 2013). Given that the world's population is expected to reach eight billion by 2025, growing demands on drinking water supplies and water for food production are evident, and competing uses of limited resources are inevitable (UNDP, 2006). Coastal waters are facing a wide variety of stress affecting both the ecosystem and human health via domestic wastewater treatment and disposal practices that may lead to the introduction of high levels of nutrients and enteric human pathogens (Newton et al., 2013). Ocean water has many uses in industry, transportation, aquaculture, for domestic purposes, and so forth. Oceans have long been used for disposal purposes. The main anthropogenic impact caused by the seasonal population rise in coastal regions during monsoon season is the elevated organic load discharged into the water bodies, which are used as sewers (Pati et al., 2014).

Groundwater is one of the most vital resources for the sustenance of humans, plants, and other forms of life. It is required in all aspects of life for producing food for agricultural activities and for energy generation. Drinking water is a major source of microbial pathogens in developing regions, although poor sanitation and food sources are integral to enteric pathogen exposure. The lack of safe drinking water and adequate sanitation measures lead to a number of diseases such as cholera, dysentery, salmonellosis, and typhoid, and every year millions of lives are claimed in developing countries. Groundwater is the main source of drinking water in the villages without any treatment. It may be contaminated by disease-producing pathogens, leachate from landfills and septic systems, careless disposal of hazardous household products, agricultural chemicals, and leaking underground storage tanks (Selvam et al., 2017). Major and trace elements above permissible limits in groundwater have direct consequences on human health and the environment. Their occurrence in groundwater is often guided by two main pathways: natural and anthropogenic. Natural pathways are associated with rock weathering, leaching, rock water interaction, and resident time. On the other hand, anthropogenic activities, such as burning of fossil fuels and discharge of industrial effluents and sewage into water bodies can contribute a significant concentration of trace elements in groundwater (Selvam et al., 2014).

In recent years, many water-borne diseases from contaminated groundwater have been reported by countries with various levels of economic development (Zhang et al., 2014; Beer et al., 2015). About 2.3 billion people worldwide suffer from diseases that are linked to water, and most of the victims are children in developing countries. Safe water would prevent around 2.5 million deaths from diarrheal diseases, 150 million cases of schistosomiasis, or 75 million cases of trachoma each year (Vignesh et al., 2016). These diseases are caused by pathogenic microorganisms of enteric origins, distributed by anthropogenic and natural processes, such as grazing, manure spreading, and uncontrolled sewage disposal (Giglio et al., 2016). The bacteriological examination of water has a special significance in pollution studies, as it is a direct measurement of the deleterious effects of pollution on human health. For assessment of water quality, microbiological water quality (Gokul et al., 2018) must also be included, in addition to the regular monitoring of physic-chemical characteristics (APHA, 2012).

Human fecal material is generally considered to be a risk to human health, as it is likely to contain human enteric pathogens (Staley et al., 2012). Fecal coliforms are generally used as principal indicators of fecal pollution in water of target catchments. These organisms are excreted in feces by man and animals, and are useful indicators to complement assessment of fecal pollution in water. Natural areas can also be a source of bacteria originating from wildlife, including birds, mammals, pets, and livestock (Griffith et al., 2006). In this chapter, we report the bacteria levels, fecal coliforms, fecal *Streptococci*, *Vibrio cholerae*, *Salmonella* sp., *Shigella* sp., and *Pseudomonas* sp., and trace metals measured from 12 stations throughout Tamilnadu coastal systems in seasonal intervals. In addition, an epidemiological survey of 2000 respondents was also conducted. This work addresses the issue of water contamination, which has long been neglected by researchers and policy makers in assessing the quality of coastal water. Hence, it is expected to assist local authorities in developing plans and policies and implementing actions to reduce pollution to acceptable levels.

2 Materials and Methods

2.1 Study Area

The Indian coastline is about 8129 km in length. The state of Tamil Nadu has a 1076 km long coastline, accounting for 17% of the total coastline, and has a number of ecologically sensitive areas, with about 15 million people living in the coastal areas. Twelve sampling stations were established for the present study (Fig. 1). They are: Chennai Marina beach (S1), Pondicherry beach (S2), Cuddalore beach (S3), Tarangabadi beach (S4), Nagapattinam (S5), Velangami beach (S6), Vetharanyam beach (S7), Thondi beach (S8), Rameshwaram (S9), Tuticuorin (S10), Tiruchendur beach (S11), and Kanayakumari beach (S12).

2.2 Sample Collection and Preparation

A total of 48 groundwater samples were collected from open wells and bore holes ranging in depth between 6 and 86 m BGL. From each station, 2 L of coastal groundwater were collected for microbiological (1 L) and trace metal analysis (1 L) during each season, for four seasons: postmonsoon (December 2014, January and February—2015), summer (March, April and May—2015), premonsoon (June, July and August), and monsoon (September, October and November—2015). Samples were collected in acid-washed polyethylene 1000-mL HDPE bottles. Each bottle was completely filled with water, taking care that no air bubbles were trapped within the water sample. Then, to prevent evaporation, the double plastic caps of the bottles were sealed. Precautions were also taken to avoid sample agitation during transfer to the laboratory. The samples were stored at a temperature below 4°C prior to analysis in the laboratory.

2.3 Isolation of Pollution Indicators and Epidemiological Survey

All the media were prepared with the addition of aged seawater, and were autoclaved properly. All plates were prepared 5 days prior to sampling. The bacterial population in different samples was estimated by the spread plating method on selective medium plates with 0.1 mL of suitable dilutions. All the media plates were incubated at 37°C at least for 24–48 h, and final counts of colonies were noted (Vignesh et al., 2016). All trials were performed in triplicate. For bacterial enumeration, spread plates were used to determine the number of colonies (CFU/volume of sample). Specific selective medium were used for enumerating the bacterial types. All the culture media were obtained from Hi-Media Pvt. Ltd., Bombay, India. Because recommended selective media were used for all organisms, specific biochemical tests were not performed for identification, and they were therefore referred to as 'like organisms (LO).' For endorsement of the pathogens, typical colonies were inoculated into rapid microbial limit test kits that were recommended for diagnostic microbiology, and supplied by Hi-Media Laboratories Limited.

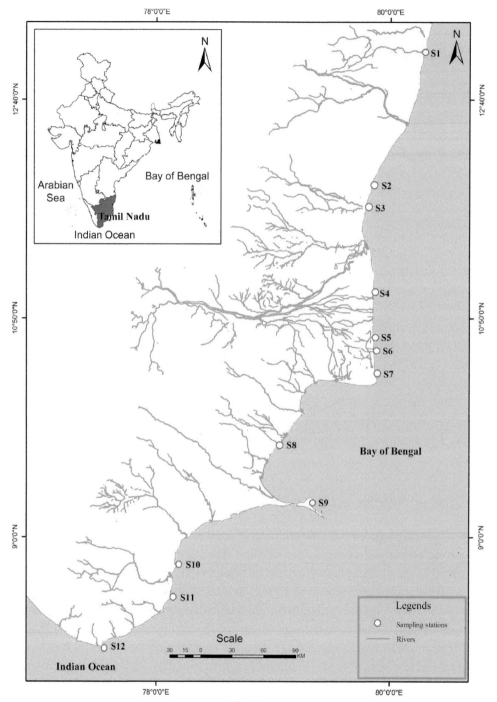

Fig. 1
Study area and sampling sites of Tamilnadu coast.

In order to properly investigate the public health implication of contaminated groundwater on the residents of the study, a survey was carried out using structured questionnaires. The demographics, habits, and occupational data of the respondents were observed in the questionnaires (Dahunsi et al., 2014).

2.4 Trace Metal Analysis

One liter of water sample was filtered through a 0.45 μm nitrocellulose membrane filter paper and adjusted to pH 2, with HNO_3 taken in a separator funnel. Then 10 mL (3%, w/v) of a freshly prepared solution of amino-pyrolidine dithiocarbamate (APDC) was added into the funnel, and the mixture was shaken by a mechanical shaker for 10 min. Further, 25 mL of methyl-isobutyl-ketone (MIBK) was added to this mixture and shaken for 15 min. The phases were allowed to separate. The top organic phase was collected. The bottom aqueous phase was again shaken with 25 mL of MIBK, and the organic phase obtained then pooled with the earlier one. The pooled organic phase was mixed with 2 mL of 50% HNO_3 and shaken vigorously for 10 min to separate the bottom acid layer (Muthukumar et al., 2017).

2.5 Antibiotic Resistant Strains

In order to evaluate the bacterial resistance to 10 types of antibiotics used in this study, disk diffusion assay was employed. Bacterial cultures were grown at 35°C for 18 ± 2 h in nutrient broth, diluted to $1 \cdots 10^7$ Cfu/mL in tempered 0.75% agar (45°C), mixed gently, and poured onto Muller-Hinton agar (MHA; Difco). After solidification, antimicrobial susceptibility test disks were applied, and plates were incubated at 35°C for 24 h. These 10 antibiotic discs represented seven different chemical structural classes of antibiotics: aminoglycosides (gentamycin), ß-lactams (amoxicillin, ampicillin, methicillin, and penicillin-G), glycopeptide (vancomycin), macrolides (erythromycin), quinolones (ciprofloxacin), tetracycline (tetracycline), and others (chloramphenicol). All these culture media and antibiotic discs were obtained from Hi-Media Pvt. Ltd., Mumbai, India. Antibiotics were dissolved in Milli-Q water prior to testing, and were diluted with the appropriate medium immediately before the tests (Vignesh et al., 2012).

3 Result and Discussion

3.1 Pollution Indicators

Total viable counts (TVC) for water samples were the highest during monsoon season, and the least during the postmonsoon season. The TVC was in the range of 9.1 $[\times 10^3]$; 5.8 $[\times 10^3]$; 6.6 $[\times 10^3]$ and 4.1 $[\times 10^3]$ mL^{-1} at in Velanganni (S6) and least ranges of 7.8 $[\times 10^2]$; 2.3 $[\times 10^2]$; 6.2 $[\times 10^2]$ and 2.1 $[\times 10^2]$ mL^{-1} were found at Vedharaniyam (S7), during monsoon season, summer, premonsoon season, and postmonsoon season, respectively. The highest abundance of all the examined groups was observed during monsoon season (Fig. 2). Compared with Vedharanyam, the Velangani showed higher fisher communities. It is one of the important tourism destinations for all seasons. The major reason is that pilgrimage places suffer from insufficient sanitation facilities and aging sanitary infrastructure, forcing many people to adopt open defecation in nearby water bodies such as rivers, lakes, ponds, and shoreline environments (Viji and Shrinithivihahshini, 2017).

In fact, the groundwater of the Tamil Nadu coast is used for several purposes in most of the regions, although the higher TVC values suggest that this practice should be avoided. The highest TVC values may be attributed to the presence of large populations residing in the coastal zone. In addition to pathogens, urine and feces contain organic matter, as well as eutrophic substances in the form of phosphorus and nitrogen compounds (Langergraber and Muellergger, 2005). Visits by the public and livestock to ocean water systems are common in developing countries, particularly in slum communities where most residents lack access to clean water. As a result, they usually obtain water for their daily needs from coastal groundwater systems that are often contaminated (Nevondo and Cloete, 1999; Obi et al., 2002). Consequently, during visits, livestock may be watered; people may bathe; waste is disposed of; shipping yards, clothes, and vehicles may be washed; and water is usually abstracted for domestic needs (Nevondo and Cloete, 1999; Mathooko, 2001; Yillia et al., 2008).

During monsoon season and summer, the counts of TC and TS were generally higher in all the sample stations. The mean concentration of TC load in seawater showed the maximum value of 2607 mL^{-1} (monsoon season) in Tuticorin (S10), and the minimum value of 100 mL^{-1} (postmonsoon season) at Vedharaniyam (S7). Total coliform was high in Tuticorin (S10) during all the seasons in groundwater, that is, 2000, 3110, 2300, and 4300 mL^{-1} during postmonsoon season, summer, premonsoon season, and monsoon season, respectively. Pollution indicator bacteria such as TC, FC, and TS are routinely examined for an understanding of the preponderance of human pathogenic bacteria (APHA, 1980). Tremendous increases in human activities, urbanization, and industrialization have disturbed the balance of the coastal environment (Selvam et al., 2014). These sources of bacterial contamination include surface runoff, and pastures and other land areas where animal wastes are deposited. Additional sources include seepage or discharge from septic tanks, sewage treatment facilities, natural soil/plant bacteria, and buckle channels.

In groundwater, counts of total *Streptococci* (TS) ranged from 240; 350; 290; and 510 mL^{-1} during the postmonsoon season, summer, premonsoon season, and the monsoon season, respectively. The fecal *Streptococci* (FS) ranged from 0 to 220 mL^{-1} and maximum populations were noticed in Velanganni (S6) (monsoon season). The presence of fecal coliforms does not always correspond to the presence of pathogens, but pathogens are rarely detected without indicators of fecal pollution. The presence of bacterial indicators of fecal contamination in different depths of groundwater aquifers clearly revealed the bacteriological

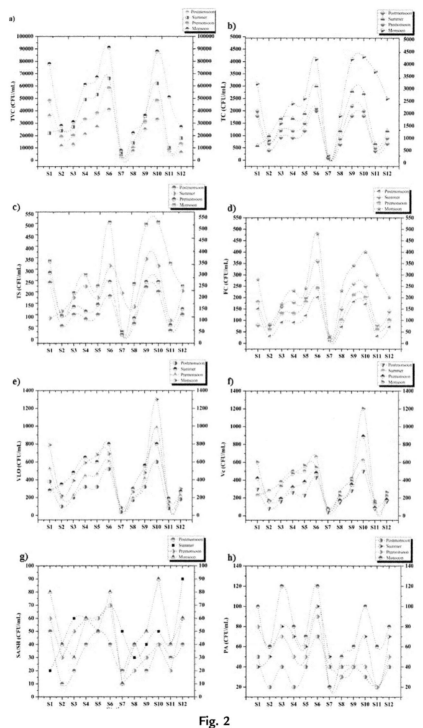

Fig. 2
Seasonal distribution of Pollution indicators (A) TVC, (B) TC, (C) TS, (D) FC, (E) VLO, (F) VC, (G) SA/SH, and (H) PA in Tamilnadu coastal groundwater.

status of the water at that site. Fecal material from swimmers, domestic animals (dogs, cattle, and horses), and waterfowl (geese, gulls, and ducks) leads to an increase of microbial loadings at beaches. Our results on groundwater quality demonstrated that FC, FS, and pathogens are widely distributed in the studied area. These waters serve as an important natural resource for drinking water supply, recreation and domestic activities for most of the shoreline/slum populations. This fact indicated potential health effects on populations using this groundwater. The summary of the health challenges reported by respondents during the survey conducted in this research is shown in Table 2. Cholera was reported by 9%–25% of the respondents, diarrhea by 5%–27% of the respondents, gastroenteritis by 33%–44% of the respondents, typhoid by 08%–24%, and dysentery was reported by 17%–23% of the respondents. Other health problems reported by less than 3% of the respondents include anemia, cancer, sleeping disorders, poor/loss of appetite, constipation, vomiting/nausea, and malaria. The presence of *Escherichia coli* is of medical importance because it is an indication of the presence of other enteric pathogens. *E. coli* is known to cause many enteric diseases, such as diarrhea. Other important pathogens identified are *Pseudomonas* sp., *Citrobacter* sp., *Klebsiella* sp., and *Enterobacter aerogenes*. *Bacillus* sp. and *Staphylococcus* sp. are known to cause gastrointestinal disorders such as diarrhea.

According to WHO (2003) standards, the studied groundwater was completely unsuitable for drinking and other domestic uses. Many concerns have been raised recently about the pathogens originating from rainfall runoff (Conboy and Goss, 2000; Ferguson et al., 2003; Smith and Perdek, 2004). Unhygienic domestic sanitation and unsafe environments lead to incidences of waterborne illness, and place children at risk of death (Ezzati et al., 2002; Guilbert, 2003). More than half of the reported waterborne disease outbreaks have been linked to contaminated groundwater (Craun et al., 1997). Ingestion of contaminated water due to the lack of hygiene and sanitation contributes to about 1.5 million child deaths per year, and around 88% of them are from diarrhea (Ezzati et al., 2002; Guilbert, 2003). During the summer, the Vibrio-like organisms (VLO) were higher in most locations, especially in Tuticorin (S10), Velangani (S6), Nagapattinam (S5), and Rameshwaram (S9); and the elevated values were observed to be 1300, 980, 800, and 600 mL^{-1}. The range of Vibrio-like organisms (VLO) in groundwater samples were 40–600; 70–800; 70–980; and 90–1300 mL^{-1} in the postmonsoon season, summer, the premonsoon season, and monsoon season, respectively. In coastal groundwater, the VC, SA, SH, and PA were in the ranges of 0–1200, 0–50, 0–90, and 40–120 mL^{-1}, respectively.

3.2 Antibiotic-Resistant Strains

In all the sampling sites, higher concentrations of pathogen counts were noted in Velanganni (S6) groundwaters than any other sites. Among the colonies isolated from Tamil Nadu coastal groundwater, 61.5% showed resistance to antimicrobials. In these strains, *E. coli* showed the highest frequencies of resistance to Penicillin (20.6%), Methicillin (8.9%), and Vancomycin (7%) (Table 1). *Salmonella* sp. showed higher resistance to penicillin (15.3%), amoxicillin (6.6%), and ampicillin (5.6%). Most of the isolated strains in water were resistant to one antibiotic (68.5%). Similar antimicrobial resistance profiles were also reported in studies

Table 1 Antibiotic resistant and multidrug resistant strains isolated from Tamilnadu coastal groundwater

Classes of Antibiotics	Antibiotics	Tamilnadu Coastal Groundwater (S1–S12) ($n = 600$)							
		E. coli		Salmonella sp.		Shigella sp.		Vibrio sp.	
		N	%	N	%	N	%	N	%
Aminoglycosides	Gentamycin	18	03	29	4.8	–		11	1.8
β-lactams	Amoxicillin	29	4.8	40	6.6	09	1.5	32	5.3
	Ampicillin	66	11	34	5.6	11	1.8	22	3.6
	Methicillin	49	8.9	19	3.1	10	1.6	28	4.6
	Penicillin	124	20.6	92	15.3	80	13.3	59	9.8
Quinolone	Ciprofloxacin	30	5	–		–		05	0.8
Glycopeptides	Vancomycin	42	7	15	2.5	67	11.16	26	4.3
Macrolides	Erythromycin	18	3	09	1.5	20	3.3	10	1.6
Tetracycline	Tetracycline	01	0.16	–		–		05	0.8
Other	Chloramphenicol	–		–		20	1.8	03	0.5

Table 2 Epidemiological health-based survey in Tamilnadu coastal communities

Parameters	S1	S2	S3	S4	S5	S6	S7	S8	S9	S10	S11	S12
Demographic												
Male (Adult)	460	629	397	402	551	585	445	389	692	549	624	684
Female (Adult)	578	540	349	360	520	482	520	410	584	482	584	640
Male (children)	490	354	410	480	504	460	481	507	645	468	528	588
Female (Children)	607	470	430	417	490	502	438	490	580	494	560	648
Health-based (%)												
Cholera	15	10	12	8	18	20	10	9	25	28	18	14
Typhoid	21	14	17	10	15	24	17	19	14	24	17	12
Diarrhea	18	21	18	16	12	19	5	20	27	24	12	19
Dysentery	12	09	05	11	21	14	11	8	20	19	14	12
Gastroenteritis	09	11	07	10	14	11	14	06	20	18	17	10
Poor appetite	22	34	19	17	20	30	24	18	31	40	17	13
Constipation	36	28	21	19	27	22	19	30	44	28	38	30
Vomiting	28	20	21	18	26	30	22	14	30	19	24	20
Malaria	19	23	32	18	41	31	18	24	21	28	17	20
Pneumonia	10	09	23	16	11	22	16	13	11	21	15	09
Skin diseases	08	04	13	05	19	26	11	08	14	09	11	13
Nausea	10	03	14	02	10	13	09	16	12	07	14	04

involving isolates from coastal environments, farmed fish, and from tap water. Koesak et al. (2012) detected bacterial resistance against ampicillin, gentamycin, erythromycin, tetracycline, and ciprofloxacin at different times. Double resistance was observed in 18.75% of the strains, and only 12.5% demonstrated resistance to 3 antimicrobials. According to Harakeh et al. (2006), the emergence of antimicrobial-resistant bacteria increases in environments where antimicrobials are indiscriminately used by the public. Multiple bacterial resistances to drugs had earlier been reported in aquaculture environments (Hatha et al., 2005). Puah et al. (2013) reported up to six different resistance patterns, and resistance to at least one antibiotic was seen in 46 isolates (98%); and multidrug resistance (to two or more drugs) in 93% of tested isolates. Resistance to multiple antibiotics can lead to occurrence of newly emerging resistant bacteria that may be transmitted to consumers, causing infections that are difficult to treat.

3.3 Trace Metal Occurrence

Elevated levels of metals (Cd, Cr, Cu, Fe, Ni, Pb, and Zn) were detected in coastal aquifer waters on the Tamil Nadu coast. The abundance of metals in the coastal water is in the order of Fe > Zn > Cu > Pb > Cr > Ni > Cd. The metal concentration is related to point/nonpoint source inflows from the urban/village/industries/rivers discharging directly into the ocean. The higher concentrations at these sites might be due to contributions of domestic sewage from town/slum areas, industrial/shipping activities, and fishing communities located along the Tamil Nadu coast. Sewage and urban waste water from towns and several settlements constitute other discharges that are negatively affecting the water quality in the coastal regions (Koukal et al., 2004). The concentrations obtained for Pb ranged from BDL to $0.585\,mg\,L^{-1}$ and were above the permissible limits of WHO, US-EPA, Canadian guideline. The highest concentration of Pb ($0.858\,mg\,L^{-1}$) was obtained from an S12 water sample during monsoon season. Adverse chronic effects may occur at $0.5-1.0\,mg\,L$ Pb. At levels greater than $0.1\,mg\,L$, possible neurological damage in fetuses and children may occur (Department of Water Affairs and Forestry (DWAF), 1993). These levels were found to be in excess of permissible limits in the sampling locations, and the direct use of water for domestic purposes without treatment could be detrimental to pregnant women and young children. Possible sources of Pb in the study area could be from domestic sewage and effluent discharge from industries, shipping activities, paints, glass works, and dense traffic emissions (Caeiro et al., 2005). Fe concentrations ranged from 0.03 to $2.85\,mg\,L^{-1}$ and were within permissible limits (Fig. 3). The highest concentration was $2.85\,mg\,L^{-1}$ reported at S10 during summer. Other significantly concentrated levels of Fe were found at: 2.3 (S9); 1.46 (S10); and $1.33\,mgL^{-1}$
(S9) during monsoon season, premonsoon season, and postmonsoon season, respectively. The concentrations of Cu ranged from 0.02 to $0.798\,mg\,L^{-1}$ and were within the permissible limits of the WHO ($2.0\,mg\,L^{-1}$), US-EPA ($1.30\,mg\,L^{-1}$), Canadian guidelines ($1.0\,mg\,L^{-1}$).

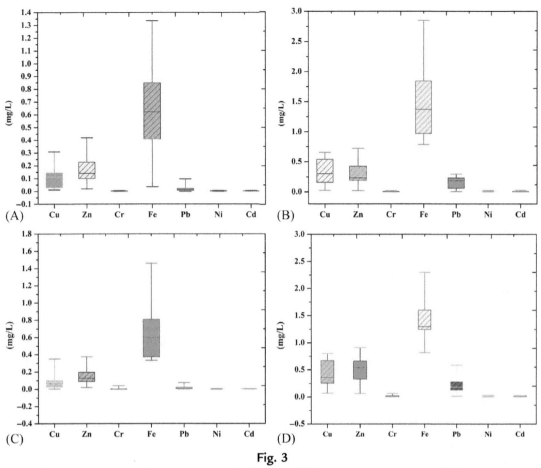

Fig. 3
Seasonal variation of trace metals distribution (A) Postmonsoon season, (B) Summer, (C) Premonsoon season and (D) monsoon season in Tamilnadu coastal groundwater.

The highest concentration of Cu (0.0.798 mg L^{-1}) was observed in the S10 station during monsoon season, and the lowest (0.02 mg L^{-1}) was in the S7 station during premonsoon season. Zn concentrations ranged from 0.02 to 0.912 mg L^{-1} and were within permissible limits. The highest concentration of Zn (0.0.912 mg L^{-1}) was obtained in S10 during monsoon season, and the lowest (0.02 mg L^{-1}) was obtained in S7 during premonsoon season. Sources of zinc are associated with urban/industrial effluents, shipyard traffic emission, fertilizers, paints, and electrical apparatus (Caeiro et al., 2005).

The concentrations of Cr, ranged from 0 to 0.065 mg L^{-1} and exceeded the permissible limits of WHO, Canadian guideline, US-EPA for drinking water. The highest Cr concentration (0.065 mg L^{-1}) was obtained in S11, and the lowest (BDL) was obtained in S2, S4, and S8, respectively. Low levels of chromium values were also found in most of the stations. Sources of chromium are associated with agro materials, the manufacture of chemicals, chrome plating, and cooling towers (McConnell et al., 1996). Concentrations of Ni ranged from 0 to 0.0307 mg L^{-1}. These concentrations were above the WHO permissible limit (0.02 mg L^{-1}) for drinking water. The highest concentration (0.0307 mg L^{-1}) was obtained from the S8 water sample during monsoon season, and the lowest concentration was obtained from S7 during postmonsoon season. The monsoon and summer seasons registered elevated levels of the metals, as compared with the premonsoon and postmonsoon seasons. The continuous increase in heavy metal contamination of coastal water is a cause of concern, as these metals have the ability to bioaccumulate in the tissues of various biotas, and may also affect the distribution and density of benthic organisms, as well as the composition and diversity of faunal communities. As a result, residents of many fishing communities suffer from serious skin diseases around coastal cities (Balasubramanian, 1999).

3.4 Correlation Matrix and Factor Analysis

Interelemental relationships can provide interesting information on the element sources and pathways. In the present study, TVC showed a strong positive correlation with TC ($r=0.858$), TS ($r=0.812$), FC ($r=0.879$), VLO ($r=0.876$), VC ($r=0.0.846$), and PA ($r=0.746$). TC with TS ($r=0.949$), FC ($r=0.972$), VLO ($r=0.754$), and VC ($r=0.737$). TS with FC ($r=0.932$), VLO ($r=0.772$), VC ($r=0.753$), PA (0.702). VLO with VC ($r=0.992$), Cu—Pb ($r=0.709$) and Pb—Cd ($r=0.705$), respectively (Fig. 4). Microbial parameters of TVC and TC showed a moderate correlation with heavy metals Cu, Zn, Fe, Ni, and Pb. Bacterial parameters were commonly associated with heavy metal parameters, confirming that the minimum level of trace metals were acting as micronutrients, and were supporting microbial growth with high significance (Vignesh et al., 2016). Principal component analysis (PCA) is the method that provides a unique solution, so that the original data can be reconstructed from the results. The first step in multivariate statistical analysis was applying PCA with an aim to group the individual parameter components by loading plots for the investigated contaminated sites. The use of PCA to evaluate water quality assessment has increased in recent years, mainly due to the need to obtain appreciable data reduction for analysis and decision. According to these results, the Eigen values of the three extracted components are greater than the one before and after the varimax rotation. As a consequence, heavy metals could be grouped into a three-component model that accounts for 73.52% of all the data variations. In the rotated component matrix, the first PC (PC1, variance of 48.6%) included Cu, Pb, Zn, Fe, and Cd (Table 3); and the second PC (PC2, variance of 20.5%) was constituted by Ni. PC1, including Cu, Pb, Zn, Fe, and Cd, can be defined as an anthropogenic component due to the high variability observed in the present study.

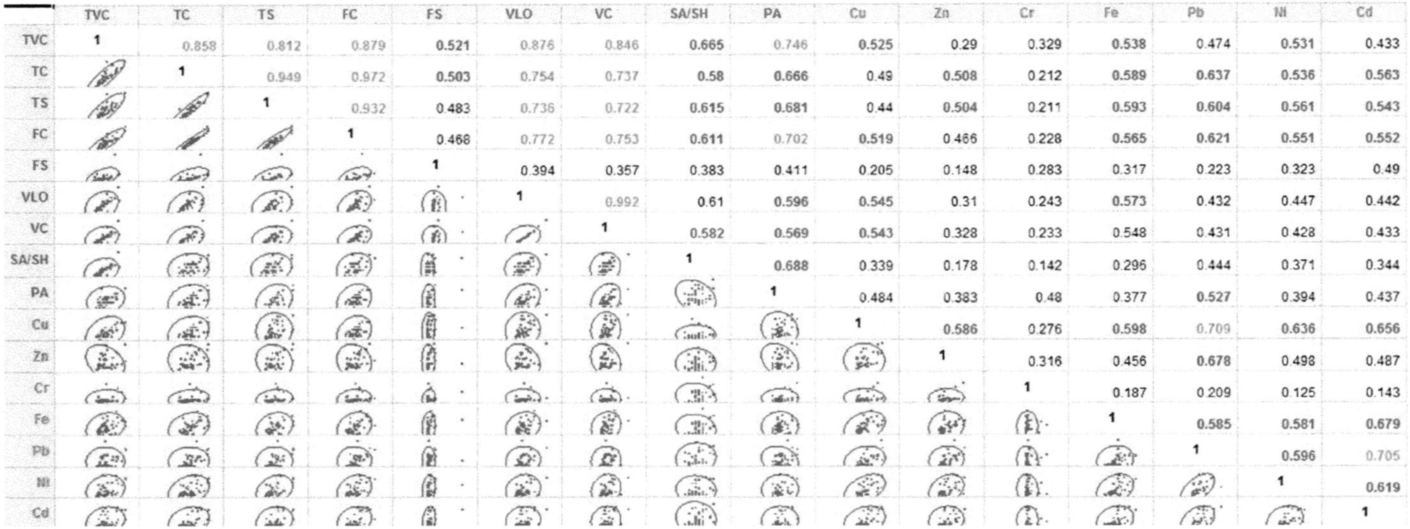

	TVC	TC	TS	FC	FS	VLO	VC	SA/SH	PA	Cu	Zn	Cr	Fe	Pb	Ni	Cd
TVC	1															
TC	0.858	1														
TS	0.812	0.949	1													
FC	0.879	0.972	0.932	1												
FS	0.521	0.503	0.483	0.466	1											
VLO	0.876	0.754	0.736	0.772	0.394	1										
VC	0.846	0.737	0.722	0.753	0.357	0.992	1									
SA/SH	0.665	0.58	0.615	0.611	0.383	0.61	0.582	1								
PA	0.746	0.666	0.681	0.702	0.411	0.596	0.569	0.688	1							
Cu	0.525	0.49	0.44	0.519	0.205	0.545	0.543	0.339	0.484	1						
Zn	0.29	0.508	0.504	0.466	0.148	0.31	0.328	0.178	0.383	0.586	1					
Cr	0.329	0.212	0.211	0.228	0.283	0.243	0.233	0.142	0.48	0.276	0.316	1				
Fe	0.538	0.589	0.593	0.565	0.317	0.573	0.548	0.296	0.377	0.598	0.456	0.187	1			
Pb	0.474	0.637	0.604	0.621	0.223	0.432	0.431	0.444	0.527	0.709	0.678	0.209	0.585	1		
Ni	0.531	0.536	0.561	0.551	0.323	0.447	0.428	0.371	0.394	0.636	0.498	0.125	0.581	0.596	1	
Cd	0.433	0.563	0.543	0.552	0.49	0.442	0.433	0.344	0.437	0.656	0.487	0.143	0.679	0.705	0.619	1

Fig. 4
Correlation coefficient between physiochemical and trace metal parameters in Tamilnadu coastal groundwater.

Table 3 Principal component analysis (PCA) for microbial and trace metal parameters of ground waters of coastal Tamil Nadu

Variables	Mean	SD	Communalities	F1	F2	F3
Microbial parameters						
TVC	33,110.02	22,489.84	0.914	0.910		
TC	1739.78	1080.45	0.859	0.815		
TS	197.10	121.66	0.831	0.808		
FC	166.95	103.40	0.874	0.835		
FS	26.08	32.62	0.367	0.681		
VLO	407.60	263.43	0.824	0.872		
VC	336.30	224.94	0.784	0.848		
SA/SH	46.08	19.37	0.581	0.735		
PA	56.73	25.82	0.758	0.650		
Trace metals						
Cu	0.279	0.235	0.697		0.780	
Zn	0.310	0.231	0.667		0.759	
Cr	0.008	0.132	0.872			0.920
Fe	1.116	0.609	0.633		0.693	
Pb	0.117	0.133	0.770		0.820	
Ni	0.008	0.008	0.630		0.725	
Cd	0.007	0.008	0.702		0.785	
Eigen value				8.807	1.817	1.140
% of variance				37.74	26.97	8.81
Cumulative %				37.74	64.71	73.52

4 Conclusion

A detailed study and critical evaluation of analytical data showed that the groundwater aquifer of the coastal study area extending from Chennai to Kanyakumari of southern Tamil Nadu is severely affected by saline water intrusion and microbial pollution. The epidemiological health-based survey indicated the incidences of cholera up to 25%, diarrhea 27%, and gastroenteritis 44%. The study also reported that MDR strains such as *E. coli* 34%, *Salmonella* sp. 26%, and *Shigella* sp. 15%, which are resistant, showed a considerable increase in coastal aquifers. The study emphasized the need to create awareness in the coastal community for sanitation. The study also emphasized the need to attempt water quality monitoring with biological indicators, along with routinely attempted physic-chemical assessments for better and more accurate assessments of quality.

References

APHA (American Public Health Association), 1980. Standard Methods for the Examination of Water and Wastewater, 15th ed. Washington, DC.

APHA, AWWA, WEF, 2012. Standard Methods for Examination of Water and Wastewater, 22nd ed. American Public Health Association, Washington 1360 pp. ISBN 978-087553-013-0.

Balasubramanian, T., 1999. In: Impacts of coastal pollution.Proceedings of Indo-British integrated coastal zone management training short-course conducted by Institute for Ocean Management, Anna University, India.

Beer, K.D., Gargano, J.W., Roberts, V.A., Hill, V.R., Garrison, L.E., Kutty, P.K., Hilborn, E.D., Wade, T.J., Fullerton, K.E., Yoder, J.S., 2015. Surveillance for waterborne disease outbreaks associated with drinking water—United States, 2011–2012. MMWR Morb. Mortal. Wkly. Rep. 64, 842–848.

Caeiro, S., Costa, M.H., Ramos, T.B., Fernandes, F., Silveira, N., Coimbra, A., Medeiros, G., Painho, M., 2005. Assessing heavy metal contamination in Sado estuary sediment: an index analysis approach. Ecol. Indic. 5, 151–169.

Conboy, M.J., Goss, M.J., 2000. Natural protection of groundwater against bacteria of fecal origin. J. Contam. Hydrol. 43 (1), 1–24.

Craun, G.F., Berger, P.S., Calderon, R.L., 1997. Coliform bacteria and water borne disease outbreaks. J. Am. Water Works Assoc. 89, 96–104.

Dahunsi, S.O., Owamah, H.I., Ayandiran, T.A., Oranusi, U.S., 2014. Drinking water quality and public health of selected communities in South Western Nigeria. Water Qual. Exp. Health 6, 143–153.

Department of Water Affairs and Forestry (DWAF), 1993. South African Water Quality Guidelines, first ed. Domestic Use, vol. 1 The Government Printer, Pretoria, South Africa.

Ezzati, M., Lopez, A.D., Rodgers, A., Vander Hoorn, S., Murray, C.J., 2002. Selected major risk factors and global and regional burden of disease. Lancet 360, 1347–1360.

Ferguson, C., Husman, A.M.D., Altavilla, N., Deere, D., Ashbolt, N., 2003. Fate and transport of surface water pathogens in watersheds. Environ. Sci. Technol. 33 (3), 299–361.

Giglio, O.D., Barbuti, G., Trerotoli, P., Brigida, S., Calabrese, A., Vittorio, G.D., Caggiano, G.L.G., Uricchio, V.F., Montagna, M.T., 2016. Microbiological and hydrogeological assessment of groundwater in southern Italy. Environ. Monit. Assess. 188, 638.

Gokul, M.S., Dahms, H.U., Muthukumar, K., Henciya, S., Kaviarasan, T., James, R.A., 2018. Multivariate drug resistance and microbial risk assessment in tropical coastal communities. Hum. Ecol. Risk Assess. https://doi.org/10.1080/10807039.2018.1447361.

Griffith, J.F., Schiff, K.C., Lyon, G.S., 2006. Microbiological water quality at non-human impacted reference beaches in southern California during wet weather. Technical report 495, Southern California Coastal Water Research Project, Westminster, CA.

Guilbert, J.J., 2003. The world health report 2002—Reducing risks, promoting healthy life. Educ. Health 16, 230–233.

Harakeh, S., Yassine, H., El-fadel, M., 2006. Antimicrobial resistance pattern of *Escherichia coli* and Salmonella strains in the aquatic Lebanese environments. Environ. Pollut. 143, 269–277.

Hatha, M., Viverkanandam, A.A., Joice, G.J., Chistol, G.J., 2005. Antibiotic resistance patterns of mobile aeromonds from farm raised fresh fish. Int. J. Food Microbiol. 98, 131–134.

Koesak, D., Borek, A., Daniluk, S., Grabowska, A., Pappelbaum, K., 2012. Antimicrobial susceptibilities of listeria monocytogenes strains isolated from food and food processing environment in Poland. Int. J. Food Microbiol. 158, 203–208.

Koukal, B., Dominik, J., Vignati, D., Arpagaus, P., Santiago, S., Ouddane, B., 2004. Assessment of water quality and toxicity of polluted Rivers fez and Sebou in the region of fez (Morocco). Environ. Pollut. 131, 163–172.

Langergraber, G., Muellergger, E., 2005. Ecological sanitation—a way to solve global sanitation problems? Environ. Int. 31, 433–444.

Mathooko, J.M., 2001. Disturbance of a Kenyan Rift Valley Rift Valley stream by the daily activities of local people and their livestock. Hydrobiologia 458, 131–139.

McConnell, R., DeMott, R., Schulten, J., 1996. Toxic contamination sources assessment: risk assessment for chemicals of potential concern and methods for identification of specific sources. Final Report. Technical Publication #09-96. Tampa Bay National Estuary Program, Tampa, FL, USA.

Muthukumar, K., Dahms, H.U., Palanichamy, S., Subramanian, G., Vignesh, S., 2017. Multi metal assessment on biofilm formation in off shore environment. Mater. Sci. Eng. C 73, 743–755.

Nevondo, T.S., Cloete, T.E., 1999. Bacterial and chemical quality of water supply in the Dertig village settlement. Water SA 25 (2), 215–220.

Newton, R.J., Bootsma, M.J., Morrison, H.G., Sogin, M.L., McLellan, S.L., 2013. A microbial signature approach to identify fecal pollution in the waters off an urbanized coast of Lake Michigan. Microb. Ecol. 65 (4), 1011–1023.

Obi, C.L., Potgieter, N., Bessong, P.O., Matsaung, G., 2002. Assessment of the microbial quality of river water sources in rural Venda communities in South Africa. Water SA 28 (3), 287–292.

Pati, S., Mihir, K., Dash, C.K., Mukherjee, B., Pokhrel, D.S., 2014. Assessment of water quality using multivariate statistical techniques in the coastal region of Visakhapatnam, India. Environ. Monit. Assess. 186, 6385–6402.

Puah, S., Puthucheary, S.D., Liew, F., Chua, K., 2013. Aeromonasaquariorum clinical isolates: Antimicrobial profiles, plasmids and genetic determinants. Int. J. Antimicrob. Agents 41, 281–284.

Selvam, S., Manimaran, G., Sivasubramanian, P., Balasubramanian, N., Seshunarayana, T., 2014. GIS-based evaluation of water quality index of groundwater resources around Tuticorin coastal city, South India. Environ. Earth Sci. 71, 2847–2867.

Selvam, S., Venkatramanan, S., Sivasubramanian, P., Chung, S.Y., Singaraja, C., 2017. Geochemical characteristics and evaluation of minor and trace elements pollution in groundwater of Tuticorin City, Tamil Nadu, India using geospatial techniques. J. Geol. Soc. India, 62–68.

Smith, J.E., Perdek, J.M., 2004. Assessment and management of watershed microbial contaminants. Crit. Rev. Environ. Sci. Technol. 34 (2), 109–139.

Staley, C., Reckhow, K.H., Lukasik, J., Harwood, V.J., 2012. Assessment of sources of human pathogens and fecal contamination in a Florida freshwater lake. Water Res. 46 (17), 5799–5812.

[UN Water] United Nations Water, 2013. Water security & the global water agenda: a UN-water Analytical brief. Available from: http://www.unwater.org/downloads/watersecurity_analyticalbrief.pdf.

[UNDP] United Nations Development Programme, 2006. Human development report 2006: beyond scarcity: power, poverty and the global water crisis. ISBN 0-230-50058-7. Available from: http://hdr.undp.org/sites/default/files/reports/267/hdr06-complete.pdf.

Viji, R., Shrinithivihahshini, N.D., 2017. An assessment of water quality parameters and survival of indicator in pilgrimage place of Velankanni, Tamil Nadu, India. Ocean Coast. Manag. 146, 36–42.

Vignesh, S., Dahms, H.U., Muthukumar, K., Vignesh, G., James, R.A., 2016. Biomonitoring along the tropical southern Indian coast with multiple biomarkers. PLoS ONE 11(12) https://doi.org/10.1371/journal.pone.0154105.

Vignesh, S., Muthukumar, K., Arthur James, R., 2012. Antibiotic resistant pathogens versus human impacts: a study from three eco-regions of the Chennai coast, southern India. Mar. Pollut. Bull. 64, 790–800.

WHO (World Health Organization), 2003. Guidelines for Safe Recreational Water Environments: Coastal and Fresh Waters. Vol. 1. Geneva.

Yillia, P.T., Kreuzinger, N., Mathooko, J.M., 2008. The effect of in-stream activities on the Njoro River—part I: stream flow and chemical water quality. Phys. Chem. Earth 33 (8–13), 722–728.

Zhang, Y., Kelly, W.R., Panno, S.V., Liu, W.T., 2014. Tracing fecal pollution sources in karst groundwater by Bacteroidales genetic biomarkers, bacterial indicators, and environmental variables. Sci. Total Environ. 490, 1082–1090.

CHAPTER 14

Chromium Fractionation in the River Sediments and its Implications on the Coastal Environment: A Case Study in the Cauvery Delta, Southeast Coast of India

S. Dhanakumar*, R. Mohanraj[†]
*Department of Environmental Science, PSG College of Arts and Science, Coimbatore, India,
[†]Department of Environmental Management, Bharathidasan University, Tiruchirappalli, India

1 Introduction

Sediment is an integral component of aquatic environments, and act as a sink of pollutants (Pejman et al., 2015; Benson et al., 2016). Among the inorganic pollutants, heavy metals are ubiquitous in nature, and gaining scientific interest due to their reactivity, lithophilic nature, toxicity, and nonbiodegradability. Metallic element pollutants have a relatively high density, and are toxic even at low concentrations. Contamination of freshwater sediment by heavy metals is a worldwide problem stemming from numerous natural and anthropogenic activities in the industrial, domestic, and agricultural sectors (Davutluoglu et al., 2011; Pokorny et al., 2015). Heavy metal contamination in sediment can affect water quality, and thus the bioassimilation and bioaccumulation of metals in aquatic organisms, resulting in long-term implications for human and ecosystem health (Ip et al., 2007).

Heavy metals in sediment may exist in different mineral forms and chemical compounds. Based on the primary accumulation mechanism of heavy metals on sediments, metals may exist in several forms, namely; (i) easily exchangeable ions, (ii) metal carbonates, (iii) oxides, (iv) sulphides, (v) organometallic compounds, and (vi) ions in crystal lattices of minerals (Weisz et al., 2000; Kuang-Chung et al., 2001). The chemical species of a heavy metal may exhibit different physical and chemical behaviors in terms of chemical interaction, mobility, bioavailability, and potential toxicity (Ibragimow et al., 2013; Wang et al., 2014). For example, metals that are adsorbed on carbonate, sulfide, and organic bonds are highly correlated to pollution and a higher risk of bioavailability (Karbassi et al., 2008). Determination of total metal concentration could not provide any reliable information about mobility, bioavailability, and toxicity of metal. It is generally accepted that partitioning of heavy metals in sediment samples provides better approximations about metal mobility, bioavailability, and environmental and human health risks (Benson et al., 2017).

The concept of elemental speciation dates back to 1954, when Goldberg introduced the concept of oxidized versus reduced, and chelated versus free metal ions of trace elements in seawater (Jain, 2004). Since then, speciation studies attracted worldwide environmentalists and chemists, and became very prominent when Tessier et al. (1979) described the sequential extraction analytical procedure for heavy metals. Thus a new practice emerged to identify various classes of species of a metal, and to determine the sum of its concentrations in each class. Identification and quantification of metals associated with predefined phases was termed as "metal fractionation," where metal portioning was performed based on its physical and chemical characteristics (Templeton et al., 2000). The metal fractionation studies are carried out through sequential extractions of operationally defined metal fractions using specific reagents at various conditions.

Chromium (Cr) is one among the 14 most noxious heavy metals, and the seventh most abundant element on earth. Among the heavy metals, Cr is one of the biochemically active transition metals. Weathering of the earth's crust is a primary natural source of Cr in surface water (Mortuza and Al-Misned, 2017). Anthropogenic sources of emissions of Cr in surface water are from municipal wastes, chemical industries, paints, leather, road run-off due to tire wear, corrosion of bushings, brake wires, and radiators (Dixit and Tiwari, 2008). In aquatic sediments, Cr (III) and Cr (VI) are two oxidation states of Cr showing varying geochemical and ecotoxicologic properties. Cr(VI) is a known carcinogen and persistent industrial contaminant. The trivalent Cr can be oxidized to the more toxic hexavalent Cr species, and may accumulate into aquatic biota through bioaccumulation and biomagnification processes (Costa and Klein, 2006; Owlad et al., 2009). Cr in excess of naturally occurring background levels are prevalent in urbanized and industrialized river sediments due to the run-off from road surfaces, combined sewer overflows, and municipal and industrial discharges (Paul et al., 2002; USEPA, 2004).

The discharge of heavy metals from riverine systems into coastal environments is a major concern today. These heavy metal contaminants readily accumulate in coastal sediments, which can also act as a repository of pollutants due to changes in environmental conditions (e.g., pH, redox conditions, and competition for absorption sites among metal ions and absorptive area), and can increase the bioavailability of metals associated with sediment fractions, ultimately leading to potential hazards to aquatic organisms (Korfali and Jurdi, 2007; Zakir and Shikazono, 2011). The Cauvery River is one of the major rivers of Peninsular India that flows across three states and drains into the Bay of Bengal. The river, on its course, receives a considerable amount of industrial effluents, untreated municipal sewage, urban runoff, and agricultural runoff (Solaraj et al., 2010; Dhanakumar et al., 2011). Although several studies focused on total heavy metal concentrations in the River Cauvery water and sediments, there is no detailed study on Cr fractionation and its possible implications on the costal environment. In this context, the present study focused on Cr fractionation and its associated impacts on costal environments through a case study in the River Cauvery, on the Southeast coast of India.

2 Regional Setting

The Cauvery River Basin covers a major part of Peninsular India over the states of Tamil Nadu, Karnataka, Kerala, and the Union Territory of Puducherry. It occupies about 2.7% of the total geographical area of India. It extends over an area of 85,626.23 sq. km with a maximum length and width of 560 and 245 km, respectively. It is bounded by the Western Ghats on the west, by the Eastern Ghats on the east and the south and by the ridges separating it from the Krishna Basin and Pennar Basin on the north. The Cauvery River Basin falls in three agro-climatic zones, which include the west coast plains and ghat region, east coast plains and hill region, and the southern plateau and hill region. The basin is divided into three agro-ecological zones, which are mainly the hot, humid eco-region with red, lateritic, and alluvium-derived soils; the hot, subhumid to semi-arid eco-region with coastal alluvium-derived soils; and the hot, semi-arid eco-region with red loamy soils. The Cauvery River Basin experiences four distinct seasons namely; winter, summer, south-west monsoon season, and north-east monsoon season. The basin is mainly influenced by the south-west monsoon season in the Karnataka and Kerala, and the north-east monsoon season in Tamil Nadu. The basin has a tropical and subtropical climate. About 21 land use/land cover classes exist in the basin. Among them, agriculture is the prime land use (66.21%), followed by forest area (20.50%). The utilizable surface water resource in the basin is 19 BCM. The average annual runoff and average annual water potential in the basin are the same as 21.36 BCM. The maximum area falls under fine soil texture (68.01%), followed by medium soil texture (17.90%). Soil erosion is a main problem in the basin, and nearly 17.93% of the total basin area is affected. Granite gneiss, biotite gneiss, sand with clay and sandstone, sand clay gravel, and sandstone are the major lithology groups found in the basin (Central Water Commission, Ministry of Water Resources, Government of India).

The Cauvery River Basin constitutes three subbasins, namely the Cauvery Upper, Cauvery Middle, and Cauvery Lower sub-basin. There are 132 watersheds, each of which represents a different tributary system whose sizes range from 362 to 991 sq. km, with the maximum number of watersheds falling in the Cauvery Middle Sub-basin. The important tributaries joining the Cauvery are Harangi, Hemavathi, Kabini, Suvarnavathi, Amaravathi, Noyyal, and Bhavani. River Cauvery, which is downstream of the Grand Anicut (Tiruchirappalli), subdivides itself into two main branches; namely, the Cauvery and Vennar System, which get further sub-divided into 36 rivers to feed the delta through a network of channels and branches, distributaries, and sub-distributaries. Annual rainfall in the delta is 1000 mm. The Cauvery delta region is the most fertile tract in the basin. The delta region is spread over 6566 km², which is 8.09% of the total area of the Cauvery Basin. It has a total geographic land area of 14.47 lakh ha, which is equivalent to 11.13% of the state area, and it lies in the eastern part of the Tamil Nadu state between 10°00′–11°30′, north latitude, and between 78°15′–79°45′ longitude. In the Delta region, the Cauvery River splits into numerous tributaries that irrigate large areas of agricultural land. At Tiruchirappalli (in the upper reaches of the delta region), three medium sized reservoirs are in operation, of which the Upper Anicut reservoir is upstream of the urban region, while the Grand Anicut and the Anaikarai are located downstream.

3 Methods and Materials

Surface sediments (top 15 cm) were collected from forty sampling sites along the River Cauvery during the dry (May 2010) and wet seasons (November 2010). The sampling sites were chosen to represent various land use practices, point source and non-point source inputs, upstream, mid-stream, and downstream consideration (Table 1). Further, the sampling sites

Table 1 List of Sampling sites and land use patterns

Sample No	Name of Station	Land Use Pattern in the Vicinity of Sampling Station
\multicolumn{3}{c}{Upper delta region}		
U1	Karur	Urban and commercial mixed
U2	Thirumukudallur	Agricultural and residential mixed
U3	Kulithalai	Agricultural and commercial mixed
U4	Upper Anicut	Predominantly agricultural area
U5	Tiruchirappalli-Mambala salai	Urban and commercial mixed
U6	Grand Anicut	Predominantly agricultural area
\multicolumn{3}{c}{Mid-delta region}		
M1	Pundi	Agricultural and residential mixed
M2	Sathanur	Agricultural and residential mixed
M3	Ammapalli	Agricultural and residential mixed
M4	Thiruperambium	Agricultural and residential mixed
M5	Anaikarai	Agricultural and commercial mixed
M6	Mannalmedu	Agricultural and residential mixed
M7	Thirukattupalai	Agricultural and residential mixed
M8	Papanasam	Agricultural and residential mixed
M9	Kumbakonam	Urban and commercial mixed
M10	Aduthurai	Agricultural and commercial mixed
M11	Mayavaram	Urban and commercial mixed
M12	Kanjanagaram	Agricultural and residential mixed
M13	Thenperambur	Agricultural and residential mixed
M14	Thiruvankadu	Agricultural and residential mixed
M15	Padakacherry	Agricultural and residential mixed
M16	Sellur	Agricultural and residential mixed
M17	Okkur	Agricultural and residential mixed
M18	Thiruvarur	Urban and commercial mixed

Table 1 List of Sampling sites and land use patterns—Cont'd

Sample No	Name of Station	Land Use Pattern in the Vicinity of Sampling Station
M19	Utharamangalam	Agricultural and residential mixed
M20	Semanallur	Agricultural and residential mixed
M21	Rajapuram	Agricultural and residential mixed
M22	Vadapadi	Agricultural and residential mixed
M23	Arasalur	Agricultural and residential mixed
M24	Krishnapuram	Agricultural and residential mixed
Estuarine region		
E1	Kollidam MID	Aqua culture
E2	Kollidam estuary	Aqua culture and
E3	Pichavaram	Recreational and Mangrove forest
E4	Killiyur	Agricultural and residential mixed
E5	Poompuhar	Recreational and agricultural mixed
E6	Nagore	Urban and commercial mixed
E7	Nagapattinam	Urban and commercial mixed
E8	Seruthur	Estuary and residential
E9	Pallingamedu	Agricultural and residential mixed
E10	Muthupet	Agricultural and residential mixed

were categorized into upper delta (U1 to U6), mid-delta (M1 to M 24), and estuarine region (E1 to E10) to differentiate upstream, midstream, and downstream sampling sites (Fig. 1). "Dry season" in the Cauvery delta region refers to February through May, during which the water flow is low or nil, whereas "wet season" implies August through November, during which 66% of the annual water discharges occur (Central Water Commission, Government of India). Because the Cauvery River Basin receives the major proportion of rainfall during these months, the water flow at this time is also substantially greater than the normal flow.

For the Cr fractionation in sediments, one gram (grain size <63 μm) from each sample was subjected to the sequential extraction scheme proposed by Tessier et al. (1979). This scheme extracts the sediment bounded metals into the following five fractions: exchangeable (F1), carbonates bound (F2), Fe—Mn oxides bound (F3), organic matter bound (F4), and residual form (F5). A supernatant solution from each extraction phase was analyzed for heavy metals using an atomic absorption spectrophotometer (GBC SenSAA Spectrometer, Australia). The standard solutions of each metal were prepared from the stock solution obtained from Fisher Scientific Company, USA. All the reagents and chemical solutions used were analytical grade. The recoveries were carried out by the addition of the standards for each element at different levels, and calibrations were regularly performed to assess the accuracy of the analytical method. Throughout the metal analysis, the recovery rate was measured in the range of 82%–106%, and the relative standard deviation of triplicate measurements was <5%. To verify the precision and accuracy of the method, blanks and internal standards were run at regular intervals. In order to evaluate the risk and metal mobility associated with fractions of Cr,

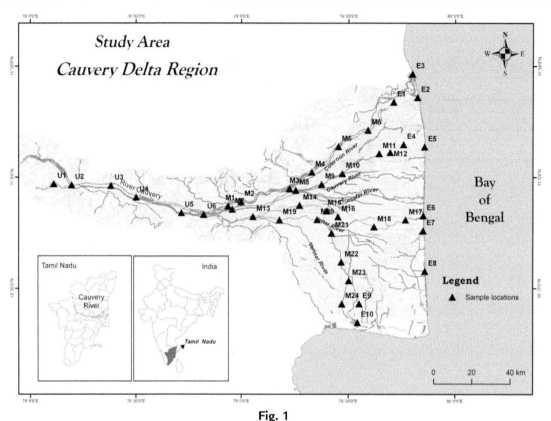

Fig. 1
Study area- River Cauvery in the delta region with sample locations.

the relative index was applied to calculate the risk-assessment code (RAC) based on the sum of exchangeable (F1) and carbonate bound fractions (F2). The mobility factor (MF) was calculated on the basis of the ratio of exchangeable (F1), carbonate-bound fraction (F2) and reducible fractions (F3) to the sum of all fractions (Kabala and Singh 2001; Olajire et al. 2003). The high MF values have been interpreted as symptoms of relatively high mobility and biological availability of heavy metals in soils (Ma and Rao 1997).

Metal fractionation data obtained from the sequential extraction were subjected to statistical analysis using SPSS 11.0 version. One way analysis of variance (*ANOVA*) was carried out to identify the significant differences in metal concentrations between the sampling locations in terms of urbanization, season, and reservoir sites, in addition to upstream and downstream differentiation.

4 Results

The Cr associated with various geochemical fractions in sediments of the river Cauvery in the delta region is presented in Fig. 2. Among the Cr fractions studied, the lithogenic fraction (F5) stands as a prime carrier for Cr. In the study area, Cr in the F5 fraction measured in

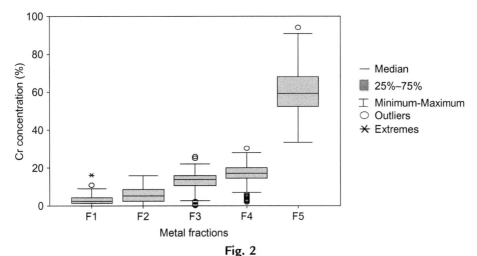

Fig. 2
Chromium fractionation in river sediments of the Cauvery delta region.

the range from 81% to 94%, 41% to 75.2%, and 33.4% to 65.1%, respectively in the upper delta, mid-delta, and estuarine regions. The Cr concentration in F4 ranged from 2.0% to 30.4%. The maximum level was observed at Thiruvankadu (M14) during the dry season (2010), and the minimum at Kulithalai (U3) during the wet season (2010). The order of Cr—F4 fraction distribution in the River Cauvery in the delta region was observed as estuarine region ≈ mid delta > upper delta.

Fraction F3 (Fe—Mn oxide matrix) was found to carry 0.5% to 26% of Cr. The maximum level of Cr in F3 was recorded at Thenperambur (M13) during the wet season (2010). Fraction F2-associated Cr was distributed in the range from 0% to 16%. The mean level of F2 fraction was notably high in the wet season (7.5%), more so than the dry season samples (3.8%). The order of Cr—F2 fraction was recorded as estuarine region > mid-delta > upper delta. An analysis of variance yielded significant variation of F2 fractions between seasons, years, and three regions. Fraction F1 was recorded in the range from 0% to 16.2%. The maximum level was recorded at Nagapattinam (E7), and the minimum at Sellur (M16) in the wet period in 2010. The mean level of F1 fraction was relatively high in the wet season (4.0%), more so than the dry season samples (2.3%). The Cr—F1 fraction was recorded in the order: estuarine region > upper delta > mid delta. A statistically significant difference in F1-associated Cr was observed between seasons, and three regions of delta and urban, suburban, and rural sites.

Overall, the order of Cr fraction observed was: residual (F5) > organic bound (F4) > Fe–Mn oxide bound (F3) > carbonate bound (F2) > exchangeable (F1). As per the risk assessment code assessment, the upper delta regions fell into the low-risk category, and about 50% of samples were recorded in the medium-risk category in the wet season (Table 2). In the case of the estuarine region, 70 and 95% samples were recorded as the medium-risk category in dry and wet season samples, respectively. In the mobility factor assessment (Table 3), the

Table 2 Correlation matrix for Chromium fractionation in Cauvery delta regions

	F1	F2	F3	F4	F5	pH	OM	CaCO$_3$	Fe-F3	Mn-F3	Fe	Mn
Dry season											
F1	1											
F2	0.44[b]	1.00										
F3	−0.09	0.28[a]	1.00									
F4	−0.33[b]	0.20	0.62[b]	1.00								
F5	−0.07	−0.60[b]	−0.84[b]	−0.83[b]	1.00							
pH	−0.13	−0.27[a]	−0.13	−0.03	0.17	1.00						
OM	−0.12	−0.01[a]	0.06	−0.01	0.01	0.14	1.00					
CaCO$_3$	−0.11	0.02	0.28[a]	0.28[a]	−0.26[a]	0.07	0.03	1.00				
Fe-F3	0.03	0.09	0.07	−0.02	−0.05	−0.10	−0.12	0.10	1.00			
Mn-F3	0.06	0.06	−0.10	−0.04	0.04	0.14	0.22[a]	−0.07	−0.19	1.00		
Fe	−0.21	−0.27[a]	−0.30[b]	−0.17	0.34[b]	0.06	0.10	−0.22[a]	0.07	−0.11	1.00	
Mn	0.23[a]	0.06	−0.03	−0.22	0.08	−0.17	−0.27[a]	−0.18	−0.14	0.05	0.06	1.0
Wet season											
F1	1											
F2	0.54[a]	1.00										
F3	0.33[a]	0.49[a]	1.00									
F4	0.26[a]	0.43[a]	0.72[a]	1.00								
F5	−0.57[a]	−0.74[a]	−0.88[a]	−0.86[a]	1.00							
pH	−0.16	−0.07	−0.20	−0.29[a]	0.25[a]	1.00						
OM	−0.04	0.12	−0.15	0.02	0.02	−0.01	1.00					
CaCO$_3$	0.30[a]	0.17	0.13	0.17	−0.22[a]	−0.06	−0.01	1.00				
Fe-F3	−0.09	0.16	−0.24[a]	−0.21	0.24[a]	0.13	−0.07	0.07	1.00			
Mn-F3	0.12	0.05	−0.02	0.16	−0.10	0.03	0.09	0.07	0.08	1.00		
Fe	−0.22[a]	−0.20	−0.10	−0.08	0.17	−0.28[a]	0.06	−0.10	0.22[a]	−0.16	1.00	
Mn	−0.01	−0.13	−0.08	−0.04	0.08	−0.04	−0.06	−0.04	0.16	−0.01	0.25[a]	1.0

[a]Correlation is significant at the 0.05 level (2-tailed).
[b]Correlation is significant at the 0.01 level (2-tailed).

Table 3 Percentage of samples falling under various classes as per risk assessment code and summary of mobility factor assessment for Chromium fractionation

Risk Class	Upper Delta		Mid-Delta		Estuarine	
	Dry Season	Wet Season	Dry Season	Wet Season	Dry Season	Wet Season
Risk Assessment Code						
No risk	–	–	8	–	–	–
Low risk	100	100	92	50	30	5
Medium risk	–	–	–	50	70	95
High risk	–	–	–	–	–	–
Very high risk	–	–	–	–	–	–
Mobility Factor						
Minimum	5.0	2.0	10.2	16.0	17.0	22.0
Maximum	16.0	12.0	27.7	36.6	34.7	45.6
Mean	9.7	8.3	17.1	26.3	26.5	33.1

maximum level of mobility (45.6%) was recorded at Nagore (E6), and the minimum level (10.2%) was found at Sathanur (M2). The order of mobility observed was: estuarine region > mid delta > upper delta. The highly mobile nature of Cr in the estuarine region may pose a risk to the aquatic biota and humanity.

5 Discussion

A major portion of Cr in sediment of River Cauvery were recorded in the lithogenic fraction F5. Residual fraction (F5) is generally measured as an immobile or environmentally unreactive fraction of metal (Chakraborty et al. 2015). The Cr—F5 fraction was observed in the order of upper delta > mid-delta > estuarine region. In the study area, Cr bound to organic matter (F4 fraction) was the second-largest fraction. Organic matter plays an important role in the distribution and dispersion of heavy metals, by mechanisms of chelation and cation exchange processes (Purushothaman and Chakrapani, 2007). The affinity of heavy metals with organic substances and their decomposition products are of great importance for the release of the metals into water.

Cr associated with non-residual fractions (F1–F4) has been used as an indicator of anthropogenic enrichment (Sutherland and Tack, 2003). Notable level of F3 fractions of Cr was recorded in a few sampling sites, which implies that Fe and Mn oxyhydroxides can co-precipitate Cr in aquatic sediment. Although Cr is not a divalent metal and does not form an insoluble metal sulfide complex, in the presence of acid-volatile sulfur, Cr can exist as Cr(III) (Berry et al., 2004; Besser et al., 2004; Becker et al., 2006).

According to Jain et al. (2010), F5 and F3 fractions are the major carriers of Cr in sediments. It has been reported that metals absorbed by Fe—Mn oxides tend to decrease in the order of Cr > Zn > Cu (Badarudeen et al., 1996). It is known that oxides are amphoteric in nature, and their charges change according to the pH of the environment. Significant differences in F3-bounded Cr were observed between seasons, in three regions of the delta and reservoir sites. Comparatively, a lower level of F2 —Cr associated with this fraction may be due to the inability of Cr^{3+} to form a precipitate or complex with carbonates (Sundaray et al., 2011). A lower level of Cr associated with the F1 fraction in most of the dry season samples was probably due to the solubility of Cr in response to the reducing conditions prevailing under anoxic environment (Singh et al., 2008). According to Chakraborty et al. (2011), organic matter is also an important binding phase for metals in sediments, in addition to the hydrous oxides of Fe and Mn. A significant proportion of Cr associated with Fe—Mn oxides and organic matter fractions suggests the scavenging role of these components in Cr fractionation in the sediment of the River Cauvery.

In Cr fractionation profile, mobile fractions such as exchangeable (F1) and carbonate-bound fractions (F2) are recorded up to 16.2 and 16%, respectively. Higher level of Cr associated with F1 and F2 fractions in a few sampling sites may cause an environmental risk. It indicates the anthropogenic enrichment of Cr likely from disposal of industrial and municipal wastewater in addition to storm water and agrochemical runoff into the river from the banks of the River Cauvery. Similar observations were also made in Serbian river sediments by Sakan et al. (2016). A statistically significant difference in exchangeable (F1) and carbonate bound fractions (F2) associated Cr observed between seasons, in three land use types of delta (urban, suburban, and rural sites), is probably due to varying levels of anthropogenic input (Table 4). It is worth mentioning that the contribution of domestic sector wastewater is 65.46% from the total wastewater disposed into the River Cauvery, in which 91.33% of the domestic wastewater is disposed as raw sewage, and only 8.67% of the domestic wastewater gets to treatment facilities before discharge. The Stanley Reservoir is located on the banks of the River Cauvery in Mettur, an industrial town in Tamil Nadu, India. The three large industries viz. Chemplast, Madras Aluminum Company, and the Mettur Thermal Power Plant, are located in Mettur. Besides these industries, several chemical industries are located on the banks of the River Cauvery in Mettur as part of the Small Industries Development Corporation (SIDCO) industrial estate. According to a report by the Department of Environment of the State Government of Tamil Nadu (2005), the estimated waste water discharge from the Salem District into the Cauvery River is 64 million liters/day.

Low levels of Cr associated with F1 fractions in most of the dry season samples are probably due to the solubility of Cr in response to the reducing conditions prevailing under an anoxic environment (Singh et al., 2008). The Cr—F1 fraction was recorded in the order: estuarine region > upper delta > mid delta. Comparatively, a lower level of Cr-associated

Table 4 Analysis of variance of Chromium fractionation in various comparisons

Cr Fraction	Comparisons	df	F	Significance
F1	Between season	1	22.4	**0.000**
	Between years	1	1.9	0.169
	Between three regions	2	38.6	**0.000**
	Between reservoir sites	5	0.8	0.542
	Between urban, suburban and rural sites	2	5.6	**0.005**
F2	Between season	1	46.7	**0.000**
	Between years	1	4.3	**0.040**
	Between three regions	2	42.5	**0.000**
	Between reservoir sites	5	2.7	0.056
	Between urban, suburban and rural sites	2	0.4	0.695
F3	Between season	1	4.43	**0.037**
	Between years	1	0.7	0.397
	Between three regions	2	92.3	**0.000**
	Between reservoir sites	5	19.1	**0.000**
	Between urban, suburban and rural sites	2	2.2	0.111
F4	Between season	1	4.09	**0.045**
	Between years	1	1.7	0.188
	Between three regions	2	118.7	**0.000**
	Between reservoir sites	5	29.3	**0.000**
	Between urban, suburban and rural sites	2	9.9	**0.000**
F5	Between season	1	6.99	**0.009**
	Between years	1	0.003	0.958
	Between three regions	2	200.9	**0.000**
	Between reservoir sites	5	28.3	**0.000**
	Between urban, suburban and rural sites	2	2.0	0.132

Bolded values are significant at the 0.05 level.

with F2 fractions may be due to the inability of Cr^{3+} to form a precipitate or complex with carbonates. Similarly, a low level of F2-Cr fractions was reported in the river estuarine sediment of the River Mahanadi, India (Sundaray et al., 2011). The bioavailability of Cr depends on its partitioning particulate phases. Favorable environmental conditions (e.g., pH, and redox potential) can increase the bioavailability of Cr associated with sediment fractions, such as inorganic solid phases, organic matter, and oxides of iron and manganese, and it ultimately leads to potential hazards to aquatic organisms. Further, it can affect biotic and abiotic components of coastal environments, and eventually can cause ill effects to human society through the food chain. A study conducted by Damodharan (2013) in Uppanar, Cuddalore, on the southeast coast of India also reported similar findings. Although a major portion of Cr is locked in the residual fraction (F5), the percentage of Cr in the nonresidual forms collectively is significant, and chances of its leaching from the sediment as a result of slight environmental perturbation cannot be ignored.

6 Conclusion

- The order of Cr fraction in terms of concentration was measured in the sequence: residual (F5) > organic bound (F4) > Fe-Mn oxide bound (F3) > carbonate bound (F2) > exchangeable (F1). Mobile fractions such as exchangeable (F1) and carbonate-bound fractions (F2) and are recorded up to 16.2% and 16%, respectively. A higher level of Cr associated with F1 and F2 fractions in a few sampling sites are may cause an environmental risk. It indicates the anthropogenic enrichment of Cr.
- A significant proportion of Cr associated with Fe—Mn oxides and organic matter fractions suggests the scavenging role of these components in Cr fractionation in the sediment of the River Cauvery. High RAC and MF values were recorded at a few sampling sites, suggesting that certain pockets of the Cauvery delta are polluted by Cr due to human influences.
- The present investigation demonstrated that the metal fractionation study can be effectively used for predicting the possible mobility of Cr and its associated implications in costal ecosystems. Remedial measures to reduce point and non-point sources of Cr should be implemented as a long-term approach.

References

Badarudeen, A., Damodaran, K.T., Sajan, K., 1996. Texture and geochemistry of the sediments of a tropical mangrove ecosystem, southwest shore of India. Environ. Geol. 27, 164–169.

Becker, D.S., Long, E.R., Proctor, D.M., Ginn, T.C., 2006. Evaluation of potential toxicity and bioavailability of chromium in sediments associated with chromite ore processing residue. Environ. Toxicol. Chem. 25 (10), 2576–2583.

Benson, N.U., Udosen, E.D., Essien, J.P., Anake, W.U., Adedapo, A.E., Akintokun, O.A., Fred-Ahmadu, O.H., Olajire, A.A., 2017. Geochemical fractionation and ecological risks assessment of benthic sediment-bound heavy metals from coastal ecosystems of the equatorial Atlantic Ocean. Int. J. Sediment Res. 32 (3), 410–420.

Benson, N.U., Asuquo, F.E., Williams, A.B., Essien, J.P., Ekong, C.I., Akpabio, O., Olajire, A.A., 2016. Source evaluation and trace metal contamination in benthic 1 sediments from equatorial ecosystems using multivariate statistical techniques. PLoS ONE 11 (6), 1–19.

Besser, J.M., Brumbaugh, W.G., Kemble, N.E., May, T.W., Ingersoll, C.G., 2004. Effects of sediment characteristics on the toxicity of chromium(III) and chromium(VI) to the amphipod, Hyalella azteca. Environ. Sci. Technol. 38 (23), 6210–6216.

Berry, W.J., Boothman, W.S., Serbst, J.R., Edwards, P.A., 2004. Predicting the toxicity of chromium in sediments. Environ. Toxicol. Chem. 23 (12), 2981–2992.

Chakraborty, P., Raghunadh Babu, P.V., Sarma, V.V., 2011. A multimethod approach for the study of lanthanum speciation in coastal and estuarine sediments. J. Geochem. Explor. 110 (1–2), 225–231.

Chakraborty, P., Ramteke, D., Chakraborty, S., 2015. Geochemical partitioning of cu and Ni in mangrove sediments: Relationships with their bioavailability. Mar. Pollut. Bull. 93 (1), 194–201.

Costa, M., Klein, C.B., 2006. Toxicity and carcinogenicity of chromium compounds in humans. Crit. Rev. Toxicol. 36 (2), 155–163.

Damodharan, U., 2013. Bioaccumulation of heavy metals in contaminated river Water-Uppanar, Cuddalore, South East Coast of India. In: Ahmad, I., Dar, M.A. (Eds.), Perspectives in Water Pollution. Croatia-European Union; InTech, Rijeka, pp. 23–34. https://doi.org/10.5772/53374.

Davutluoglu, O.I., Seckin, G., Ersu, C.B., Yilmaz, T., Sari, B., 2011. Heavy metal content and distribution in surface sediments of the Seyhan River, Turkey. J. Environ. Manag. 92 (9), 2250–2259.

Dhanakumar, S., Mani, U., Murthy, R.C., Veeramani, M., Mohanraj, R., 2011. Heavy metals and their fractionation profile in surface sediments of upper reaches in the Cauvery river delta, India. Int. J. Geol. Earth Environ. Sci. 1 (1), 38–47.

Dixit, S., Tiwari, S., 2008. Impact assessment of heavy metal pollution of Shahpura Lake, Bhopal, India. Int. J. Environ. Res. 2 (1), 37–42.

Ibragimow, A., Walna, B., Siepak, M., 2013. Physico-chemical parameters determining the variability of actually and potentially available fractions of heavy metals in fluvial sediments of the middle ODRA River. Arch. Environ. Protect. 39 (2), 3–16.

Ip, C.C., Li, X.D., Zhang, G., Wai, O.W., Li, Y.S., 2007. Trace metal distribution in sediments of the Pearl River estuary and the surrounding coastal area, South China. Environ. Pollut. 147 (2), 311–323.

Jain, C.K., Gurunadha Rao, V.V.S., Prakash, B.A., 2010. Meal fractionation study on bed sediments of Hussainsagar Lake, Hyderabad, India. Environ. Monit. Assess. 166 (1–4), 57–67.

Jain, C.K., 2004. Metal fractionation study on bed sediments of river Yamuna, India. Water Res. 38, 569–578.

Kabala, C., Singh, B.R., 2001. Fractionation and mobility of copper, lead and zinc in soil profiles in the vicinity of a copper smelter. J. Environ. Qual. 30, 485–492.

Karbassi, A.R., Monavari, S.M., Gh, R., Nabi, B., Nouri, J., Nematpour, K., 2008. Metal pollution assessment of sediment and water in the Shur River. Environ. Monit. Assess. 147, 107–116.

Korfali, S.I., Jurdi, M., 2007. Assessment of domestic water quality, case study, Beirut, Lebanon. Environ. Monit. Assess. 135, 241–251.

Kuang-Chung, Y., Li-Jyur, T., Shih-Hsiuns, C., Shien-Tsong, H., 2001. Correlation analyses binding behavior of heavy metals with sediment matrices. Water Res. 35, 2417–2428.

Ma, L.Q., Rao, G.N., 1997. Chemical fractionation of cadmium, copper, nickel, and zinc in contaminated soils. J. Environ. Qual. 26 (1), 259–264.

Mortuza, M.G., Al-Misned, F.A., 2017. Environmental contamination and assessment of heavy metals in water, sediments and shrimp of Red Sea coast of Jizan, Saudi Arabia. J. Aquat. Pollut. Toxicol. 1 (1), 1–8.

Olajire, A.A., Ayodele, E.T., Oyedirdar, G.O., Oluyemi, E.A., 2003. Levels and speciation of heavy metals in soils of industrial southern Nigeria. Environ. Monit. Assess. 85, 135–155.

Owlad, M., Aroua, M.K., Daud, W.A.W., Baroutian, S., 2009. Removal of hexavalent chromium-contaminated water and wastewater: a review. Water Air Soil Pollut. 200 (1), 59–77.

Paul, J.F., Comeleo, R.L., Copeland, J., 2002. Landscape metrics and estuarine sediment contamination in the mid-Atlantic and southern New England regions. J. Environ. Qual. 31, 836–845.

Pejman, A., Nabi, G., Ardestani, M., 2015. A new index for assessing heavy metals contamination in sediments: A case study. Ecol. Indic. 58, 365–373.

Pokorny, P., Pokorny, J., Dobicki, W., Senze, M., Kowalska, G.M., 2015. Bioaccumulations of heavy metals in submerged macrophytes in the mountain river Biala Ladecka (Poland, Sudety Mts.). Arch. Environ. Protect. 41 (4), 81–90.

Purushothaman, P., Chakrapani, G.J., 2007. Heavy metals fractionation in Ganga River sediments, India. Environ. Monit. Assess. 132, 475–489.

Sakan, S., Popović, A., Škrivanj, S., Sakan, N., Đorđević, D., 2016. Comparison of single extraction procedures and the application of an index for the assessment of heavy metal bioavailability in river sediments. Environ. Sci. Pollut. Res. 23 (21), 21485–21500.

Singh, A.P., Srivastava, P.C., Srivastava, P., 2008. Relationships of heavy metals in natural lake waters with physico-chemical characteristics of waters and different chemical fractions of metals in sediments. Water Air Soil Pollut. 188 (1–4), 181–193.

Solaraj, G., Dhanakumar, S., Murthy, K.R., Mohanraj, R., 2010. Water quality in select regions of Cauvery delta river basin, southern India, with emphasis on monsoonal variation. Environ. Monit. Assess. 166, 435–444.

State of Environment of Tamil Nadu (2005). Department of Environment, Government of Tamilnadu.

Sundaray, S.K., Nayak, B.B., Lin, S., Bhatta, D., 2011. Geochemical speciation and risk assessment of heavy metals in the river estuarine sediments—a case study, Mahanadi Basin, India. J. Hazard. Mater. 186, 1837–1846.

Sutherland, R.A., Tack, F.M.G., 2003. Fractionation of cu, Pb, and Zn in certified reference soils SRM 2710 and SRM 2711 using the optimized BCR sequential extraction procedure. Adv. Environ. Res. 8 (1), 37–50.

Templeton, D.M., Ariese, F., Cornelis, R., Danielson, L.G., Muntau, H., Van Leeuwen, H.P., Lobinski, R., 2000. Guidelines for terms related to chemical speciation and fractionation of elements, definitions, structural aspects, and methodological approaches. Pure Appl. Chem. 72 (8), 1453–1470.

Tessier, A., Cambell, P.G.C., Bission, M., 1979. Sequential extraction procedure for the speciation of particulate traces metals. Anal. Chem. 51, 844–851.

USEPA (United States Environmental Protection Agency), 2004. Sewer Sediment and Control. Office of Research and Development.

Wang, L., Yuan, X., Zhong, H., Wang, H., Wu, Z., Chen, X., Zeng, G., 2014. Release behavior of heavy metals during treatment of dredged sediment by microwave-assisted hydrogen peroxide oxidation. Chem. Eng. J. 258, 334–340.

Weisz, M., Polyak, K., Hlavay, J., 2000. Fractionation of elements in sediment samples collected in rivers and harbors at Lake Balaton and its catchment area. Microchem. J. 67, 207–217.

Zakir, H.M., Shikazono, N., 2011. Environmental mobility and geochemical partitioning of Fe, Mn, co, Ni and Mo in sediments of an urban river. J. Environ. Chem. Ecotoxicol. 3 (5), 116–126.

CHAPTER 15

Seasonal Variations of Groundwater Geochemistry in Coastal Aquifers, Pondicherry Region, South India

S. Chidambaram*,[†], S. Pethaperumal[‡], R. Thilagavathi*, V. Dhanu Radha[†], C. Thivya[§], M.V. Prasanna[¶], K. Tirumalesh[∥], Banaja Rani Panda*

*Department of Earth Sciences, Annamalai University, Chidambaram, India, [†]Water Research Center, Kuwait Institute for Scientific Research, Kuwait City, Kuwait, [‡]State Groundwater Unit and Soil Conservation, Department of Agriculture, Pondicherry, India, [§]Department of Geology, University of Madras, Chennai, India, [¶]Department of Applied Geology, Faculty of Engineering and Science, Curtin University, Miri, Malaysia, [∥]SOE, IAD, Bhabha Atomic Research Center, Mumbai, India

1 Introduction

Groundwater is considered to be a valuable resource on earth, as it serves the basic needs of human endeavors; however, it is governed by quality and availability (Raza et al., 2017). Water-related issues are intensified due to deficiency (Farid et al., 2014; Raza et al., 2017) and improper management. These problems are aggravated in coastal regions, as the demands exceed the naturally renewable supply in the region (Janshidzadeh and Mirbagheri, 2011). The reasons for variation in the quality of groundwater degradation is mainly due to the chemical and biochemical interactions in the aquifer matrix that enhance the dissolved inorganic and organic constituents, seawater intrusion into aquifers, and anthropogenic activities particularly in arid and semiarid regions (Vengosh, 2005, Chidambaram et al., 2009a; Thilagavathi et al., 2017, Thivya et al., 2013; Prasanna et al., 2011). The anthropogenic activities that directly or indirectly influence the groundwater are considered as major hazards in coastal regions (Srinivasamoorthy et al., 2011).

An assessment of groundwater quality for irrigation is determined by various techniques including, geochemistry (Ordookhani et al., 2012; Ishaku et al., 2011), statistics (Goyal and Chaudhary, 2010), and geographic information systems (Venkateswaran and Vediappan, 2013). Adhikari et al. (2012) evaluated the quality of groundwater for the suitability of various purposes using geographic information systems. Nosrat and Mogaddm (2010) and Sarala (2012) assessed the impacts of groundwater quality using irrigational quality parameters. Overexploitation and modern commercial agriculture were found to be the major causes that affected the groundwater resources in southern India (Thilagavathi et al., 2014a, b, 2016a, b).

The Pondicherry region includes a multilayered aquifer where the groundwater chemistry study was carried out to understand the spatial and temporal variability by Pethaperumal et al. (2008); Thilagavathi et al. (2012) attempted to identify groundwater quality and its suitability for domestic use by comparing the concentrations of selected water quality parameters. The lack of maintenance of irrigation tanks in the Pondicherry region has led to the increase in dependency of groundwater resources for irrigation. Apart from this, domestic water requirements and industrial usage are also being met from the groundwater resources. Overexploitation of groundwater resources for agriculture and the seawater intrusion of coastal aquifers are irreversible (Chidambaram et al., 2003). The present study was attempted to identify the ground water quality, geochemistry, and the isotopic signature of the groundwater in the Pondicherry region.

2 Study Area

The study area is located on the east coast of India (Fig. 1). It is enclosed by north latitudes 11° 45′ and 12° 03′ and east longitudes 79° 37′ and 79° 53′ and forms parts of Survey of India topographical maps Nos.58 M/9, M/13 and 57 P/12 and P/16. The region is bounded by the Bay of Bengal on the east, and the remaining sides by lands of the Cuddalore and Villupuram districts. The region extends up to 293 km^2, consisting of 179 villages. Gingee and Pennaiyar are the major rivers in the Pondicherry region (Fig. 1). The Gingee River traverses diagonally from northwest to southeast, and the Pennaiyar River forms the southern border of the Pondicherry region. The climate of the Pondicherry region is humid and tropical. The mean temperature ranges between 22°C and 33°C, which is at maximum during the months of May and June. Coastal plain, alluvial plain, and uplands are the predominant physiographic divisions in the study region. Of the study area, 86% is occupied by the agricultural lands, followed by the water bodies and built-up lands. Most of the Pondicherry region is covered by sedimentary formations ranging in age from Cretaceous to Recent (Table 1). The sedimentary formations occur in almost the entire region, and are represented by Cretaceous, Paleocene, Mio-Pliocene, and Quaternary formations.

The general strike of the Cretaceous and Paleocene formations trend NE-SW, with a gentle dip, ranging from 2° to 5° toward the southeast. The Cuddalore formation has the same strike, and shows a dip up to 10° SE. The Cretaceous and Paleocene formations forming inliers have been exposed due to the denudation of the overlying Cuddalore formation. The low angle fault trending in a NNE-SSW direction is inferred at Muthrapalayam. Groundwater is in unconfined to semiconfined condition, with a water level fluctuation of 1.8 to 34 m bgl during premonsoon season, and 1.6 to 37.6 m bgl during the postmonsoon season (Chidambaram et al., 2011; Pethaperumal et al., 2008).

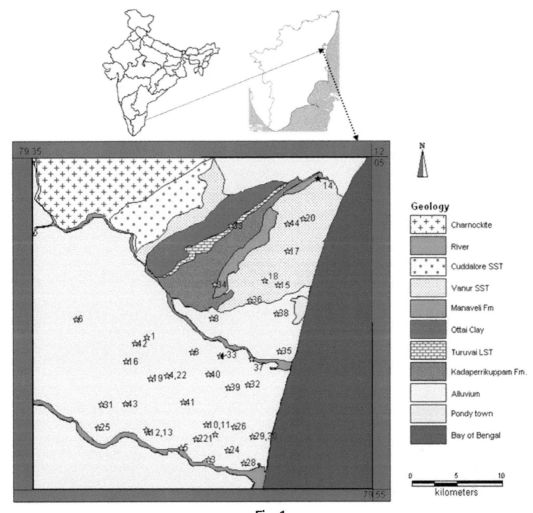

Fig. 1
Sampling locations over the Geology map of the study area.

2.1 Methodology

The samples were collected with respect to aquifer types, and depending upon their spatial coverage (Fig. 1). A total of 98 water samples were collected; 54 during premonsoon season (May 2007), and 44 during postmonsoon season (January 2008). Sampling and analysis were carried out using standard procedures (Ramesh and Anbu, 1996; APHA, 1998). Five hundred mL of water samples from tube wells tapping specific aquifers were collected in polyethylene bottles. Then, they were sealed and brought to the laboratory for analysis, stored properly (4°C), and filtered with 0.45-μm filter paper before analysis. The pH, total dissolved solids (TDS), and electrical conductivity (EC) of the water samples were measured in the field using a portable

Table 1 General stratigraphic succession of the study area (Nair and Rao, 1971)

Era	Period	Formations	Lithology
Quaternary	Recent	Alluvium Laterite	Sand, clay, silt, Kankar and gravel laterite
Tertiary		Mio-Pliocone Cuddalore Formation	Pebbly and gravelly and coarse-grained sandstones with minor clay and siltstones with thin seams of lignite
Unconformity			
	Paleocene	Manaveli Formation	Yellow and yellowish brown, gray calcareous siltstone and claystone and shale with thin bands of limestone
		Kadapperi-kuppam	Yellowish white to dirty white sandy, hard fossiliferous limestone calcareous formation sandstone and clays
Unconformity			
Mesozoic	Upper cretaceous	Turuvai limestone	Highly fossiliferous limestone, conglomerate at places, calcareous sandstone and clays
		Ottai clay stone	Grayish to grayish green claystones, silts with thin bands of sandy limestone and fine-grained calcareous sandstone
		Vanur sandstones	Quartzite sandstones, hard, coarse grained, occasionally feldspathic or calcareous with minor clays
	Lower cretaceous	Ramanathapuram Formation (unexposed)	Black carbonaceous silty clays and fine to medium-grained sands with bands of lignite and medium to coarse-grained sandstones
Unconformity			
Archaean		Eastern Ghat Complex	Charnockite and biotite-hornblende gneiss

water-analysis kit. The method for the ions analyzed in the present study is given in the Table 2. The error percentage of the analysis was carried out by calculating the percentage using the following calculation (Freeze and Cherry, 1979). Our results have shown an error limit from ±1% to ±10%.

$$\text{Error Percentage}(\%) = \frac{(TZ^+ - TA^-)}{(TZ^+ + TA^-)} \times 100$$

Table 2 Details of Sample collection and analysis method of groundwater

	(A) Sample Collection Details			
	Aquifer Types			
Monsoon	Alluvial aquifer	Upper Cuddalore Sandstone	Lower Cuddalore Sandstone	Cretaceous
Premonsoon (N=54)	16	9	19	10
Postmonsoon (N=44)	11	9	15	9
(B) Chemical Analysis of Groundwater				
Parameter	Units	Instrument (Make)	Reagents	Reference
Ca^{2+}	mg/L	EDTA titrimetric	EDTA, sodium hydroxide, and murexide (Rankem)	APHA (1998)
Mg^{2+}	mg/L	EDTA titrimetric	EDTA, sodium hydroxide, Ammonia buffer, and Erichrome black-T (Rankem)	APHA (1998)
Na^+	mg/L	Flame photometer (Elico)	Sodium chloride (NaCl), KCl, and calcium carbonate ($CaCO_3$) (Rankem)	APHA (1998)
K^+	mg/L	Flame photometer (Elico)	NaCl, KCl, and $CaCO_3$ (Rankem)	APHA (1998)
HCO_3^-	mg/L	Titrimetric	Hydrosulfuric acid (H_2SO_4), phenolphthalein, and methyl orange (Rankem)	APHA (1998)
Cl^-	mg/L	Titrimetric	Silver nitrate, potassium chromate (Rankem)	APHA (1998)
SO_4^{2-}	mg/L	UV-vis double beam spectrophotometer (Elico)	Glycerol, HCl, ethyl alcohol, NaCl, Barium chloride, sodium sulphate (Rankem)	Eaton et al. (1995)

(Continued)

Table 2 Details of Sample collection and analysis method of groundwater—Cont'd

	(A) Sample Collection Details			
		Aquifer Types		
Monsoon	Alluvial aquifer	Upper Cuddalore Sandstone	Lower Cuddalore Sandstone	Cretaceous
NO_3	mg/L	Spectrophotometer (Systronics)	Brucine-sulfanilic acid, potassium nitrate, and H_2SO_4 (Rankem)	APHA (1998)
H_4SiO_4	mg/L	UV-vis double beam Spectrophotometer (Elico)	H_2SO_4, ammonium mlybdate, sodium hydroxide, oxalic acid, ascorbic acid, sodium metasilicate (Rankem)	Eaton et al. (1995)
PO_4	mg/L	UV-vis double beam Spectrophotometer (Elico)	H_2SO_4, potassium antimony tartarate, ammonium molybdate, ascorbic acid, potassium dihydrogen phosphate (Rankem)	Eaton et al. (1995)
Br and I	mg/L	Consort ion electrode (Thermo Orion)		Pethaperumal et al. (2008)
$\delta^{18}O$ and δD	‰	Finnigan Deltaplus Xp, Thermo Electron Corporation, Bermen, Germany		Coplen (1993), Thivya et al. (2015)
Tritium	TU (Tritium Unit)	Quantulus 1220 spectrophotometer		Tirumalesh (2012)

2.2 Results and Discussion

The minimum, maximum, and average values for the chemical composition of groundwater are given in Tables 3 and 4. The cation dominance follows the order $Na^+ > Ca^{2+} > Mg^{2+} > K^+$ and that of anions is $HCO_3^- > Cl^- > SO_4^{2-}$. The trend remains the same in both seasons.

2.3 Spatial Distribution of EC

Electrical conductivity increases with the number of ions in solution. The spatial data of electrical conductivity gives a general trend of the anions and cations present in water. The EC during the premonsoon season ranges from 209 to 2360 μS/cm. Lower EC was observed along

Table 3 Chemical concentration of groundwater samples during premonsoon of the maximum, minimum, and average (all values are in mg/L except, EC in µS/cm and pH)

		pH	EC	TDS	Na	K	Ca	Mg	Cl	HCO$_3$	SiO$_2$	PO$_4$	SO$_4$	NO$_3$	Br	I
Alluvium	Max	8.82	2360.00	1780.00	892.30	6.30	56.00	105.60	497.00	390.40	250.00	12.40	18.00	23.30	10.90	0.71
	Min	6.66	309.00	215.00	116.20	3.10	24.00	2.40	53.17	73.20	30.00	0.90	3.00	0.60	1.28	0.06
	Avg	7.55	1020.50	829.19	550.71	3.61	40.50	28.80	171.80	263.83	118.00	8.05	7.13	7.64	3.93	0.27
Upper Cuddalore	Max	7.50	1530.00	1154.00	795.50	4.70	48.00	26.40	336.77	219.60	300.00	12.90	10.50	39.90	7.01	0.52
	Min	5.89	209.00	106.50	45.70	3.00	16.00	0.00	53.17	48.80	10.00	0.05	2.50	3.12	0.01	0.09
	Avg	6.55	547.67	422.61	270.39	3.36	28.89	6.67	122.10	107.09	113.89	5.96	5.39	13.19	2.49	0.22
Lower Cuddalore	Max	7.84	1440.00	1175.00	706.40	4.40	56.00	55.20	248.15	366.00	330.00	15.00	11.50	22.80	5.70	0.56
	Min	6.64	349.00	228.00	161.30	3.10	16.00	2.40	35.45	97.60	20.00	0.10	3.00	0.28	1.01	0.04
	Avg	7.32	666.00	511.58	411.82	3.31	30.53	20.97	96.09	214.46	145.53	7.23	5.50	4.27	2.36	0.25
Cretaceous	Max	8.58	2020.00	1760.00	883.00	11.50	84.00	81.60	531.75	786.80	230.00	9.60	15.00	72.60	12.70	2.08
	Min	7.14	699.00	543.00	130.60	3.10	32.00	7.20	70.90	0.00	0.00	0.05	4.00	1.76	1.95	0.06
	Avg	7.67	1109.20	950.30	552.22	4.22	50.80	29.28	207.38	299.50	93.00	6.90	8.30	17.18	4.79	0.45

Table 4 Chemical concentration of groundwater samples during postmonsoon of the maximum, minimum, and average (all values are in mg/l except, EC in µS/cm and pH)

		pH	EC	TDS	Ca	Mg	Na	K	Cl	HCO$_3$	SO$_4$	H$_4$SiO$_4$	PO$_4$	NO$_3$	Br	I
Alluvium	MAX	7.50	1942.86	1360.00	88.00	67.20	344.83	18.20	389.95	500.20	72.05	124.00	0.80	6.25	1.06	1.88
	MIN	6.76	641.43	449.00	24.00	0.00	39.08	13.10	88.63	207.40	40.50	53.25	0.05	0.80	0.03	0.17
	AVG	7.11	1494.03	1045.82	44.00	17.02	216.72	16.39	215.92	379.31	48.35	88.73	0.32	2.88	0.40	0.65
Upper Cuddalore	MAX	7.61	1241.43	869.00	44.00	19.20	275.86	17.90	319.04	573.40	81.65	170.00	0.70	18.90	6.59	1.49
	MIN	5.74	144.00	100.80	8.00	0.00	114.94	13.10	88.63	61.00	43.50	71.00	0.05	0.00	0.01	0.16
	AVG	6.91	704.41	493.09	26.22	8.00	185.39	15.98	198.91	222.31	50.04	99.31	0.38	5.95	0.87	0.52
Lower Cuddalore	MAX	7.92	1157.14	810.00	44.00	48.00	218.39	18.80	248.14	317.20	224.00	171.75	0.90	5.04	0.98	1.16
	MIN	6.46	324.29	227.00	8.00	0.00	68.97	11.50	53.18	146.40	12.00	3.00	0.00	0.29	0.00	0.16
	AVG	7.11	582.29	407.60	21.87	10.24	144.83	16.68	132.43	206.59	61.29	61.80	0.41	0.89	0.18	0.37
Cretaceous	MAX	8.02	1505.71	1054.00	60.00	38.40	229.89	18.60	248.14	414.80	49.00	101.75	0.80	17.70	0.28	3.21
	MIN	6.62	531.43	372.00	12.00	0.00	68.97	11.80	53.75	170.80	41.00	48.00	0.05	0.92	0.01	0.09
	AVG	7.24	1200.79	840.56	36.00	13.87	149.65	15.61	143.83	310.42	45.42	81.36	0.39	4.57	0.14	0.91

the eastern, northeastern, and northwestern boundaries of the study area, whereas higher EC was observed in the central, western, and southern part of the study area, which may be mainly due to sea water intrusion (Fig. 2A).

The spatial distribution of EC during the postmonsoon season represents a range from 144 – 1942 µS/cm (Fig. 2B). It is interesting to note that the regions with higher drainage intensity and covered with alluvium formations have greater EC, which is mainly due to dissolution and leaching in the study area. But the trend remains the same with respect to formations studied. One limitation of using EC as an indicator of TDS is that EC does not respond to the presence of uncharged dissolved substances such as silica, a common weathering product from igneous rock.

2.3.1 Piper classification

The trilinear piper diagram is the common method to determine the geochemical facies (Piper, 1944), and to determine the evolution of groundwater.

Premonsoon season

Two major types of groundwater were identified by the Piper facies during the premonsoon season: Na-HCO_3 type and Na-Cl type (Fig. 3A). All the Upper Cuddalore Sandstone aquifer samples fall in the Na-Cl type, along with a few representations of Cretaceous and Alluvial aquifers. The Na-HCO_3 facies are represented by all the Lower Cuddalore Sandstone aquifer samples, and the majority of the Alluvial and Cretaceous aquifers samples.

Postmonsoon season

A majority of the samples fall in different types of facies, outlined as follows:

1. Ca-HCO_3 facies
2. Na-HCO_3 facies
3. Na-Cl facies
4. Ca-Cl facies

The dominant facies in all the aquifers are Na-Cl and Ca-HCO_3 facies (Fig. 3B). During postmonsoon season, few samples of Alluvial and Cretaceous aquifers fall in Ca-HCO_3 facies and that of Lower Cuddalore Sandstone in Ca-Cl facies. It's to be noted that there is a significant variation of cation in the postmonsoon season, and variation of the anion in the premonsoon season. There are a few compositional evolutionary pathways; namely,

1. Ca-HCO_3→Na-HCO_3→Na-Cl (ion exchange)
2. Ca-HCO_3→Na-HCO_3 (ion exchange)

    → Na – Cl (saturation and removal/dilution)

3. Ca-Cl→Na-Cl (ion exchange)

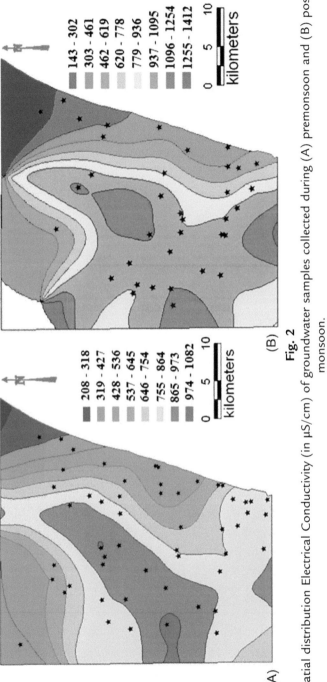

Fig. 2
Spatial distribution Electrical Conductivity (in µS/cm) of groundwater samples collected during (A) premonsoon and (B) post monsoon.

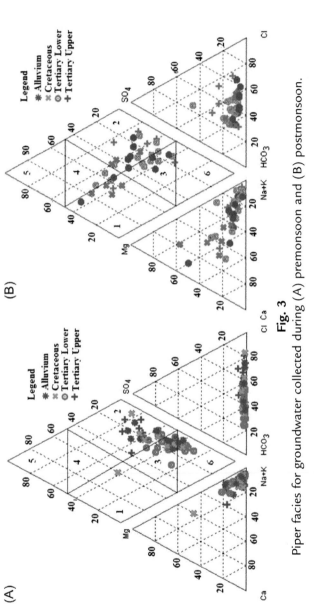

Fig. 3 Piper facies for groundwater collected during (A) premonsoon and (B) postmonsoon.

2.4 Chadda's Hydrogeochemical Process Evaluation

The hydrochemical processes suggested by Chadha (1999) are indicated in each of the four quadrants of the graph (Fig. 4A and B). These are broadly summarized as:

Field 1—Ca-HCO$_3$ type recharging waters
Field 2—Ca-Mg-Cl type reverse ion-exchange waters
Field 3—Na-Cl type end-member waters (sea water)
Field 4—Na-HCO$_3$ type base ion-exchange waters

Field 1 (recharging water) waters enter into the ground from the surface, and carry dissolved carbonate in the form of HCO$_3$ and the geochemically mobile Ca. Field 2 (reverse ion-exchange) waters are less easily defined and less common, but represent groundwater where Ca+Mg is in excess to Na+K, either due to the preferential release of Ca and Mg from mineral weathering of exposed bedrock, or possibly reverse base cation-exchange reactions of Ca+Mg into solution, and subsequent adsorption of Na onto mineral surfaces. Field 3 (Na-Cl) waters are typical of seawater mixing, and are mostly constrained to the coastal areas. Field 4 (Na-HCO$_3$) waters form a wide band between the western part of the study area and the seacoast, possibly representing base exchange reactions or an evolutionary path of groundwater from Ca-HCO$_3$ type fresh water to Na-Cl mixed sea water where Na-HCO$_3$ is produced by ion-exchange processes.

It was found that almost all the samples of the Upper Cuddalore Sandstone aquifer of the premonsoon season fall in the sea water mixing zone (Fig. 4A). There were few representations of Alluvial aquifer samples and Cretaceous aquifers in this field (Field 3). The majority of the Lower Cuddalore Sandstone, Alluvial, and Cretaceous aquifer samples were noted in the cation exchange region (Field 4). The postmonsoon season (Fig. 4B) also exhibits a similar trend, but there is a reduction in Field 3 (seawater intrusion), and with representations of samples from the Cretaceous aquifer, Alluvial, and Lower Cuddalore Sandstone aquifers in the recharge region (Field 1).

Fields 3 and 4 have more representation of samples in both seasons, of which more representation of the samples was in Field 4. The impact of the monsoon is also noted by the shift of samples from 3 to the Fields 1 and 4 during the postmonsoon season.

2.5 Isotope Studies

Seven groundwater samples from the study area namely, Alluvium (1), Upper Cuddalore (3), Lower Cuddalore (1), and Cretaceous (2) were collected for isotope analysis. The samples from the Upper Cuddalore Sandstone and the Alluvial aquifers fall near the meteoritic water line, and that of the Alluvial aquifer was very close to the LMWL (The LMWL for Tamil Nadu State) (Chidambaram et al., 2009), which exhibits an equation of $\delta D = 7.8941 * \delta^{18}O \pm 10.385$. It is

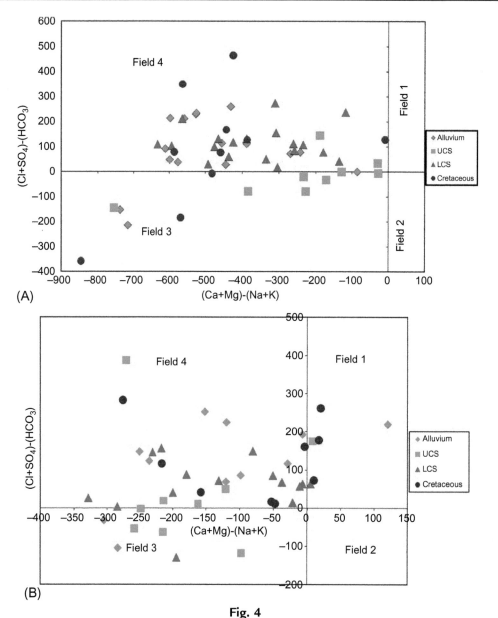

Fig. 4
Chadda's plot of process evaluation for samples collected during (A) premonsoon and (B) postmonsoon.

also noted that the heavier isotopes are more in the Alluvial aquifer samples which may be due to the precipitation near the shore, or that of saline intrusion (Prasanna et al., 2009). Samples of the Cretaceous and Lower Cuddalore Sandstone aquifers fall away from the meteoric line which may be due to the processes of weathering of rocks/sediments, or evaporation before recharging from other sources.

2.5.1 $\delta^{18}O$

Depletion of groundwater reflects a bias in seasonality of recharge, and the nature of summer precipitation. The study area is a flat terrain with minimal topographical variation, and all the alluvial samples were from coastal groundwater. The recharge in arid regions exhibits high evaporation, which is a highly fractionating process (Clarke and Fritz, 1997). Evaporative enrichment in the falling droplets beneath the cloud base was effective during warm and dry months when the amount of rainfall is small. This partially evaporated rain is characterized by relatively higher $\delta^{18}O$ values. The isotopic values (Fig. 5) of the samples reflect that $\delta^{18}O$ ranges from -2 to -6.91 permil, exhibiting recharge in the comparatively warmer period.

2.5.2 Tritium

Only qualitative interpretations can be made with tritium values. Recharge is sporadic, and shallow groundwater can have a wide range of ages for coastal and low latitude regions (Clarke and Fritz, 1997).

<0.8 TU	Submodern-recharged prior to 1952
0.8 to ~2 TU	Mixture between submodern and recent recharge
2–8 TU	Modern (<5–10 years)
10–20 TU	Residual "bomb" ^3H present
>20 TU	Considerable component of recharge from the 1960s or 1970s

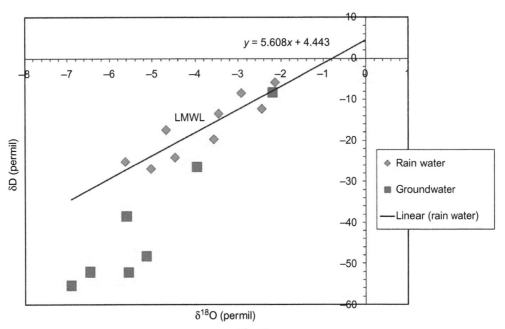

Fig. 5
Comparison of groundwater samples with LMWL.

The study (Fig. 6) reveals the groundwater of different types.

Type A: Old groundwater with low electrical conductivity.
Type B: Modern groundwater with high electrical conductivity.

Type A
The groundwater of this group has tritium units <0.8, and they are submodern waters recharged prior to 1952. These samples fall in the Cuddalore Sandstone aquifer (Upper and Lower). The electrical conductivity of these samples were <176 µS/cm, indicating that the locally recharged groundwater existed with long residence time. It also reveals that the recharge process of this area is less active, and interaction between the groundwater and the aquifer matrix is less.

Type B
The groundwater of this type is comparatively younger, with >2 TU, and accordingly designated as modern water (Clark and Fritz, 1997). These samples are found in the Alluvial and Cretaceous aquifers. The electrical conductivity of these samples ranges from 246 to 456 µS/cm. Thus, the residence time of this groundwater was estimated to be relatively short after infiltration. Moreover, the groundwater flow system was thought to be more active than type A.

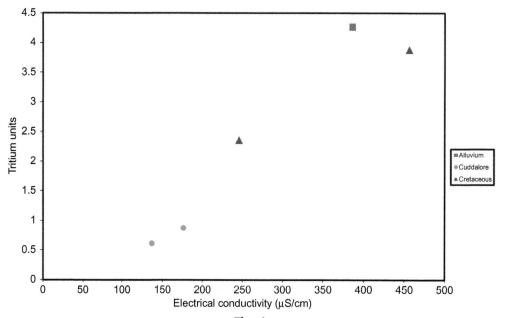

Fig. 6
Relationship of EC with Tritium.

2.6 Statistical Analysis

2.6.1 Correlation

Pearson's correlation coefficient gives the degree of correlation, as well as the direction of the correlation. During premonsoon season, correlation exists between Na-SO_4: Na-Ca: Na-HCO_3: Na-pH: Ca-HCO_3: Ca-Br: Br-K: Cl-K: and SO_4-Br. Good correlation exists between Na-Br; and Br-Cl and Na-Cl exhibit good to poor positive correlations and ionic contributions from anthropogenic activities, especially fertilizers. The contribution of HCO_3, H_4SiO_4 and K indicate feldspar weathering from the country rock/sediments. Cl exhibits a good correlation with Mg, and Ca indicates leaching of secondary salts.

Postmonsoon

Mg
|
Ca — Cl
|
HCO_3— Na — H_4SiO_4

Premonsoon

pH — SO_4
 \\ /
  Na ⇌ Cl
 / \\
HCO_3 Br
  \\ | \\
   Ca   K

The postmonsoon analysis exhibits the correlation between Mg-Ca: Ca-HCO_3: Ca-Cl: and HCO_3-Ca.

2.6.2 Factor analysis

Principal component analysis (PCA) was used to reduce the dimensionality of a data set by explaining the correlation among many variables in terms of a smaller number of underlying factors (principal components), without losing much information (Jackson, 1991; Meglen, 1992). The Eigen value gives a measure of the significance of the factor: the factors with the highest Eigen values are the most significant. Eigen values of 1.0 or greater are considered significant (Kim and Mueller, 1987) (Tables 5 and 6). Varimax rotation (Knudson et al., 1977)

Table 5 Rotated component matrix for premonsoon

	1	2	3	4
Na	0.67	0.56	−0.11	0.24
K	0.84	−0.23	0.11	0.00
Ca	0.34	0.73	0.19	0.15
Mg	0.13	0.35	0.45	0.43
Cl	0.87	0.36	0.09	0.07
HCO_3	0.12	0.87	−0.10	−0.01
H_4SiO_4	−0.08	0.14	−0.64	−0.16
PO_4	−0.06	0.37	0.33	0.05
SO_4	0.37	0.22	0.30	0.57
NO_3	0.15	0.16	0.81	−0.05
Br	0.89	0.30	0.13	0.20
I	−0.02	−0.10	0.10	0.79
pH	0.23	0.33	−0.38	0.56

Table 6 Rotated component matrix for postmonsoon

	1	2	3	4	5
Mg	0.58	−0.07	0.45	−0.08	−0.26
Ca	0.91	0.01	0.13	0.06	−0.13
K	−0.01	0.19	0.19	0.71	−0.14
Na	−0.01	0.56	−0.21	−0.15	0.69
Cl	0.70	0.20	0.02	0.05	0.20
HCO_3	0.79	0.59	−0.20	0.08	0.05
PO_4	−0.01	−0.21	0.01	−0.15	0.03
SO_4	−0.18	−0.13	0.20	0.41	0.56
H_4SiO_4	−0.09	0.70	−0.02	0.17	−0.53
NO_3	0.13	0.30	0.62	−0.07	−0.20
Br	−0.20	0.05	0.26	−0.75	0.11
I	0.37	0.17	0.14	−0.06	0.02
pH	0.06	0.06	−0.84	−0.08	−0.12

has also been applied to find more clearly defined factors that could be more easily interpreted. Liu et al. (2003) classified the factor loadings as "strong," "moderate" and "weak," corresponding to absolute loading values of >0.75, 0.75–0.50, and 0.50–0.30, respectively.

2.6.3 Premonsoon season

During premonsoon season, four factors were extracted. Factor 1 explains 23.5% of variance, and represents Na, K, Cl, and Br (Table 5). The higher loading on Na$^+$ in the factor seems to be related to two processes (Papatheodorou et al., 2006):

(i) sea water intrusion (Thilagavathi et al., 2017).
(ii) the application of rock salt (halite and potassium salt) used in cesspools.

The association of these ions indicates the saline nature of groundwater. This factor is represented by the samples along the southern part of the study area in the Alluvial aquifer, and along the central part of the region of the Upper Cuddalore Sandstone aquifer, indicating the predominance of sea water intrusion as the factor controlling the geochemistry of the region. Factor 2 explains 18% of the variance, and is represented by Na, Ca, and HCO_3. These could be the result of weathering of plagioclase feldspar, which is common in the rocks/soils of the entire alluvial aquifer (Chidambaram et al., 2011). Factor 3 explains 13% of variance, and is represented by NO_3. This factor is also represented in the postmonsoon samples as the third factor, but with less percent of variance explained. Loading of nitrate in both the seasons reveals that anthropogenic contamination plays a significant role in groundwater geochemistry. High nitrate concentrations in the study area can be explained by agricultural fertilizers, domestic sewage, septic tanks, and other anthropogenic activities, such as residential water softeners, and septic tanks (Vengosh and Keren, 1996; Wayland et al., 2003). The fourth factor, represented by SO_4, I, pH explains 12% of variance. This may be due to the dissolution and leaching of salt precipitates during the process of infiltration (Pethaperumal et al., 2008).

2.6.4 Postmonsoon season

Factor analysis of the postmonsoon season samples exhibit five different sets (Table 6) of factor loading, explaining 64% of total variance. Factor 1 explains 19% of variance, and is represented by Mg, Ca, Cl, and HCO_3. This factor exhibits the dominance recharge, followed by the ion exchange process. This factor is represented in the southern, central, and northern parts of the study area. It is to be noted that this factor sprawls in the regions covered by the Alluvial aquifer. Factor 2 explains 13% of variance, and is represented by silicate weathering (H_4SiO_4, Na, and HCO_3). Factor 3 explains 11.9% of variance, and is represented by NO_3 and interpreted as an anthropogenic factor. Factor 4 explains 10% of variance, and is represented by positive loading of K and negative loading of pH and Mg. From this factor it can be interpreted that an increase of hydrogen and potassium in the aqua system, along with a decrease of Mg indicating the cation exchange processes. K^+ is also a common constituent of soluble fertilizers, or could come from livestock excrement. Factor 5 explains 9.8% variance, and includes Na and SO_4. This may be due to pollution or anthropogenic activities. Factor loadings of Na, SO_4, exhibit the solubility of the parameters in acidic sewage/effluents as responsible for their inverse relationship with pH (Panda et al., 2006).

Five factors are represented during the postmonsoon season, reflecting the hydrogeochemical complexity during this season. It is also noted that all the five factors put together contribute only 64% of the (TDV). This indicates that there are many other factors with lesser Eigenvalues contributing to the hydrogeochemistry of the groundwater during the postmonsoon season.

3 Conclusion

The study reveals the influence of monsoon season on the groundwater of the region, as the spatial distribution pattern of EC remains almost the same, with changes in values indicating dilution during POM. Further significant variation in the cations and the anions were noted with respect to season, as the predominant facies of PRM was NaCl, and that of POM was $NaHCO_3$. Seawater intrusion was revealed to be the predominant factor during PRM, and representation of recharge was noted in POM. Weathering was observed to be induced after the monsoon season with the ion exchange process. The anthropogenic factor was found to be represented in both seasons, indicating the influence of domestic sewage, agriculture, and industrial effluents. The sea water intrusion was identified to be represented in Alluvium and Upper Cuddalore formations. Isotope studies reveal that the sea water intrusion and the natural recharge are significant in the alluvial aquifers. The depleted samples represented inland groundwater. The older groundwater was found to be less contaminated than the youngest water, and it was also revealed that this is due to less circulation of groundwater in the Lower Cuddalore formation.

Hence, the study concludes that there are seasonal influences in the groundwater chemistry, chiefly along the southeast part of the study area, in both Alluvial and Upper Cuddalore aquifers.

References

Adhikari, K., Chakraborty, B., Gangopadhyay, 2012. Assessment of irrigation potential of groundwater using water quality index tool. Environ. Res. J. 6 (3), 197–205.

APHA, 1998. Standard Methods for the Examination of Water and Wastewater, 19th edn APHA, Washington, DC.

Chadha, D.K., 1999. A proposed new diagram for geochemical classification of natural waters and interpretation of chemical data. Hydrgeol. J. 7 (5), 431–439.

Chidambaram, S., Ramanathan, A.L., Srinivasamoorthy, K., 2003. Lithological influence on the groundwater chemistry—Periyar district a case study. Conference on Coastal and Freshwater Issues, p. 173.

Chidambaram, S., Prasanna, M.V., Srinivasamoorthy, K., Manivannan, R., Ramanathan, A.,.L., Tirumalesh, K., Vasu, K., Hameed, S., Anandhan, P., Warrier, U.K., Johnsonbabu, G., 2009. A study on the factors affecting the stable isotopic composition in precipitation of Tamil Nadu, India. Hydrol. Process. https://doi.org/10.1002/hyp.7300.

Chidambaram, S., Senthil Kumar, G., Prasanna, M.V., John Peter, A., Ramanathan, A.L., Srinivasamoorthy, K., 2009a. A study on the hydrogeology and hydrogeochemistry of groundwater from different depths in a coastal aquifer: Annamalai Nagar, Tamilnadu, India. Environ. Geol. 57 (1), 59–73.

Chidambaram, S., Karmegam, U., Prasanna, M.V., Sasidhar, P., Vasanthavigar, M., 2011. A study on hydrochemical elucidation of coastal groundwater in and around Kalpakkam region, Southern India. Environ. Earth Sci. 64, 1419–1431. ISSN 1866-6280.

Clark, I.D., Fritz, P., 1997. Environmental Isotopes in Hydrogeology. Lewis Publishers, New York.

Coplen, T.B., 1993. Uses of environmental isotopes. In: Alley, W.M. (Ed.), Regional GroundWater Quality. Van Nostrand Reinhold, New York, pp. 227–254.

Eaton, A.D., Clesceri, L.S., Greenberg, A.E., 1995. Standard Methods for the Examination or Water and Wastewater, 19th ed. APHA, AWWA, WEF, Washington DC.

Farid, M., Irshad, M., Fawad, M., Ali, Z., Eneji, A.E., Aurangzeb, N., Ali, B., 2014. Effect of cyclic phytoremediation with different wetland plants on municipal wastewater. Int. J. Phytoremediation 16, 572–581.

Freeze, A.R., Cherry, J.A., 1979. Groundwater. Prentice-Hall Inc, Englewood Cliffs, p. 604.

Goyal, S.K., Chaudhary, B.S., 2010. GIS based study of Spatio-temporal changes in groundwater depth and quality in Kaithal district of Haryana, India. J. Ind. Geophys. Union 14 (2), 75–87.

Ishaku, J.M., Ahmed, A.S., Abubakar, M.A., 2011. Assessment of groundwater quality using chemical indices and GIS mapping in Jada area, Northeastern Nigeria. J. Earth Sci. Geotech. Eng. 1 (1), 35–60. ISSN: 1792-9040 (print), 1792-9660 (online).

Jackson, J.E., 1991. A User's Guide to Principal Components. Wiley, New York.

Janshidzadeh, Z., Mirbagheri, S.A., 2011. Evaluation of groundwater quantity and quality in the Kashan Basin. Central Iran Desal 270, 23–30.

Kim, J.O., Mueller, C.W., 1987. Factor Analysis: Statistical Methods and Practical Issues, Sage University Paper Series on Quantitative Applications in the Social Sciences, series no 07–014. Sage Publications, Beverly Hills.

Knudson, E.J., Duewer, D.L., Christian, G.D., Larson, T.V., 1977. Application of factor analysis to the study of rain chemistry in the Puget Sound region. In: Kowalski, B.R. (Ed.), Chemometric: Theory and Application. ACS Symposium Series, Washington, DC, pp. 80–116.

Liu, C.W., Lin, K.H., Kuo, Y.M., 2003. Application of factor analysis in the assessment of groundwater quality in a blackfoot disease area in Taiwan. Sci. Total Environ. 313, 77–89. https://doi.org/10.1016/S0048-9697(02)00683-6.

Meglen, R.R., 1992. Examining large databases: a chemometric approach using principal component analysis. Mar. Chem. 39, 217–237.

Nair KM, Rao VP (1971) Result of shallow drilling in the area north of Pondicherry, unpublished ONGC field season report of 1969-1970.

Nosrat, A., Mogaddm, A.A., 2010. Assessment of groundwater quality and its suitability for drinking and agricultural uses in the Oshnavieh area, Northwest of Iran. J. Environ. Protect. 1, 30–40.

Ordookhani, K., Amiri, B., Saravani, A., 2012. Assessment of ground water quality using geographic information system.Parsian Plain International Conference on Environment, Agriculture and Food Sciences (ICEAFS'2012) Phuket (Thailand).

Panda, U.C., Sundaray, S.K., Rath, P., Nayak, B.B., Bhatta, D., 2006. Application of factor and cluster analysis for characterization of river and estuarine water systems—a case study: Mahanadi River (India). J. Hydrol. 331, 434–445.

Papatheodorou, G., Demopoulou, G., Lambrakis, N., 2006. A long-term study of temporal hydrochemical data in a shallow lake using multivariate statistical techniques. Ecol. Model. 193, 759–776. https://doi.org/10.1016/j.ecolmodel.2005.09.004.

Pethaperumal, S., Chidambaram, S., Prasanna, M.V., Verma, V.N., Balaji, K., Ramesh, R., Karmegam, U., Paramaguru, P., 2008. A study on groundwater quality in the Pondicherry region. ECO—Chronicle 3 (2), 85–90.

Piper, A.M., 1944. A graphic procedure in the geochemical interpretation of water-analyses. Trans. Am. Geophys. Union 25, 914–923.

Prasanna, M.V., Chidambaram, S., Shahul Hameed, A., Srinivasamoorthy, K., 2009. Study of evaluation of groundwater in Gadilam Basin using hydrogeochemical and isotope data. Environ. Monit. Assess. https://doi.org/10.1007/s 10661-009-1092-5.

Prasanna, M.V., Chidambaram, S., Shahul Hameed, A., Srinivasamoorthy, K., 2011. Hydrogeochemical analysis and evaluation of groundwater quality in the Gadilam river basin, Tamil Nadu, India. Aust. J. Earth Sci. 120 (1), 85–98.

Ramesh, R., Anbu, M., 1996. Chemical Methods for Environmental Analysis—Water and Sediments. Macmillan Publisher, India, p. 161.

Raza, M., Hussain, F., Lee, J.-Y., Shakoor, M.B., Kwon, K.D., 2017. Groundwater status in Pakistan: a review of contamination, health risks, and potential needs. Crit. Rev. Environ. Sci. Technol. https://doi.org/10.1080/10643389.2017.1400852.

Sarala, C., 2012. Assessment of groundwater quality for irrigation use. Int. J. Res. Eng. Technol. ISSN: 2319-1163.

Srinivasamoorthy, K., Vasanthavigar, M., Vijayaraghavan, K., Chidambaram, S., Anandhan, P., Manivannan, R., 2011. Use of hydrochemistry and stable isotopes as tools for groundwater evolution and contamination investigations. Geosci 1 (1), 16–25.

Thilagavathi, R., Chidambaram, S., Prasanna, M.V., Thivya, C., Singaraja, C., 2012. A study on groundwater geochemistry and water quality in layered aquifers system of Pondicherry region, southeast India. Appl Water Sci. https://doi.org/10.1007/s13201-012-0045-2.

Thilagavathi, R., Chidambaram, S., Thivya, C., Prasanna, M.V., Pethaperuamal, S., Tirumalesh, K., 2014a. A study on the behaviour of Total carbon and dissolved organic carbon in Groundwaters of Pondicherry region, India. Int. J. Earth Sci. Eng. 7, 1537–1550.

Thilagavathi, R., Chidambaram, S., Thivya, C., Prasanna, M.V., Singaraja, C., Tirumalesh, K., Pethaperumal, S., 2014b. Delineation of natural and anthropogenic process controlling hydrogeochemistry of layered aquifer sequence. Proc. Natl. Acad. Sci. 84 (1), 95–108.

Thilagavathi, R., Chidambaram, S., Thivya, C., Prasanna, M.V., Tirumalesh, K., Pethaperumal, S., 2016a. Dissolved organic carbon in multi-layered aquifers of Pondicherry region (India): spatial and temporal variability and relationships to major ion chemistry. Nat. Resour. Res. https://doi.org/10.1007/s11053-016-9306-3.

Thilagavathi, R., Chidambaram, S., Pethaperuamal, S., Thivya, C., Rao, M.S., Tirumalesh, K., Prasanna, M.V., 2016b. An attempt to understand the behaviour of dissolved organic carbon in coastal aquifers of Pondicherry region, South India. Environ. Earth Sci. 75, 235.

Thilagavathi, R., Chidambaram, S., Thivya, C., Prasanna, M.V., Tirumalesh, K., Pethaperumal, S., 2017. Assessment of groundwater chemistry in layered coastal aquifers using multivariate statistical analysis. Sustain. Water Resour. Manag 3, 55–69.

Thivya, C., Chidambaram, S., Singaraja, C., Thilagavathi, R., Prasanna, M.V., Jainab, I., 2013. A study on the significance of lithology in groundwater quality of Madurai district, Tamil Nadu (India). Environ. Dev. Sustain. 15, 1365–1387.

Thivya, C., Chidambaram, S., Rao, M.S., Thilagavathi, R., Prasanna, M.V., Manikandan, S., 2015. Assessment of fluoride contaminations in groundwater of hard rock aquifers in Madurai district, Tamil Nadu (India). Appl Water Sci. https://doi.org/10.1007/s13201-015-0312-0.

Tirumalesh, K., 2012. Characterization of Groundwater in the Coastal Aquifers of Pondicherry Region Using Chemical, Isotopic and geochemical Modelling Approaches (Unpublished Ph. D. thesis). Bhabha Atomic Research Centre, Mumbai.

Vengosh, A., 2005. Salinisation and saline environments. In: Sherwood Lollar, B., Holland, H.D., Turekian, K.K. (Eds.), Environmental Geochemistry. Treatise on geochemistry, vol. 9. Elsevier. pp. 333–365.

Vengosh, A., Keren, R., 1996. Chemical modifications of groundwater contaminated by recharge of treated sewage effluent. Contam. Hydrol. 23, 347–360.

Venkateswaran, S., Vediappan, S., 2013. Assessment of groundwater quality for irrigation use and evaluate the feasibility zones through Geospatial Technology in Lower Bhavani Sub Basin, Cauvery River, Tamil Nadu, India. Int. J. Innov. Technol. Explor. Eng. 3 (2). ISSN: 2278-3075.

Wayland, K., Long, D., Hyndman, D., Pijanowski, B., Woodhams, S., Haack, K., 2003. Identifying relationships between base flow geochemistry and land use with synoptic sampling and R-mode factor analysis. J. Environ. Qual. 32, 180–190.

CHAPTER 16

Anthropogenic Influence of Heavy Metal Pollution on the Southeast Coast of India

R. Rajaram, A. Ganeshkumar
Department of Marine Science, Bharathidasan University, Tiruchirappalli, India

1 Introduction

The coastal zone is an area where interaction of the land and ocean take place. The coastal and marine ecosystems such as bays, mangroves, salt marshes, estuaries, beaches, and coral reefs provide buffer zones and filtering systems for the coastal ecosystems (Venkataraman, 2007; Kathiresan and Rajendran, 2005) that accumulate and redistribute sediment, water, and nutrients for the sustenance of coastal, marine, land, and avian fauna and flora. Coastal zones represent 18% of the Earth's total surface, and it is estimated that around 44% of the world's population directly or indirectly depends on coastal zones for their livelihoods (Brown et al., 2014; Crossland et al., 2005). During the past few decades, terrestrial and coastal ecosystems experienced ever-increasing pressures from various anthropogenic activities (Salomidi et al., 2012), often to the detriment of their own survival. According to the WWF (World Wildlife Fund) (2017), more than 80% of marine pollution came from the direct influence of land-based activities, which includes oil spills, fertilizers, solid garbage, sewage disposal, and toxic chemicals (heavy metals, POP, and PCP). Halpern et al. (2007, 2008) estimated that around 17 different anthropogenic drivers cause severe changes in coastal ecosystems. In normal circumstances, the majority of pollutants from runoff and direct discharge are retained in coastal ecosystems because of their natural buffering and filtering action (Liang et al., 2016; Sharifinia et al., 2016).

Metallic elements are present in dissolved and particulate form in the Earth's surface, and are added to natural ecosystems from many natural processes, including weathering of rocks, volcanic fallout, and so forth (Callender, 2003). Extraneous catastrophic events such as cyclones, tsunamis, and so forth also add exotic sediments that may contain metallic ingredients (Liang et al., 2016; Unnikrishnan et al., 2015; IPCC, 2013). Superimposed on these natural accumulations are the contaminations from anthropogenic activities, especially land use changes in the inland and coastal regions. Attendant with these changes of urbanization, industrialization (Thevenon et al., 2011; Nriagu, 1990, 1979), agriculture, commercial and recreational usages, and so forth (Sharifinia et al., 2016), are the enhanced surface and subsurface flow and mobility of heavy metals that increase phenomenally over time, especially in coastal ecosystems due to their inherent buffering and filtering traits.

Loss of coastal and marine biodiversity has been the major concern for the past few decades (Vikas and Dwarakish, 2015). India is among the world's megabiodiversity countries, and is listed 17th among the 34 hotspots (Marchese, 2015). The ~8000 km-long Indian coastline has an exclusive economic zone of 2.02 million km^2 and encompasses a wide range of ecologically sensitive coastal ecosystems, including estuaries and mangroves (Venkataraman et al., 2012). Although the approximately 250 million people living within the coastal zones of India may be a driver of economic growth, (Brown et al., 2014; Sivakumar, 2012), it may also sound caution bells for the quality of coastal ecosystems as this large population may exert pressure through activities in terms of urbanization, effluent disposal, commercial and recreational activities, and so forth. As a result, there are several direct and indirect influences arising from different types of socio-economic and other developmental activities that impact coastal and marine biodiversity across the country (Kulkarni et al., 2018; Sivakumar, 2013).

Metals are essential components found on the Earth's surface, and in the atmosphere and aquatic ecosystems (Rajaram et al., 2017). However, when exceeded, they pose a severe threat to the natural equilibrium, and affect the natural environment. In recent years, elevated levels of metals have been observed in various environments. Apart from the lithogenic sources, major anthropogenic activities, including domestic and industrial effluents to rivers, sewage sludge disposal, application of pesticides, runoff from fertilized agricultural and aquatic fields (Ganeshkumar et al., 2018; Rajaram et al., 2017; Mathivanan and Rajaram, 2014a) cause elevated levels of metals in natural environments. India has huge natural water resources, including 14 major, 55 minor, and several 100's of small rivers fulfilling the daily water needs of the people (Kulkarni et al., 2018). As could be observed elsewhere, these natural rivers and other aquatic ecosystems are also used as a means of waste disposal (Ramkumar, 2007; MoEF, 2012). In India, only 9% of wastewater gets properly treated, and the rest is directly discharged into various water bodies that ultimately reach coastal zones. According to the Virtus Global Partners (VGP) report, Indian waste water increases 10 to 12% every year (Virtus Global Partners, 2008).

Tamil Nadu is the southernmost state in India and it has long coastal line around 1075 km covering 13 districts. Among them, the coast lines of Ramanathapuram and Nagapattinam occupy 40% of the total area (Ramesh et al., 2008). The primary productivity of the Tamil Nadu coastal zone is high due to the presence of coastal ecosystems such as the Gulf of Manner, Muthupet and Pitchavaram mangroves, and Pulicat Lake. The Gulf of Manner is one among the marine national parks recognized by UNESCO with unique marine biodiversity resources on the Indian subcontinent (Sacratees and Karthigarani, 2008). Situated among these biodiversity resources are several socio-economic, commercial, and recreational areas, including ports, fishing harbors, industries such as nuclear and thermal power plants, refineries, mining locations, electroplating facilities, fertilizer and chemical producers, and so forth (Rajaram et al., 2017; Mathivanan and Rajaram, 2014a; Ramesh et al., 2008). The coastal zone of Tamil Nadu is home to 5500 major industries in coastal districts, among which about 2500 are in the coastal zone itself. In addition, all the 13 coastal districts are affected by domestic and industrial waste water, from industries such as textiles, tanning, paper and pulp, thermal and nuclear power stations, oil refiners, and petrochemical production. Several previous studies (Rajaram et al., 2017, 2013; Balakrishnan et al., 2015; Natesan et al., 2014; Priya et al., 2014; Dhinesh et al., 2014; Mathivanan and Rajaram, 2014b) have raised the issue of documenting heavy metal pollution in the coastal ecosystems. This study was aimed at the estimation of concentrations of heavy metals such as Cu, Cd, Pb, and Zn from water and sediment samples, and was designed to determine the source of heavy metals in the coastal tracts of Tamil Nadu, South India.

2 Study Area

Three sites were selected based on the polluted and nonpolluted status of the coastal environment. Among the coastal zones in Tamil Nadu, Cuddalore, and Tuticorin are well developed industrial areas characterized by the discharge of effluents with various pollutants, including heavy metals into adjacent river systems. Muthupet does not have any significant activity in the vicinity. In addition, it was recognized as a pristine environment by the government of Tamil Nadu.

The Muthupet mangrove-lagoon (Lat 10°25′N: Long 79°39′E) is situated at the southernmost end of the Cauvery delta between the Nagapattinam and Thiruvarur districts, and connected to Palk Bay (Fig. 1). Many distributaries, namely, Paminiyar, Koraiyar, Kandankurichanar, Kilaithangiyar, and Marakkakoraiyar discharge into the Muthupet Lagoon. In addition to the lagoon, Mangrove plants are distributed across the wetland dominated by *Avicenna marina* and *Rhizophora mucronata*. Northeast monsoon season is the major contributor of freshwater into the lagoon during October through December via various drainage channels after draining agricultural soils, mangrove swamps, and aquaculture ponds (Ganeshkumar et al., 2018; Rajaram et al., 2017, 2013; Mathivanan and Rajaram, 2014b; Thilagavathi et al., 2011).
A negligible level of freshwater influence was observed from February through September. Three stations, including the confluence, river, and lagoon areas were established for sampling within Muthupet. The sediments in the lagoon are clayey silt, and the depth reaches a maximum of 3 m. This mangrove swamp and the creeks are the thoroughfares of commercial fishing boats (Ganeshkumar et al., 2018; Rajaram et al., 2017).

Tuticorin is located between the latitudinal and longitudinal extensions of 8°40′–8°55′ N and 78°0′–78°15′ E along the southeastern part of Tamil Nadu (Fig. 1). There are many major and

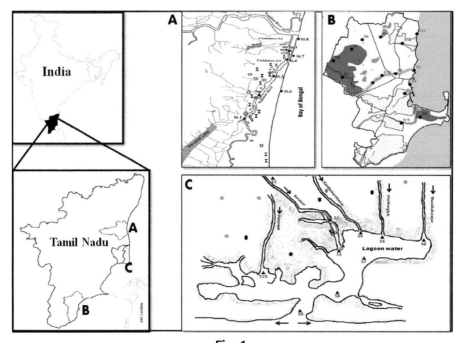

Fig. 1
Sampling sites. (A) Cuddalore Coast, (B) Tuticorin Coast, and (C) Muthupet mangroves.

minor industries, including refinery; aquaculture; metal processing; and chemical, pharmaceutical, and fertilizer manufacture; thermal power; and heavy water plants in addition to vast expanses of saltpans in and around the Tuticorin (Jayaraju et al., 2009; Asha et al., 2010; Karunanidhi et al., 2017; Selvam et al., 2017). According to an estimate (Asha et al., 2010), about 10.4 MLD (millions of liters per day) effluents are discharged from these industries. Apart from these, aquaculture activity alone generates about 91.2 MLD of effluents (Asha et al., 2010). Together, these effluents cause degradation of water quality in this area (Jayaraju et al., 2009; Asha et al., 2010).

Cuddalore (Lat. 11°43′N: Long. 79°49′E) is among the well-developed industrial areas of Tamil Nadu, located along the coast line and on the banks of the Uppanar Estuary (Fig. 1). There are numerous industries, in addition to an industrial complex SIPCOT housing 55 major and minor industries (State Industrial promotion Corporation of Tamil Nadu Limited) located along the Uppanar Estuary. The untreated waste water from the SIPCOT industrial complex and domestic waste water from the town of Cuddalore are directly discharged into the Uppanar Estuary (Mathivanan and Rajaram, 2013, 2014a, b). Given this background, the sampling stations and their sources of pollution are identified (Mathivanan and Rajaram, 2014a; Rajaram et al., 2005) for the quantification of toxic heavy metals in water and sediment samples.

3 Materials and Methods

3.1 Collection and Preservation of Samples

Sampling was performed for a year (2015–16) for four seasons (summer, premonsoon, monsoon, and postmonsoon) in the three sites for the collection of surface water and sediment. In Muthupet, sediment samples were collected using acid-cleaned PVC pipes (Ganeshkumar et al., 2018; Rajaram et al., 2017) up to 2 cm depth. In Tuticorin and Cuddalore, surface sediments were collected through stainless-steel forceps (Mathivanan and Rajaram, 2014a). 1 L of water sample was collected for heavy metal extraction in acid-cleaned polyethylene containers. 1 mL of concentrated HNO_3 was added to preserve the water samples for heavy metal extraction. Collected water samples were refrigerated and stored at 4°C, and the sediment samples were shade dried in the laboratory, and finally pulverized before acid digestion and subsequent analysis.

3.2 Analysis of Heavy Metals in Water and Sediment Samples

Water samples were filtered through a millipore filtration unit (0.45 μ) to remove the suspended solids. A 10 mL of 1% APDC solution (ammonium pyrrolidine dithio carbamate) and 25 mL MIBK (methyl isobutyl ketone) was added into the separatory funnel with 1 L of seawater and vigorously shaken for 15 min. After the aqueous phase was removed, the organic layer was collected separately. The procedure was repeated until complete extraction of the organic layer. The collected organic layer was digested with 50% HNO_3, and finally made up into 25 mL with millipore filtered water (Rajaram et al., 2017).

The heavy metal extraction was carried out by the modified method (Rajaram et al., 2017). One gram of finely ground sediment sample was digested with 10 mL of an acid mixture including HNO_3, H_2SO_4, and $HCLO_4$ in the ratio of 5:2:1 at 60°C on a hot plate. A few drops of hydrochloric acid were added to allow the complete digestion of the sample. Finally, the dried and digested samples were made up into 25 mL with millipore water and filtered using Whatman No.1 paper for analysis of heavy metals.

All the extracted samples were subjected to Atomic Absorption Spectrophotometer (AA7000-Shimadzu Corporation, Japan) analysis. The working wavelength for the heavy metals was 213.85 nm for Zn, 324.75 nm for Cu, 228.80 nm for Cd, and 217.00 nm for Pb. Quality assurance and percentage of recovery (Table 1) testing relied on the control of blanks and yield for chemical procedures (Ganeshkumar et al., 2018). The blanks were prepared with ultrapure water. Standardized reference materials were used during analysis. Triplicate analysis was performed for each sample. Prepared standard solutions were evaluated by linear model regression to maintain the precision of the instrument throughout the analysis.

Table 1 Quality control and quality assurance of instrumental methods

Metal	True (ppm)	Concentration Detected (ppm)	%R	r
Cd	0.50	0.59	118	
	1.00	1.05	105	1.00
	2.00	1.95	97.5	
Cu	0.50	0.46	92	
	1.00	1.03	103	0.99
	2.00	1.99	99.5	
Pb	0.50	0.67	134	
	1.00	0.99	99	0.99
	2.00	1.82	91	
Zn	0.50	0.49	98	
	1.00	1.12	112	0.99
	2.00	1.95	97.5	

%R = recovery percentage, r = linear regression.

3.3 Multivariate Statistical Analysis

Principal component analysis (PCA) was performed to reduce the data. Cluster analysis (CA) was used to identify the groups or clusters of similar stations based on the correlation between the metal concentrations from different study areas (Mathivanan and Rajaram, 2014b; Sparks, 2000). Similarities were examined by performing hierarchical cluster analysis (HCA) by the Ward method in order to identify the distances between samples. Most correlated stations formed the same groups with lesser distances (Birth, 2003). At last, the results were plotted and the dendrogram was developed. All the statistical calculations were performed with Past 3.0 (Windows 7.0) software, and other datasets were processed with Microsoft Excel (Windows 7.0).

4 Results

4.1 Variation of Heavy Metal Concentration in Water Samples

The distribution of heavy metals in the water samples varied depending upon the seasons premonsoon > postmonsoon > monsoon by the following order: Pb > Zn > Cu > Cd (Table 2). Cd concentration in water was the least among all the metals analyzed. It ranged between 0.07 and 2.25 ppm. The minimum and maximum levels of Cd were recorded on the Tuticorin coast during the monsoon and summer seasons, respectively. In summer, all three stations exhibited higher concentrations. Copper is an essential metal that plays a crucial role in both terrestrial and aquatic life. In general, the highest concentration of Cu in water was observed during the postmonsoon season at Tuticorin, and the least during premonsoon season at Cuddalore (Table 2). The highest mean values of Cu were observed in all the seasons at Tuticorin.

Table 2 Annual variation of heavy metal concentration (ppm) in water samples of three stations

Season	Station		Cu	Cd	Pb	Zn
Premonsoon	Cuddalore	Range	0.25–0.89	0.21–0.78	0.31–2.03	0.48–1.14
		Mean	0.56±0.20	0.45±0.18	1.07±0.65	0.78±0.25
	Tuticorin	Range	1.45–4.46	0.09–1.97	16.47–28.43	1.15–8.89
		Mean	2.77±1.08	0.74±0.77	21.63±4.42	3.82±2.73
	Muthupet	Range	1.01–1.65	0.05–0.35	6.31–8.12	1.86–5.99
		Mean	1.28±0.22	0.14±0.09	7.28±0.55	4.93±1.34
Monsoon	Cuddalore	Range	0–5.47	1.41–2.8	3.97–9.17	0.49–1.52
		Mean	1.54±1.75	2.14±0.55	6.33±1.55	0.88±0.37
	Tuticorin	Range	2.09–7.49	0.01–0.17	20.63–22.43	1.69–6.75
		Mean	4.3±1.77	0.07±0.05	21.57±0.6	3.24±1.8
	Muthupet	Range	0.91–1.81	0.08–0.2	7.11–8.19	0–0.20
		Mean	1.41±0.25	0.12±0.03	7.65±0.34	0.04±0.07
Postmonsoon	Cuddalore	Range	0.29–3.13	0.85–1.96	0.65–2.88	0.32–2.6
		Mean	1.0±0.95	1.31±0.45	1.74±0.86	1.03±0.85
	Tuticorin	Range	1.82–8.53	0.04–1.53	17.48–22.36	1.04–9.23
		Mean	4.71±2.2	0.47±0.54	20.59±1.66	4.09±2.77
	Muthupet	Range	0.5–4.43	0.06–1.80	11.82–15.16	0–9.78
		Mean	2.66±1.55	0.44±0.53	13.36±1.25	2.04±3.07
Summer	Cuddalore	Range	0.27–2.63	0.96–1.90	1.62–7.21	0.27–0.98
		Mean	1.02±0.78	1.33±0.33	3.51±1.74	0.59±0.23
	Tuticorin	Range	1.59–10.02	0.12–4.52	16.89–22.09	2.33–7.22
		Mean	3.81±2.9	2.25±1.85	20.04±1.73	4.58±1.96
	Muthupet	Range	0.79–1.85	0.37–3.72	13.78–17.87	2.3–12.91
		Mean	1.26±0.33	2.16±1.45	15.49±1.17	5.69±3.17

It was observed that the Pb concentration in water samples ranged from 1.07 to 21.63 ppm, with the highest level at Tuticorin and lowest in Cuddalore during premonsoon season. Also, elevated levels of Pb were observed in all the seasons at Tuticorin, and it has recorded more than 20 ppm were recorded during the study period (Table 2). The Pb concentration in the study area has shown the following order: premonsoon > monsoon > postmonsoon > summer. These results confirmed that the elevation of Pb in water was observed during the rainy seasons.

Zinc oxide, or zinc acrylate, is used as a biocide to control fouling organisms, but mostly it is combined with copper as an enhancer (Andrew Turner, 2010). The concentration of Zn in water samples ranged between 1.89 and 57.95 ppm, with a maximum (57.95 ppm) during summer at Tuticorin, and a minimum in Cuddalore (Table 2).

4.2 Variation of Heavy Metal Concentration in Sediment Samples

Seasonal variation of heavy metal concentrations in sediment samples are shown in Table 3. Among the analyzed metals, Pb was abundant in sediment when compared with all other metals (Cu, Cd, and Zn). A maximum concentration of Pb (60.54 ppm) was found in Cuddalore during the monsoon season, followed by Muthupet and Tuticorin during the postmonsoon and

Table 3 Annual variation of heavy metal concentration (ppm) in sediment samples of three stations

Season	Station		Cu	Cd	Pb	Zn
Premonsoon	Cuddalore	Range	0.42–1.57	0.38–1.30	1.36–2.72	0.86–4.69
		Mean	0.97±0.43	0.75±0.36	1.97±0.51	2.16±1.45
	Tuticorin	Range	0.42–54.25	0.01–0.23	6.59–15.05	7.79–53.78
		Mean	13.51±14.07	0.09±0.09	11.95±2.81	26.35±12.09
	Muthupet	Range	4.46–15.68	0.3–0.57	2.9–4.64	4.41–39.18
		Mean	10.88±3.55	0.45±0.10	4.01±0.54	27.5±11.09
Monsoon	Cuddalore	Range	0.49–11.71	2.26–5.30	37.33–82.81	1.22–5.78
		Mean	4.7±4.2	3.27±1.07	60.54±17.03	3.44±1.80
	Tuticorin	Range	1.88–49.82	0.08–0.36	12.99–18.06	13.99–79.17
		Mean	14.3±13.81	0.15±0.09	14.5±1.25	34.98±18.41
	Muthupet	Range	11.22–20.59	0.00	8.98–12.5	14.32–32.72
		Mean	16.83±2.74	0.00	10.83±1.31	26.69±7.00
Postmonsoon	Cuddalore	Range	1.13–3.79	1.8–2.85	0.94–8.12	1.2–4.05
		Mean	2.5±1.09	2.24±0.40	3.55±2.68	2.47±0.98
	Tuticorin	Range	1.41–53.53	0.01–0.21	8.34–20.44	6.79–85.96
		Mean	17.56±15.73	0.1±0.06	15.11±2.42	48.14±23.04
	Muthupet	Range	6.14–17.52	0.06–0.45	15.67–21.12	11.32–36.59
		Mean	11.5±3.37	0.25±0.10	18.33±1.57	27.15±8.25
Summer	Cuddalore	Range	2.83–12.64	1.69–3.40	4.4–10.31	0.72–3.99
		Mean	6.62±3.63	2.43±0.61	7.67±2.04	1.89±1.18
	Tuticorin	Range	6.86–67.6	0.31±0.23	14.77–19.76	30.12–97
		Mean	24.65±14.03	0.16±0.10	16.97±1.45	57.95±20.89
	Muthupet	Range	0.24–33.54	0.27–29.47	0.06–15.67	0.45–36.59
		Mean	16.44±13.81	1.66±0.14	8.15±6.73	18.79±14.83

summer seasons, respectively. Whereas, the minimum concentration of 1.97 ppm was found during premonsoon season at Cuddalore (Table 3).

Zn was the second most abundant metal in sediment samples. It was found to be higher and lower in the summer season, and it ranged between 1.89 and 57.95 ppm. During the entire study period, Zn in the sediment sample of Tuticorin was higher than Muthupet and Cuddalore. The major sources of Zn in the coastal zones included Chemical and allied products; fabricated metal products; pulp, paper, and paper products; iron and steel; textile mill products; electronic parts and devices; rubber and plastic products; municipal wastewater treatment plants; domestic and some industrial sources; road runoff; and fertilizers and pesticides (Naito et al., 2010).

High Cu content was found in Tuticorin samples (24.65 ppm) during the summer season, followed by the postmonsoon, monsoon, and premonsoon seasons (Table 3). In contrast, Muthupet has more Cu, after Tuticorin. Rajaram et al. (2017) observed elevated level of Cu in Muthupet sediment because of boating activities, Cu-based antifouling paints, and use of copper oxychloride and copper hydroxide as agricultural fungicides. Cu and Zn are applied over the surface of marine vessels to prevent the attachment of marine organisms, but controlled leaching may occur by causing an elevation of metal concentration in closed marine ecosystems such as bays, estuaries, and harbors. Antifouling paint particles collected from boat maintenance facilities had elevated levels of Cu (23–380 ppm) and Zn (14–160 ppm), along with elevated levels of organic boosters such as chlorothalonil, dichlofluanid, and irgarol 1051 (Parks et al., 2010). During monsoon season, the maximum concentration of Cd was found in Cuddalore (3.27 ppm), followed by other seasons; whereas in monsoon season, the least concentration of 0 ppm was found in Muthupet. Cd was emitted in the atmosphere through natural (volcanic activities) and manmade activities, including drying and roasting of ores, Ni-Cd batteries, fossil fuel combustion, and cement manufacturing (Kazantzis, 1987).

5 Source Identification

PCA was performed to find the sources of heavy metals in the study area. Table 4 displays the loadings values, eigen values, and cumulative variance; and Fig. 2 presents the biplot of principal components 1 and 2 for four heavy metals. These two factor loadings explain 74.75%. Principal component 1 shows positive loadings of Cu and Zn, and negative loadings of Cd and Pb.

Principal component 2 is positively loaded with Cd and Pb, and explains a variance of 25.92%. The biplot shows the interrelationship between seasons and heavy metal concentrations (Fig. 2). The Cd and Pb are many-folds higher than the previous studies, and the Earth's crust values (Taylor and McLennan, 1985, 1995). It is inferred that these elevated levels are not due to natural accumulation, and certainly due to anthropogenic activities.

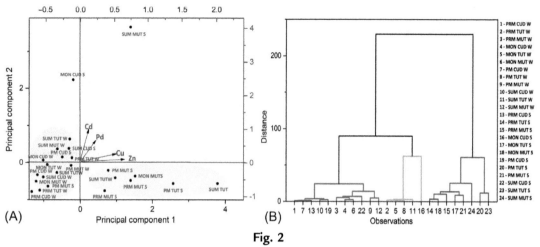

Fig. 2
(A) Principal component analysis and (B) Cluster analysis. Multivariate statistical analysis of samples from the southeast coast of India.

5.1 Analysis of Interrelationship Between Heavy Metal Concentrations

Two main cluster nodes, namely, clusters A and B, are recognizable while processing the metal concentration in water and sediment samples of the study area (Fig. 2). Cluster A was subdivided into cluster A1, containing most of the water samples of the Cuddalore and Muthupet mangroves, and cluster A2 had water samples from Tuticorin. A unique dendrogram pattern in cluster A that grouped the polluted and nonpolluted stations into a separate cluster suggests an anthropogenic impact on sediment and water chemistry; especially metal contamination from industrial, agricultural, and other nonnatural sources. The maximum concentration of 57.95 ppm was comparatively twofold less than the permissible level of sediment quality (Tables 4 and 5) (Harikrishnan et al., 2017a; Vignesh et al., 2016; Keshav and Achyuthan, 2015; Damodharan, 2013).

Table 4 Factors and rotated principle components of four heavy metals from the samples of study area (strongly positively loaded values are highlighted in bold letters)

Element	PC 1	PC 2	PC 3	PC 4
Cu	**0.71**	0.04	−0.09	−0.70
Cd	0.05	**0.88**	−0.45	0.16
Pb	0.12	**0.44**	0.89	0.03
Zn	**0.70**	−0.18	−0.03	0.69
Eigenvalue	1.95	1.04	0.97	0.04
Percentage of variance	49	26	24	1
Cumulative (%)	49	75	99	100

Table 5 Mean concentration of heavy metals in different industrial and estuary region of India and world coastal line

Study Area	Cd	Cu	Pb	Zn	Reference
Muthupet	0.57	20.59	21.35	39.18	Rajaram et al. (2017)
Uppanar, Cuddalore	0.44	5.96	6.00	11.53	Ayyamperumal et al. (2006)
Bay of Bengal, off Chennai	5.24	677	NA	60.39	Raju et al. (2010)
Ennmore	6.58	506.21	NA	126.83	Muthuraj and Jayaprakash (2007)
Pitchavaram Estuary	6.60	43.40	11.2	93.00	Ramanathan et al. (1999)
Coleroon Estuary	8.60	4.30	4.60	60.00	Ramanathan et al. (1999)
Korampallam Creek, Tuticorin	5.29	98.1	67.38	190.35	Magesh et al. (2013)
Palk Strait, Southeast Coast of India	0.40	69.1	19.52	244.2	Kasilingam et al. (2016)
Pallikaranai Marsh	1.60	101.5	36.6	149.9	Jayaprakash et al. (2010)
Cuddalore Coast	NA	59.2	24.0	89.7	Keshav and Achyuthan (2015)
Karaikal Coast	3.90	NA	0.80	3.00	Lakshmanasenthil et al. (2013)
Kalpakkam Coast	0.31	44.00	28.0	163.0	Selvaraj et al. (2004)
Vaigai River Estury	1.01	44.73	69.34	164.2	Paramasivam et al. (2015)
Cuddalore SIPCOT	1.60		1.38	1.50	Mathivanan et al. (2010)
Petrochemical Industrial City in Xinjiang, China	0.74	64.22	32.56	123.32	Wang et al. (2016)
Swan River, Islamabad, Pakistan	2.79	15.89	15.48	60.32	Irum et al. (2017)
Industrial Area of Surat, Gujarat, Western India	NA	110	NA	109.4	Krishna and Govil (2007)
Pali Industrial Area, Rajasthan, India	NA	93.7	57.05	357.5	Krishna and Govil (2004)
Gebze Industrial Area, Turkey	4.41	95.88	246.0	632.0	Yaylalı-Abanuz (2011)
Industrial Areas Uttar Pradesh, India	NA	42.9	38.30	159.3	Gowd et al. (2010)
Cuddalore	2.36	1.74	10.80	1.66	Present study
Tuticorin	10.70	0.50	17.80	22.89	Present study
Muthupet	7.78	0.65	10.64	14.10	Present study

NA, not applicable.

6 Discussion

The results are indicative of the continuous influx of pollution into sediment and water, and the dilution effects of rainwater during the monsoon seasons in all the studied sites. The elevated level of Cd in coastal water is due to the release of municipal waste water containing metal residues, untreated or partially treated effluents from plating and galvanizing industries, and several other industrial activities (Rajaram et al., 2017; Birth, 2003). While higher Cu in the vicinity of Tuticorin can be expected due to the presence of the major copper processing industry there, the occurrence of higher levels of Cu in relatively pristine coastal regions such as Muthupet is unexpected, and may explain the distal nonpoint sources such as discharges from agricultural, domestic, and fishery-boating, and antifouling agent usage as sources (Ashok Kumar et al., 2011). The concentrations recorded exceed the permissible limit for aquatic life by UNEPA and WHO (Table 6). The reported values were fivefold higher than sediment quality guidelines (Table 5); but around tenfold less than the reported values (Table 4) (Karunanidhi et al., 2017; Thilagavathi et al., 2011).

Occurrences of untreated effluents into the pristine ecosystem brought through the distributaries draining urban and industrial regions, and the enhanced influx of pollutants during monsoon season are observed in this study, and the effects are several-fold higher in industrial areas such as the Uppanar Estuary (Mathivanan and Rajaram, 2013). The observed maximum concentrations of Pb were above the permissible level, as per the Canadian Environmental Quality Guidelines (Table 6), and a previous report by Damodharan (2013) (Table 7) from the Cuddalore industrial area.

PCA has shown natural occurrences of essential metals Cu and Zn. However, Cu was several-fold higher than the background values of the Earth's crust (Taylor and McLennan, 1985, 1995). Together, it indicated that the principal component 1 was influenced by both anthropogenic and lithogenic sources (Rajaram et al., 2017; Mathivanan and Rajaram, 2013). Claisse and Alzieu (Bird et al., 1996; Claisse and Alzieu, 1993) have concluded that boat traffic

Table 6 Permissible level of heavy metals in water and sediment (ppm) by international agencies

Component	International Agencies	Cu	Cd	Pb	Zn	Reference
Water	Aquatic life	0.007	0.002	0.01	0.086	Ganeshkumar et al. (2018)
	US EPA Aquatic life protection	0.004	0.005	0.014	3.6	Ganeshkumar et al. (2018)
	WHO	2	0.003	0.01	3	Ganeshkumar et al. (2018)
Sediment	CEQG 2001	18.7	0.7	30.2	124	CEQG 2001

US EPA, US Environmental Production Agency; *WHO*, World Health Organization; *CEQG*, Canadian Environmental Quality Guidelines.

may be a significant source of estuarine metals through the use of zinc in anodes and copper in antifouling paints. A significant level of copper emission was observed from antifouling paint applied on recreational vessels for managing fouling organisms (Schiff et al., 2004). Antifouling paints contain a Cu-based biocidal pigment that includes cuprous oxide and cuprous thiocyanate (Turner, 2010).

The cluster analysis results showed that Muthupet also acts as a sink for various heavy metals by showing the strong correlation with other stations. The concentrations of heavy metals in sediment samples were several-fold higher than the water samples. As sediment contain higher organic matter content, which has a tendency to accumulate heavy metals (Luoma and Bryan, 1981; Hollert et al., 2003; Bettinetti et al., 2003; Ramkumar, 2003, 2007), higher concentrations of heavy metals in sediment than water samples are explained. Occurrences of all the polluted and nonpolluted sites together in a cluster provide affirmative evidence for point and nonpoint sources of pollution, as well as the accumulative character of the coastal zones.

The results of metal concentrations as documented by this study were compared (Table 7) with other coastal zones and esturarine regions of the world. It can be observed that metal concentration on the southeast coast is several-fold less than the previous studies, namely, the Bay of Bengal (Raju et al., 2010), Ennmore (Jayaprakash et al., 2010), the Pitchavaram Estuary (Ramanathan et al., 1999), Korampallam Creek, Tuticorin (Magesh et al., 2013), Palk Strait (Kasilingam et al., 2016), Vaigai Estuary (Paramasivam et al., 2015), the petrochemical industrial city in Xinjiang, China (Wang et al., 2016), Swan River, Islamabad, Pakistan (Irum et al., 2017), the industrial area of Surat, Gujarat (Krishna and Govil, 2007), the Pali industrial area, Rajasthan (Krishna and Govil, 2004), the Gebze industrial area, Turkey (Yaylalı-Abanuz, 2011), and industrial areas in Uttar Pradesh, India (Gowd et al., 2010). Heavy metal concentrations in this study appear to be slightly higher than the values reported in the following studies: Coleroon Estuary (Ramanathan et al., 1999) and Uppanar Estuary, Cuddalore (Mathivanan and Rajaram, 2014b; Ayyamperumal et al., 2006). The sources of contamination in various sites are inclusive of electroplating, power generation, industrial waste production, and smelting (Wang et al., 2016), vehicle lubricating oil, burning of fossil fuels (Yesilonis et al., 2008), electric energy transport, industrial wastes (Hu et al., 2014), industries such as tanneries, refractories, fertilizers, paints (Gowd et al., 2010), the release of agricultural wastages (Behera et al., 2013), pulp, paper, and fertilizer (Caeiro et al., 2005). However, only a few of the studies highlighted the low level of heavy metals in both sediment and the edible part of marine organisms (Ong et al., 2013a, b).

7 Conclusion

Coastal zones are the repositories of economically important flora and fauna; however, degradation causes extreme impacts on aquatic organisms, as well human beings. Increased industrialization along the coastal zones of India is the prime factor contributing to the loss of

Table 7 Previous report of mean concentration of heavy metals on the southeast coast of India

Study area	Cu	Cd	Pb	Zn	Reference
Palk Bay, Southeast Coast of India	19.42	7.05	11.52	17.52	Sithu et al. (2016)
Kallar Estuary, Tuticorin Coast	27.43	3.61	29.11	320.00	Magesh et al. (2013)
Korampallam Creek, Tuticorin Coast	98.10	5.29	67.38	190.38	Magesh et al. (2013)
Punnakayal Estuary, Tuticorin Coast	30.98	10.40	28.13	231.00	Magesh et al. (2013)
Tuticorin Coast	NA	NA	34.50	72.40	Jayaraju et al. (2009)
Tuticorin Coast, Gulf of Mannar	8.84	0.16	3.12	3.59	Jonathan et al. (2004)
Gulf of Mannar, India	NA	0.20	16.00	73.00	Jonathan and Ram-Mohan (2003)
Tuticorin	10.70	0.50	17.80	22.89	Present study
Cuddalore OT	ND	3.60	6.10	29.00	Harikrishnan et al. (2017b)
Cuddalore Coast Line	2.57	0.95	2.49	16.47	Vignesh et al. (2016)
Cuddalore Coast	NA	59.20	24.00	89.70	Keshav and Achyuthan (2015)
Uppanar River, Cuddalore Coast	191.50	36.05	98.50	201.38	Damodharan (2013)
Cuddalore SIPCOT	1.60		1.38	1.50	Mathivanan et al. (2010)
Uppanar, Cuddalore	0.44	5.96	6.00	11.53	Ayyamperumal et al. (2006)
Cuddalore	2.36	1.74	10.80	1.66	Present study
Muthupet Mangrove	0.57	20.59	21.35	39.18	Rajaram et al. (2017)
Muthupet, India	5.41	0.18	3.45	7.28	Balakrishnan et al. (2015)
Muthupet, India	58.63	0.74	41.52	209.02	Usha Natesan et al. (2014)
Muthupet, India	22.00	1.70	NA	29.00	Priya et al. (2014)
Muthupet, India	NA	48.58	16.59	628.00	Thilagavathi et al. (2011)
Muthupet, India	NA	2.32	NA	210.10	Ashok Kumar et al. (2011)
Muthupettai, India	NA	11.35	30.00	15.89	Janaki Raman et al. (2007)
Muthupet	7.78	0.65	10.64	14.10	Present study

NA, not applicable; ND, not detectable.

India's mega-biodiversity. This investigation documented the concentrations of heavy metals in water and sediment samples from the coastal industrial city and relatively unaltered coastal ecosystem on the southeast coast of India. The results showed the excessive distribution of nonessential elements in the coastal ecosystem through unrestricted anthropogenic activities when compared with the previous reports. During the past decade, Muthupet was recognized as one among the biologically important sites by the state government of Tamil Nadu. This study clearly indicated point and nonpoint distal sources of contaminants that pollute relatively secluded pristine coastal environments, which in turn, is a cause of worry. It also necessitates strict implementation of coastal regulation acts and practices for environmental nourishment and sustenance.

Acknowledgments

The authors are thankful to the University Grants Commission, Govt. of India (Ref. No.F. No. 36-55/2008 (SR) Dated 24-03-2009), Ministry of Earth Sciences, Govt. of India (Ref. No.: MoES/36/OOIS/Extra./9/2013) and Gulf of Mannar Marine Biosphere Reserve Trust, Govt. of Tamil Nadu (Ref. No. GOMBRT/C.No. 1008/2013/DO dated 28/03/2014) for financial support to carry out these works. The authors are also thankful to the authorities of Bharathidasan University for providing the necessary facilities.

References

Asha, P.S., Krishnakumar, P.K., Kaladharan, P., Prema, D., Diwakar, K., Valsala, K.K., Bhat, G.S., 2010. Heavy metal concentration in sea water, sediment and bivalves off Tuticorin. J. Mar. Biol. Assoc. India 52 (1), 48–54.

Ashok Kumar, S., Rajaram, G., Manivasagan, P., Ramesh, S., Sampathkumar, P., Mayavu, P., 2011. Studies on hydrographical parameters, nutrients and microbial populations of Mullipallam Creek in Muthupettai mangroves (southeast coast of India). Res. J. Microbiol. 6, 71–86.

Ayyamperumal, T., Jonathan, M., Srinivasalu, S., Armstrong-Altrin, J., Ram-Mohan, V., 2006. Assessment of acid leachable trace metals in sediment cores from River Uppanar, Cuddalore, Southeast coast of India. Environ. Pollut. 143 (1), 34–45.

Balakrishnan, T., Sundaramanickam, A., Shekhar, S., Balasubramanian, T., 2015. Distribution and seasonal variation of heavy metals in sediments of Muthupet Lagoon, Southeast Coast of India. J. Ecol. Eng. 16, 49–60.

Behera, B.C., Mishra, R.R., Patra, J.K., Sarangi, K., Dutta, S.K., Thatoi, H.N., 2013. Impact of heavy metals on bacterial communities from mangrove soils of the Mahanadi Delta (India). Chem. Ecol. 29, 604–619.

Bettinetti, R., Giarei, C., Provini, A., 2003. A chemical analysis and sediment toxicity bioassays to assess the contamination of the River Lambro (northern Italy). Arch. Environ. Contam. Toxicol. 45, 72–80.

Bird, P., Comber, S.D.W., Gardner, M.J., Ravenscroft, J.E., 1996. Zinc inputs to coastal waters from sacrificial anodes. Sci. Total Environ. 181, 257–264.

Birth, G., 2003. A scheme for assessing human impacts on coastal aquatic environments using sediments. In: Woodcoffe, C.D., Furness, R.A. (Eds.), Coastal GIS. Wollongong University Papers in Centre for Maritime Policy, 14, Wollongong.

Brown, S., et al., 2014. Shifting perspectives on coastal impacts and adaptation. Nat. Climatic Change 4, 752–755.

Caeiro, S., Costa, M.H., Ramos, T.B., Fernandes, F., Silveira, N., Coimbra, A., Medeiros, G., Painho, M., 2005. Assessing heavy metal contamination in Sado Estuary sediment: an index analysis approach. Ecol. Indic. 5, 151–169.

Callender, E., 2003. Heavy metals in the environment—historical trends. In: Holland, H.D., Turekian, K.K. (Eds.), Treatise on Geochemistry.In: vol. 99. pp. 67–105.

Claisse, D., Alzieu, C., 1993. Copper contamination as a result of antifouling paint regulations? Mar. Pollut. Bull. 26, 395–397.

Crossland, C.J., et al., 2005. The coastal zone—a domain of global interactions. In: Crossland, C.J., Kremer, H.H., Lindeboom, H.J., Marshall Crossland, J.I., Le Tissier, M.D.A. (Eds.), Coastal Fluxes in the Anthropocene. Global Change. The IGBP SeriesSpringer, Berlin/Heidelberg.

Damodharan, U., 2013. Bioaccumulation of heavy metals in contaminated River Water-Uppanar, Cuddalore, South East Coast of India. Perspect. Water Pollut. https://doi.org/10.5772/53374.

Dhinesh, P., Rajaram, R., Mathivanan, K., Vinothkumar, S., Ramalingam, V., 2014. Heavy metal concentration in water and sediment samples of highly polluted Cuddalore coast, southeastern India. Int. J. Curr. Res. 6, 8692–8700.

Ganeshkumar, A., Rajaram, R., Vinothkumar, S., Rameshkumar, S., Mathivanan, K., 2018. Flow of toxic metals in food-web components of tropical mangrove ecosystem, southern India. Hum. Ecol. Risk Assess. Int. J. 24 (5), 1367–1387.

Gowd, S.S., Reddy, M.R., Govil, P., 2010. Assessment of heavy metal contamination in soils at Jajmau (Kanpur) and Unnao industrial areas of the Ganga Plain, Uttar Pradesh, India. J. Hazard. Mater. 174 (1–3), 113–121.

Halpern, B.S., Selkoe, K.A., Micheli, F., Kappel, C.V., 2007. Evaluating and ranking the vulnerability of global marine ecosystems to anthropogenic threats. Conserv. Biol. 21, 1301–1315.

Halpern, B.S., Walbridge, S., Selkoe, K.A., Kappel, C.V., Micheli, F., 2008. A global map of human impact on marine ecosystems. Science 319, 948–952.

Harikrishnan, N., Ravisankar, R., Chandrasekaran, A., Gandhi, M.S., Kanagasabapathy, K., Prasad, M., Satapathy, K., 2017a. Assessment of heavy metal contamination in marine sediments of East Coast of Tamil Nadu affected by different pollution sources. Mar. Pollut. Bull. 121 (1–2), 418–424.

Harikrishnan, N., Ravisankar, R., Gandhi, M., Kanagasabapathy, K., Prasad, M.V., Satapathy, K., 2017b. Heavy metal assessment in sediments of east coast of Tamil Nadu using energy dispersive X-ray fluorescence spectroscopy. Radiat. Prot. Environ. 40, 21–26.

Hollert, H., Keiter, S., Konig, N., Rudolf, M., Ulrich, M., Braunbeck, T., 2003. A new sediment contact assay to assess particulate bound pollutants using zebrafish (Danio rerio) embryos. J. Soil. Sediment. 3, 197–207.

Hu, Y.N., Wang, D.X., Wei, L.J., 2014. Heavy metal contamination of urban topsoils in a typical region of Loess Plateau, China. J. Soil. Sediment. 14 (5), 928–935.

IPCC, 2013. In: Stocker, T.F. et al., (Ed.), Summary for policymakers. Climate Change 2013: The Physical Science Basis, Contribution of Working Group I to the Fifth Assessment of the Intergovernmental Panel on Climate Change. Cambridge University Press, Cambridge, p. 27.

Irum, P., Muhammad, A., Raza, S.S., Iffat, N., Brian, Y., Safia, A., 2017. Heavy metal contamination in water, soil and milk of the industrial area adjacent to Swan River, Islamabad, Pakistan. Hum. Ecol. Risk Assess. Int. J. https://doi.org/10.1080/10807039.2017.1321956.

Janaki Raman, D., Jonathan, M.P., Srinivasalu, S., Armstrong-Altrin, J.S., Mohan, S.P., Ram-Mohan, V., 2007. Trace metal enrichments in core sediments in Muthupet mangroves, SE coast of India: application of acid leachable technique. Environ. Pollut. 145, 245–257.

Jayaprakash, M., Urban, B., Velmurugan, P.M., Srinivasalu, S., 2010. Accumulation of total trace metals due to rapid urbanization in microtidal zone of Pallikaranai marsh, South of Chennai, India. Environ. Monit. Assess. 170, 609–629.

Jayaraju, N., Sundara Raja Reddy, B.C., Reddy, K.R., 2009. Heavy metal pollution in reef corals of Tuticorin coast, southeast coast of India. Soil Sediment Contam. 18 (4), 445–454.

Jonathan, M.P., Ram-Mohan, V., 2003. Heavy metals in sediments of the inner shelf off Gulf of Mannar, southeast coast of India. Mar. Pollut. Bull. 46, 263–268.

Jonathan, M.P., Ram-Mohan, V., Srinivasalu, S., 2004. Geochemical variations of major and trace elements in recent sediments, off the Gulf of Mannar, southeast coast of India. Environ. Geol. 45, 466–480.

Karunanidhi, K., Rajendran, R., Pandurangan, D., Arumugam, G., 2017. First report on distribution of heavy metals and proximate analysis in marine edible puffer fishes collected from gulf of Mannar marine biosphere reserve, South India. Toxicol. Rep. 4, 319–327.

Kasilingam, K., Suresh Gandhi, M., Krishnakumar, S., Magesh, N.S., 2016. Trace element concentration in surface sediments of Palk Strait, southeast coast of Tamil Nadu, India. Mar. Pollut. Bull. 111, 500–508.

Kathiresan, K., Rajendran, N., 2005. Mangrove ecosystem in the Indian Ocean region. Ind. J. Marine Sci. 1, 104–113.

Kazantzis, G., 1987. Cadmium. In: Fishbein, L., Furst, A., Myron, A. (Eds.), Genotoxic and Carcinogenic Metals: Environmental and Occupational Occurrence and Exposure. Advances in Modern Environmental Toxicology. vol. 11. Princeton Scientific Publishing Co., Princeton, NJ.

Keshav, N., Achyuthan, H., 2015. Late Holocene continental shelf sediments, off Cuddalore, East coast, Bay of Bengal, India: geochemical implications for source-area weathering and provenance. Quat. Int. 371, 209–218.

Krishna, A.K., Govil, P.K., 2004. Heavy metal contamination of soil around Pali industrial area, Rajasthan, India. Environ. Geol. 47 (1), 38–44.

Krishna, A.K., Govil, P.K., 2007. Soil contamination due to heavy metals from an industrial area of Surat, Gujarat, Western India. Environ. Monit. Assess. 124, 263–275.

Kulkarni, R., Deobagkar, D., Zinjarde, S., 2018. Metals in mangrove ecosystems and associated biota: a global perspective. Ecotoxicol. Environ. Saf. 153, 215–228.

Lakshmanasenthil, S., Vinothkumar, T., Ajith Kumar, T.T., Marudhupandi, T., Veettil, D.K., Ganeshamurthy, R., Ghosh, S., 2013. Harmful metals concentration in sediments and fishes of biologically important estuary, Bay of Bengal. J. Environ. Health Sci. Eng. 11, 33–39.

Liang, P., Wu, S.C., Zhang, J., Cao, Y., Yu, S., Wong, M.H., 2016. The effects of mariculture on heavy metal distribution in sediments and cultured fish around the Pearl River Delta region, South China. Chemosphere 148, 171–177.

Luoma, S.N., Bryan, G.W., 1981. A statistical assessment of the form of trace metals in oxidized estuarine sediments employing chemical extractants. Sci. Total Environ. 17, 165–196.

Magesh, N., Chandrasekar, N., Kumar, S.K., Glory, M., 2013. Trace element contamination in the estuarine sediments along Tuticorin coast—Gulf of Mannar, southeast coast of India. Mar. Pollut. Bull. 73 (1), 355–361.

Marchese, C., 2015. Biodiversity hotspots: a shortcut for a more complicated concept. Global Ecol. Conserv. 3, 297–309.

Mathivanan, K., Rajaram, R., 2013. Anthropogenic influences on toxic metals in water and sediment samples collected from industrially polluted Cuddalore coast, Southeast coast of India. Environ. Earth Sci. 68, 487–497.

Mathivanan, K., Rajaram, R., 2014a. Tolerance and biosorption of cadmium (II) ions by highly cadmium resistant bacteria isolated from industrially polluted estuarine environment. Ind. J. Geo Marine Sci. 43 (4), 580–588.

Mathivanan, K., Rajaram, R., 2014b. Isolation and characterization of the cadmium resistant bacteria from industrially polluted coastal ecosystem, Southeast coast of India. Chem. Ecol. 30, 622–635.

Mathivanan, V., Prabavathi, R., Prithabai, C., Selvisabhanayakam, 2010. Analysis of metals concentration in the soils of SIPCOT Industrial Complex, Cuddalore, Tamil Nadu. Toxicol. Int. 17 (2), 102–105.

MoEF (2012) Performance Audit of Water Pollution in India, Comptroller and Auditor General of India Report No. 21 of 2011-2012.

Muthuraj, S., Jayaprakash, M., 2007. Distribution and enrichment of trace metals in marine sediments of Bay of Bengal, off Ennore, south-east coast of India. Environ. Geol. 56, 207–217.

Naito, W., Kamo, M., Tsushima, K., Iwasaki, Y., 2010. Exposure and risk assessment of zinc in Japanese surface waters. Sci. Total Environ. 408 (20), 4271–4284.

Natesan, U., Madan Kumar, M., Deepthi, K., 2014. Mangrove sediments a sink for heavy metals? An assessment of Muthupet mangroves of Tamil Nadu, southeast coast of India. Environ. Earth Sci. 72 (4), 1255–1270.

Nriagu, J.O., 1979. Global inventory of natural and anthropogenic emissions of trace metals to the atmosphere. Nature 279, 409–410.

Nriagu, J.O., 1990. Global metal pollution: poisoning the biosphere? Environ Sci Policy Sustain Dev 32, 7–33.

Ong, M.C., Menier, D., Shazili, N.A.M., Kamaruzzaman, B.Y., 2013a. Geochemical characteristics of heavy metals concentration in sediments of Quiberon bay waters, South Brittany, France. Orient. J. Chem. 29, 39–45. ISSN: 0970-020 X.

Ong, M.C., Menier, D., Shazili, N.A., Effendy, A.W.M., 2013b. Levels of trace elements in tissue of Ostrea edulis and Crassostrea gigas from Quiberon Bay, Brittany, France. J. Fish. Aquat. Sci. 8 (2), 378–387.

Paramasivam, K., Ramasamy, V., Suresh, G., 2015. Impact of sediment characteristics on the heavy metal concentration and their ecological risk level of surface sediments of Vaigai river, Tamilnadu, India. Spectrochim. Acta A Mol. Biomol. Spectrosc. 137, 397–407.

Parks, R., Donnier-Marechal, M., Frickers, P.E., Turner, A., Readman, J.W., 2010. Antifouling biocides in discarded marine paint particles. Mar. Pollut. Bull. 60 (8), 1226–1230.

Priya, K.L., Jegathambal, P., James, E.J., 2014. Trace metal distribution in a shallow estuary. Toxicol. Environ. Chem. 96, 579–593.

Rajaram, R., Srinivasan, M., Rajasegar, M., 2005. Seasonal distribution of physic-chemical parameters in effluent discharge area of Uppanar estuary, Cuddalore, South-east coast of India. J. Environ. Biol. 26, 291–297.

Rajaram, R., Sumitha Banu, J., Mathivanan, K., 2013. Biosorption of Cu (II) ions by indigenous copper-resistant bacteria isolated from polluted coastal environment. Toxicol. Environ. Chem. 95, 590–604.

Rajaram, R., Ganeshkumar, A., Vinothkumar, S., Rameshkumar, S., 2017. Multivariate statistical and GIS-based approaches for toxic metals in tropical mangrove ecosystem, southeast coast of India. Environ. Monit. Assess. 189, 288.

Raju, K., Vijayaraghavan, K., Seshachalam, S., Muthumanickam, J., 2010. Impact of anthropogenic input on physicochemical parameters and trace metals in marine surface sediments of Bay of Bengal off Chennai, India. Environ. Monit. Assess. 177 (1–4), 95–114.

Ramanathan, A.L., Subramanian, V., Ramesh, R., Chidambaram, S., James, A., 1999. Environmental geochemistry of the Pitchavaram mangrove ecosystem (tropical), Southeast coast of India. Environ. Geol. 37 (3), 223–233.

Ramesh, R., Nammalwar, P., Gowri V.S. (2008). Database of coastal information of Tamil Nadu. Report Submitted to Environmental Information System (ENVIS) Centre Department of Environment, Government of Tamil Nadu: 1–133.

Ramkumar, M., 2003. Geochemical and sedimentary processes of mangroves of Godavari delta: implications on estuarine and coastal biological ecosystem. Ind. Jour. Geochem. 18, 95–112.

Ramkumar, M., 2007. Spatio-temporal variations of sediment texture and their influence on organic carbon distribution in the Krishna estuary. Ind. Jour. Geochem. 22, 143–154.

Sacratees, J., Karthigarani, R., 2008. Environment Impact Assessment. APH Publishing, p. 10. ISBN 81-313-0407-8.

Salomidi, M., Katsanevakis, S., Borja, A., Braeckman, U., Damalas, D., 2012. Assessment of goods and services, vulnerability, and conservation status of European seabed biotopes: a stepping stone towards ecosystem-based marine spatial management. Mediterr. Mar. Sci. 13, 49–88.

Schiff, K., Diehl, D., Valkirs, A., 2004. Copper emissions from antifouling paint on recreational vessels. Mar. Pollut. Bull. 48 (3–4), 371–377.

Selvam, S., Venkatramanan, S., Sivasubramanian, P., et al., 2017. Geochemical characteristics and evaluation of minor and trace elements pollution in groundwater of Tuticorin city, Tamil Nadu, India using geospatial techniques. J. Geol. Soc. India 90, 62–68.

Selvaraj, K., Ram Mohan, V., Szefer, P., 2004. Evaluation of metal contamination in coastal sediments of the Bay of Bengal, India: geochemical and statistical approaches. Mar. Pollut. Bull. 49, 174–185.

Sharifinia, M., Mahmoudifard, A., Namin, J.I., Ramezanpour, Z., Yap, C.K., 2016. Pollution evaluation in the Shahrood River: do physico-chemical and macroinvertebrate-based indices indicate same responses to anthropogenic activities? Chemosphere 159, 584–594.

Sithu, S.G.D., Kuppusamy, B., Subramaniyan, M., 2016. Sediment geochemistry with population of recent benthic ostracoda in Pak Bay, Southeast coast of India. Geosci. J. 20, 199–207.

Sivakumar, K., 2012. Marine biodiversity conservation in India. Bio Div Newslett 2 (2), 10–12.

Sivakumar, K., 2013. Coastal and marine biodiversity protected areas in India: challenges and way forward. In: Venkataraman, K., Sivaperuman, C., Raghunathan, C. (Eds.), Ecology and Conservation of Tropical Marine Faunal Communities. Springer, Berlin/Heidelberg.

Sparks, T., 2000. Statistics. In: Ecotoxicology. vol. 320. Wiley, Chichester.

Taylor, S.R., McLennan, S.M., 1985. The Continental Crust: Its Composition and Evolution. Blackwell, Oxford.

Taylor, S.R., McLennan, S.M., 1995. The geochemical evolution of the continental crust. Rev. Geophys. 33, 241–265.

Thevenon, F., Guedron, S., Chiaradia, M., Loizeau, J.L., Pote, J., 2011. (Pre-) historic changes in natural and anthropogenic heavy metals deposition inferred from two contrasting Swiss Alpine lakes. Quat. Sci. Rev. 30, 224–233.

Thilagavathi, B., Raja, K., Bandana Das, K., Saravanakumar, A., Vijayalakshmi, S., Balasubramanian, T., 2011. Heavy metal distribution in sediments of Muthupettai Mangroves, South East Coast of India. J. Ocean Univ. China 10, 385–390.

Turner, A., 2010. Marine pollution from antifouling paint particles. Mar. Pollut. Bull. (41), 159–171.

Unnikrishnan, A.S., Nidheesh, A.G., Lengaigne, M., 2015. Sealevel-rise trends off the Indian coasts during the last two decades. Curr. Sci. 108 (5), 966–971.

Venkataraman, K., 2007. In: Coastal and Marine Wetlands in India.Proceeding of Taal 2007: The 12th World Lake Conference, pp. 392–400.

Venkataraman, K., Raghunathan, C., Sreeraj, C.R., Raghuraman, R., 2012. Guide to the Dangerous and Venomous Marine Animals of India. pp. 1–98.

Vignesh, S., Dahms, H., Muthukumar, K., Vignesh, G., James, R.A., 2016. Biomonitoring along the tropical southern Indian coast with multiple biomarkers. PLoS One 11(12).

Vikas, M., Dwarakish, G., 2015. Coastal pollution: a review. Aquatic Proc. 4, 381–388.

Virtus Global Partners, 2008. Indian Water and Wastewater Treatment Market Opportunities for US Companies. .

Wang, W., Lai, Y., Ma, Y., Liu, Z., Wang, S., Hong, C., 2016. Heavy metal contamination of urban topsoil in a petrochemical industrial city in Xinjiang, China. J. Arid. Land 8 (6), 871–880.

WWF (World Wildlife Fund), 2017. Marine Problems: Pollution. Retrieved July 12, 2017, from, wwf.panda.org/about_our_earth/blue_planet/problems/pollution/.

Yaylalı-Abanuz, G., 2011. Heavy metal contamination of surface soil around Gebze industrial area, Turkey. Microchem. J. 99 (1), 82–92.

Yesilonis, I.D., Pouyat, R.V., Neerchal, N.K., 2008. Spatial distribution of metals in soils in Baltimore, Maryland: role of native parent material, proximity to major roads, housing age and screening guidelines. Environ. Pollut. 156 (3), 723–731.

Further Reading

Canadian Environmental Quality Guidelines Canadian Council of Ministers of the Environment, 1999, updated 2001.

Jayaraju, N., Reddy, B.C., Reddy, K.R., 2008. Metal pollution in coarse sediments of Tuticorin coast, southeast coast of India. Environ. Geol. 56 (6), 1205–1209.

Soundarapandian, P., Premkumar, T., Dinakaran, G.K., 2009. Studies on the physico-chemical characteristics and nutrients in the Uppanar estuary of Cuddalore, south east coast of India. Curr. Res. J. Biol. Sci. 1 (3), 102–105.

PART 5

Coastal Zone Management Concepts and Applications

CHAPTER 17

Adaptation Strategies to Address Rising Water Tables in Coastal Environments Under Future Climate and Sea-Level Rise Scenarios

Alex K. Manda[*,†,‡], Wendy A. Klein[*,†,‡]

[*]Department of Geological Sciences, East Carolina University, Greenville, NC, United States,
[†]Institute for Coastal Science and Policy, East Carolina University, Greenville, NC, United States,
[‡]Coastal Resources Management Program, East Carolina University, Greenville, NC, United States

1 Introduction

Globally, approximately 3.3 billion people live near the coastline (Cahoon et al., 2009, http://www.oceansatlas.com), all of whom are at risk from climate-induced marine inundation (where sea water replaces previously dry land), and saltwater intrusion (where sea water replaces freshwater in the subsurface) (e.g., Guha and Panday, 2012; Langevin and Zygnerski, 2013; NC Coastal Resources Commission, 2010; Fitzgerald et al., 2008). Previous work on the vulnerability of coastal aquifers to various external stressors indicates that groundwater extraction and climate change may have significant impacts on potable water resources. Many researchers have illustrated that saltwater intrusion in coastal zones can be caused by groundwater extractions, particularly in densely populated coastal regions (e.g., Barlow and Reichard, 2010). The extraction of groundwater in these settings may have a bigger influence on saltwater intrusion than sea-level rise (Ferguson and Gleeson, 2012). In areas where groundwater extraction is minimal, sea level rise may play a larger role in triggering saltwater intrusion into coastal freshwater aquifers (Werner and Simmons, 2009; Ketabchi et al., 2016).

Whereas the effects of climate change and sea-level rise on saltwater intrusion are well-known, few researchers (e.g., Hoover et al., 2017; Rotzoll and Fletcher, 2013; Habel et al., 2017) have documented the influence of climate change and sea-level rise on potential impacts of groundwater inundation in coastal regions. Also known as groundwater flooding, groundwater inundation is the process by which groundwater pools on the ground surface as a consequence of the water table intersecting the ground surface. Groundwater inundation may therefore occur after storm events if the water table is close to the ground's surface, or the phenomenon may occur as a result of an increase in the elevation of the water table owing to sea-level rise (Rotzoll and Fletcher, 2013). This process is likely to occur in shallow unconfined aquifers that are close to the coast.

Using a density-dependent, three-dimensional numerical groundwater flow model, Masterson and Garabedian (2007) are among the few to show that water tables in shallow coastal aquifers would intersect the land's surface under various sea-level rise scenarios. In an ecohydrological study, Masterson et al. (2013) ran numerical simulations that revealed that extensive groundwater inundation of the land's surface would occur on barrier islands with a shallow depth under various sea-level rise scenarios.

Understanding the impact of climate change on the environment will therefore be important because sea-level rise projections of 0.2–1.4m by 2100 (NC Coastal Resources Commission, 2010; Jevrejevaet al., 2012; NRC, 2012; Rahmstorf et al., 2012; Horton et al., 2014) suggest that sea-level rise will not only cause marine inundation, but will lead to a reduction in freshwater storage and/or elevation of water tables in surficial aquifers along the coast (Manda et al., 2014; Masterson et al., 2013; Rotzoll and Fletcher, 2013) (Fig. 1).

Shallow water tables and groundwater inundation will in turn lead to severe but gradual changes to the environment that include loss of dry land, expansion of wetlands, persistently saturated soils, changes to salinity and contamination of estuaries, changes and losses to fragile habitats, and deterioration of important infrastructure and associated services (e.g., septic systems, building foundations, roads, etc.). Other effects of rising water tables in coastal regions include increased evaporation and groundwater discharge, and decreased infiltration and drainage (Rotzoll and Fletcher, 2013). These effects are in addition to corrosion of subsurface structures that may be exposed to rising salinity levels.

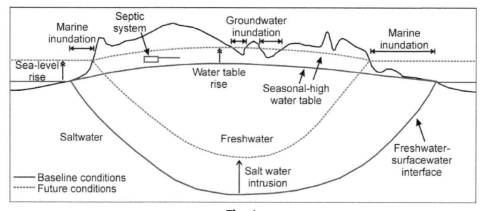

Fig. 1
Schematic illustrating groundwater inundation, marine inundation, and saltwater intrusion under a sea-level rise scenario.

Dune dominated barrier islands with a large human footprint are particularly vulnerable to groundwater inundation under a sea-level rise scenario because the dune and swale topography may provide an environment for groundwater inundation to dominate at the expense of marine inundation (e.g., Manda et al., 2014). On many barrier islands, human development (e.g., roads, parking lots, houses, tennis courts, shopping plazas, etc.) serve to retard the natural evolution of barrier islands over time. Additionally, if local communities actively protect the beach through hard (e.g., by building groins, breakwaters, jetties, and sea walls) and soft stabilization (i.e., beach nourishment) approaches, then the natural transport of sand will be disturbed, leading to a largely stationary barrier island that is restricted from moving laterally and vertically when the sea level changes.

An increase of impervious surfaces has precipitated storm water flooding problems in many coastal communities. To address this concern, coastal residents have employed various remediation/adaptation approaches (e.g., bioretention facilities, rain gardens, storm water wetlands, cisterns, permeable pavements, etc.) (Fig. 2) that encourage infiltration of surface water into the groundwater system (Dietz, 2007). Such low impact development (LID) strategies are encouraged because of relatively low cost, ease of use, and effectiveness. However, some of these approaches may not be appropriate under future environmental conditions where the water table may be close to or above the land's surface.

Fig. 2
Examples of low-impact development strategies (A) stormwater wetland, (B) bioretention facility, (C) permeable pavement, and (D) rainwater cistern.

It is therefore paradoxical that good management practices involving LID approaches to mitigate storm water runoff may exacerbate future problems of shallow water tables and/or groundwater inundation on barrier islands. The future challenge to reduce storm water flooding while decreasing groundwater infiltration will involve assessing the effectiveness of current strategies while developing and testing new and innovative approaches to deal with shallow water tables and/or groundwater inundation. This discussion ought to start sooner rather than later.

2 Current Options

The Intergovernmental Panel on Climate Change (IPCC) has recommended three adaptive strategies for managing the effects of sea-level rise resulting from climate change (Dronkers et al., 1990). Although the alternatives of (1) retreat, (2) accommodation, and (3) protection, as recommended by the IPCC, relate more to marine inundation than groundwater inundation, and are designed specifically for coastal erosion, the suggestions are viable options for initiating short- and long-term strategies for addressing shallow water tables and/or groundwater inundation. Unfortunately, there is not a "one size fits all" solution, as the impacts of shallow water tables and/or groundwater inundation may vary considerably by region, intensity, and time scales (Nicholls et al., 2011). Although none of the alternatives is ideal, as there are social, environmental, and economic compromises involved with each scenario, these options provide good starting points for addressing the problem.

Retreat, in the context of groundwater inundation, would allow the groundwater to rise in response to sea-level rise in the future. The human response would involve an inland movement away from the inundated areas. Because abandonment of important existing infrastructure and services may become necessary under this response, retreat is a costly socio-economic solution. In other cases, retreat may not even be feasible as barrier islands are typically long and narrow land masses. Where possible, retreat would be extremely disruptive to societies, as families, cultures, and traditions would have to be dislocated. Another disadvantage of retreat is that it has the potential to discriminately displace a large proportion of the population based on language, race, religion, and/or socio-economic status. Loss of property through condemnation, resettlement costs, rebuilding of infrastructure, and the elimination of land for future development would be other effects of retreat. Consequently, retreat would also be a barrier to economic development as noncondemned properties would diminish in value.

Accommodation allows continued use of the land, while making no attempt to alter the potential for groundwater inundation. Accommodation options include land use changes, future limits on land use, and land use plans to incorporate site selection and design of infrastructure based on future climate change and sea-level scenarios. Engineering solutions include such approaches as elevating structures, altering existing septic or sewer systems, retrofitting infrastructure, and rebuilding roads and bridges. Thoughtful planning for future development with LID techniques such as installing rain gardens and rain barrels, cisterns, and other storm water collection projects, vegetated rooftops, and rain gutter disconnects and reroutes may help to ameliorate the severity of flooding and groundwater inundation, but preserve the current socio-economic way of life for barrier island residents and businesses.

Protection against groundwater inundation entails utilizing engineering strategies that allow current use to continue with relatively few disruptions to the way of life of humans. This approach is, however, a costly and continuing alternative that may involve managed or perpetual pumping of groundwater, and aquifer storage and recovery (e.g., Werner et al., 2013, http://www.pitwatch.org). The impacts of sea level rise and groundwater inundation vary by region and time scales (Nichols, et al., 2007); however, elevation highly influences the probability for groundwater inundation and the reduction of aquifer thickness (Cahoon, et al. 2009). Unfortunately, human development has, in many coastal regions, hindered the natural processes of island building. In a well-intentioned attempt to stave off the impacts of coastal erosion, developers and coastal planners have enacted strategies (e.g., building hard structures) that essentially starve coastal areas of needed sediment for island migration. At the same time, while human intervention impedes vertical island building, low-lying coastal areas and swales cannot shift or grow vertically, adding to the problem of groundwater inundation caused by sea-level rise. Because inundation occurs at approximately the same rate as marine inundation, elevation must keep pace with rising water tables.

The first approach to addressing the problem of groundwater inundation is to understand the problem on both local and regional scales by integrating population dynamics, land use, and infrastructure data with inundation and erosion models. The threat of groundwater inundation will be managed by taking advantage of new and available technologies, strategies, and development opportunities. No one strategy will eliminate the social, environmental, and economic impacts of extensive groundwater inundation. Perhaps the best and most credible alternative to solve the paradox of reducing storm water flooding while decreasing groundwater infiltration for barrier islands and other low-lying areas will be a combination of accommodation and protection. Accommodation is the most realistic option for preserving the economic stability of an area, while protection is the best strategy for maintaining the integrity of the aquifer. If viewed through an innovative lens, management alternatives present an impetus for new economic opportunities, including recreational development, desalination, aquifer storage and recovery, and new water uses and disposal technologies. In order to develop a groundwater management policy that will be acceptable to all stakeholders, it is crucial that the solution be determined with the involvement of all stakeholders through a fair and just process; governmental institutions on all levels, industry and the community at large (e.g., Manda and Klein, 2014). Communities in low-lying coastal regions are likely to face multiple onslaughts from the sea owing to climate change and sea-level rise. Resource managers in the coastal zone therefore ought to start to develop tools and strategies to combat not only marine inundation and saltwater intrusion, but groundwater inundation. Although these strategies may require resourceful thinking, resource managers, policy makers, and coastal communities will be well-served if they starting considering such options sooner rather than later. The consequences of not having a plan to address the potential impacts of groundwater inundation may be too costly for communities in coastal regions.

3 Conclusions

The report of the Coastal Zone Management Subgroup of the IPCC (Dronkers et al., 1990) stresses the importance of acting proactively before the consequences of sea level rise can no longer be mitigated. This is especially imperative in low-lying areas such as barrier islands and other coastal areas. Although the societal costs cannot be overstated, groundwater inundation is frequently overlooked when the management of sea-level rise is considered. Solutions to the paradox of reducing storm water flooding, coastal erosion, and island transformation resulting from sea-level rise, while decreasing infiltration to the water table, will demand understanding and integrating the issues impacting both the natural and human systems. Many coastal communities are not well-informed about the multitiered threat posed by sea-level rise, and are therefore poorly prepared. Successful coastal zone planning and management will necessitate educating coastal communities and others most affected by the hazards about the risks of groundwater inundation and sea-level rise. Engaging the public in potential solutions and rewards of a preemptive coastal management approach will encourage coastal communities to preserve their ways of life and livelihoods, while giving them a measure of control over their future security.

References

Barlow, P.M., Reichard, E.G., 2010. Saltwater intrusion in coastal regions of North America. Hydrgeol. J. 18, 247–260.

Cahoon, D.R., Williams, J., Gutierrez, B.T., Anderson, K.E., Thieler, R., Gesch, D.B., 2009. The Physical Environment. In: Titus, J.G. (Ed.), Coastal Sensitivity to Sea-Level Rise: A Focus on the Mid-Atlantic Region. U.S, Synthesis and Assessment Product 4.1 Reportby the U.S. Climate Change Science Program and the Subcommittee on Global Change Research, pp. 9–84.

Dietz, M.E., 2007. Low impact development practices: a review of current research and recommendations for future directions. Water Air Soil Pollut. 186, 351–363.

Dronkers, J., Gilbert, J., Butler, L., Carey, J., Campbell, J., James, E., McKenzie, C., Misdorp, R., Quin, N., Ries, K., 1990. Strategies for Adaptation to Sea Level Rise, Report of the IPCC Coastal Zone Management Subgroup. (Intergovernmental Panel on Climate Change).

Ferguson, G., Gleeson, T., 2012. Vulnerability of coastal aquifers to groundwater use and climate change. Nat. Clim. Chang. 2, 342–345.

Fitzgerald, D.M., Fenster, M.S., Argow, B.A., Buynevich, I.V., 2008. Coastal impacts due to sea-level rise. Annu. Rev. Earth Planet. Sci. 36, 601–647. https://doi.org/10.1146/annurev.earth.35.031306.140139.

Guha, H., Panday, S., 2012. Impacts of sea-level rise in a coastal community of South Florida. J. Am. Water Resour. Assoc. 48 (3), 510–529.

Habel, S., Fletcher, C.H., Rotzoll, K., El-Kadi, A.I., 2017. Development of a model to simulate groundwater inundation induced by sea-level rise and high tides in Honolulu, Hawaii. Water Res. 114, 122–134.

Hoover, D.J., Odigie, K.O., Swarzenski, P.W., Barnard, P., 2017. Sea-level rise and coastal groundwater inundation and shoaling at select sites in California, USA. J. Hydrol. Reg. Stud. 11, 234–249.

Horton, B.P., Rahmstorf, S., Engelhart, S.E., Kemp, A.C., 2014. Expert assessment of sea-level rise by AD 2100 and AD 2300. Quat. Sci. Rev. 84, 1–6.

Jevrejeva, S., Moore, J.C., Grinsted, A., 2012. Sea level projections to AD2500 with a new generation of climate change scenarios. Global Planet. Change 80-81, 14–20.

Ketabchi, H., Mahmoodzadeh, D., Ataie-Ashtiani, B., Simmons, C.T., 2016. Sea-level rise impacts on seawater intrusion in coastal aquifers: review and integration. J. Hydrol. 535, 235–255.

Langevin, C.D., Zygnerski, M., 2013. Effects of sea-level rise on salt water intrusion near a coastal well field in southeastern Florida. Ground Water 51, 781–803.

Manda, A.K., Klein, W., 2014. Rescuing degrading aquifers in the Central Coastal Plain of North Carolina (USA): just process, effective groundwater management policy and sustainable aquifers. Water Resour. Res. 50, 5662–5677. https://doi.org/10.1002/2013WR015242.

Manda, A.K., Sisco, S., Mallinson, D., Griffin, M., 2014. Relative role and extent of marine and groundwater inundation on a dune-dominated barrier island under sea-level rise scenarios. J. Hydrol. Proc. https://doi.org/10.1002/hyp.10303.

Masterson, J.P., Garabedian, S.P., 2007. Effects of sea-level rise on ground water flow in a coastal aquifer system. Ground Water 45, 209–217. https://doi.org/10.1111/j.1745-6584.2006.00279.x.

Masterson, J.P., Fienen, M.N., Thieler, E.R., Gesch, D.B., Gutierrez, B.T., Plant, N.G., 2013. Effects of sea-level rise on barrier island groundwater system dynamics—ecohydrological implications. Ecohydrology. https://doi.org/10.1002/eco.1442.

National Research Council, 2012. Sea-Level Rise for the Coasts of California, Oregon, and Washington: Past, Present, and Future (Board on Earth Sciences and Resources, Ocean Studies Board, 2012).

NC Coastal Resources Commission, 2010. North Carolina sea-level rise assessment report: N.C. Coastal Resources Commission's Science Panel on Coastal Hazards, N.C. Division of Coastal Management, 6–12.

Nicholls, R.J., Marinova, N., Lowe, J.A., Brown, S., Vellinga, P., de Gusmao, D., Hinkel, J., Tol, R.S.J., 2011. Sea-level rise and its possible impacts given a 'beyond 4°C world' in the twenty-first century. Phil. trans. R. Soc. 369, 161–181. https://doi.org/10.1098/rsta.2010.0291.

Rahmstorf, S., Foster, G., Cazenave, A., 2012. Comparing climate projections to observations up to 2011. Environ. Res. Lett. 7, 1–5. https://doi.org/10.1088/1748-9326/7/4/044035.

Rotzoll, K., Fletcher, C.H., 2013. Assessment of groundwater inundation as a consequence of sea-level rise. Nat. Climate 3, 477–481.

Werner, A.D., Simmons, C.T., 2009. Impact of sea-level rise on seawater intrusion in coastal aquifers. Ground Water 47, 197–204. https://doi.org/10.1111/j.1745-6584.2008.00535.x.

Werner, A.D., Bakker, M., Post, E.A., Vandenbohede, A., Lu, C., Ataie, B., Simmons, C., Barry, D.A., 2013. Seawater intrusion processes, investigation and management: Recent advances and future challenges. Adv. Water Resour. 51, 3–26.

Further Reading

Nicholls, R.J., 1995. Coastal megacities and climate change. Geo J 37, 369–379.

Nicholls, R.J., Cazenave, A., 2010. Sea-level rise and its impact on coastal zones. Science 328, 1517–1520.

Taylor, R.G., Scanlon, B., Doll, P., Rodell, M., va Beek, R., Wada, Y., Longuevergne, L., Leblanc, M., Famigelietti, J.S., Edmunds, M., Konikow, L., Green, T.R., Chen, J., Taniguchi, M., Bierkins, M.F.P., MacDonald, A., Fan, Y., Maxwell, R.M., Yechieli, Y., Gurdak, J.J., Allen, D.M., Shamsudduha, M., Hiscock, K., Yeh, P., J-F Holman, I., Treidel, H., 2013. Ground water and climate change. Nat. Clim. Chang. 3, 322–329.

CHAPTER 18

Analytic Hierarchy Process to Weigh Groundwater Management Criteria in Coastal Regions

Wendy A. Klein[*,†,‡], Alex K. Manda[†,‡]
[*]*Coastal Resources Management Program, East Carolina University, Greenville, SC, United States,*
[†]*Department of Geological Sciences, East Carolina University, Greenville, SC, United States,* [‡]*Institute for Coastal Science and Policy, East Carolina University, Greenville, SC, United States*

1 Introduction

Sustainable use of potable water resources is imperative to global health (Gleick, 1998; Hunter et al., 2010; United Nations, 2007; World Health Organization, 2004). However, despite years of management interventions, use of many groundwater resources has become unsustainable due to overexploitation and contamination arising from human population growth, urban sprawl, industrial advancement, agricultural development, and climate change (Taylor et al., 2013; Hughes et al., 2011). Globally, population growth, low rates of groundwater recharge and high groundwater extractions have worked together to severely degrade water quality and diminish levels in groundwater reservoirs (Gleick, 1998; Rosegrant et al., 2002; Green et al., 2011; Cosgrove and Rijsberman, 2014). Although reversal of this trend is possible with management decisions that consider the multidimensions of groundwater, there is currently no consensus on the importance of various criteria for the sustainable use and equitable allocation of groundwater resources, particularly those in coastal regions.

The literature has shown how the perceptions of criteria for managing groundwater resources have evolved over time. For example, early philosophies for managing groundwater resources were based on the concept of "safe yield," which was defined as the "limit to the quantity of water which can be withdrawn regularly and permanently without dangerous depletion of the storage reserve" (Lee, 1915). The definition of safe yield was later expanded to include economic considerations (Meinzer, 1923). Although these definitions focus on long-term aquifer health, they were developed prior to sophisticated knowledge about aquifer drawdown, dewatering, and salt water intrusion (Conkling, 1946; Fetter, 1988; Schwartz and Zhang, 2003), processes that are important in coastal areas where large numbers of the world's population reside.

Following Theis's (1935, 1940), pioneering work on the analysis of aquifer characteristics, groundwater management criteria have proceeded to include the infringement on water rights (Banks, 1953), and the economic disadvantages and social impacts of pumping (Conkling, 1946; Freeze and Cherry, 1979). Other researchers have advocated for adoption of economic theories of optimization and exhaustible resources (Mays, 2013) and adaptive management (e.g., Gleeson et al., 2010; Klein et al., 2014) as viable strategies for implementing groundwater policy. Although authoritative bodies such as the Intergovernmental Panel on Climate Change (IPCC, 2014) and the United Nations (United Nations, 1997) have stressed the importance of applying multicriteria analysis and interdisciplinary research that incorporate qualitative and quantitative data from the natural and social sciences in studies of groundwater policy, it is still unclear how water professionals perceive and apply these criteria in their efforts to manage coastal groundwater resources. Understanding these perceptions is important, because the choice of criteria and the assessment of the relative importance for those criteria are critical determinants to creating new and innovative management options (Hajkowicz and Collins, 2007; Chen et al., 2010) that would safeguard water resources for future generations.

Criteria on which groundwater management is based impact the efficient and sustainable use of groundwater; however, because not all criteria are of equal importance (Hajkowicz and Collins, 2007), it is not currently clear which criteria are the most important for managing groundwater resources. This chapter seeks to evaluate the relative importance of these groundwater management criteria using the Analytic Hierarchy Process (AHP), a comparative weighting procedure increasingly being used to assess water resource planning (Calizaya et al., 2010; Cabrere et al., 2011; Panagoloulos et al., 2012; Li and Sun, 2017). In this research, AHP is used to assess how groundwater professionals perceive the importance of various criteria affecting groundwater management. Addressing this objective has implications for managing groundwater resources in coastal regions, especially when multiple criteria are considered. Coastal regions are particularly sensitive areas because these regions not only have large population centers, but they also harbor fragile ecosystems that are threatened by a myriad of coastal processes, including saltwater intrusion and tropical storms. The use of the AHP in determining the relative importance of groundwater criteria is a novel application in groundwater research in coastal regions.

2 Methods

2.1 Establishing Criteria for Groundwater Management

The criteria for assessing groundwater management are drawn from international guidelines that regulate the nonnavigational use of water that were adopted by the International Law Association (ILA) in 2004. Known as the Berlin Rules on Water Resources (ILA, 2004), these rules stipulate the relevant factors to be considered when determining equitable and reasonable use of water. The Berlin Rules were developed to refine the Helsinki Rules on the Uses of International Rivers (ILA, 1966; Salman, 2007), and the United Nations Convention on the Law of Non-Navigational Uses of International Watercourses (United Nations, 1997). The revisions account for changes in international environmental, international human rights, and the humanitarian law (ILA, 2004). In contrast to previous conventions, the Berlin Rules considered principles and rules that specifically applied to groundwater. Thus, the Berlin Rules list nine factors that affect the equitable and reasonable use of groundwater resources. These factors served as a starting point for developing the seven pertinent groundwater management criteria that were appropriate for this study. The seven criteria that were borne out of the Berlin Rules are: (a) hydrogeologic and aquifer characteristics (i.e., attributes that influence that flow of water in the subsurface), (b) socioeconomic demands and needs of a community, (c) the population dependent on groundwater resources in a given area, (d) the volume of groundwater available in an aquifer at a given time, (e) the availability of alternate water sources to develop, (f) financial capability of communities to develop alternate water resources, and (e) political motivation and support to develop alternate water resources (Table 1).

These criteria are particularly applicable to the study, as coastal zones have very specific policy challenges. Coastal regions face complex variations in aquifer characteristics (e.g., salt water intrusion) due to the land-sea interface; seasonal demands, needs and financial inputs; and coastal population clusters, often surrounded by rural communities, both of which require available and secure freshwater resources. In addition, global climate change will exacerbate a continual change to theses dynamic, interacting systems.

Table 1 The seven groundwater management criteria that were used to explore the perceptions of water professionals

Code	Criterion
C1	Hydrogeologic/aquifer characteristics of the area of interest
C2	Socio-economic demands and needs of the community
C3	Population dependent on the groundwater resources
C4	Available groundwater volume to use for the population's demands and needs
C5	Availability of alternate water sources to develop for the population's demands and needs
C6	Financial resources to develop alternate water resources
C7	Political motivation and support to develop alternate water resources

2.2 Survey Instrument

To assure relevance to those who influence the strategic planning for water resources (i.e., water managers and water resource advisors), a survey instrument was designed to explore the perceptions of water professionals with respect to the seven groundwater management criteria as shown in Table 1. Survey questions targeted practicing water professionals working in different locations and work sectors.

Surveys were circulated using a random sampling approach via paper and online avenues. Paper copies of the surveys were dispensed during a professional water resource meeting held in the state of North Carolina in the United States in 2016. Approximately 50 people attended the meeting, during which 36 surveys were distributed, completed, collected, and tabulated by the researchers. Additional surveys were administered online ($n = 100$), via the American Institute of Professional Geologists (AIPG) web portal. Information on the age range and experience of the respondents was not collected. The 136 responses represent an 8% margin of error, at the 95% confidence level (assuming a total population of 10,000 people). Institutional review board approval was acquired prior to administering the surveys.

The survey format consisted of carefully crafted questions that were worded to avoid ambiguity or confusion. The survey was designed to consider the seven criteria introduced in Table 1. These criteria were assessed via a standard, Likert-type survey, with responses on a scale of one through five (where 1 = not at all important, 2 = slightly important, 3 = moderately important, 4 = very important, and 5 = extremely important) (Likert, 1932). Typically, Likert-type scales of 1–5 provide the respondents with answer sensitivity, indicating a relative strength of each response and avoiding extreme responses of "agree" or "disagree." The five-point scale is easy to understand, and accurately captures respondents' opinions (Alharbi and Sayed, 2017) without diminishing the response rate due to frustration (Babakus and Mangold, 1992). For each question, respondents were also given the opportunity to comment or elaborate on each question. Other information was collected to better understand the characteristics of the sample population.

The respondents of the survey were asked to self-identify (a) the location where they predominantly worked (i.e., in the United States and/or internationally), and (b) the sector in which they worked. The researchers later grouped the work locations in United States by the US census bureau regions (i.e., Northeast, Midwest, South, and West). An answer of "National" indicates that the respondent worked in more than one US census region, whereas, "International" represents those respondents who worked in the United States and other countries outside of the United States, or worked exclusively outside of the United States. An answer of "No response" indicates that the respondent did not identify the work sector. The work sectors that were self-identified included academia (identified as personnel from institutions of higher education), government agencies (which included federal, state, local, or tribal agencies), industry, nongovernmental/nonprofit agencies, or other. Respondents included 18 academics, 30 from government, 48 from industry, 5 from Non-Governmental Organizations (NGO), and 34 from other sectors. The majority of the "other" category was composed of professionals in the engineering and environmental consulting fields (hydrogeologic consulting—12, environmental consulting—10, planning/permitting/inspection—3, well field development—2, retired—2, oil, gas, minerals—2, not specified—2, elected official—1). The "others" category are grouped together as an individual sector for the analysis. The participants responded voluntarily and received no compensation.

2.3 Analysis

A mixed methods approach was used to evaluate the survey results, as the responses were in the form of both qualitative and quantitative data (Creswell and Clark, 2007; Clark et al., 2008; Denzin and Lincoln, 2013). Likert-type results were analyzed as quantitative and ordinal data, the short responses (i.e., work location and sector) were considered quantitative and nominal data, whereas any additional comments were treated as qualitative data. Only surveys in which each criterion was assessed were used in the study. A lack of a response for a criterion was considered as one less response in the statistical analysis.

Basic descriptive statistics were generated in SPSS and R studio statistical software programs. The data that were acquired were ordinal and nominal, and not normally distributed. As a result, the assessed scores (in the case of the criteria analysis) have no relative difference between each value, and the mean was deemed inappropriate for analyses. Therefore, the Kruskal-Wallis test was used to explore if there were significant differences in how different groups of water professionals perceived the importance of the seven groundwater management criteria. The goal of this exercise was to determine if there was a difference between the perceptions of water professionals from different work sectors, and whether there were significant differences between the perceptions of water professionals from different locations at the 95% confidence level. The null hypothesis that was tested was that the medians of all variables were equal.

The scaled Likert-type data that were collected from the participant survey were converted to a ratio scale using the AHP to allow for a comparison of the intensity of each criterion. The AHP is useful for comparing different kinds of criteria (e.g., physical and social processes) and assigning priorities by using pairwise comparisons of criteria to assess the relationships between those criteria (Saaty, 1980, 1990, 2008; Saaty and Vargas, 1991). The comparisons were derived from the values acquired from the Likert-type survey, and in this case, represent the relative perceptions of the water professionals with regard to important groundwater management criteria. The AHP was used to develop criteria weights from paired comparisons by considering the relationships and variations in judgment between the multiple criteria from the many surveyed water professionals at the same time.

The protocol for determining weights of criteria consisted of first identifying criteria that the authors deemed were important for effective groundwater management (Fig. 1). Then, a Likert survey was administered to quantitatively determine how various stakeholders perceived the importance of the selected criteria. The geometric means from the scores of the Likert survey (expressed as a score between 1 and 5 on a scale of importance) were thereafter calculated and placed in a matrix. The first row and the first column in the resulting matrix comprised the headers. Because each header was a criterion that was identified by the authors, the resulting matrix was a 7×7 square matrix. Another matrix was then created by computing the ratio of the geometric means of scores of any two criteria (e.g., geometric mean of scores from criterion 1 to geometric mean of scores from criterion 2, geometric mean of scores from criterion 1 to geometric mean of scores from criterion 3, and so on and so forth). Therefore, each cell in the matrix was populated with a value that was based on the comparison between the raw Likert results for any two criteria.

The geometric means from the Likert survey were then converted to a "Saaty intensity scale" (Saaty, 1980). The Saaty intensity scale (Table 2) represents the relative importance between any two criteria on a nine-point scale. Table 3 illustrates the relationships utilized to convert the Likert-type scale to the Saaty scale, and shows the inverse relationships in the Saaty intensity scale. A value of 9 or 1/9 signifies that one criterion is nine times more important than the other. For example, if, for one criterion, the geometric mean from the scores of the Likert survey is equal to 1, then this value represents the largest possible difference between criteria, as the maximum possible score is 5 (i.e., 1/5). The value of 1 out of a possible score of 5, represents an "unquestionable support of the importance of one criteria over the

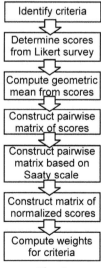

Fig. 1
The protocol for assessing criteria weights utilizing the Analytic Hierarchy Process.

Table 2 The fundamental scale of values representing the intensities of judgments between two groundwater management criteria

Intensity of Importance	Definition	Description
1/1	Equal importance	Two Criteria are of equal importance
2, 1/2	Weak importance	
3, 1/3	Moderate importance	Moderate importance of one criteria over the other
4, 1/4	Moderate plus importance	
5, 1/5	Strong importance	Strong Importance of one criteria over the other
6, 1/6	Strong plus importance	
7, 1/7	Very strong importance	Very strong importance of one criteria over the other, evidence based
8, 1/8	Very, very strong importance	
9, 1/9	Extreme importance	Unquestionable or demonstrated support of the importance of one criteria over the other

From Saaty, T.L., Vargas, L.G. 2006. Decision Making with the Analytic Network Process: Economic, Political, Social and Technological Applications with Benefits, Opportunities, Costs and Risks, Kluwer Academic Publisher, Dordrecht.

Table 3 The conversion of the Likert scale to the Saaty fundamental scale representing the intensities of judgments between two groundwater management criteria

Likert Importance	Saaty Scale
1/5 or 5/1	9 or 1/9
2/5 or 5/2	7 or 1/7
3/5 or 5/3	5 or 1/5
4/5 or 5/4	3 or 1/3
5/5	1

other" and is therefore assessed a value of 9 or 1/9 on the Saaty scale (Saaty, 2008). This process converts a fixed scale, with no measurable distance between the scores, to one in which the criteria have measurable relative values to each other (Saaty, 2008).

After a pairwise comparison matrix of the Likert data (Table 4) were converted to the Saaty Intensity Scale (Table 5), the pairwise comparison matrix was adjusted to create a normalized matrix (Table 6). This was performed by summing the values in each column of the matrix in Table 5 and dividing the values in each cell by the total of the column to yield normalized values (the sum of the normalized values in each column must equal one). The weights for the criteria are then determined by computing the average values from each row that represents a criterion in the normalized matrix (Table 6).

Table 4 Pairwise comparison matrix of the seven water management criteria using scores from a 5-point Likert scale

	C1	C2	C3	C4	C5	C6	C7
C1	1.00	0.82	0.87	0.91	0.82	0.76	0.67
C2	1.22	1.00	1.05	1.11	1.00	0.92	0.81
C3	1.15	0.95	1.00	1.05	0.95	0.87	0.77
C4	1.10	0.90	0.95	1.00	0.90	0.83	0.73
C5	1.22	1.00	1.05	1.11	1.00	0.92	0.81
C6	1.32	1.09	1.15	1.21	1.09	1.00	0.88
C7	1.50	1.23	1.30	1.37	1.23	1.13	1.00

Table 5 Pairwise comparison matrix converted to a Saaty fundamental scale

	C1	C2	C3	C4	C5	C6	C7
C1	1	3	2	2	3	2	4
C2	1/3	1	1	1/2	1	2	3
C3	1/2	1	1	1	2	2	3
C4	1/2	2	1	1	2	3	4
C5	1/3	1	1/2	1/2	1	2	3
C6	1/2	1/2	1/2	1/3	1/2	1	2
C7	1/4	1/3	1/3	1/4	1/3	1/2	1

Table 6 Normalized pairwise comparison matrix of the seven groundwater management criteria on a Saaty fundamental scale

	C1	C2	C3	C4	C5	C6	C7	Weight
C1	0.29	0.34	0.32	0.36	0.31	0.16	0.20	0.28
C2	0.10	0.11	0.16	0.09	0.10	0.16	0.15	0.12
C3	0.15	0.11	0.16	0.18	0.20	0.16	0.15	0.16
C4	0.15	0.23	0.16	0.18	0.20	0.24	0.20	0.19
C5	0.10	0.11	0.08	0.09	0.10	0.16	0.15	0.11
C6	0.15	0.06	0.08	0.06	0.05	0.08	0.10	0.08
C7	0.07	0.04	0.05	0.04	0.03	0.04	0.05	0.05

Note that judgment scoring in the pairwise comparison is not always consistent (Saaty, 1980, 1990, 2008; Saaty and Vargas, 1991, 2006; Chen et al., 2010). Therefore, prior to accepting the weights, a degree of consistency or a Consistency Ratio was determined by computing the ratio of a Consistency Index (CI) to a Random Consistency Index (RI). This check assures that the subjective score from the Likert survey and the associated Saaty comparisons are consistent.

The CI indicates the consistency of judgment in the pairwise matrix and is given by:

$$\text{CI} = (\lambda - n)/(n - 1) \tag{1}$$

where λ is the product of the reciprocal of the normalized values for each criterion in the matrix and the average of each row in the matrix (Saaty, 1990), and n is the number of criteria. The RI is a randomly generated pairwise comparison matrix using the same 1–9 relative importance scale of a sample of 500 randomly generated matrices (Saaty, 1990). A ratio of less than 0.1 indicates a satisfactory degree of consistency, allowing for the computed weights to be accepted (Saaty, 1990). The reader is referred to Saaty (1990) for a complete description of the method, as well as the RI reference table used in the process.

3 Results and Discussion

As accessible potable water resources become threatened in coastal regions, adequate and accurate decision-making methods are needed to develop strong, equitable, and effective groundwater management strategies. Consideration of the many criteria involved with water management has proven to be a successful, integrative, and comprehensive tool that respects the myriad of physical and socio-economic dimensions of water management, and yields well informed and enforceable strategies adaptable to the long-term sustainability of aquifers (Hajkowicz and Collins, 2007; Hajkowicz and Higgins, 2008). However, it has been unclear how to best assess weights to criteria, and whether there is consensus among water professionals about which criteria are the most important for managing groundwater, and, how the degree of importance varies among these criteria.

The results from the survey that queried water professionals regarding their perception about the relative importance of seven physical and socio-economic criteria that impact how groundwater is managed and allocated indicate that responses pertaining to all the criteria, excluding political will, had median scores of 4 or 5 on a Likert scale (Table 7). These results illustrate that water professionals valued all of the criteria (excluding political will) as either important or extremely important. Of all the criteria, hydrogeologic, and aquifer characteristics (C1), and the available groundwater volume (C4) were perceived by the participants to have been extremely important, with median scores that were equal to 5 on the Likert scale. In contrast, political will (C7) had a lower median score of 3 on the Likert scale, suggesting that the survey participants perceived this criterion to be moderately important.

Socio-economic demands and needs (C2), population (C3), alternate water sources (C5), and financial capability (C6) had median scores of 4 (very important) on the Likert scale. For completeness, and as an illustration of how the descriptive statistics compare to each other, the geometric mean, median, mode, and variance are included in Table 7. The geometric mean and the mode align with the patterns displayed by the medians (the statistics for hydrogeologic and aquifer characteristics (C1) and the available groundwater volume (C4) are generally higher than the statistics for the other criteria). Although viewed as a confirmation of the validity of the criteria choices, the raw Likert scores give little indication of the relative importance of the

Table 7 Summary statistics illustrating the relative importance of seven groundwater management criteria

	C1 Hydrogeologic and Aquifer Characteristics	C2 Socioeconomic Demands and Needs	C3 Population	C4 Available Groundwater	C5 Alternate Sources	C6 Financial Capability	C7 Political Will
Geometric mean	4.5	3.7	3.9	4.1	3.7	3.4	3
Median	5	4	4	5	4	4	3
Mode	5	4	5	5	4	4	3
Variance	0.67	0.9	0.9	0.87	1.08	0.95	1.48

criteria, and the data do not allow for direct and quantifiable comparisons of criteria. The results from the AHP (see the following) resolve this issue.

The respondents to the questionnaire indicated that their primary work responsibilities were in 46 states of the United States and 22 other countries. For the respondents that primarily worked in the United States, approximately 12% worked in the Northeast, 16% in the Midwest, 23% in the West, and 49% in the South. Results from the Kruskal-Wallis test reveal that there were no significant differences in the perceptions of water professionals from different geographic locations (Fig. 2). This result suggests that there was general agreement about the level of importance for groundwater management criteria, regardless of geographic location.

Approximately 4% of the respondents self-identified their work sector as nongovernmental organization/nonprofit, 13% as academic, 22% as government, 36% as industry, and 25% as "other." The results indicate that there is little variation in the distribution of responses from groundwater professionals to all criteria on the basis of work sector. Comparable to the results from the responses about various work locations, the results from the Kruskal-Wallis test of significance reveal that there were no significant differences in the perceptions of water professionals from different work sectors (Fig. 3). This result suggests that there was general agreement about the level of importance for groundwater management criteria, regardless of the sector in which the respondents work.

As shown in the boxplots (Figs. 2 and 3), there is very little difference in the responses from either group, illustrating that water professionals from different locales and work sectors have a common perception of important groundwater management criteria. In all cases, the P value is greater than 0.05, indicating weak evidence against the null hypothesis. These results suggest that, because neither the location (Fig. 2), nor work sector (Fig. 3) impacted the answers of the water professionals as to the importance of the various criteria, the criteria chosen seem to be of universal importance among the sample population.

A check of the consistency ratio following the creation of a matrix with the Saaty (1980) intensity scale shows that the comparison judgments are consistent, as the computed ratio is 0.03. The judgments are consistent because the research was designed to avoid requiring the water professionals make direct comparisons of criteria. Allowing water professionals to make direct comparisons of criteria would have resulted in inconsistent comparisons, as judgments by different individuals would likely have resulted in different weight values. The average values that represent the weights of all the criteria are therefore acceptable because the consistency ratio is less than 0.1.

The Likert-type survey was originally employed to prevent the water professionals from having to directly assess weights to the criteria. Having water professionals directly assess weights has proven to be biased and unreliable (Goldstein, 1990; Hajkowicz and Higgins, 2008; Korhonen et al., 2013), and often giving inaccurate and misleading results (Rowe and Pierce, 1982). Using the

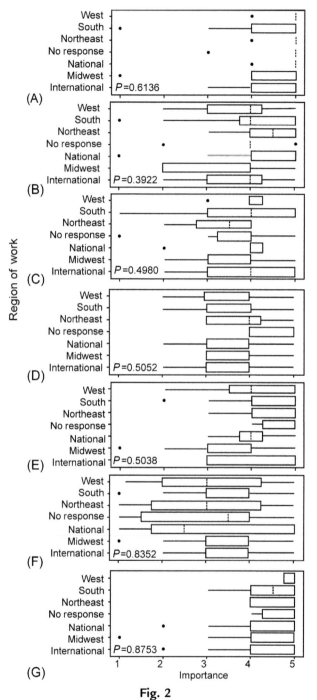

Fig. 2

Box plots illustrating how groundwater professionals perceived the importance of Criteria 1–7 by location (A) C1—hydrogeologic and aquifer characteristics, (B) C2—socio-economic demands and needs, (C) C3—population dependent on the groundwater resources, (D) C4—available groundwater volume, (E) C5—availability of alternate water sources, (F) C6—financial resources, and (G) C7—political motivation and support. *Thick vertical lines* in each box plot represents the median. The *dots* represent outliers.

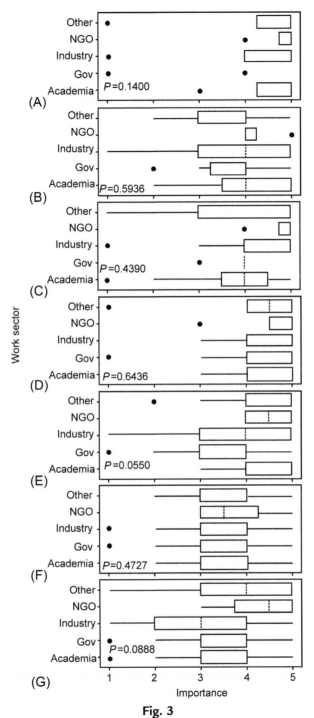

Fig. 3

Box plots illustrating the variations of groundwater professional's perceptions to Criteria 1–7, by work sector (i.e., Other, NGO (Non-Governmental Agency or Nonprofit), Industry, Gov (Governmental Agency), and Academia). (A) C1—hydrogeologic and aquifer characteristics, (B) C2—socio-economic demands and needs, (C) C3—population dependent on the groundwater resources, (D) C4—available groundwater volume, (E) C5—availability of alternate water sources, (F) C6—financial resources, and (G) C7—political motivation and support. *Thick vertical lines* in each *box* plot represents the median. The *dots* represent outliers.

AHP, the authors used the geometric means from Likert surveys to compute the importance of each criterion and how each criterion compares with other criteria. As such, the researchers assigned larger weights to the more important criteria, as determined through the AHP.

The weights that were derived through the AHP (Table 6) were compared with the weights acquired from two traditional techniques in order to determine how well the AHP performed in establishing weights for groundwater management criteria (Table 8). The traditional techniques are (a) a ranking based on median scores from Likert surveys, and (b) a simple ratio of the sum of scores for each criterion to the sum of all criteria scores from Likert surveys. Had a ranking approach based on median scores from Likert surveys been used, the rankings would have revealed that hydrogeologic and aquifer characteristics (C1) and available groundwater (C4) would have had the highest ranking, with social and economic demands and needs (C2), population (C3), availability of alternate water sources (C5), and financial capability (C6) being of intermediate ranking. Political will (C7) would have had the lowest ranking. Had the traditional simple ratio method been used, the weights would have had little variation in the ranges (between 11 and 17). However, when the AHP is utilized, the hydrogeologic and aquifer characteristics criterion (C1) is perceived to be the most important, followed by available groundwater (C4), population (C3), socio-economic demands and needs (C2), alternate water sources (C5), financial capability (C6), and political will (C7), respectively (Table 8). Although the relative importance of the criteria (i.e., rankings) is the same with the simple ratio method and the AHP, the AHP reveals that the hydrogeologic and aquifer characteristics criterion has a weight of approximately 1.5 times more than the criterion with the next largest weight (available groundwater (C4)) and approximately 6 times the weight of the criterion with the least weight (political will (C7)) (Table 8). These characteristics are not evident in the ranking and simple ratio techniques.

Political will was perceived to be of low importance, yet 26% of those who responded with "optional" comments mentioned regulatory or political concerns. Although this may indicate the respondents' recognition of a need for, and difficulty of working within the constraints of the political system, there appears to be a disdain for the importance of political will, and an unwillingness to accept the relevance of political will to groundwater management. For

Table 8 Comparison of weights from AHP and traditional techniques

Criterion	Median Ranking	Simple Ratio	AHP
C1	1	16.64	28.16
C2	3	13.98	12.43
C3	3	14.63	15.86
C4	1	15.73	19.33
C5	3	14.12	11.30
C6	3	12.96	8.18
C7	7	11.94	4.75

example, a respondent from the West stated that there is no real political will, including that of the Governor of the US state of California, to provide other sources for drinking water resources (e.g., desalination). The respondent reasoned that this was an economic decision based solely on costs for construction and maintenance. Furthermore, the respondent added that either water agencies or the "first users" own all groundwater in California, and the courts have ruled that the state cannot regulate the consumptive use by the "first users," inferring that the political authorities are either unwilling or unable to affect change. Despite such comments that highlight the structural rigidities of groundwater management, it is interesting that water professionals, regardless of location, placed a low importance on political will.

Regulatory framework, existing legislation, and regulations weighed heavily with water professionals, especially those in the West and the Northeast. Many mentioned the futility of changing the existing regulatory structure or previous ownership claims. "Who claimed ownership first? Who has legal access to the resource? Does the current owner's use impact other current users' abilities to use the resource? Does the new proposed use impact current users' abilities to use the resource?" "Can management work within existing laws and regulations or are changes required? If the latter, are such changes politically feasible?" It is interesting to note that although the respondents highlighted several structural impediments to managing groundwater, the water professionals highly valued hydrogeologic and aquifer characteristics (C1) as the most important factor to consider when evaluating water management strategies (weight=28.16%), with political will (C7) as the least important consideration (weight=4.75%) (Tables 7 and 8).

The criteria and associated weights used in this study should not be considered as rigid constructs to be used across all spatial and temporal scales, but as guidelines for groundwater management assessments that are particularly applicable when considering environmental changes, as well as responses of humans to those changes (e.g., adaptation to environmental change, subsidence, threats from saltwater intrusion, etc.). Thus, the criteria and weights presented in this study may change depending on who uses water, what the water is used for, and time periods over which water management issues are to be considered. This is because water demands fluctuate with changes in ecological (e.g., as a consequence of climate change), political, regulatory, religious, cultural, and technological aspects of water use (Gleick, 1998).

The data from the survey represents an incomplete picture of important groundwater management criteria. There was agreement across work sectors and work locales that minimized the importance of political will. This common disdain for the necessities of working within the political system may be a hindrance to developing new management approaches that need to be promoted and funded by political entities. The surveyed water professionals ostensibly have difficulty in seeing beyond their own disciplines and fields of interest, with an apparent disconnect between political realities and the belief that hydrology is by far the most important consideration in crafting management schemes. This result has important implications for consensus on aquifer management. On regional scales, water professionals are likely to agree on large-scale strategies. Agreement on the importance of each criterion may encourage professionals to view management on an aquifer-wide scale, crossing geopolitical boundaries, with considerations of the unique hydrogeologic problems facing coastal systems, thereby crafting "big picture" approaches. Perhaps the charge for water professionals is the establishment of the vision, creation of performance standards, establishment of monitoring networks, and the creation of specific hydrogeologic tactics to manage the aquifers. However, because the water professionals do not acknowledge the importance of political will, and daily management, coordination and oversight might be most effective if left on a local stakeholder scale. The reality is that sustainable management strategies in coastal regions are developed and moved forward through political will, for without political support, funding, permits, and supporting management infrastructure are difficult, if not impossible, to obtain or maintain. Until water professionals understand and reconcile the multidimensional nature of water management, the hydrogeologic issues in coastal regions will be difficult to tackle (Kay and Adler, 2005).

4 Shortcomings

Likert-type scaled surveys are typically employed to explore attitudes and perceptions (Likert, 1932; Dittrich et al., 2007). The respondents answer a series of questions in which they provide their perceptions of the relative strength of importance to various statements, and although it is a common method to ascertain attitudinal information, there is no quantitative measure of the difference between respondents' answers. The Likert-type scale was employed to allow respondents to assess their perception of each criterion without regard for the other criteria. As such, the importance of one criterion did not diminish the importance of the others.

The criteria introduced in this research study are intended as a preliminary position for management discussions, but are not intended to constitute an exclusive list. Other criteria suggested by respondents included water budgets, available infrastructure, competing current and future uses, ownership, quality and treatment, conservation, sustainability, water rights, and impacts to ecology. These are important considerations in coastal regions. In this study, omission of criteria that may be deemed important for sustainability by other water professionals does not necessarily lessen the importance of any other criterion. Because each criterion was scored independently, the resulting scores did not influence the impressions of the respondents toward the importance of other criteria (Hajkowicz & Higgins, 2008). For example, a score of 5 (extremely important) on one criterion, does not negate the importance of another criterion. It should be noted, however, that if new criteria were added to the list presented in this study, then the relative weights of the criteria that were computed herein would change. This should be taken into account if researchers are to include additional criteria to the ones suggested in this chapter.

5 Conclusions

The novelty and value of this research study is that it addresses questions about how groundwater management criteria in coastal regions should be analyzed, how the perceptions vary by water professionals, and how the degree of importance varies among the criteria. The use of the AHP in establishing weights of groundwater management criteria is also a novel contribution. Although the results of the study reveal that that the respondents to the survey perceived hydrogeologic and aquifer characteristics, as well as the available groundwater volume in the area of interest to be of the highest importance, it is important to note that the hydrogeologic and aquifer characteristics were 1.5 times as important as the available groundwater volume. Here, political will was perceived to have the least importance, although many respondents commented on its relevance and implications.

Coastal regions are complex and interconnected, creating groundwater concerns most effectively managed when strategies are adapted to the appropriate spatial and temporal scales. Groundwater issues impacting the local coastline areas are different than those issues viewed at a more regional scale. The results of the survey of water professionals highlight the need for management on the various scales. Those surveyed represent professionals working on different scales. They also represent professionals working domestically and internationally. When given a common set of criteria for managing groundwater, water professionals agree on the most and least important criteria. The results also show that the perceptions of water professionals concerning the importance of groundwater management criteria were similar across work sector and work locale.

The focus on hydrogeologic criteria and downplaying of socio-political criteria may be a function of the demographics of the respondents. Largely, the water professionals who were surveyed in this study consisted of geologists and engineers. Missing from this study are other stakeholders such as mayors, council members, financial officers, and other decision makers whose responsibilities it is to improve, update, or expand existing water sources for coastal communities. Thus, future studies might involve a comparison of the perceptions of water professionals with other stakeholder groups. For example, water purveyors and members of water boards could contribute ideas on data sharing and the willingness to participate in collaborative projects. Also, end users could be surveyed to unravel their interest and willingness to take part in conservation measures that minimize how much water is used during various water consumption activities. It is envisioned that multidimensional, mixed-methods approaches such as the AHP and the textual analysis presented in this chapter will contribute to the sustainable use of coastal groundwater resources to meet current and future demands of water by providing means for various stakeholders to focus on the most important criteria for groundwater management.

Acknowledgments

This work was supported by funding from East Carolina University. The authors thank the American Institute of Professional Geologists for their assistance in distributing the criteria survey.

References

Alharbi, S.H., Sayed, O.A., 2017. Measuring services quality: Tabuk municipal. British J. Econ. Manag. Trade 17 (2), 1–9.
Babakus, E., Mangold, G., 1992. Adapting the SERVQUAL scale to hospital services: an empirical investigation. Health Service Res. 26, 767–780.
Banks, H.O., 1953. Utilization of underground storage reservoirs. T Am Soc Civ Eng 118, 220–234.
Cabrere, E., Cobacho, R., Vicent, E.-G., Aznar, J., 2011. Analytical hierarchical process (AHP) as a decision support tool in water resources management. J Water Supply Res T. J. 60 (6), 343–351.
Calizaya, A., Meixner, O., Bengtsson, L., Berndtsson, R., 2010. Multi-criteria decision analysis (MCDA) for integrated water resources management (IWRM) in the Lake Poopo Basin, Bolivia. Water Resour. Manag. 24, 2267–2289.
Chen, Y., Yu, J., Khan, S., 2010. Spatial sensitivity analysis of multi-criteria weights in GIS-based land suitability evaluation. Environ Modell Soft 25, 1582–1591.
Clark, V.L., Creswell, J.W., Green, D.O., Shope, R.J., 2008. Mixing quantitative and qualitative approaches: an introduction to emergent mixed methods research. In: Hess-Biber, S.N., Leavy, P. (Eds.), Handbook of Emergent Methods. Guilford, New York, pp. 363–387.
Conkling, H., 1946. Utilization of ground-water storage in stream system development. T Am Soc Civ Eng 3, 275–305.
Cosgrove, W.J., Rijsberman, F.R., 2014. World Water Vision: Making Water Everybody's Business. Routledge, London.
Creswell, J.W., Clark, V.L.P., 2007. Designing and Conducting Mixed Methods Research. Sage Publ, Thousand Oaks.
Denzin, N.K., Lincoln, Y.S., 2013. The Landscape of Qualitative Research: Theories and Issues. Sage Publ, Thousand Oaks.
Dittrich, R., Francis, B., Hatzinger, R., Katzenbeisser, W., 2007. A paired comparison approach for the analysis of sets of Likert-scale responses. Stat. Model. 7 (1), 3–28.
Fetter, C.W., 1988. Applied Hydrogeology. Merrill Publishing Company, Columbus, OH, pp. 450–452.
Freeze, R.A., Cherry, J.A., 1979. Groundwater. Prentice Hall, Englewood Cliffs, NJ, pp. 364–365.
Gleeson, T.J., Vander Steen, M.A., Sophocleous, M., Taniguchi, M., Alley, W.M., Allen, D.M., Zhou, Y., 2010. Groundwater sustainability strategies. Nat. Geosci. 3, 378–379.
Gleick, P.H., 1998. Water in crisis: paths to sustainable water use. Ecol. Appl. 8 (3), 571–579.
Goldstein, W.M., 1990. Judgements of relative importance in decision making: global vs local interpretations of subjective weight. Organ Behav Hum Dec 47 (2), 313–336.
Green, T.R., Taniguchi, M., Kooi, H., Gurdak, J.J., Allen, D.M., Hiscock, K.M., Treidel, H., Aureli, A., 2011. Beneath the surface of global change: impacts of climate change on groundwater. J. Hydrol. 405 (3), 532–560.
Hajkowicz, S., Collins, K., 2007. A review of multiple criteria analysis for water resource planning and management. Water Resour. Manag. 1553–1566.
Hajkowicz, S., Higgins, A., 2008. A comparison of multiple criteria analysis techniques for water resource management. Eur. J. Oper. Res. 184, 255–265.
Hughes, C.E., Cendón, D.I., Johansen, M.P., Meredith, K.T., 2011. Climate change and groundwater. In: JAA, J. (Ed.), Sustaining Groundwater Resources. International Year of the Planet, Springer Science+Business Media, pp. 97–118.
Hunter, P.R., MacDonald, A.M., Carter, R.C., 2010. Water supply and health. PLoS Med. 7 (11), 1–9.
International Law Association, (ILA), 1966. In: The Helsinki rules on the uses of the waters of international rivers. *Report of the Fifty-Second Conference* (Helsinki). International Law Association, London.
International Law Association, (ILA), 2004. In: The Berlin rules on water resources. *Report of the Seventy-First Conference* (Berlin). International Law Association, London.
International Panel on Climate Chane (IPCC), 2014. Climate Change 2014: Synthesis Report. Contribution of Working Groups I, II and III to the Fifth Assessment Report of the Intergovernmental Panel on Climate Change

[Core Writing Team, R.K. Pachauri and L.A. Meyer (Eds.)], IPCC, Geneva, Switzerland, 151 pp. http://www.ipcc.ch/report/ar5/syr/. Accessed September 2018.

Kay, R., Adler, J., 2005. Coastal Planning and Management. Taylor and Francis, New York.

Klein, W.A., Manda, A.K., Griffin, M.T., 2014. Refining management strategies for groundwater resources. Hydrgeol. J. 22, 1727–1730.

Korhonen, P.J., Silvennoinen, K., Wallenius, J., Öörni, A., 2013. A careful look at the importance of criteria and weights. Ann. Oper. Res. 211, 565–578.

Lee, C.H., 1915. The determination of safe yield of underground reservoirs of the closed basin type. T Am Soc Civ Eng 78, 148–151.

Li S, Sun A (2017) IOP Conf. Ser: Earth and Environ Sci, 51.

Likert, R., 1932. A technique for the measurement of attitudes. Arch Sci Psychol 410, 1–55.

Mays, L.W., 2013. Groundwater sustainability: Past, present, and future. Water Resour. Manag. 27, 4409–4424.

Meinzer OE (1923) Outline of ground-water hydrology, with definitions. *USGS Water Suppl. Pap.*, 494, 55–56.

Panagoloulos, G.P., Bathrellos, G.D., Skilodimou, H.D., Martsouka, F.A., 2012. Mapping urban water demands using multi-criteria analysis and GIS. Water Resour. Manag. 26, 1347–1363.

Rosegrant, M.W., Cai, X., Cline, S.A., 2002. Global Water Outlook to 2025. Averting an Impending Crisis. IWMI, Colombo.

Rowe, M.D., Pierce, B.C., 1982. Sensitivity of the weighted summation decision method to incorrect application. Socioecon. Plann. Sci. 16 (4), 173–177.

Saaty, T.L., 1980. The Analytical Hierarchy Process: Planning, Priority Setting, Resource Allocation. McGraw Hill, New York.

Saaty, T.J., 1990. How to make a decision: the hierarchy process. Euro J Op Res 48, 9–26.

Saaty, T.L., 2008. Decision making with the analytic hierarchy process. Int. J. Serv. Sci. 1 (1), 83–98.

Saaty, T.L., Vargas, L.G., 1991. Prediction Projection and Forecasting. Springer Science+Business Media, New York, pp. 1–26.

Saaty, T.L., Vargas, L.G., 2006. Decision Making with the Analytic Network Process: Economic, Political, Social and Technological Applications with Benefits, Opportunities, Costs and Risks. Kluwer Academic Publisher, Dordrecht.

Salman, S.M.A., 2007. The Helsinki rules, the UN watercourses convention and the Berlin rules: perspectives on international water law. Water Resour. Dev. 23 (4), 625–640.

Schwartz, F.W., Zhang, H., 2003. Fundamentals of Groundwater. John Wiley & Sons, Inc, New York, NY, pp. 337–338.

Taylor, R.G., Scanlon, B., Döll, P., Rodell, M., van Beek, R., Wada, Y., Longuevergne, L., Leblanc, M., Famigliette, J.S., Edmunds, M., Knoikow, L., Green, T.R., Chen, J., Taniguchi, M., Bierkens, M.F.P., MacDonald, A., Fan, Y., Maxwell, R.M., Yechieli, Y., Gurdak, J.J., Allen, D.M., Shamsuddha, M., Hiscock, K., Yeh, P.J.F., Holman, I., Treidel, H., 2013. Ground water and climate change. Nat. Clim. Change 3, 322–329.

Theis CV (1935) The relation between the lowering of the piezometric surface and the rate and duration of discharge of a well using ground-water storage: American Geophysical Union Transactions, 16th Annual Meeting, vol 16 (pt. 2), p. 519–524.

Theis, C.V., 1940. The source of water derived from wells-essential factors controlling the response of an aquifer to development. Civ. Eng. 10 (5), 277–280.

United Nations (UN), 1997. General Assembly resolution 51/229 (1997). In: Convention on the law of the non-navigational uses of international watercourses. General Assembly of the United Nations, pp. 4–6. http://www.thewaterpage.com/UN_Convention_97.html. [(Accessed 10 December 2014)].

United Nations (UN), 2007. The Millennium Development Goals Report. United Nations, New York. Available: http://www.un.org/millenniumgoals/pdf/mdg2007.pdf. [(Accessed May 2016)].

World Health Organization (WHO), 2004. Guidelines for Drinking-Water Quality, third ed. Vol. 1. WHO, Geneva.

Further Reading

Global Water Partnership, 2000. Integrated Water Resources Management. TAC Background Papers 4.

CHAPTER 19

Interlinking of Rivers as a Strategy to Mitigate Coeval Floods and Droughts: India in Focus With Perspectives on Coastal Zone Management

Mu. Ramkumar*, David Menier[†], K. Kumaraswamy[‡], M. Santosh[§,¶], K. Balasubramani[||], Rathinam Arthur James[#]

*Department of Geology, Periyar University, Salem, India, [†]Géosciences Océan Laboratory UMR CNRS 6538, University of South Brittany, Vannes Cedex, France, [‡]Department of Geography, Bharathidasan University, Tiruchirappalli, India, [§]School of Earth Sciences and Resources, China University of Geosciences Beijing, Beijing, P.R. China, [¶]Department of Earth Sciences, University of Adelaide, Adelaide, SA, Australia, [||]Department of Geography, Central University of Tamil Nadu, Thiruvarur, India, [#]Department of Marine Sciences, Bharathidasan University, Tiruchirappalli, India

1 Introduction

India is the seventh largest country in the world, covering an area of 3.29 million km^2. Located exclusively in the northern and eastern hemispheres, the mainland of India stretches from latitude 8°4′ north to 37°6′ north, and from longitude 68°7′ east to 97°25′ east of Greenwich. It extends about 3214 km N-S and about 2933 km E-W. India has a coastline of 6100 km on the main land and an additional 1417 km long coastline comprising the Lakshadweep Andaman and Nicobar Island groups. Physiographically, the mainland constitutes several distinct units, namely, the Great Mountains of the North, the Indo-Gangetic Plain, the Peninsular Plateau, the Coastal Plains, the Thar Desert, and the island groups. Along with these diverse physiographic features are the river systems emanating from glacial melt water, monsoons, and a mix of both, that produce distinct climatic-physiographic-ecosystems of their own.

According to the Geological Survey of India (http://www.portal.gsi.gov.in/portal/), the flood plains of the rivers Ravi, Sutlej, Yamuna-Sahibi, Gang, Gandak, Gaggar, Teesta, Kosi, Brahmaputra, Mahananda, Mahanadi, Damodar, Mayurakshi, Godavari, and Sabarmati and their tributaries and distributaries are prone to flooding. Most of the flood affected areas lie in the Ganga Basin, Brahmaputra Basin, the northwestern river basins comprising the Jhelum,

Chenab, Ravi, Sutlej, Beas and Ghagra, the Peninsular river basins comprising the Tapti, Narada, Mahanadi, Baitarani, Godavari, Krishna, Pennar, and the Kaveri. Owing to flooding, an estimated area of 7.181 million hectares of land, and the population, and infrastructures are affected (Ramkumar, 2009), and there seems to be a reduction in area and intensity of flooding from the north toward the south. Ironically, the interval between successive occurrences of droughts is longer in the northern region, and shows a distinct reduction toward the south. Owing to its unique geography and rainfall pattern, the Indian subcontinent experienced severe droughts all through documented history. According to the India Meteorological Department, the country experiences drought once every 4 years. Geographically, the repetition of droughts increases from north toward the south: Assam-once in 15 years; West Bengal, Madhya Pradesh, the Konkan region of Maharashtra, Bihar, and Orissa-once in 5 years; the southern Karnataka, eastern Uttar Pradesh, and Vidharba regions-once in 4 years; Gujarat, eastern Rajasthan and western Uttar Pradesh-once in 3 years; Tamil Nadu, Jammu, and Kashmir, western Rajasthan, and the Telangana regions—once in two and half years. These observations, along with the data on temporal occurrences of flooding and drought, often show their occurrences almost simultaneously (Suresh and Ramkumar, 2009) and/or in successive years, such as in the years 1860, 1861, 1917, 1918, 1941, 1942, 1971, 1972, 1987, 1988, and so forth. Despite being home to one of the ancient civilizations adept with well-planned cities, higher-educational institutions, and a thriving agrarian climate; the bludgeoning population, industrialization, urbanization, and haphazard utilization of natural resources led India into a situation of coexistent ironies, in terms of economy, ecology, and so forth. On a geological note, coeval droughts in one part of the country, while another part of the country reels under flooding, often bring the populace, and rescue and remedial personnel to their knees, through causing loss, and irreparable damage to lives, infrastructure, and agriculture (Suresh and Ramkumar, 2009).

India has ironic, dual problems related to water: its dependence on highly variable population density, and monsoon-rainfall creates an imbalance of water availability. For example, according to a Ministry of Water Resources report, in 2010 the average per capita availability of water in the Ganga-Brahmaputra-Meghna system was 20,136 cubic meters per year, compared with 263 cubic meters in the Sabarmati Basin. Among the populace of this country, one in three reels under drought, while one in eight is affected by flood! Given these scenarios, linking glacier-monsoon fed perennial rivers of the north with that of monsoonal rivers of south has been proposed as a measure to effectively mitigate coeval floods and droughts in India. On completion, the mammoth program of interlinking of rivers (ILR) will be the world's largest irrigation infrastructure project, linking 37 rivers through 30 links through the establishment of about 15,000 km of new canals and 3000 dams of various sizes. The program has two components: the Himalayan rivers component, with 14 links, and the peninsular component, with 16 links, which will transport 33 and 141 trillion L of water, respectively, per year. The estimated irrigation potential to be created is 34 million hectares of land, while this project is expected to provide drinking water to the five metro areas and 101 districts. The expected hydro power irrigation is 34,000 MW. In addition, flood control, navigation, drinking water supply, salinity control, and so forth, are also envisaged through this program.

Rivers have been prime sources of sustenance for mankind since the advent of civilization, and humans have continued reaping the benefits provided by rivers for centuries, without understanding much on how the river ecosystem functions and maintains its vitality (Naiman and Bilby, 1998; Subramanian, 2002). The rivers play(ed) significant roles in the provision of water for domestic, agricultural, and industrial activities (Ayivor and Gordon, 2012), and generate(d) sediments and nutrients for the sustenance of the natural ecosystem. Traditionally, the river basins are treated as treasure troves for natural resources, but human needs have taken precedence over environmental concerns (Triedman, 2012). In view of the prevailing favorable conditions for habitation, cultivation, and industry, the explosive growth of the human population and resultant pressure on the natural environment is high in river basins (Zarea and Ionus, 2012). Within a river basin, the deltaic and coastal regimes, being at the receiving basin, act as an interface between fluvial, oceanographic, atmospheric, and anthropogenic dynamics. These traits make this region ecologically fragile and susceptible to environmental deterioration very easily, even by changes in the catchment (pollution, siltation, flooding, etc.), and ocean (inundation, erosion, accretion, etc.) and atmospheric equilibrium. Any change in the factors that influence the equilibrium results in recognizable changes in the system, including adverse impacts such as flooding, erosion (Walling, 1999), and desertification, which in turn, may cause a loss of critical resources that provide sustenance to the human race such as land, agricultural produce, and other commercial activities. In addition, it also alters the nutrient availability in the deltaic region, and contributes to the proliferation of exotic species in coastal regions, that in turn adversely affect the people who depend on the normal natural processes in downstream regions (e.g., Wu et al., 2008). In a classic review, Barrow (1998) stated that except for the most arid and cold regimes, the world's landscape can be divided into distinctly mappable river basins of various scales (large, medium, and small) and the river basins themselves can be subdivided into upper, middle, and lower basins, based on hydrologic and geomorphic characteristics. Because each river basin acts as a holistic system in tune with the climatic, geological, and anthropogenic interactions, any study on the river system should understand the dynamics of the river basin as a whole. A river basin is a basic geographic and climatological unit within which the vagaries of natural processes act and manifest at different spatio-temporal scales. However, even if juxtaposed, no two river basins respond to natural processes in a similar way, and thus, each river basin is unique. Hence, any developmental activity or conservation effort has to be designed and implemented unique to each river basin. At this juncture, linking the diverse river systems that are distinct in terms of climate, geomorphology, geology, structure, land use, source of water, monsoon systems, biodiversity, human populace, and so forth, as envisaged in the ILR seems unprecedented. Additionally, the program poses constraints on the assessment of probable environmental, ecological, and other feedback, and strategizing remedial measures. Recognizing this, we explore the interlinked nature of river systems and coastal stability, and probable impacts, and suggest necessary mitigation measures with reference to the proposed ILR.

2 The History of ILR Program

The plan to interlink India's rivers is not new, and was proposed in 1858 by a British irrigation engineer, Arthur Thomas Cotton. While this proposal died its natural death due to colonial powers, there were similar proposals in the history of independent India; first in the name of the National Water Grid Project in 1972, and second, in the name of the Garland Canal Project in 1977, which were also relegated to oblivion for various reasons.

Though these proposals were under discussion during every major catastrophic flood and drought, serious discussions were initiated during a 3-day conference, "Jal Manthan," organized in New Delhi from November 20–22, 2014. The conference deliberated about issues related to the "Interlinking of Rivers Program." Later, a 2-day conference, "Jal Manthan-2," was organized in New Delhi from February 22–23, 2016, and discussions were focused on Water Management, Coordination between Centers and States, Water Conservation, Innovations in Water Governance, Interlinking of Rivers, and so forth. In a ruling during 2002, the Supreme Court of India directed the Government to interlink the rivers within 10 years, and later in 2012, the Court directed the Government to constitute a committee for ILR. Accordingly, a Cabinet Note on ILR was submitted to the Union Cabinet on July 16, 2014, which approved the constitution of the Special Committee on ILR on July 24, 2014. Later, subcommittees for comprehensive evaluation of various studies and reports, system studies for identification of alternative plans, restructuring of the National Water Development Agency (NWDA), and negotiations and agreements between states were formed. A Committee on Intra-State River Links was constituted by the ministry on March 12, 2015, followed by formation of a Task Force for Interlinking of Rivers on April 13, 2015.

Then, ILR was declared as a program of national importance to ensure greater equity in the distribution of water by enhancing the availability of water in drought prone and rainfed areas. Under this program, the Ministry of Water Resources has identified 30 links (Fig. 1), out of which 14 are under the Himalayan Rivers Component, and 16 are under the Peninsular Rivers Component. The nodal agency, NWDA, has received 46 proposals of intrastate links from 9 States, namely, Maharashtra, Gujarat, Jharkhand, Orissa, Bihar, Rajasthan, Tamil Nadu, Karnataka, and Chhattisgarh, and as of March 2015, out of these 46 proposals, prefeasibility reports for 35 intrastate links were completed. On July 6, 2017, India's ambitious plan to interlink rivers to achieve greater equity in the distribution of water in the country reached an important milestone as the water from the Godavari met the fourth-longest river, the Krishna. The two became the first of 30 rivers to be part of the ILR program.

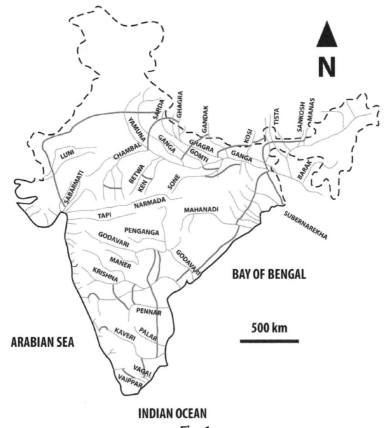

Fig. 1
River systems of India and the distribution of proposed links.

3 Regional Setting and Dynamics

3.1 Geology of the Indian Subcontinent

The Indian subcontinent (Fig. 1) is a storehouse of ~3.8 Ga of geological history spanning from the Eoarchean period to the Recent period, assembling crustal blocks of various ages at different times along major orogenic belts (Santosh et al., 2014, 2015), and involving extensive volcanism, plutonism, metamorphism, sedimentation, and tectonic deformation. The basement rocks of the Indian Peninsular Shield are dominantly composed of granites and granitic gneisses. The exposed Precambrian basement comprises four major cratons, namely, the Dharwar, Aravalli, Singhbhum, and Bastar (Figs. 2 and 3). In the southernmost part of the Peninsula, south of the Dharwar Craton, is a series of crustal blocks ranging in age from Archean to latest Neoproterozoic, welded together by intervening suture zones, and it is termed the Southern Granulite Terrane (Santosh et al., 2015 and references therein). Sedimentary successions of Proterozoic intracontinental basins unconformably overlie the basement rocks.

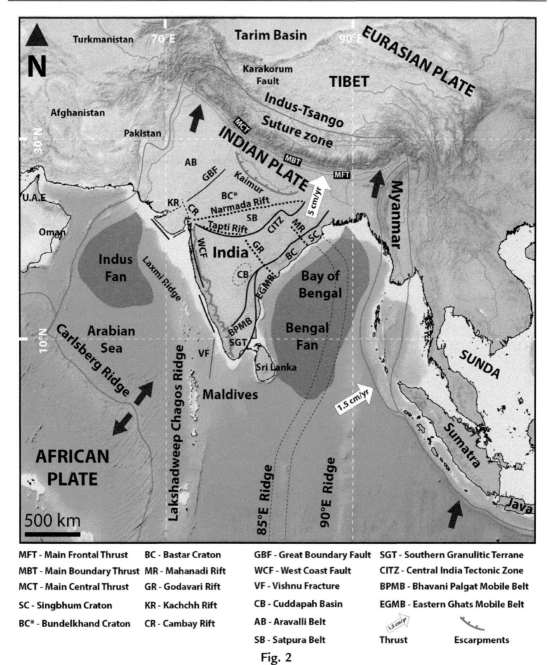

Fig. 2

Location and regional structural trends of Peninsular India and its environs. *After Ramkumar, M., Menier, D., Manoj, M.J., Santosh, M., 2016. Geological, geophysical and inherited tectonic imprints on the climate and contrasting coastal geomorphology of the Indian peninsula. Gondw. Res. 36, 52–80.*

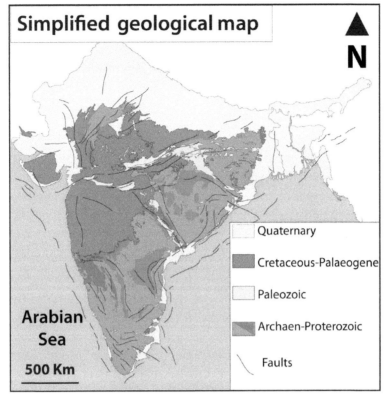

Fig. 3
Major geological provinces of the Peninsular India. *Modified from the Geology of India map, Geological Survey of India. After Ramkumar, M., Menier, D., Manoj, M.J., Santosh, M., 2016. Geological, geophysical and inherited tectonic imprints on the climate and contrasting coastal geomorphology of the Indian peninsula. Gondw. Res. 36, 52–80.*

More or less unmetamorphosed Proterozoic deposits occur in the Cuddapah Basin, and Late Paleozoic-Jurassic sedimentary deposits are preserved in the Narmada-Son, the Mahanadi, and the Godavari (Fig. 2) rift valleys. Basins related with intra- to intercontinental rifting of India, namely the Gondwana basins and those of the east and west coasts, record the Phanerozoic history of India from the Permian to Cenozoic periods (Ramkumar et al., 2016). The basement rocks are covered in the north by Indo-Gangetic alluvium and in the west central region by Deccan flood basalts. Except where exposed, all these older formations are concealed in their most part by Quaternary relict alluvium, calcrete-ferricrete, and Holocene-Recent alluvium and coastal sediments (Fig. 3). The coastal zone of the peninsular region south of the Narmada-Son rift can be broadly classified into Eastern Continental Margin (ECM) and Western Continental Margin (WCM). Except the southern margin (where granulites crop out on the surface exclusively), Precambrian, Proterozoic, and Gondwana deposits occur between the ECM and the WCM.

The ECM is a 2600 km long rifted passive margin evolved through separation of India-Antarctica-Australia from the Gondwana supercontinent during the Late Jurassic to Early Cretaceous periods. Four stages of magmatism, that is, Early Cretaceous (Rajmahal equivalents—Kerguelen and Crozet plumes—Bastia et al., 2010; Radhakrishna et al., 2012a), post-Albian, K/Pg (Deccan magmatism), and post-Eocene have influenced tectonic processes and structural styles. During its evolution, the Bay of Bengal (BOB) lithosphere witnessed two major hotspots, namely, the Kerguelen and the Crozet, resulting in the emplacement of linear N-S trending aseismic 90°E and the 85°E ridges, respectively (Fig. 2). The east coast of India is geomorphologically and tectonically segmented due to subcrustal distinctions (Lal et al., 2009; Bastia and Radhakrishna, 2012) into Bengal/Ganges, Mahanadi, Godavari, Krishna, and Kaveri basins/deltas. Each of these subcrustal blocks behaved differently and mimicked the inherent subcrustal mosaic during the Gondwana period, and carved out for itself an exclusive basin of the Mesozoic-Cenozoic age (Lal et al., 2009). All the east coast basins are connected to the open sea in the east, and are bordered on the west by highlands and the discontinuous Eastern Ghats. These basins have also developed into deltaic systems since the Neogene era (Ramkumar, 2003b), and have been shaping the east coast of India.

The 2200 km long WCM consists of Precambrian granites, Mesozoic sediments, Deccan flood basalts, and Paleocene to Recent sediments. The sedimentary basins of the WCM consist of thinned continental crust overlain by volcanogenic, volcaniclastic, and terrigenous sediments. The WCM is a rifted passive continental margin (Gombos Jr. et al., 1995). Prominent among major structural features are the West coast fault, Panvel flexure, Cambay rift, Kachchh rift, (Fig. 4) Laxmi and Laccadive-Chagos ridges, and Vishnu Fracture (Fig. 2). Block faulting owing to extensional tectonics, thinning, and shear rifting, and uplift of Western Ghats affected the coastal tract, and manifested through several major lineaments. The edge of the continental shelf of the west coast is remarkably straight, representing a fault line that formed/reactivated during Late Pliocene time (Jayalakshmi et al., 2004). Several basement arches trending perpendicular to the west coast divide the shelf region into sedimentary basins, namely, the Kachchh, Saurashtra, Mumbai (Ratnagiri), Konkan, and Kerala basins along the western offshore of India.

The rift-drift events related to the separation of Madagascar during the mid-Cretaceous and the Seychelles in the Late Cretaceous era from India when the Indian Plate moved above the Marion Plume (Raval and Veeraswamy, 2003) gave rise to the formation of major offshore rift basins (Storey et al., 1995). In the northernmost part, the Kutch and the Saurashtra basins are separated by a southwesterly plunging basement high, termed as the Saurashtra Arch (Biswas, 1987). The major conjugate rift systems, the Narmada and the Cambay, cross each other to the south of the Saurashtra Peninsula in the Surat offshore and form a deep Surat depression (Fig. 4). According to Biswas (1987), the northern part of the west coast was the first to be subjected to continental rifting since the Late Triassic, giving rise to three pericontinental Mesozoic marginal rift basins in the onshore such as the Kutch, the Narmada, and the Cambay

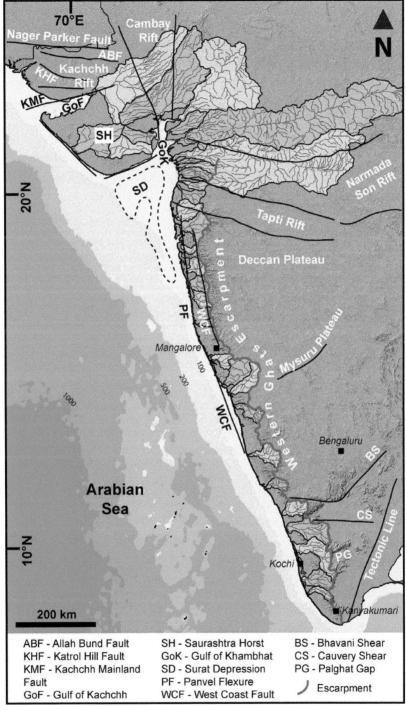

Fig. 4
Location, distribution, and drainage morphology, major structural trends, and Western Ghats Escarpment along the west coast of India. *After Ramkumar, M., Menier, D., Manoj, M.J., Santosh, M., 2016. Geological, geophysical and inherited tectonic imprints on the climate and contrasting coastal geomorphology of the Indian peninsula. Gondw. Res. 36, 52–80.*

basins. The subsurface geology and depositional and tectonic history of the Konkan-Kerala (KK) offshore have a genetic linkage to the denudational and uplift history of the Western Ghats (Gunnell and Radhakrishna, 2001).

3.2 Geomorphology of India

The geomorphology of Peninsular India can be broadly classified into eastern and western coastal zones (Fig. 2), including WGE (Fig. 4), Mysuru, Deccan, Malwa, and Bastar plateaus (Fig. 5) and southwestern high mountains (Fig. 2). There are two more escarpments, namely, the Kaimur and the Narmada, that also emerged as a result of postcollision reactivation of Gondwanan failed rifts.

The 2600 km long east coast of Peninsular India from Sundarbans to Kanyakumari (Fig. 2) consists of river delta systems, vast coastal plains, and extensive fluvial and marine sedimentary deposits,. and so forth, indicative of the prograding nature of the coast. Most of the major easterly and southeasterly flowing rivers, namely, Ganga, Mahanadi, Godavari, Krishna, and Kaveri, and a few minor rivers, such as Pennar, Palar, Tamiraparani, and Vaigai, have developed deltaic platforms at their mouths. Despite this commonality, the deltaic systems of these rivers differ considerably from each other as a function of riverine influx, wave, tide, and a variety of morphological, climatological, and environmental parameters (Ramkumar et al., 2016). These major rivers discharge large quantities of sediments to the deltaic system and to the offshore regions as well, building one of the largest (Tripathy et al., 2011) and thickest subaqueous fans of the world, the Bengal fan. The east coast has been prograding since inception, as evidenced by the prevalent cyclic sedimentation and subsidence, and a general thickening of Mesozoic, Tertiary, and Quaternary basinfill from onland toward offshore. Large lagoons/brackish lakes, namely, the Chilka, Kolleru, Pulicat, and Muthupet, among many others, also occupy the coastal plain. The BOB, along the eastern coast, receives huge volumes of river discharge from large fluvial systems, namely, Ganga-Brahmaputra, Mahanadi, Krishna, Godavari, and Kaveri. In addition, relatively smaller rivers, namely, the Subarnarekha, Baitarani, Vamsadara, Rishikulya, Ponnaiyar, Gadilam, Adyar, Vellar, Vaigai, and Thamiraparani also drain the plains of the east coast. Coastal streams, especially in the west, are short and episodic.

The west coast is essentially limited by the WGE, which also forms a prominent orographic feature with a linear relief zone (Sukhtankar, 1995) along the western fringe of the Indian Peninsula. It came into existence as a result of separation of Madagascar-Seychelles from Greater India around 88–80 Ma. It extends from the Tapi valley located north of Mumbai to the southern tip of Kanyakumari for a distance of about 1600 km. The plateau-escarpment is very prominent in the northern part. The average elevation of the escarpment is around 1200 m. The escarpment is bounded on the landward side by plateaus of >800 m elevation along most of its length (Fig. 4). Many embayments along the length of the escarpment, due to the influence of

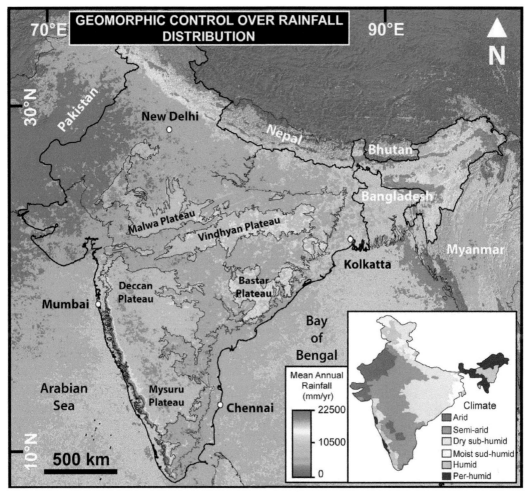

Fig. 5
Spatial distribution of 2600, 500, 100 m topographic contours, Western Ghats Escarpment and mean annual rainfall. Note that the zones of highest rainfall, including the second highest annual rainfall (~5000 mm/year), are located along the Western Ghats, signifying its orographic effect. The 2600 m contour defines the escarpment, across which drastic changes in the rainfall distribution could be discerned. This observation, together with the delineation of a 500 m contour that defines the distribution of plateaus, the 100 m contour that delineates the pre-Quaternary terraces, and general alignment of all these either on or in proximity with inherited structures (Ramkumar et al., 2016). Fig. 5 shows the climatic zones of India that generally follow the topography, suggestive of geomorphic control, which in turn has been influenced since inception by a subcrustal plume directed by tectonic events (Ramkumar et al., 2016, 2017). Together, the intricately linked, self-regulating tectonics- topography-climate interactions and feedback (Ramkumar, 2015; Menier et al. 2016, 2017) are indicated. *After Ramkumar, M., Menier, D., Manoj, M.J., Santosh, M., 2016. Geological, geophysical and inherited tectonic imprints on the climate and contrasting coastal geomorphology of the Indian peninsula. Gondw. Res. 36, 52–80*

headward erosion of westerly-flowing streams occur, among which, the 80 km wide Palghat Gap is the largest (Campanile et al., 2007). The WGE acts as a major catchment divide, between the few, large, east-flowing river systems of Godavari, Krishna, and Kaveri and the numerous, short rivers flowing toward the west. The Western Ghats is the wettest region in the Indian Peninsula, with annual rainfall up to 5000 mm (Fig. 5). The highest peak, Anaimudi, [2695 m above mean sea level (amsl)] is located at its southern extremity. Wide semiarid plains are located north of Surat, and Great and Little Ranns along the coast form the northern limits of the west coast. The Saurashtra coast has a crenulated rocky shoreline with channels, submerged islands, sand bars, dunes, mudflats, and saltpans. Further south, Late Paleocene to Late Eocene, pre-Quaternary, and Quaternary planation surfaces of lateritic platforms and Quaternary-Recent wave-cut terraces with eroded cliffs and steep descents further west were identified to have genetic links with the uplift and emergence of the WGE and neotectonics (Sukhtankar, 1995; Prasad et al., 1998; Nair, 1999). The 30–50 km wide Konkan Coast is undulating lowland. It encompasses numerous hills and detached ridges amid the lowland area characterized by flat shores with long sandy spits. The southern part of the Konkan Coast is rugged and has high hills and elevated plateaus cut by numerous creeks. Further south, wide estuaries mark the coast around Goa. These extend west by coastal plains with a topographic elevation of 10 m (amsl) interspersed by hills. The short and fast flowing nature of the rivers resulted in formation of estuaries and lakes along the crenulated coastline amid narrow discrete zones of beach plains.

The coastal landscape near Mangalore in the Netravati valley has frequent river mouths, creeks, promontories, and cliffs. Toward the south, the lowland is relatively wide ($\sim$70 km). Three broadly parallel belts of landform can be recognized in this region, namely, coastal plains (up to 30 m amsl), a 20 km wide erosional platform (between 30 and 60 m amsl), and isolated, denuded hillocks. Toward the south, the 5–35 km wide coast encompasses many estuaries, backwaters, lagoons, lakes, sand dunes, and rocky cliffs, and documents the Pleistocene-Recent sea level and climatic fluctuations (Jayalakshmi et al., 2004). Low laterite plateaus and foothills occur east of the alluvial coast in this region. The west coast terminates near the rocky outcrops at Kanyakumari.

In the northernmost part of the west coast, two distinct sets of rivers, the Rupen, Saraswati, and Banas in the NW part, flow into the Rann of Kachchh, while the major rivers draining the central and southern coast, namely, Narmada, Tapi, Sabarmati, and Mahi, discharge into the Gulf of Khambhat. Down south, many rivers, namely, Saraswati, Bhadar, Rupen, and Machundri originate in the central plateau region of Saurashtra and meander in a radial pattern through plains to meet the Arabian Sea. Only a few rivers, such as the Khari, Nagmati, and Luni flow in the Kachchh. These rivers show highly varied drainage patterns (Fig. 4), depending on the bedrock characteristics (Kar, 1993; Prasad et al., 1998) such that the patterns range from braided in basalt, dendritic in upper reaches, and straight in limestone terrain. A few of the rivers, namely, Konkan, Pinjal, Vaitarna, Bhatsai, and Ulhas drain a relatively narrow strip of land (<100 km width). Although the general slope is westward, a number of rivers and streams also have north-south alignment in certain stretches, displaying structural control. Further south, the rivers Netravati, Gurupur, Gangoli, Hangarkatta, Sharavathi, Aganashini, Gangavali, Kalinadi Kumaradhara, Nandini, Shambavi, Swarna, Zuari, and Mandovi drain the coastal area. The coastal plain of Kerala is drained by a large number of rivers, and their tributaries originate from the Western Ghats. The main rivers originating from the Western Ghats are Chandragiri, Kuppam, Beypore, Bharatapuzha, Periyar, Meenachil, Pampa, and Manimala. The rivers, namely, Periyar, Achankovil, Manimala, Muvattupuzha, Kallada, and Ittikara run for small distances and debouch into the Vembanad or Ashtamudi Lakes.

3.3 Tectono-Geomorphic Evolution of India and the Cenozoic-Recent River Basins

A summary of the geologic, geophysical, geomorphic, and climatic studies on the Indian subcontinent form a recent synthesis by Ramkumar et al. (2016), and is presented herein.

From the Gondwana supercontinent, initial rifting of India occurred along preexistent Precambrian faults during the Permo-Triassic period to create Gondwana basins, but failed to completely separate the continent (Biswas, 2005; Roy, 2004). Actual separation of India (continental splitting or pull-apart phase) from the Gondwana supercontinent commenced at around 128–130 Ma. The rifting was aligned parallel to the structural trend of the Eastern Ghats, which defines the configuration of the eastern continental margin, and partly, the present coastline of eastern India (Gombos Jr. et al., 1995). It appears from the structural, stratigraphic, and sediment infill that this Early Cretaceous rifting commenced from the Bengal Basin (Banerji, 1984) and proceeded south (present day), as similar rifts, synrift, and postrift sedimentation occur in all the sedimentary basins, except in the case of the Kaveri Basin, wherein shearing-rifting (Radhakrishna et al., 2012a, b) commenced during Barremian at about 120–118 Ma (Ramkumar et al., 2004, 2011). As could be visualized from the structural styles and sedimentary fills of the east coast basins, multiple episodes of basin differentiation in terms of normal faulting and resultant horst-graben structures occurred, and continued from the Early Cretaceous-Cenozoic period, progressively shifting the principal loci of deposition from the cratonic interior toward the continental margin. Around 88–80 Ma, Madagascar-Seychelles separated from Greater India (Melluso et al., 2009), forming WCM, a coast-parallel deep crustal fault (west coast fault/WCF—refer Fig. 2), and a series of parallel faults that initiated the formation of WGE. Passing of the Indian plate over the Reunion hotspot during 68–64 Ma led to a continental flood basalt volcanic episode, and the Panvel flexure (Hooper et al., 2010). Studies suggest that plume upwelling resulted in uplift of the Indian plate (Gunnel et al., 2003; Singh et al., 2007) and was tilted east (Gunnell and Radhakrishna, 2001), contributing to the rise of WGE (Harbor and Gunnell, 2007), inception of orographic effect (Gunnell, 1998), water divide along the peaks (Kale, 2014; Kale and Vaidyanadhan, 2014), birth of precursors to the east-flowing modern rivers, and supply of sediments and water to the BOB. However, the rivers flowing to the north of Narmada-Son and debouching at the Arabian Sea are an exception, suggestive of the limiting of uplift and tilt by the Narmada-Son rift. The occurrences of escarpments along and north of the Narmada-Son rift (Narmada and Kaimur escarpments—Fig. 2) suggest subcrustal lithospheric uplift related to a northerly and easterly tilt, instead of the traditional easterly tilt of the Indian plate. After the rift-drift of Madagascar during 88–80 Ma, the rift-drift of Seychelles during 68–64 Ma from the WCM, followed up by northerly drift and the collision of India with Eurasia around ~54 Ma and an anticlockwise rotation, with subduction under the Tibetan plate (Lippert et al., 2014), and ridge push from Carlsberg and Central Indian ridges (Banerjee et al., 2001) all contributed to the tilt-related tectono-geomorphic evolution of the subcontinent. The emergence of Western Ghats and the easterly tilt of the Indian Peninsula along with Himalayan uplift due to the collision of Eurasian-Indian plates during the Miocene era resulted in abundant clastic supply, leading to the start of deltaic deposition by the Ganges-Brahmaputra, Mahanadi, Krishna-Godavari, and Kaveri Rivers all over the east coast (Lal et al., 2009). The emergence of WGE, dated to be Late Cenozoic-Pleistocene, extends for about 1600 km all along the western margin. The Miocene-Recent tectonic uplift in the Raan of Kutchchh (Biswas, 1982, 1987, 2005) could be cited as evidence for the easterly tilt, driven by continued rifting since the separation of India-Madagascar at around 88 Ma (Melluso et al., 2009; Chatterjee et al., 2013). At the macrogeomorphic level, together with the structural trends and geological information, it can be surmised that the Indian subcontinent as a whole experiences thrust and tilt from the south due to its continued northerly flight and collision with Eurasia since rifting from Gondwanaland, together with rift and tilt toward the NE from the west. This is applicable essentially to the region south of the Narmada-Son Lineament.

The macrogeomorphic subdivisions of Peninsular India are such that a 2600 m topographic contour delineates the WGE, a 500 m contour the plateaus, a 100 m contour the maximum seaward limit of Neogene- pre-Quaternary terraces, a 30 m contour the Pleistocene terraces, a 10 m contour defines the Holocene delta heads, and only the regions below a 5 m contour are active delta lobes. As shown in Figs. 2–5, all these delineated zonal boundaries are, without exception, either aligned or located along, or in close proximity to underlying Precambrian, Permo-Triassic, Late Jurassic-Early Cretaceous, Late Cretaceous, or Neogene-Quaternary-Recent structural trends, suggestive of a continuum of topographic/geomorphic evolution under primary tectonic control, and ongoing landscape evolution.

3.4 Climate

The Indian Ocean is the only tropical ocean bounded by a landmass entirely to the north. In the boreal summer, this unique geographical setting allows for a large differential heating between the Asian subcontinent and the ocean to the south that drives the most dramatic monsoonal wind system in the world (Chaitanya et al., 2014). Driven by transequatorial pressure differences and coupled with the insolation difference between the Northern and Southern subtropical

Hemispheres, the Indian monsoons produce intense spatio-temporal variations. These monsoon systems are also influenced by El Niño, La Niña, the Indian Ocean dipole, and Walker circulation in the equatorial Pacific, and cover a large region spreading from Africa to Southeast Asia, including India (Chauhan et al., 2013).

India receives about 4000 km^3 of precipitation annually, including snowfall; out of which, about 3000 km^3 is from SW (June to September) and NE (October to December) monsoonal rainfall (Kumar et al., 2005). The SW monsoon season covers the entire peninsula, except the southeastern extremity, and accounts for ~80% of the total rainfall (http://nwda.gov.in; http://india-wris.nrsc.gov.in). Orographic influence is dominant in the distribution of rainfall in the SW season, as the prevailing winds blow at right angles against the Western Ghats. The monsoonal precipitation is highly spatio-temporally heterogeneous, and produces a variety of climatic zones (Fig. 5), from evergreen tropical rainforests, humid, semihumid, and to hot-dry climes that constitute India, one of the wettest and heaviest snowfall regions of the world, to the hottest deserts. Thus, the Indian Peninsula is climatically zoned in terms of annual precipitation and temperature. For example, the sparsely vegetated (mainly grass and shrub cover) peneplain region of the uplifted Mysuru plateau with semiarid tropical climate (mean annual rainfall <1000 mm) grades within a few tens of kilometers into the Western Ghats, with mean annual precipitation of >4000 mm, mean annual temperature of 25°C, and lush vegetative cover in a subhumid climatic setting. The SW monsoon accounts for 75% of the annual rainfall, 73% of the water discharge, and 85% of the annual sediment transport (Rao, 1975). During SW monsoon season, water vapor collected on the ocean's surface by powerful southwesterly winds is flushed over India and the BOB. A large fraction of the runoff to the ocean occurs during or shortly after the SW monsoon season, and contributes to freshwater flux into the BOB in equal proportion with rainfall over the ocean (Chaitanya et al., 2014). After the end of SW monsoon season, the NE monsoon season sets in, and it is characterized by a relatively dry season with less rainfall. The annual mean rainfall distribution along the coastal tracts of the country (Fig. 5) is in contrast with the west and east coasts. Along the west coast, it ranges from 300 mm in the northern and northwestern parts, to 2500 mm in the southern part, with a maximum of >5000 mm in some areas of the Western Ghats, forming the second highest rainfall in the country. Along the east coast, the rainfall distribution is almost reversed, with higher rainfall of 600–1000 mm in the northern parts, and relatively less in the southern parts, within the range of 200–600 mm.

As explained in the tectono-geomorphic evolution section, India was an Island Continent, akin to present day Australia. Since the breakup from the Gondwana Supercontinent, and after one of the extraordinary drift rates (Ramkumar et al., 2017), it started colliding with the Eurasian plate from 54 Ma onward, which gave rise to the evolution of the Himalayan Ranges. Since 88–80 Ma, the Madagascar-Seychelles started separating from India, and was followed by the eastern-northern tilt of India that gave rise to the WGE. The monsoon system of India is, thus, the result of the emergence of these two orographic barriers, entrenched by approximately 30 Ma in the north and 0.8 Ma in the west.

3.5 Coastal Hydrodynamics

The east coast in the BOB experiences two different current directions, durations, and intensities (Fig. 6) with a net northward littoral sediment transport (Sanil Kumar et al., 2006; Kunte et al., 2013), a phenomenon also evidenced by northward growing spits of all the deltas of the east coast (Ramkumar, 2003b; Jana et al., 2014), few of which are >21 km long and ~1 km wide (Ramkumar, 2000). The littoral/longshore current, or Coramandal Coastal Current (CCC), along the coastal margin (Fig. 6) reverses twice a year under the influence of alternating monsoons. It flows northward from February to September, and southward from October to January. Heavy discharges from the Ganges-Brahmaputra and BOB, as a result of the intense SW monsoon season, result in 100 km wide freshwater lens floating over saline water. This current slowly withdraws from the north of Kanyakumari during the middle of September, and

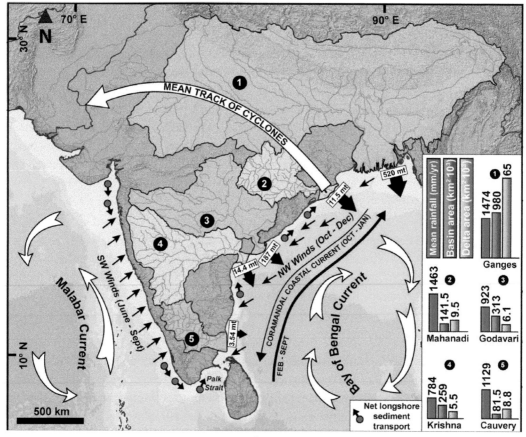

Fig. 6
Drainage characteristics and coastal hydrodynamics of the east and west coasts. *After Ramkumar, M., Menier, D., Manoj, M.J., Santosh, M., 2016. Geological, geophysical and inherited tectonic imprints on the climate and contrasting coastal geomorphology of the Indian peninsula. Gondw. Res. 36, 52–80.*

flows northward during the rest of the year (Chaitanya et al., 2014). The deeper part of the BOB experiences clockwise circulation (BOB Current or BBC—Fig. 6), seemingly driven by the thermal differential between the Arabian and Indian coastal regions, having contrasting atmospheric and sea surface temperatures. The Arabian Sea in the west coast experiences anticlockwise currents (Sanil Kumar et al., 2006; Malabar Current or MC—Fig. 6), probably driven by prevailing winds and wave refraction (Avinash et al., 2013). Both the coastal and deeper regions of the Arabian Sea show a net southerly sediment transport (Kunte, 1994; Kunte and Wagle, 1993, 2000; Sanil Kumar et al., 2006). Owing to the general topographic slope and the absence of any major riverine influx, the southerly sediment transport continues unabated, often extending into the Palk Strait region (Fig. 6) of the east coast (Sanil Kumar et al., 2006).

Wind speeds at 10 m (amsl) height range from 39 to 50 m/s along both the coasts; however, higher speeds are associated along the east coast than the west coast (Sanil Kumar et al., 2006). A wide range of wave heights (0.2–16 m) are observed along the Indian coastline. During fair-weather seasons, the wave heights are centered around 0.5–2.2 m and are controlled by bottom topography, owing to the highly varied sediment delivery by the rivers (quantum, rate, and loci) and the structural control over the continental margin (distance from coast and slope) and resultant dynamics in the breaker zone. Wave influence on coastal zones and deltas varies considerably due to the coastal morphology and nearshore bathymetry (Anthony, 2015). During fair-weather seasons, waves approaching the east coast are always at an angle (Ramkumar, 2000, 2003b), a common feature observable in coasts experiencing predominant swells.

The Indian Ocean is bordered on the north entirely by a landmass that enables the BOB to act as a nursery for cyclones and storms (Allison et al., 2003), and supercyclones almost annually, accompanied by ~12 m waves. Hence, shoreline morphology is significantly influenced by waves, particularly associated with inclement weather. As deltas, lowlands, and/or tidal creeks cover the east coast, the onshore-offshore sediment transport dynamics are also influenced by the cyclones. As most of the coastal zones are located within a mean sea level to 10 m amsl (Ramkumar et al., 2001a, b; Allison et al., 2003), the inland reach of these cyclones (that always accompany monsoons, which in turn brings in excessive discharge) is attenuated (Ramkumar, 2009). The fair-weather seasons, as they are devoid of significant riverine influx and powerful currents or waves, could not rework the storm-resultant morphologies (Anthony, 2015). This phenomenon preserves the storm-generated morphologies along the coast, as observed elsewhere (e.g., Simarro et al., 2015). Thus, all along the coast, behind the berm zone toward land, extensive low lands, in the form of bays, back waters, and lagoons, namely, Pichavaram, Pulicot, Kolleru, and Chilika, are formed (Ramkumar, 2003b; Ramkumar et al. 2016). All the deltaic systems of the east coast are reported (Ramkumar et al., 2016) to have possessed similar lowlands during prehistoric times as well, suggestive of the control exercised by the storm-influenced coastal morphologies also in the geologic past. The wave climate of the west coast varies regionally and has significantly affected the coastal morphology in terms of extensive backwaters in the south, cliff and wave-cut platforms in the middle, and extensive tidal flats in the north. Similar to the east coast, the west coast is also subjected to seasonal-annual recurrence of cyclones and storms; however, the effects are not as dramatic as in the east due to the generally raised coastal lands.

Both the coasts experience diurnal and seasonal tidal asymmetries and show an increase of tidal range from the south to north. While a range of 0.2–4 m is observed from the Kanyakumari-Ganges mouth, a range of 0.2–10 m is observed along the west coast. In the funnel-shaped estuarine mouths of Ganges and Rann of Kachchh, that are located in northern extremes of east and west coast, respectively, geomorphic amplification of tidal range is observed. While the tidal waters inundate up to 100 km from coast in the Ganges, it ranges up to 160 km in the Kachchh.

4 Discussion

Understanding geological and geomorphological phenomena is fundamental to the essential needs of modern civilization, and sustaining the human race at large. Since the advent of life on our planet, particularly the human race, this knowledge has helped in evolution and survival. The human evolution commenced with the acquiring of knowledge to find and use chert and flint to light fire and to shape large rock fragments as stone tools to chop meat, and to make spearheads, and so forth. Since then, many major steps in evolutionary development and the significant advancement of civilization relied on acquiring additional knowledge on the Earth and Earth's materials. For example, the change from nomads to settlers (finding suitable caves for dwelling—Stone Age), the hunter-gatherer life-style to cultivation (finding suitable landscapes and perennial freshwater sources for agriculture); learning to find suitable ores for metal extraction and production of tools for livelihood (the Iron Age), establishment of village-fiefdoms to kingdoms (learning to produce alloys from multiple ores and minerals—the Bronze Age), sophistication of life and the advent of modern man (Copper Age—learning to extract precious metals and making jewelry for adorning deities and others), learning to locate and extract fossil fuels, large-scale mining of other metals and ores (Industrial era), and finally, the computerized era (modification natural minerals to produce chips, production of nuclear energy, etc.) can be cited as evidence for the coeval nature of human evolution and advancement of exploration methods of ores and minerals. Collectively, these can be collated to state that geoscience is the mother of all sciences (Ramkumar, 2017), and an ability to adopt and/or learn to live with the rhythms of natural geological cycles/equilibrium should be the basis for human endeavors that may ensure safety and sustenance (Ramkumar et al., 2015). This is the reason why our ancestors and forefathers utilized this effectively and formulated their lifestyles in tune with natural geological processes and conditions. Failing to stick to this often leads to the detriment of human endeavors, not to mention the catastrophic responses. Numerous examples could be cited supporting this bare, crude reality! Within this underlying construct, we place our perspectives.

4.1 Spatio-Temporal Variations of Precipitation

India is home to unique river systems that drain through highly contrasting climatic zones, geological formations, and topographic features. Similar to many other rivers elsewhere, the river systems in India have uniqueness in terms of drawing their waters predominantly from glacial meltwater (Ravi, Peas, etc.), glacial and SW monsoon (Ganges, Brahmaputra), exclusively SW monsoons (Narmada, Tapti), SW and NE monsoons (Kaveri, Krishna, Godavari), exclusively NE monsoons (Palar, Vaigai, Vellar), exclusively during storm weather related cloud bursts (Marudaiyar), and so forth. These characteristics, together with short-duration, high-intensity monsoonal precipitation enforce high-spatial distribution patterns of water sources, thereby resulting in coeval floods and droughts. Ironically, high-precipitation regions are also regions of drinking water scarcity! Furthermore, though there are many perennial rivers, the surplus water either floods the plains to wreak havoc annually and/or debauches into the sea. The quantity of water that drains into the sea is so high to initiate river-in-sea (Chaitanya et al., 2014) (Fig. 6).

As could be visualized from Fig. 5, the annual mean rainfall of India (1176 mm/year for the period between the years 1901–2014, Source: https://data.gov.in/catalog/all-india-area-weighted-monthly-seasonal-and-annual-rainfall-mm) is higher than the world average of 1050 mm/year (Source: http://www.physicalgeography.net/fundamentals/8g.html). It is perceivable that except certain rain shadow regions, India receives copious rainfall. Nevertheless, notwithstanding the natural geological-geomorphic-climatic factors, other factors, including, but not limited to paucity of storage structures, improper planning, land use, agriculture practices, and so forth, among others, might be the reasons for coeval flooding and drought conditions. In this case, it is simply common sense to store, divert, and utilize the surplus water to deficit regions. The role of juxtaposed, high disparity geological, topographic, and other characteristics, such as rainforest-dry deciduous regions of WGE and Mysuru Plateau, Kaimur escarpment-Thar desert, and so forth, also factor in the coeval flood-drought situations. Added with these are the political situations, wherein water being state subject, the states located in catchment regions constructed dams to the detriment of riparian states, which also causes water woes. The absence of proper planning and management of water storage/augmentation structures can also be another factor that contributes to the current state of affairs. For example, in the Tamil Nadu State, there are 39,202 surface water storage tanks, with storage capacity of 390 TMC, which is too large against the storage capacity of river dams, which is 243 TMC. Despite this available installed capacity, this state reels under perennial drought.

These observations bring forth three paradigms: first, superimposed on the natural conditions, improper, unscientific practices of water storage, transfer, and utility contribute to coeval flood-drought; second, bludgeoning population, enhancement of areas under irrigation, and infrastructure development, all without proper planning and water resource management, result in intensive pressure on available resources; and third, water being state subject, there exists a disconnect between catchments and riparian regions within a river basin in terms of planning and execution of water storage structures and practices, and the absence of proper land use planning, and the absence of scientific methods of surface and subsurface water resource nourishment. In order to obviate these problems, nationalization of rivers could bring all the planning, execution, monitoring, maintenance of water storage and transfer structures, which in turn provide for comprehensive, holistic planning and management. Second, transfer of surplus water from flood-prone rivers to water deficits and dry regions through ILR seems to be the remedial measure. To be precise, the solutions are: nationalize, and interlink rivers! However, these are not simple and straightforward solutions, and are wrought with many impediments. The impediments are examined herein.

4.2 Political Impediments on Nationalization and ILR

While the ILR program is about linking rivers in India, its neighbors are watching how it unfolds with apprehension. The ILR program's Himalayan component envisages construction of reservoirs on the principal tributaries of the Ganga and the Brahmaputra in India and Nepal, and involves transfer of water from the eastern tributaries of the Ganga to the west, apart from linking the Brahmaputra to the Ganga and the Ganga to the Mahanadi. As the Ganga and Brahmaputra are Transboundary Rivers, India's proposed engineering of their waters would impact Nepal and Bhutan, where these rivers originate, and Bangladesh, the lower riparian country whose economy and agriculture depend almost exclusively on the waters of these rivers. Bangladesh, meanwhile, is deeply apprehensive over the diversion of water from the Ganga's tributaries upstream, and the Brahmaputra and Teesta rivers to the Ganga, as this would reduce water flows into its territory, increasing salinity of the water, rendering the soil unfit for cultivation, and resulting in the desertification of large parts of the country.

4.3 Uniqueness of River Basins and Geological Impediments

The geologic history of the Earth is replete with repetition due to multiple tectonic cycles. The present landscape of India is largely built on Precambrian basement rocks and structures that were repeatedly reactivated all through the entire past, until recently. As seen in the structural information of India (Figs. 2 and 3), in comparison with Fig. 7, it is explicit that the continent remains to be seismically active, essentially along the structural trends. All the sedimentary basins and continental margins of India are either exclusively shaped and or delimited by these structures. The deltaheads of all the rivers of the east coast of India are all located either on or east of the tectonic line (Fig. 2), which is also seismically active.

As explained in the previous sections, the east coast deltas have commenced significant aggradation and progradation since 54 Ma, and prograde so, by delivering floodwaters and sediments all along the onland-offshore regions of the eastern continental margin. Almost all the rivers of India were perennial until the historic past, and were known to have overflown annually, building vast flood plains and coastal tracts. Should the ILR be implemented, the existing

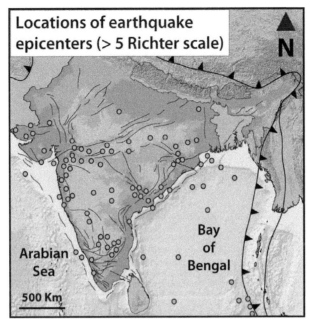

Fig. 7

Map showing major earthquake epicenters. Only earthquakes of >5 magnitude on the Richter scale since 1800 BP are depicted in the figure. Note that all of these are located at or in the vicinity of major structural trends. Interestingly, most of these structures are Precambrian shears, faults, and rifts that were reactivated during Proterozoic and/or Paleozoic eras, and again episodically during Late Jurassic, Upper Cretaceous, Neogene, Pleistocene, Holocene, and Recent periods. *After Ramkumar, M., Menier, D., Manoj, M.J., Santosh, M., 2016. Geological, geophysical and inherited tectonic imprints on the climate and contrasting coastal geomorphology of the Indian peninsula. Gondw. Res. 36, 52–80.*

equilibrium between sealevel-shoreline-sediment influx will be significantly modified, which in all probability will affect the coastal land use, resources, population, and infrastructure, among other issues. These, in addition to the potential enhancement of geohazards, include seismicity, as most of the deltaheads are aligned either on the tectonic line or adjoining it.

This commonality of rivers notwithstanding, each river basin and delta system is unique. For example, the Ganges delta is dominated by tides, the Brahmaputra by floods and tides, the Mahanadi by cyclones and storms (Fig. 6), the Krishna by fluvial flow, the Godavari by waves, and the Kaveri by costal currents. These differences not only resulted in varied morphologies of the respective deltas, but also in terms of geo-bio-oceanographic interactions as well. The nature and type of feedback that may result from any change in their characteristics, especially in terms of enhancement/reduction/diversion of huge quantities of water and sediment taken up in a large, continental scale, as proposed in ILR, are unknown and unprecedented. This singularly makes the ILR program scary. This unknown threat looks gloomier when viewed with the statistic that though India might have only 2.4% of the world's land area, it sustains 8% of the world's biodiversity.

4.4 Self-Regulating River Dynamics and Plausible Impacts of ILR

The rivers are dynamic geomorphic expressions; the occurrence, distribution, sprawl, morphology, direction, and course of which are transient on short-long term temporal scales, and respond to tectonics, slope, lithology, climate, hydrology, land use, vegetation, and so forth, in the catchment to deltaic regions, and are additionally influenced by oceanographic and other factors at the lower reaches, and most importantly, they respond at the rate that these factors evolve/change. It means river systems show the shortest response times to any of the factors that interact with it. Also, from a geological timescale to present conditions, rivers have demonstrated their ability to self-regulate their flow paths. They also have a memory to return to their old courses. Examples galore in this regard include Kosi River flooding associated with the recapture of its paleochannel in the year 2006. The river, due to its dwindling flow volumes was flowing in a course, returned to its older path/course when flow volumes changed following intense monsoonal precipitation.

In order to explain the long-short term topographic and channel dynamics, the example of the River Kaveri is presented herein.

The River Kaveri has existed since the breakup of the Gondwana continent. Until about 66 Ma, it was debauching in the Perambalur-Ariyalur regions. The collision of the Indian continent with Eurasia, and following the emergence of Himalaya in 54 Ma, was accompanied by shifting the river course akin to a pendulum swing from north toward the southeast, east, and northeast; and it currently flows east. The apex of this swing remained more or less located at Trichy, on the east of the tectonic line (Fig. 2). The current location and course of this river is controlled by the flow volume. Significant changes in the flow volumes during seasonal and catastrophic floods have invariably resulted in overflow of the river, and changes in course as well. There are historically documented floods from this river during the years 1198, 1257, 1924, 1954, 1965, 1977, 1979, 1983, 1999, 2000, and most recently in 2005. An interesting fact is, the bank breach occurred more or less in the same location during all these flood events, which kindles the memory of the river! It also suggests that, through understanding the river dynamics, unique to each of the river basins, appropriate river regulatory measures, including geomorphic, hydrological, land use reinforcements/modifications need to be taken up, should the flooding menace need to be controlled, to which ILR is one such effective procedure.

4.5 Interlinked Nature of River Basin Dynamics and the Potential Threat by ILR

Meteoric precipitation, the landscape, rivers, river basins and deltas are all bounties of nature. As with any resource, these also are not bountiful everywhere, and never follow a uniform distribution pattern. Similarly, the snowfall and rainfall are not uniform everywhere, and do not precipitate uniformly throughout the year. They change geographically, seasonally, and even have short-long term cyclic changes. These changes are the norm of nature. As long as the human race accepts this and learns to live with this norm, we will be better equipped to deal with these conditions.

Rivers, flood plains, and deltas were and are the benefactors of human evolution since time immemorial. Next to coastal regions, river banks and plains are the preferred locations of human habitations and infrastructure developmental sites. The close relationships between population expansion and industrialization, and urbanization and infrastructure development are well known. Invariably, the population expansion and resultant changes in Earth systems, particularly in river basin topography, land use, and so forth, induce changes in physical, chemical, and biotic cycles of riverine environmental systems. In addition, the rivers are unmindfully utilized as an outlet for wastes that emanate from the anthropogenic activities. As reviewed in the introduction section, the landscape and land use changes induce a chain of geological-geomorphic-geochemical-biotic changes in the river system. For example, when natural vegetation cover is modified/removed, it encourages soil erosion, nutrient recycling, and so forth, not only in the vicinity, but also many hundreds of kilometers downstream, into the deep marine ecosystems.

The river water is not a commodity; rather, it is a catalyst, supplier of nutrients, microecosystem, transporter, food, and a habitat. Each of the river systems and water contained in it promote unique biodiversity, dependent on the climatic zonation, geoenvironmental conditions, quantity, and characteristics of flow (Subramanian, 2000). If this is the case, ILR essentially attempts to homogenize the system in its entirety, which may turn out to be endangering, if not driving the system into catastrophe! The Indian subcontinent has uniquely distributed endemic fauna and flora that are adapted to local environmental-ecological-geological conditions. For example, the flood plains of Brahmaputra, home to the endangered rhinoceros and water buffaloes, depend entirely on the floodwater-loving flora for food and sustenance. Similarly, the gharials, otters, and dolphins of the Ganges are endemic, and depend exclusively on the estuarine tidal waters. Once the "surplus" flood waters of these two major rivers are diverted to "deficit" regions through ILR, then the quantum of water, sediment, and nutrients contained in it, together with the floral distribution, coastline, coastal ecosystem, littoral circulation, and all that is dependent on the enormous quantities of water-sediment-nutrient will be affected, irreversibly! This means, ILR will not solve the surplus-deficit problem, but rather, will aggravate the existing environmental deterioration, and also introduce a hitherto unknown environmental-ecological-biotic crisis. This is not to argue against the ILR. The guiding principle is, and should be, that developmental activities and utilization of natural resources are two sides of the same coin. In this context, when the already executed ILR programs in other parts of the world, or even within India (for example, the Ken-Betwa program, Krishna-Godavari program) should be viewed through the prism of the fact that interlinking river systems within the same climatic zones introduces minimal interference into the natural processes and homogenization of biotic realms. The term "minimal interference" with reference to the ILR starts from determination of "ecological flow" (sensu Robson et al., 2005) in terms of quantity, rate, location, distribution, and so forth. The ecological flows of surplus, as well as deficit, river systems need to be determined on a basin scale, which should form the basis for ALL other modification plans. An example could be cited to emphasize the importance of this: The very occurrence and location of the Thar Desert augments the evapotranspiration that catalyzes wind movement to storm the Himalayan orographic barrier, wherein descent of cooler winds due to snowy mountains creates monsoonal rainfall. The meltwater and rainfall are collected in the catchments to flow via might river systems, transporting water and sediments to the deltaic plains and deep into the BOB. These sediments and water provide sustenance to the thriving of river-coastal-marine ecosystems, including the sprawling mangrove swamps. The mangrove ecosystem, associated aquatic, land, and avian fauna form important links in the food and habitats and provide ecosystem services (Robertson and Duke, 1987), which in turn will get modified, if the seasonal quantum, rate, and distribution of water and sediment are modified by ILR. When the importance, relative roles, and quantum of contribution of these are not yet fully understood, introducing another unknown will invariably result in detriment.

The geosphere, biosphere, hydrosphere, and atmosphere are all intertwined, and any change in any one of them will invariably promote changes/reequilibration in the others as well. An equilibrium between the natural geosphere-biosphere existed until the advent of the human race, especially until the evolution of settlers. Until then, the occurrence, distribution, and other vagaries of nature were controlled exclusively by natural cycles and processes. Since then, and particularly since the concept-practice of "everything is for human" and "human race at the top of biopyramid," there has been a growing disequilibrium and detrimental anthropogenic intervention into the natural processes and cycles. Since the advent of life on the earth, there were six major and about 30 minor-moderate scale extinctions that occurred entirely due to natural causes, and in no circumstance did the organisms ever contribute to these causes. Should there be another extinction/catastrophe, there will be one unprecedented example, that is, the causative organism will be human.

The flow in the river is not water alone; river water sources sediments from the terrain in which it flows. It transfers sediments from its source to their final place of rest, through four major modes: rolling, saltation, suspension, and dissolved load (Fig. 8). The mode, quantum, type, and other characteristics of the sediments so sourced are dependent on the bedrock characteristics, slope, lithology, channel morphology, volume, and flow characteristics. These are then transported downstream, deposited in the lower reaches, building deltas and also dumping deep into the sea. Thus, the downstream regions, including deep marine systems and their ecological and geoenvironmental stability, are controlled by the river flows.

Superimposed on these general characteristics, the uniqueness of each river basin, especially the lithological distribution, offers uniqueness to each of the river waters, over which the entire ecosystem and biodiversity are dependent. In addition, as explained in previous

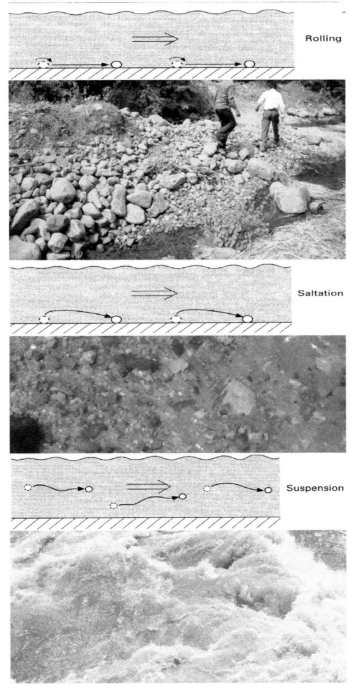

Fig. 8
Modes of Sediment Transport.

sections, the river deltas of India are controlled by unique oceanographic-climatic-fluvial processes: the Ganges-tides; the Brahmaputra-tide-flood; Mahanadi-storms and cyclones; the Krishna-fluvial flood; Godavari-waves; the Kaveri-littoral current, due to which, these deltas exhibit unique morphology, sediment characteristics, flora and fauna, and so forth. If so, changing the river flow and sediment characteristics, which ultimately would result due to ILR, will inevitably homogenize these systems and wipe out endemic habitats, ecosystems, and biodiversity. Added with this threat is the uniqueness of sprawl of the river basins of India, which are limited within individual climate zones, either from catchments to confluence, or at least deltaic regions (Figs. 1 and 4–6). The linked canals of ILR located within each geological province (Fig. 3) and climatic zone (Fig. 5), should be minimal; if not, not-so-drastic ecological shocks could be expected.

4.6 Probable Impacts on Deltaic and Coastal Regions by ILR

Location, morphology, and stability of the shoreline is a dynamic equilibrium between fluvial-oceanographic physical processes, and is also controlled by sediment influx, vegetation cover, and so forth. When surplus river water is transferred to deficit river systems, it may immediately initiate a transient chain of responses by geomorphic-hydrologic-biotic systems. In addition, the probable influx of different types of sediments, nutrients, and microbes would encourage shift/modify habitats, and the proliferation of opportunistic colonizers at the expense of natives. Regionally, the anthropogenic share in phosphate flux rates is distinctively higher, and reaches levels 10–100 times greater than preindustrial levels. The high rates of phosphate loading in these regions severely affect ecosystems, favoring opportunistic species and causing widespread retreat of formerly indigenous species (Föllmi, 1996). Without having a contingent plan to tackle this, the more important ecological crisis, ILR, may spell doom, not only in the river basin, but also in coastal and marine ecosystems.

First, a visible and large-scale modification that may result from ILR is change in land use and vegetal cover. The availability of a variety of floodplain environments in large lowland river valleys provides opportunities and constraints for a host of human activities (Hudson et al., 2006). One of the visible signs of pressure on the natural environment is the change in land use/land cover, which often, if not most probably, results as a function of population growth (Qi and Luo, 2006). While drastic impacts on water resources as a direct result of changes in land use has been seen previously (for example, Coulthard and Macklin, 2001; Albinus et al., 2008; Jessel and Jocobs, 2005; Clapcott et al., 2011; Ayivor and Gordon, 2012), the interlinked nature of land use-surface and atmospheric temperature, and finally, the impact on short-long term climatic changes have also been reported recently (Sabajo et al., 2017). Hesslerova et al. (2012) are of the opinion that the higher human activity, particularly the conversion of natural land cover into land use, increases the surface temperature, which results in water loss on a basin scale. The impacts of change of natural land cover (Wu et al., 2008) by urban expansion is detrimental to the local and regional hydrology, and water quality (Li et al., 2008; Koch et al., 2011), and increases the nonpoint sources of pollution (Tang et al., 2005). The impacts of human activities translate into catchment erosion (Walling, 1999), sedimentation within river channels, surface and ground water pollution, increased levels of turbidity and acidity, and mean dissolved oxygen and nutrient loads (Ayivor and Gordon, 2012; Memarian et al. 2013). When compared with other ecosystems, the deltaic regions support a high population density, and provide sustenance to fauna and flora. It is common knowledge that rapid population growth in deltaic regions inevitably brings in recreational, commercial, and industrial use of these lands and waters, causing environmental deterioration.

Ramkumar (2003a) emphasized the link between land use dynamics with that of coastal environments, particularly the mangrove swamps. The coastal deltaic lands serve as repositories of rich biodiversity, and as nurseries to endemic avian, aquatic, and other species, including mangroves (Robertson and Duke, 1987; Srivastava and Farooqui, 2017). A firm link between nutrient accumulation in natural sedimentary environments, particularly in estuaries and adjoining regions, and the transport of nutrients to coastal marine regions, as a function of land use change is known (Ramkumar, 2003b, 2004a, b, 2007; Ramkumar et al., 2001a, b). As the coastal environments are home to a vast array of primary producers, and act as nurseries, which in turn are dependent on the physical and chemical characteristics of the bed-sediments, significant alteration of these traits by industrial contaminants affects the food chain and primary producers. A study by Solaraj et al. (2010) in the Delta regions of the Kaveri River Basin, Southern India, reported high concentrations of dissolved salts, organic pollutants, and phosphate exceeding the permissible levels at certain sampling stations as a result of agricultural runoff from the vicinity, sewage, and industrial effluents.

Qi and Luo (2006) observed the land use changes on a basin scale in the Heihe River Basin, and found that the observed changes caused severe environmental problems in terms of surface water runoff change, decline of groundwater table, and degeneration of surface water and groundwater quality, land desertification and salinization (Zhang and Zhao, 2010), and vegetation degeneracy. Glendenning and Vervoort (2010) have evaluated the effect of rain water harvesting structures on a catchment-basin scale in terms of water balance models, intensity of surface run off, water variability in storage capacity, and so forth, to suggest suitable eco-friendly developmental activities. The nexus between agricultural land use, effluent disposal, and the quality of river water have been examined by Tafangenyasha and Dzinomwa (2005), and the results have shown that the land use change imposes beneficial, as well as detrimental, effects on the natural environment and biodiversity.

Geohazards occur everywhere, and no region of this Earth is safe. With the opening of the global economy, and expansion of domestic, industrial, commercial, recreational, and aesthetic land use in every available piece of land by the human race, the resulting intervention of natural geological processes has led to loss of life and property. Owing to their vast expanses of monotonously low altitude topography, the lower reaches of the river basins are home to many geohazards. However, the unscientific anthropogenic developmental activities in the deltaic regions often aggravate the intensities of geohazards in terms of susceptibility to flooding, coastal inundation, river and coastal erosion and sedimentation, tsunami, storms, cyclones, and quick sand. As these areas are thickly populated, the resultant damages often reach catastrophic levels, even during the slightest change in the existing equilibrium of coastal geological processes (Jayanthi, 2009).

While the developmental activities could not be contained for the sake of nonintervention with geological processes, well thought-out planning and development with minimal intervention can thwart many a potential hazardous events. However, the very nature of the geohazards being unpredictable, it would be advisable to classify the land and water regions that are being and/or planned to be utilized, according to vulnerability levels to various potential geohazards, and not to engage in potential catalytic activities. In addition, should there be any eventuality, being ready with contingent plans for relief and rescue operations would minimize the potential loss.

Tectonics, climate, relative sea level and sediment influx are the four factors that determine delta-coast-river length, morphology, and accommodation space of depositional systems, which in turn exert control on occurrence, distribution, and diversity of habitats and lives in them. On a geological time scale, these factors and systems have learned to survive; however, implementation of ILR is truly a perturbation that the natural systems may not be able to cope with. Hence, determination of not only uniqueness of each river basin, but also the rate at which its physical and biological systems could cope with change has to be determined before actual implementation of ILR. Additionally, the anthropogenically aggravated climate change, coupled with the natural geohazard zones, need to be considered when a major program of ILR on a continental scale is initiated. If not, geohazards could result.

In the context of aggravation of naturally occurring geohazards, the location of link channels and resultant modified topographic loading, the seismic hazard zonation, occurrences of major active faults, and their inheritance from the geological past to the present need to be considered. As explained in the previous section on geotectonic-geomorphic evolution, the east coast basins of India are all under the influence of seismically active major tectonic structures, and the deltaheads of all the east flowing rivers are all invariably located east of N-S running tectonic structures (Figs. 2 and 3). In addition, historic occurrences of major seismic events within the past few centuries, either located on this and other major structures, have trended (Fig. 7). As the deltaic and coastal regions are already subject to potential geohazards (Rao, 1998), aggravation by structural, topographic, and toploading effects that may result from ILR need to be considered, potential feedback should be evaluated, and remedial/preventive/mitigation efforts should be studied, and then implemented as necessary.

In this regard, three recent incidents are to be brought to consideration, not to scare or predict doom; but to remind people of the potential hazard that may result from natural and anthropogenic intervention into deltaic and coastal topographic/geoenvironmental processes. First, NASA has reported that the meteoric precipitation and flooding of the Houston region has made the region subside up to 2 cm. Second, within days, there were major seismic events in Mexico. These two may or may not be related; however, had the cyclone-flood-subsidence prompted an already active major fault line to an 8.2 Richter scale catastrophic seismic event, which was followed up by >6 events? The answer may be yes or no, but needs evaluation. When there is an unknown in the geodynamic process, it is always safer and more prudent to act, especially when the safety and sustenance of human lives and infrastructure are under scrutiny. Third, the close coupling between tectonics, topographic evolution, and catastrophic flooding has been recently demonstrated by the studies of Menier et al. (2017). Their study demonstrated the geologically-very short term feedback between tectonics-climate-fluvial flooding, and should serve as a guiding principle for mitigating flooding hazards in regions hitherto considered "deficit" as well. This anticipated aggravation of geohazard potential is more likely in the east coast, for two reasons: one, unlike the west coast, where rocky shorelines are the norm; the east coast has vast, sandy coastal plains that are within 0–10 m topographic contours, and reel under the onslaught of annual flooding, cyclones, and tsunamis (Ramkumar, 2009).

River flows do not debauch directly into the sea; rather, they pass through estuaries, wherein the saline water is mixed and delivered into the sea during low tides. The dynamics, rate, and quantum are all controlled by a unique combination of factors including, but not limited to, geomorphology, climate, seasonal changes, lithology, slope, structure, type, and distribution of vegetation, land uses, dams, industries, settlements, population, and agricultural practices. The water and sediment delivered into the estuary undergo a variety of changes, and only latter reach the coast-deep sea. The tidal changes that differ between west and east coasts also show a wide range of changes from a few tens of cm (microtidal) to a few meters (macrotidal). These tidal variations permit incursion of seawater from confluence-shorelines into the river channel-estuary that range from a few km (Kaveri estuary), about 45 km (Krishna and Godavari estuaries), about 100 km (Ganges-Brahmaputra), and up to 160 km (Rann of Kutchchh). It also signifies the extent of influence of oceanographic processes deep into the inland regions. Any alteration of the natural river flow conditions by ILR (either reduction of flow in transferring "surplus" or enhancement in "deficit" river systems), may change the estuarine dynamics, and thereby coastal processes, requiring a proper study for each of the river basins to assess potential changes in deltaic-estuarine-coastal geo-biodynamics. There are studies galore documenting estuarine sediment-nutrient-biotic interactions and dynamics (eg., Zwolsman, 1994; Ramkumar, 2003a, 2004a, 2007), inherited structures (Singer et al., 2008; Menier et al. 2016; Ramkumar et al., 2016), and the impact of changing flow conditions by damming, shoreline changes, land use, and so forth (e.g., Naiman and Bilby, 1998; Walling, 1999; Tong and Chen, 2002; Tang et al., 2005; Zimmermann et al., 2006; Okazava et al., 2010; Zhang and Zhao, 2010; Zhang et al., 2011).

4.7 Knowledge of Hazard Mitigation by Ancestors

Taming the natural geohazards, including river flooding, has been one of the endeavors our ancestors were adept at. They were aware of the irregularity in seasonal meteoric precipitation, and resultant flooding that wreaked havoc in agricultural lands, human settlements, and ports of yore. There is documented evidence of storing river water for irrigation and drinking purposes through damming, construction of sluices in surface water bodies in order to protect bank breach without affecting storage capacity, establishment of large ponds and linking them with man-made canals so that flood waters were stored and utilized in a river basin (with which erratic, short-duration meteoric precipitation do not wreak havoc, but have been utilized all throughout the year in order to mitigate coeval drought and flooding). They were also aware of the uniqueness of each river basin, and adopted different strategies of mitigating drought-floods in different river basins. For example, reinforcement of banks of the River Vaigai from breaching has been documented in ancient vedic texts that date back a few 1000 years, reinforcement of bank and diversion of direction of flow of the River Kaveri during the 11th century, known from stone inscriptions of the Srirangam Temple, construction of a gravity dam (Grand Anicut-see Fig. 9A) to slow down the flood flow in the River Kaveri by the Karikala Chola (http://www.historyfiles.co.uk/KingListsFarEast/IndiaCholas.htm; Rao, 1967; Spencer, 1984; it is also to be stated this dam is the oldest, yet in service gravity dam in the world), as documented in ancient scripts dating back to about 2000 years ago and validated through modern scientific methods (Chitra, 2003). Notwithstanding these incidents and endeavors of historic people and kings of yore, a specific scientific fact has to be learned from these; that is, though mitigation of flooding and drought and utilization of excessive meteoric precipitation for lean seasons were the endeavors, the ancestors employed different techniques for different river basins. For example, the rivers Kaveri and Vaigai were perennial and were prone to flooding seasonally. These were impediments for irrigation and human habitation, and caused loss of life, infrastructure, and agriculture in both the deltaic regions, which necessitated remedial measures. The measures undertaken in each of the riparian regions were, entirely different, according to the topography, lithology, slope, and hydrology.

Owing to the vast expanse of monotonously flat topography and slope, the peak flow of flooding in the River Kaveri occurred suddenly and from the dealtahead. The channel had shallow, low topographic banks, loose unconsolidated sand as the bedrock of the bank, and hence to tackle flooding and bank breach and to avoid sudden occurrences of peak flow, the distribution of flood waters into a large area, quickly, without permitting the flood level to rise at or close to bank level were necessary. Thus came into existence >39,000 diversion canals that branch out from major distributories of the River Kaveri (Fig. 9A). These are distributed such that, from the Grand Anicut, the Kaveri River branches off into two distributary channels, namely, the Kaveri and the Vennar. They branch further into 36 channels whose total length is 1607 km. These, in turn, branch off into 2988 channels running for a length of about 18,395 km (Kandaswamy, 1986). Except for the major paleochannels and distributary channels, all other canals are man-made, over the span of the past few centuries! Clearly this is a manifestation of thorough knowledge on river dynamics, geomorphology, meteorology, lithology, and the method of geohazard mitigation over and above water resource management practices of the ancient human race. Whereas the River Vaigai, despite having high-intensity, short-duration meteoric precipitation, similar to the Kaveri River Basin; the deltaic region is comparatively

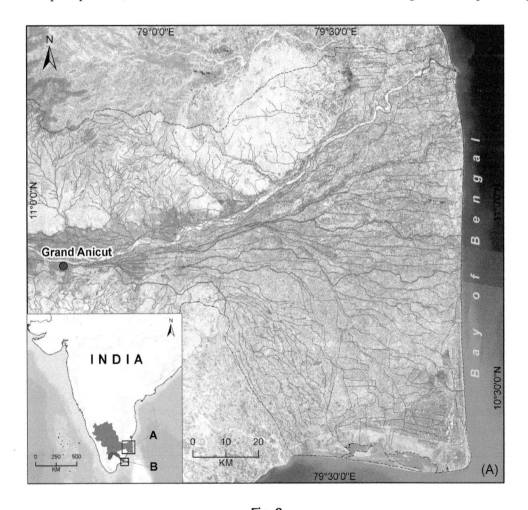

Fig. 9
(A) The Kaveri River Basin and distribution of innumerable man-made canals, and
Continued

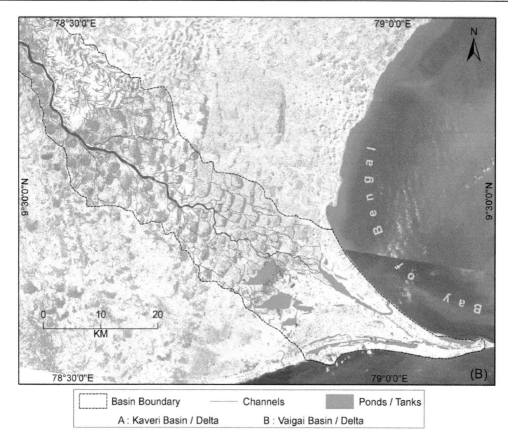

Fig. 9—cont'd
(B) the Vaigai River Basin and distribution of innumerable man-made ponds/tanks.

smaller, with slope, lithology, and topography that were highly varied and required another combination of hazard-resource management. Accordingly, our ancestors, including the local populace and kings of yore constructed thousands of ponds, dug link canals for all of them, and finally connected them with carefully chosen, historically vulnerable locations of trunk channels of the River Vaigai (Fig. 9B). The flood waters were allowed to be diverted through the link channels to be stored first in the lowest reaches of the delta. When the tanks reached the safety level, they were regulated through traditional sluices, then the flood waters were diverted to and stored in tanks aligned/located at the next upper reaches. This process was repeated into the next upper reaches. Thus came into being the thousands of interlinked pond systems in the Vaigai River Basin.

The ILR is not a novel concept, but has been in practice as a strategy to mitigate floods and droughts since historic times. However, each river basin is unique in terms of geomorphology, geology, hydrology, vegetation, and biodiversity, and hence, requires a unique management strategy for resource management and hazard mitigation. There is no single solution for all the river basins, and one method effective for a basin may not work for other basin. Hence, designing and implementing a unique combination of methods, suitable for local conditions is the best practice, especially when the stakes are high, and more specifically, when a major, continental scale program is envisaged.

5 Recommendations for Sustainable Development

Two major reasons are cited for nationalization and ILR. It is purported that these two actions would thwart water resource imbalance and minimize/avoid droughts and floods. As most of the rivers are either transnational and/or traverse interstate, the developmental activities, land use, and agricultural practices of regions located at the catchments affect the riparian regions, which in turn can be negated, and the coeval nature of droughts and floods in different parts of the country could be mitigated through transfer of the surplus to deficit regions. Among the two actions proposed, nationalization is of paramount importance, as it entails equal availability of resources, and it may not amount to interfering with the rights of states. In addition, such an action would ensure implementation of internationally accepted and practiced procedures of making available the water resources at the riparian region first. However, the monitoring and enforcement directorates have to be established and become operational in order to implement the objectives impartially.

Three facts/issues need to be considered before making the call on ILR: first, it is suggested that ILR is the solution to mitigate flooding in Ganges-Brahmaputra, and coeval drought in the peninsular region. India has more than the world's average of rainfall; however, the rainfall is not uniform in all the regions, and the rainy seasons are intense and short, which, without exception, results in drought and flood simultaneously in different parts of India. Hence, transfer of surplus to deficit regions and storage of excess water for round-the-year utility are necessary. Second, the northern rivers, the Ganges and the Brahmaputra, debauch enormous quantities of water and sediments into the Bay of Bengal, which forms a coastal river traveling all the way up to Tuticorin. It means that these two rivers have linked themselves with the peninsular regions already. Though the deltas of the east coast share a genetic relationship, their evolutionary history is unique to each of them. At this juncture, it is important to consider this: the sources, and thus characteristics, of waters of Himalayan rivers and peninsular rivers are different; the geology, geomorphology, hydrology, and land use dynamics of each of the deltas are different; however, the debauched waters-sediments of all these Himalayan and Peninsular rivers are homogenized all along the coastal tracts. It all signifies minimal changes in river hydrology, and riparian zone ecology, even if ILR is implemented through the transfer of surplus Himalayan waters to Peninsular rivers. Nevertheless, as it seems from the geographic locations of linked canals, as proposed in the program document, linking of rivers will be effected at or near the delta heads and trunk rivers. As these are the sensitive zones of river systems, despite homogenization at coastal regions, due care and scientific studies based on linking are necessary. The river Brahmaputra encompasses one of the largest sprawls of flood plains, which were built largely due to historic and geological recurrences of floods. The river basin is shared by Bangladesh, China, India, and so forth, among which India and Bangladesh are located in the riparian region, due to which the threat of floods, and damages to infrastructure, agriculture, and lives are greater in these countries. The northern neighbor, China, appears to consider combatting flooding as a tactical war strategy, and has tested their strategies in the recent past. At this juncture, in addition to mitigating droughts and floods, plans and resources are also required to handle strategic flooding.

Recognizing the geological reality, it is plain that ILR is already in place all along the eastern continental margin of India. The ILR being planned by the Government of India is not to be implemented in either the catchment or the upper reaches of the river basin; rather, mostly in the downstream and deltahead regions. As these are the ecologically sensitive, geologically active, and vulnerable regions to geohazards, and also hosts to thick populations and biodiversity, care and meticulous planning are essential before implementation of ILR. In-course monitoring and after-effect assessment-remedy-reclamation efforts are necessary; without them, the program may spell doom. More specifically, the following may be taken as guidelines for planning, executing, and monitoring mechanisms for the mammoth ILR program.

- It may appear necessary to have ILR in place given the recurrence, magnitude, and impact of coeval droughts and floods; however, once finalized, implementaton of ILR should not, in any case, be an isolated program. It is necessary to prepare a masterplan inclusive of comprehensive ILR for the whole of India, including adjoining neighboring countries, incorporating in it the national waterways grid, flood and drought management, geohazard mitigation, environmental management, ecological sustenance, a social forestry scheme, development of reserved forests, wildlife, agriculture protection and mangement, land use management, and so forth. Based on these, a comprehensive plan document with location and measure specifics and milestones tenable for 3, 5, 10, 25, and 50 years, followed by unbiased assessment-based remedial, reclamation, and development activities, with measurable attributes should be prepared, and only on this basis should the program be implemented.
- Though utilization of natural resources and their equitable distribution through human intervention are commonly accepted methods, the underlying principle in these endeavors should be sustainable to maintain optimum balance of ecological equilibrium.
- The natural geological processes maintain a pace of their own and have the ability to maintain self-controlled equilibrium. Any human intervention should be minimal, and also at the pace of natural cycles. Any changes introduced should conform to the limits of restoration by natural processes.
- Within the ambit of a comprehensive masterplan, instead of fewer largescale alterations, multiple, small scale modifications should be planned and executed, in addition to

- alternatives instead of alterations, which in turn should be data-research based instead of presumptions.
- As the river systems of India form their own climatic systems, the responses of which may differ uniquely, establishment of mechanisms to monitor source-sink dynamics, and climate-tectonics-landscape-ecology feedbacks, based on which design, implementation, and monitoring ecological, geological, geohazard, and land use changes for impact mitigation, reduction, andreclamation is necessary. Efforts should be made to include all the stakeholders involving inhabitants, scientists, administrators, and so forth in this mechanism.
- Documentation of traditional methods of flood-drought and other geohazard mitigation, practiced in each of the river basins, with implementation of them wherever appropriate.
- A hierarchical environmental baseline database at the local, regional, and continental scale should be created and incorporated with short and long-term data. Creation and maintanance of this database should be one of the functions of the mechanism suggested herein.
- The link channels should be designed and erected on the basis of least environmental disturbance, rather than economic feasibility.
- Identification of paleochannel courses and utilization of their tracts wherever possible to locate the link channels.
- The link channels should also be integrated with augmentation of subsurface water resources, recharging groundwater aquifers, and so forth, not exclusively to handle drought/flood situations, and or to enhance areas of irrigation.
- Instead of linking the rivers directly through major longer channels, interspersing them with check dams, ponds, and filteration ponds will restrict homogeneization of micronutrients and microorganisms, and thwart conductance of pathogens from one system to other.
- The ecological flow of each river (both donor and receptor) should be estimated and maintained, and only on this basis the quantum of transfer (supply and receipt), timing, location, frequency, and so forth should be planned and implemented, not on the basis of requirement or availability.

Acknowledgments

Shri.K.Vijayaragavan, a noted historian, Srirangam, is acknowledged for drawing the attention of MR on the stone inscriptions and temple documents that provided documentary evidence of past flooding events in the Kaveri deltahead, and flood-control measures undertaken by kings and seers thousands of years ago that initiated interest in the ILR program.

References

Albinus, M.P., Makalle, J.O., Bamutaze, Y., 2008. Effects of land use practices on livelihoods in the transboundary sub-catchments of the Lake Victoria Basin. Afr J Environ Sci Tech 2, 309–317.

Allison, M.A., Khan, S.R., Goodbred Jr., S.L., Kuehl, S.A., 2003. Stratigraphic evolution of the late Holocene Ganges–Brahmaputra lower delta plain. Sediment. Geol. 155, 317–342.

Anthony, E.J., 2015. Wave influence in the construction, shaping and destruction of river deltas: a review. Mar. Geol. 361, 53–78.

Avinash, K., Deepika, B., Jayappa, K.S., 2013. Evolution of spit morphology: a case study using a remote sensing and statistical based approach. J. Coast. Conserv. 17, 327–337.

Ayivor, J.S., Gordon, C., 2012. Impact of land use on river systems in Ghana. West African Jour. App. Ecol. 20, 83–95.

Banerjee, P.K., Vaz, G.G., Sengupta, B.J., Bagchi, A., 2001. A qualitative assessment of seismic risk along the peninsular coast of India, south of 19oN. J. Geodyn. 31, 481–498.

Banerji, R.K., 1984. Post-Eocene biofacies, palaeoenvironments and palaeogeography of the Bengal Basin, India. Palaeogeogr. Palaeoclimatol. Palaeoecol. 45, 49–73.

Barrow, C.J., 1998. River basin development planning and management: a critical review. World Dev. 26, 171–186.

Bastia, R., Radhakrishna, M., 2012. Continental Margin of India. Elsevier, Amsterdam, p. 417.

Bastia, R., Radhakrishna, M., Das, S., Kale, A.S., Catuneanu, O., 2010. Delineation of the 85°E ridge and its structure in the Mahanadi Offshore Basin, eastern continental margin of India (ECM), from seismic reflection imaging. Mar. Pet. Geol. 27, 1841–1848.

Biswas, S.K., 1982. Rift basins in the western margin of India and their hydrocarbon prospects. Bull. Am. Assoc. Pet. Geol. 66, 1497–1513.

Biswas, S.K., 1987. Regional tectonic framework, structure and evolution of the western marginal basins of India. Tectonophysics 135, 307–327.

Biswas, S.K., 2005. A review of structure and tectonics of Kutch Basin, western India, with special reference to earthquakes. Curr. Sci. 88, 1592–1600.

Campanile, D., Nambiar, C.G., Bishop, P., Widdowson, M., Brown, R., 2007. Sedimentation record in the Konkan–Kerala Basin: implications for the evolution of the Western Ghats and the Western Indian passive margin. Basin Res. 20, 3–22.

Chaitanya, A.V.S., Lengaigne, M., Vialard, J., Gopalakrishna, V.V., Durand, F., Kranthikumar, C., Amritash, S., Suneel, V., Papa, F., Ravichandran, M., 2014. Salinity measurements collected by fishermen reveal a "river in the sea" flowing along the eastern coast of India. J. Am. Meteorol. Soc. https://doi.org/10.1175/BAMS-D-12-00243.1.

Chatterjee, S., Goswami, A., Scotese, C.R., 2013. The longest voyage: tectonic, magmatic, and paleoclimatic evolution of the Indian plate during its northward flight from Gondwana to Asia. Gondw. Res. 23, 238–267.

Chauhan, M.S., Sharma, A., Phartiyal, B., Kumar, K., 2013. Holocene vegetation and climatic variations in Central India: a study based on multiproxy evidences. J. Asian Earth Sci. 77, 45–58.

Chitra, K., 2003. Tank and Irrigation Systems: An Engineering Analysis. (Ph. D. dissertation) submitted to the Indian Institute of Technology-Madras, Chennai, p. 182.

Clapcott, J., Young, R., Goodwin, E., Leathwick, J., Kelly, D., 2011. Relationships between multiple land-use pressures and individual and combined indicators of stream ecological integrity. DIC Res Dev Ser 326, 57.

Coulthard, T.J., Macklin, M.G., 2001. How sensitive are river systems to climate and land use changes? A model-based evaluation. J Quater Sci 16, 347–351.

Föllmi, K.B., 1996. The phosphorus cycle, phosphogenesis and marine phosphate-rich deposits. Earth Sci. Rev. 40, 55–124.

Glendenning, C.J., Vervoort, R.W., 2010. Hydrological impacts of rainwater harvesting (RWH) in a case study catchment: the Arvari River, Rajasthan, India. Part 1: field scale impacts. Agr Water Manage 98, 331–342.

Gombos Jr., A.M., Powell, W.G., Norton, I.O., 1995. The tectonic evolution of western India and its impact on hydrocarbon occurrences: an overview. Sediment. Geol. 96, 119–129.

Gunnel, Y., Gallagher, K., Carter, A., Widdowson, M., Hurford, A.J., 2003. Denudation history of the continental margin of western peninsular India since the early Mesozoic—reconciling apatite fission-track data with geomorphology. Earth Planet. Sci. Lett. 215, 187–201.

Gunnell, Y., 1998. Passive margin uplifts and their influence on climatic change and weathering patterns of tropical shield regions. Global Planet. Change 18, 47–57.

Gunnell, Y., Radhakrishna, B.P., 2001. Sahyadri, the Great Escarpment of the Indian Subcontinent. Memoirs of Geological Society of India 47. Geological Society of India, Bangaluru, p. 1053.

Harbor, D., Gunnell, Y., 2007. Along-strike escarpment heterogeneity of the Western Ghats: a synthesis of drainage and topography using digital morphometric tools. J. Geol. Soc. India 70, 411–426.

Hesslerova, P., Chmelova, I., Pokorny, J., Sulcova, J., Kropfelova, L., Pechar, L., 2012. Surface temperature and hydrochemistry as indicators of land cover functions. Ecol. Eng. 49, 146–152.

Hooper, P., Widdowson, M., Kelley, S., 2010. Tectonic setting and timing of the final Deccan flood basalt eruptions. Geology 38, 839–842.

Hudson, P.F., Colditz, R.R., Aguilar-Robledo, M., 2006. Spatial relations between floodplain environments and landuse-land cover of a large lowland tropical river valley: panuco basin, Mexico. Environ. Manag. 38, 487–503.

Jana, A., Biswas, A., Maiti, S., Bhattacharya, A.K., 2014. Shoreline changes in response to sea level rise along Digha coast, eastern India: an analytical approach of remote sensing, GIS and statistical techniques. J. Coast. Conserv. 18, 145–155.

Jayalakshmi, K., Nair, K.M., Kumai, H., Santosh, M., 2004. Late Pleistocene–Holocene palaeoclimatic history of the southern Kerala Basin, Southwest India. Gondw. Res. 7, 585–594.

Jayanthi, M., 2009. Mitigation of geohazards in coastal areas and environmental policies of India. In: Ramkumar, M. (Ed.), Geological Hazards: Causes, Consequences and Methods of Containment. New India Publishers, New Delhi, pp. 141–147.

Jessel, B., Jocobs, J., 2005. Land use scenario development and stakeholder involvement as tools for watershed management within the Havel River basin. Limnologica 35, 220–233.

Kale, V.S., 2014. Geomorphic history and landscape of India. In: Kale, V.S. (Ed.), Landscapes and Landforms of India. Springer-Verlag, Heidelberg, pp. 25–37.

Kale, V.S., Vaidyanadhan, R., 2014. The Indian Peninsula: geomorphic landscapes. In: Kale, V.S. (Ed.), Landscapes and Landforms of India. Springer-Verlag, Heidelberg, pp. 65–78.

Kandaswamy, P.K., 1986. Irrigation development in Tamil Nadu. Bhagirath 22, 67–73.

Kar, A., 1993. Neotectonic influences on morphological variations along the coastline of Kachchh, India. Geomorphology 8, 199–219.

Koch, J., Onigkeit, J., Schaldach, R., Alcamo, J., Köchy, M., Wolff, H., Kan, I., 2011. Land use scenarios for the Jordan River region. Inter. Jour. Sustain. Water. Environ. Sys. 3, 25–31.

Kumar, R., Singh, R.D., Sharma, K.D., 2005. Water resources of India. Curr. Sci. 89, 794–811.

Kunte, P.D., 1994. Sediment transport along the Goa-North Karnataka coast, western India. Mar. Geol. 118, 207–216.

Kunte, P.D., Wagle, B.G., 1993. Determination of net shore drift direction of central west coast of India using remotely sensed data. J. Coast. Res. 9, 811–822.

Kunte, P.D., Wagle, B.G., 2000. The beach ridges of India: a review. J. Coast. Res. 42, 174–183.

Kunte, P.D., Alagarsamy, R., Hursthouse, A.S., 2013. Sediment fluxes and the littoral drift along Northeast Andhra Pradesh coast, India: estimation by remote sensing. Environ. Monit. Assess. 185, 5177–5192.

Lal, N.K., Siawal, A., Kaul, A.K., 2009. Evolution of East Coast of India—a plate tectonic reconstruction. J. Geol. Soc. India 73, 249–260.

Li, S., Gu, S., Liu, W., Han, H., Zhang, Q., 2008. Water quality in relation to land use and land cover in the upper Han River basin, China. Catena 75, 216–222.

Lippert, P.C., van Hinsbergen, D.J.J., Dupont-Nivet, G., 2014. Early cretaceous to present latitude of the central proto-Tibetan plateau: a paleomagnetic synthesiswith implications for Cenozoic tectonics, paleogeography, and climate of Asia. In: Nie, J., Horton, B.K., Hoke, G.D. (Eds.), Toward an Improved Understanding of Uplift Mechanisms and the Elevation History of the Tibetan Plateau. Geological Society of America Special Paper 507.https://doi.org/10.1130/2014.2507(01).

Melluso, L., Sheth, H.C., Mahoney, J.J., Morra, V., Petrone, C.M., Storey, M., 2009. Correlations between silicic volcanic rocks of the St Mary's islands (southwestern India) and eastern Madagascar: implications for late cretaceous India–Madagascar reconstructions. J. Geol. Soc. London 166, 283–294.

Memarian, H., Tajbakhsh, M., Balasundaram, S.K., 2013. Application of SWAT for impact assessment of land use/land cover change and best management practices: a review. Inter. Jour. Advance. Earth. Environ. Sci. 1, 36–40.

Menier, D., Estournès, G., Manoj, M.J., Ramkumar, M., Briend, C., Siddiqui, N., Traini, C., Pian, S., Labeyrie, L., 2016. Relict geomorphological and structural control on the coastal sediment partitioning, north of Bay of Biscay. Z. Geomorphol. 60, 67–74.

Menier, D., Manoj, M.J., Pubellier, M., Sapin, F., Delcaillau, B., Siddiqui, N., Ramkumar, M., Santosh, M., 2017. Landscape response to progressive tectonic and climatic forcing in NW Borneo: Implications for geological and geomorphic controls on flood hazard. Nature Geosci. Rep. 7https://doi.org/10.1038/s41598-017-00620-y.

Naiman, R.J., Bilby, R.E., 1998. River ecology and management in the Pacific coastal ecoregion. In: Naiman, R.J., Bilby, R.E. (Eds.), River Ecology and Management: Lessons from the Pacific Coastal Ecoregion. Springer, New York, pp. 1–22.

Nair, M.M., 1999. Quaternary coastal geomorphology of Kerala. Ind. J. Geomorph. 4, 51–60.

Okazava, H., Yamamoto, D., Takeuchi, Y., 2010. Influences of riparian land use on nitrogen concentration of river water in agricultural and forest watersheds of northeastern Hokkaido. Jpn Int J Environ Rural Dev 1, 1–2.

Prasad, S., Pandarinath, K., Gupta, S.K., 1998. Geomorphology, tectonism and sedimentation in the Nal region, western India. Geomorphology 25, 207–223.

Qi, S., Luo, F., 2006. Land use change and its environmental impact in the Heihe River basin, arid northwestern China. Environ. Geol. 50, 535–540.

Radhakrishna, M., Rao, G.S., Nayak, S., Bastia, R., Twinkle, D., 2012a. Early cretaceous fracture zones in the BOB and their tectonic implications: constraints from multi-channel seismic reflection and potential field data. Tectonophysics 522-523, 187–197.

Radhakrishna, M., Twinkle, D., Nayak, S., Bastia, R., Rao, G.S., 2012b. Crustal structure and rift architecture across the Krishna–Godavari Basin in the central eastern continental margin of India based on analysis of gravity and seismic data. Mar. Pet. Geol. 37, 129–146.

Ramkumar, M., 2000. Recent changes in the Kakinada spit, Godavari delta. J. Geol. Soc. Ind. 55, 183–188.

Ramkumar, M., 2003a. Geochemical and sedimentary processes of mangroves of Godavari delta: implications on estuarine and coastal biological ecosystem. Ind. Jour. Geochem. 18, 95–112.

Ramkumar, M., 2003b. Progradation of the Godavari delta: A fact or empirical artifice? Insights from coastal landforms. Jour. Geol. Soc.Ind. 62, 290–304.

Ramkumar, M., 2004a. Dynamics of moderately well mixed tropical estuarine system, Krishna estuary, India: Part IV Zones of active exchange of physico-chemical properties. Ind. Jour. Geochem. 19, 245–269 (ISSN: 0970-9088).

Ramkumar, M., 2004b. Dynamics of moderately well mixed tropical estuarine system, Krishna estuary, India: part III nature of property-salinity relationships and quantity of exchange. Ind. Jour. Geochem. 19, 219–244.

Ramkumar, M., 2007. Spatio-temporal variations of sediment texture and their influence on organic carbon distribution in the Krishna estuary. Ind. Jour. Geochem. 22, 143–154.

Ramkumar, M., 2009. Flooding—a manageable geohazard. In: Ramkumar, M. (Ed.), Geological Hazards: Causes, Consequences and Methods of Containment. New India Publishers, New Delhi, pp. 177–190.

Ramkumar, M., 2015. Discrimination of tectonic dynamism, quiescence, and third order relative sea level cycles of the Cauvery Basin, South India. Ann. Geol. Penins.Balk. (76), 19–45.

Ramkumar, M., 2017. Internal 'oil war' in the Asian giant: Politics and environmental activism. Rec.t Adv. Petrochem. Sci. 2, 555590.

Ramkumar, M., Rajani Kumari, V., Pattabhi Ramayya, M., Gandhi, M.S., Bhagavan, K.V.S., Swamy, A.S.R., 2001a. Dynamics of moderately well mixed tropical estuarine system, Krishna estuary, India: part I Spatio-temporal variations of physico-chemical properties. Ind. Jour. Geochem. 16, 61–74.

Ramkumar, M., Pattabhi Ramayya, M., Rajani Kumari, V., 2001b. Dynamics of moderately well mixed tropical estuarine system, Krishna estuary, India: part II flushing time scales and estuarine mixing. Ind. Jour. Geochem. 16, 75–92.

Ramkumar, M., Stüben, D., Berner, Z., 2004. Lithostratigraphy, depositional history and sea level changes of the Cauvery Basin. South India. Ann. Geol. Penins.Balk. 65, 1–27.

Ramkumar, M., Stüben, D., Berner, Z., 2011. Barremian-Danian chemostratigraphic sequences of the Cauvery Basin, South India: implications on scales of stratigraphic correlation. Gondw. Res. 19, 291–309.

Ramkumar, M., Kumaraswamy, K., Mohanraj, R., 2015. Land use dynamics and environmental management of River Basins with emphasis on deltaic ecosystems: Need for integrated study based development and nourishment programs and institutionalizing the management strategies. In: Ramkumar, M., Kumaraswamy, K., Mohanraj, R. (Eds.), Environmental Management of River Basin Ecosystems. Springer-Verlag, Heidelberg, pp. 1–20. https://doi.org/10.1007/978-3-319-13425-3-1.

Ramkumar, M., Menier, D., Manoj, M.J., Santosh, M., 2016. Geological, geophysical and inherited tectonic imprints on the climate and contrasting coastal geomorphology of the Indian peninsula. Gondw. Res. 36, 52–80.

Ramkumar, M., Menier, D., Manoj, M.J., Santosh, M., Siddiqui, N.A., 2017. Early Cenozoic rapid flight enigma of the Indian subcontinent resolved. Geosci. Front. 8, 15–23. https://doi.org/10.1016/j.gsf.2016.05.004.

Rao, H., 1967. The Srirangam Temple: Art and Architecture. Sri Venkateswara University Historical Series 8p. 252.

Rao, K.L., 1975. India's Water Wealth. Orient Longman, p. 255.

Rao, G.K., 1998. Geo-hazard with oil and gas production in Krishna-Godavari basin. Curr. Sci. 74, 494.

Raval, U., Veeraswamy, K., 2003. India–Madagascar separation: breakup along a preexisting—mobile belt and chipping of the craton. Gondw. Res. 6, 467–485.

Robertson, A.I., Duke, N.C., 1987. Mangroves as nursery sites: comparisons of the abundance of fish and crustaceans in mangroves and other nearshore habitats in tropical Australia. Mar. Biol. 96, 193–205.

Robson, B., Austin, C., Chester, E., 2005. Environmental water allocation required to sustain macroinvertebrate species in ephemeral streams. River Riparian Lands Manage Newslett 29, 13–15.

Roy, A.B., 2004. The Phanerozoic reconstitution of Indian shield as the aftermath of breakup of the Gondwanaland. Gondw. Res. 7, 387–406.

Sabajo, C.R., Le Maire, G., June, T., Meijide, A., Rousard, O., Knohl, A., 2017. Expansion of oil palm and other cash crops causes an increase of the land surface temperature in the Jambi province in Indonesia. Biogeosciences 14, 4619–4635.

Sanil Kumar, V., Pathak, K.C., Pednekar, P., Raju, N.S.N., Gowthaman, R., 2006. Coastal processes along the Indian coastline. Curr. Sci. 91, 530–536.

Santosh, M., Yang, Q.Y., Shaji, E., Tsunogae, T., Ram Mohan, M., Satyanarayanan, M., 2014. Oldest rocks from peninsular India: evidence for hadean to Neoarchean crustal evolution. Gondw. Res. 29https://doi.org/10.1016/j.gr.2014.11.003.

Santosh, M., Yang, Q.Y., Shaji, E., Tsunogae, T., Ram Mohan, M., Satyanarayanan, M., 2015. An exotic Mesoarchean microcontinent: the Coorg Block, southern India. Gondw. Res. 27, 165–195. https://doi.org/10.1016/j.gr.2013.10.005.

Simarro, G., Guillén, J., Puig, P., Ribó, M., Iacono, C.L., Palanques, A., Muñoz, A., Durán, R., Acosta, J., 2015. Sediment dynamics over sand ridges on a tideless mid-outer continental shelf. Mar. Geol. 361, 25–40.

Singer, M.B., Aalto, R., James, L.A., 2008. Status of the lower Sacramento valley flood-control system within the context of its natural geomorphic setting. Nat Hazards Rev 9, 104–115.

Singh, K., Radhakrishna, M., Pant, A.P., 2007. Geophysical structure of western offshore basins of India and its implication to the evolution of the Western Ghats. J. Geol. Soc. India 70, 445–458.

Solaraj, G., Dhanakumar, S., Rutharvel Murthy, K., Mohanraj, R., 2010. Water quality in select regions of Cauvery Delta river basin, southern India, with emphasis on monsoonal variation. Environ. Monit. Assess. 166, 435–444.

Spencer, G.W., 1984. Heirs apparent: fiction and function in Chola mythical genealogies. Ind. Econ. Soc. History 21, 415–432.

Srivastava, J., Farooqui, A., 2017. Fluctuations in relative sea level and mangroves in the Coleroon estuary since ~4.6 ka: a palynological study from SE coast of India. Geophytology 47, 11–26.

Storey, M., Mahoney, J.J., Saunders, A.D., Duncan, R.A., Kelley, S.P., Coffin, M.F., 1995. Timing of hotspot-related volcanism and the breakup of Madagascar and India. Science 267, 852–855.

Subramanian, V., 2000. Water: Quantity—Quality Perspective in South Asia. Kingston International Publication, Surrey, p. 210.

Subramanian, V., 2002. A Textbook in Environmental Science. Alpha Science International Ltd, Oxford, p. 252.

Sukhtankar, R.K., 1995. An evolutionary model based on geomorphologic and tectonic characteristics of the Maharashtra coast, India. Quat. Int. 26, 131–137.

Suresh, R., Ramkumar, M., 2009. An Introduction to Drought. In: Ramkumar, M. (Ed.), Geological Hazards: Causes, Consequences and Methods of Containment. New India Publishers, New Delhi, pp. 155–163.

Tafangenyasha, C., Dzinomwa, T., 2005. Land-use impacts on river water quality in lowveld sand river systems in south-East Zimbabwe. Land Use Water Resour Res. 5 (3.1–3.10).

Tang, Z., Engel, B.A., Pijanowski, B.C., Lim, K.J., 2005. Forecasting land use change and its environmental impact at a watershed scale. J. Environ. Manage. 76, 35–45.

Tong, S.T.Y., Chen, W., 2002. Modeling the relationship between land use and surface water quality. J. Environ. Manage. 66, 377–393.

Triedman, N., 2012. Environment and ecology of the Colorado River basin. In: The 2012 State of the Rockies Report Card. pp. 89–107.

Tripathy, G.R., Singh, S.K., Bhushan, R., Ramaswamy, V., 2011. Sr-Nd isotope composition of the bay of Bengal sediments: impact of climate on erosion in the Himalaya. Geochem. J. 45, 175–186.

Walling, D.E., 1999. Linking land use, erosion and sediment yields in river basins. Hydrobiologia 410, 223–240.

Wu, X., Shen, Z., Liu, Z., Ding, X., 2008. Land use/cover dynamics in response to environmental and socio-political forces in the upper reaches of Yangtse River, China. Sensors 8, 8104–8122.

Zarea, R., Ionus, O., 2012. Land use changes in the Basca Chiojdului River basin and the assessment of their environmental impact. Forum Geogr. 11, 36–44.

Zhang, T., Zhao, B., 2010. Impact of anthropogenic land-uses on salinization in the Yellow River delta, China: using a new RS-GIS statistical model. Int Arch Photogram Remote Sens Spat Inf Sci 38, 947–952.

Zhang, T., Zeng, S., Gao, Y., Ouyang, Z., Li, B., Fang, C., Zhao, B., 2011. Assessing impact of land uses on land salinazation in the Yellow River delta, China using an integrated and spatial statistical model. Land. Policy 28, 857–866.

Zimmermann, B., Elsenbeer, H., de Moraes, J.M., 2006. The influence of land-use changes on soil hydraulic properties: implications for runoff regeneration. For. Ecol. Manage. 222, 29–38.

Zwolsman, J.J.G., 1994. Seasonal variability and biogeochemistry of phosphorus in the Scheldt estuary, Southwest Netherlands. Estuar. Coast. Shelf Sci. 39, 227–248.

CHAPTER 20

Utility of Landsat Data for Assessing Mangrove Degradation in Muthupet Lagoon, South India

Subbarayan Saravanan*, R. Jegankumar[†], Ayyakkannu Selvaraj*, Jacinth Jennifer J*, Parthasarathy K.S.S. [‡]

Department of Civil Engineering, NIT Tiruchirappalli, Tiruchirappalli, India, [†]Department of Geography, Bharathidasan University, Tiruchirappalli, India, [‡]Department of Applied Mechanics and Hydraulics, NIT Surathkal, Mangalore, India

1 Introduction

Mangroves can be defined as an association of trees and shrubs forming the dominant vegetation in tidal and saline wetlands along with the tropical and subtropical coasts (Tomlinson, 1986; Ricklefs and Schuter, 1993). Worldwide, there are 123 countries containing mangroves. Mangrove forests cover an area of approximately 160,000 km^2 all over the world, including the most significant forest areas of Malaysia, India, Bangladesh, Brazil, Venezuela, Nigeria, and Senegal (Giri and Muhlhausen, 2008; Alongi, 2009). Mangroves are the dominant vegetation of >70% of tropical and subtropical coastlines around the world. Mangroves grow in river deltas, estuarine complexes, and coasts in tropical and subtropical regions throughout the world. They also inhabit the shorelines and islands in sheltered coastal areas with locally variable topography and hydrology (Lugo and Snedaker, 1974). There is a need for global management of these ecosystems. Mangrove forests support various industries including, but not limited to, fisheries, timber, tourism, and coastal communities (Carter et al. 2015; Khairuddin et al., 2015). Mangroves act as a wall against cyclones, coastal erosion, and they provide nursery grounds for commercially important fish, prawns, and crabs (Alongi et al., 1992). The 20th century brought significant improvement in the field of coastal zone management by various programs and organizations such as the Ramsar Convention, International Tropical Timber Organization (ITTO), Earth Summit, Food and Agriculture Organization (FAO), and International Society for Mangrove Ecosystems (ISME) (Clark, 1992).

The coastal zone of the mainland of India and Andaman and the Nicobar Islands are covered with extensive and diverse mangrove forests. In India, mangroves are found extensively in the coastal regions of Sundarbans, Mahanadi, and the Godavari Delta, Pichavaram, the Muthupet Lagoon, Goa, Mumbai, and part of Kutch Bay. In India, the mangroves' spread has decreased from 6740 to 4460 sq. km (Thomas et al., 2017). Recent developments have posed a threat to the mangrove habitat due to the intrusion of anthropogenic activities in the mangrove environment (Dasgupta and Shaw, 2013). Several studies (Gamble and Fischer, 1957; Cornwell, 1937; Champion and Seth, 1968; Caratinal et al. 1973; Krishna Murthy et al. 1981; Gogate, 1984) have documented the mangrove forest composition and ecology in India. A study by Arunachalam et al. (2011) reveals that the agricultural area and the extent of settlements have increased, and the water bodies and other land use have decreased in the coastal region of Tamilnadu from 2000 to 2009. Selvam (2006) have studied the changes in the mangrove forest in the Pichavaram region due to the 2004 tsunami, and analyzed the coastal zone bordered with mangroves and other shelterbelts, such as casuarina plantations protected from the tsunami of 2004. Five hundred ha of mangrove forest were destroyed from 1935 to 1970. However, the vegetation cover was restored after analyzing various causes of degraded mangroves. Based on multispectral data, earlier studies imply that almost 1514 ha of mangrove forest was present during 1988 in the Vedaranniyam wetlands, on the southeast coast of India, where four species were dominant; namely, *Avicennia marina, Acanthus ilicifolius, Aeglceras corniculatum,* and *Excoecaria agallocha* (Ponnambalam et al., 2012). Ajithkumar et al., 2008, found six species of mangroves, namely, *A. marina, E. agallocha, A. corniculatum, A. ilicifolius, Suaeda monoica,* and *S. maritima*.

Poonthip et al. (2008) studied the impact of the 2004 Tsunami on the mangroves on the Thailand coast using remote sensing and GIS techniques, and analyzed the spatial pattern of damages with temporal images and the behavior of the mangrove forest during the disaster. Based on Landsat images in the Giao Thuy District, Vietnam Beland et al. (2007) documented the negative impact of shrimp culture on the environment causing wetland deterioration, thereby affecting the mangroves. Using GIS and various image processing techniques, researchers adopted numerous algorithms to study mangrove forests all over the world. Son et al. (2015) *used object-based image analysis techniques to extract the mangroves in the Ca Mau Peninsula, Vietnam, and obtained an overall accuracy of 82%.* Son et al. (2016) *carried out mangrove extraction in the Can Gio Biosphere Reserve, South Vietnam, using* tasseled cap transformation (TCT), and an unmixing model technique, and obtained an overall accuracy of 90%. Liu et al. (2008) developed a decision tree method to integrate multitemporal Landsat TM data and ancillary GIS data for identifying mangroves in the Pearl River Estuary, and obtained an average accuracy of 84%. Zhang et al. (2016) combined LiDAR data with four vegetation indices from Landsat TM imagery to detect changes in mangrove cover in South Florida due to seasonal changes, and concluded that the Normalized Difference Moisture Index (NDMI) performed best in identifying both seasonal and event-driven episodic changes. Satapathy et al. (2007) exploited multispectral data to detect changes in mangrove cover for a period of 12 years (1992–2004) in the Godavari Estuary, where anthropogenic interference destroyed an area of about 1250 ha of mangroves.

Change detection is a powerful tool to visualize, measure, and understand a trend in mangrove ecosystems (Dan et al., 2016). This chapter aims to evaluate the changes in the mangroves from 1999 to 2016, adopting supervised classification techniques with the benefits obtained from the indices such as NDVI and NDWI, and to analyze the underlying reasons behind the variations of the mangrove habitat in the Mullipallam Muthupet creek system on the southeastern coast of India.

2 Study Area

The 1076 km long Tamilnadu coastline is situated on the southeast coast of the Indian peninsula, and forms a part of the Coromandel Coast of the Bay of Bengal and the Indian Ocean. It includes the lower reaches of the Cauvery River and its delta. The delta region covers the central and east-central part of Tamilnadu, India. The coastal zone of Tamilnadu is very narrow, except in the Vedaranyam-Muthupet stretch, where extensive mudflats are present. The study area under investigation is the Muthupet region (Lat. 10°18′13″ to 10°20′71″N; and Long. 79°30′90″ to 79°34′87″E) along the coastal zone of the Bay of Bengal and Palk Strait (Fig. 1). The area is drained by the distributaries of the Cauvery River, including the River Paminiyar, Kilathangiyar, Korayar, Marakkakorayar, and the Kandankurichanar channel, namely Paminiyar, Koraiyar, Kandankurichanar Kilathangiyar, and Marakkakoraiyar. This region contains 13,000 ha of mangroves of 19 different species, in which *A. marina* is dominant (State of Forest Report, 1999). Uniform thick vegetation cover is found throughout the coast, with the bulk of the mangroves consisting of *Avicennia officinalis, A. alba*, and *A. marina* species. The vertical structure of the forest shows two distinct stories, that is, top and ground. *A. ilicifolius, Rhizophora mucronata, R. conjugata, Brugueria gymnorrhiza, E. agallocha, E. marina, Avicennia sp, Sonneratia apetala*, and so forth, form a top story, whereas, ground vegetation cover was dominated by *Suaeda nudiflora*.

3 Methodology

3.1 Image Preprocessing

Preprocessing of satellite imagery is essential for performing the change detection technique. The steps involved are (1) preprocessing of remotely sensed satellite imagery, (2) computation of spectral indices such as NDVI and NDWI, and exploiting them in the LU/LC classification using the maximum likelihood classification algorithm.

The zero cloud cover satellite dataset of Landsat TM, Landsat-7 ETM+, and Landsat OLI data of the study area were downloaded Then they were used for computation of DN values to at-sensor radiance,

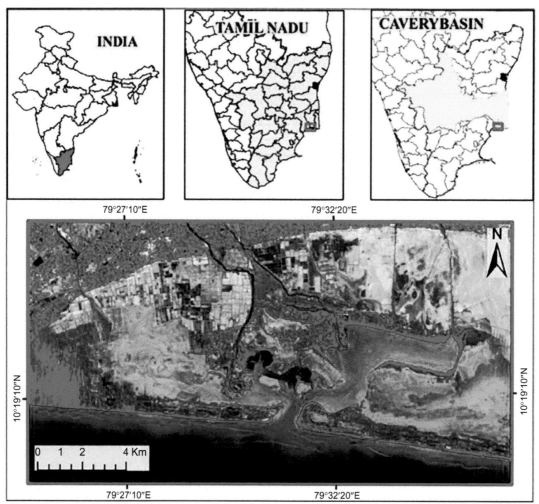

Fig. 1
Location map of the Study Area.

$$L_\lambda = \frac{L\max_\lambda - L\min_\lambda}{QCAL\max - QCAL\min} * (DN - QCAL\min) + L\min_\lambda \quad (1)$$

L_λ = Spectral radiance at the sensor (watts/m square* Ster* μm).
DN = The quantised calibrated pixel value in DN.
$L\min_\lambda$ = The spectral radiance scaled to QCALmin (watts/m square* Ster* μm).
$L\max_\lambda$ = The spectral radiance scaled to QCALmax (watts/m squared* Ster* μm).
QCALmin = The minimum quantized calibrated pixel value in DN = 1.
QCALmax = The maximum quantized calibrated pixel value in DN = 255.
Lmax and Lmin from the metadata file available with the Landsat image.

3.2 NDVI Map Generation

In this study, the Normalized Difference Vegetation Index (NDVI) was used to distinguish the vegetated regions from other surface types, especially water and land concerned in this study. The NDVI is an indicator of the vegetation that uses the red and near-infrared bands of the electromagnetic spectrum. NDVI was computed to analyze the presence of live green vegetation in the target region. Theoretically, NDVI values range from -1 to 1. NDVI values can be obtained by the following Eq. (2):

$$\text{NDVI} = \frac{NIR - R}{NIR + R} \qquad (2)$$

In this equation, Band 4 and Band 3 define the NIR and RED bands of the electromagnetic spectrum in Landsat TM and ETM+, whereas it is Band 5 and Band 4 that describe the NIR and RED band of the electromagnetic spectrum in Landsat 8 OLI/TIRS. In this study, ArcGIS 10.4 software was used to calculate NDVI.

3.3 NDWI Map Generation

The Normalized Difference Water Index (NDWI) was utilized to depict the open water features and upgrade their visibility in the remotely sensed digital image. In creek and lagoon geomorphologies, the water body and vegetation are susceptible to being misclassified; therefore, NDWI helps to delineate the water body from the other surface features. It functions on the earth surface reflectance of the green band and near-infrared band to highlight the water features from the other features of the Earth's surface. An appropriate threshold factor has to be specified for density slicing to delineate the water features. When the threshold is 0, NDWI >0 is water features, and NDWI <0 is classified as other earth features.

3.4 Land Use and Land Cover Classification

The land use and land cover categories are classified using the supervised classification method. For this, four LANDSAT images (30 m spatial resolution) from 1999, 2008, 2013, and 2016 were used. The supervised classification involved two steps: first, collection of data from the field survey; and the second step involved the exploitation of the collected data and assigning the training classes to classify the images for the study. The maximum likelihood classification algorithm served as the best algorithm for classification. The area was classified accordingly into seven classes: creek, mudflats, dense mangrove, sparse mangrove, vegetation, sand dunes, and deep water.

3.5 Land Use/Cover Change Detection

The study employed the postclassification change detection technique, performed in ArcGIS 10.4. The required amount of ground truth data has been incorporated in training the classes to improve classification accuracy. The statistics regarding the variations were obtained using image differencing techniques, using the ArcGIS 10.4 software.

4 Results and Discussion

The study adopted Landsat images of four different years, which covered nearly two decades. The spectral indices such as NDVI and NDWI extracted the regions of the vegetation cover and water bodies from other Earth features, respectively. By adopting the delineated features of the water and vegetation, a supervised classification was carried out to predict the change detection of the mangroves in the study area over the years. From the classification results obtained, the background analysis was carried out to understand the underlying reasons behind the trend in variation.

4.1 Spectral Indices

4.1.1 Normalized difference vegetation index

The value of NDVI varied between -1.0 and $+1.0$. The healthy vegetation has low red-light reflectance and high near-infrared reflectance that produce high NDVI values. The mounting amount of the positive NDVI values indicates the increase in the amounts of green vegetation. The NDVI values near zero and decreasing negative values indicate nonvegetated features, such as barren surfaces (rock and soil), water, snow, ice, and clouds. In this study, the NDVI data were rescaled to 0–255 to store the results in unsigned 8-bit data. Fig. 2 shows the difference of NDVI during different years.

4.1.2 Normalized difference water index

There are two remote sensing-derived indices concerning liquid water; one is used to monitor the changes in the water content of leaves by using NIR and SWIR band; the other is used to track the changes concerning water content in water bodies by using Green and NIR wavelength. NDWI varies between -1 to 1, which is based on the leaf water content, but also on the vegetation type and cover. A higher value of NDWI indicates higher vegetation water content. The lower value of NDWI shows low vegetation water content. In this study, Green and NIR wavelength were used to derive the NDWI to classify the clear boundary between vegetation and water body. This result was then incorporated as an input criteria to choose in the training classes for supervised classification. The derived NDWI images for the different years are shown in Fig. 3.

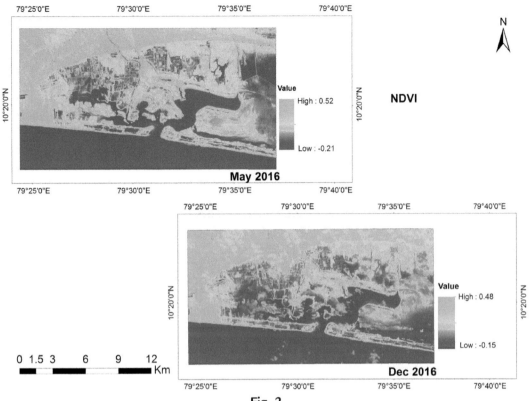

Fig. 2
NDVI of Muthupet Mangrove forest.

4.2 Change Detection

The classified image resulted with seven different classes such as creeks, deep water, dense mangroves, mud flats, sand dunes, sparse mangroves, and vegetation. The analysis revealed that the significant land use in this study was dedicated to mudflats and creeks. Fig. 4 shows the various land use and landcover changes from 1999 to 2016.

The selected study area occupies 13,343 ha, which contains the Muthupet mangrove forest, other vegetation, and mudflats; among which, the dense mangrove occupied 1152 ha (8.5%) in 1999, and shrank to 986 ha (7.3%) in 2008 (as shown in Fig. 5) due to the increase in the salinity of the water, aquaculture farms, and shoreline erosion. The total area of dense mangrove degraded about 40.3 ha from 2013 to 2016. The dense mangroves decreased by 1.29%, 1%, and 0.3% during the years 1999–2008, 2008–2013, and 2013–2016, respectively (Fig. 4). The sparse mangroves in the study area were about 716 ha in 1999. This increased to about 988 ha in 2016 due to restoration activities. About 191 ha of sparse mangroves increased from 1999 to 2008. The sparse mangroves increased by 1.53%, 0.32%, and 0.28% during the years

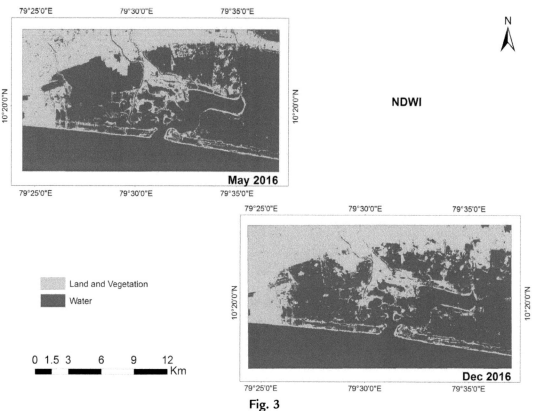

Fig. 3
NDWI of Muthupet Mangrove forest.

1999–2008, 2008–2013, and 2013–2016, respectively. The degradation of mangrove forest was due to conversion into shrimp farms, the introduction of hypersaline conditions by the reduction in the inflow of fresh water, and also due to the discharge of sewage from the shrimp farms. Other vegetation significantly increased from −0.77% to 2.25% between the years 2013 to 2016. The area under vegetation also profoundly increased during the year 2016 by 8%. Figs. 6 and 7 depict the rate of change in the land use/land cover (LU/LC) during the years 1999 to 2016. Table 1 illustrates the percentage of change in the area of the various LU/LC classes.

The shallow/creek water slightly decreased by 1%, 1.5%, and 0.8% during the years 1999–2008, 2008–2013, and 2013–2016, respectively. The width and depth of the lagoon mouth shrank due to siltation from the sea. In some regions of the lagoon, the depth of the water is not high during the high tides. The sprawl of sand dunes increased by 0.52% during 1999–2008, and subsequently shrank by 0.4% and 0.08% during the years of 2008–2013 and 2013–2016, respectively, because the sand dunes were occupied by plantations.

Mudflats consist of clay and silt with a meager percentage of sand. Tidal currents play a significant role in the formation of mudflats. The mudflats shrank by 1.8%, 0.7%, and 2.7%

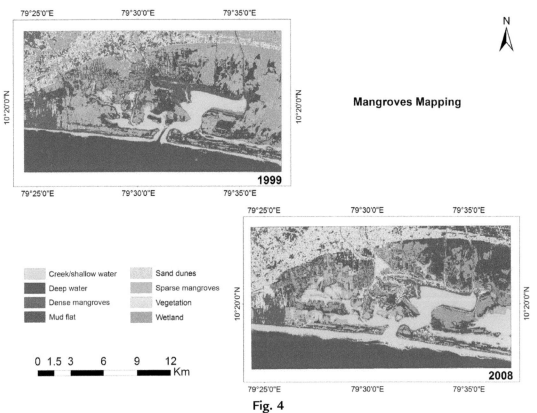

Fig. 4
Percentage of Land Use and Land Cover changes 1999–2008.

from the years 1999–2008, 2008–2013, and 2013–2016, respectively, due to the expansion of saltpans and prawn farms, and the intrusion of seawater into the mudflat region due to the 2004 tsunami.

Ponnambalam et al. (2012) used LANDSAT data for determination of degradation of mangroves in the Mullipalayam Creek, and found that 165.4 ha of mangrove cover degraded in about 16 years (1997–2007); when compared with their study, approximately 166 ha of mangroves deteriorated from 1999 to 2008. Santhi et al. (2008) used IRS LISS-III+PAN merged data for wetland mapping in the Vedaranyam region; their result showed that wetlands shrank from 2003 to 2005.

The analysis done through field visits helped in reasoning out the factors behind the degradation of mangroves and sparse mangroves. Human intervention and the encroachment of human activities in the region served as the reasons behind the increase in the salinity of the water. To reverse this degradation, various restoration techniques were introduced that resulted in the rise of sparse mangrove cover from 1999 to 2016. In the years after 2004, there seemed to be an aggressive increase in the agricultural activities in the region, which paved the way for further

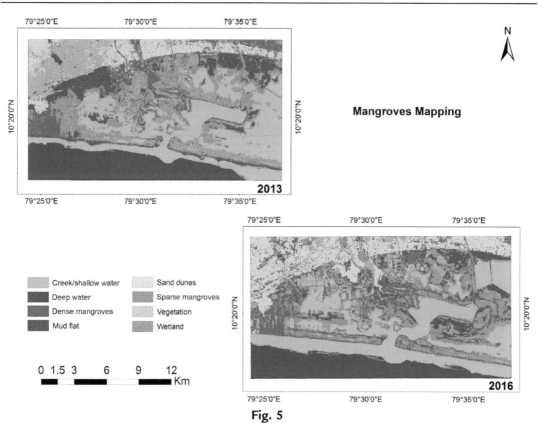

Fig. 5
Percentage of Land Use and Land Cover changes 2013–2016.

degradation in the mangrove cover. Although these anthropogenic activities exerted a great deal of influence, few rehabilitation activities have been carried out in the region to retain the mangrove cover.

5 Conclusions

The classification results from the multispectral data (Landsat) for the years between 1999 and 2016 enabled classification of seven different classes of LU/LC: creek, deep water, dense mangrove, mudflat, sand dunes, sparse mangrove, and vegetation.

There was degradation of 40.3 ha of dense mangrove from 2013 to 2016, and 135.5 ha from 2008 to 2013, and 166 ha from 1999 to 2008 due to high salinity, coastal erosion, the intrusion of saltpans, aquaculture farmlands, and human activity. The area of sparse mangrove increased by 38.2 ha between 2013 and 2016, by 42.7 ha from 2008 to 2013, and by 191.3 ha from 1999 to 2008 due to restoration activity. Other vegetation significantly increased by 298 ha from the year 2013–2016.

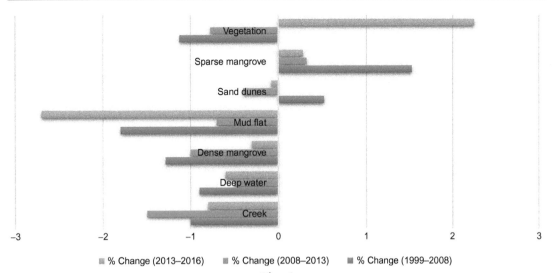

Fig. 6
Mangrove Mapping for the years 1999 and 2008.

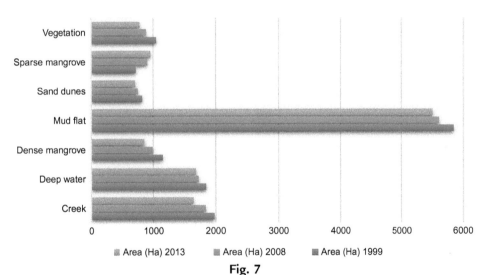

Fig. 7
Mangrove Mapping for the years 2013 and 2016.

As Muthupet mangrove wetland is situated at the end of the Cauvery Basin, freshwater reaching this region is decidedly less, owing to the construction of many dams upstream of the Cauvery River, which affects the mangrove environment. This study confirms that there is a need for proper protection activities for mangrove management and restoration programs, failing which there will be a significant loss in profitable fisheries and local economies, as well as increased erosion and shoreline uncertainty in numerous coastal communities.

Table 1 LULC Statistics for the Muthupet Lagoon

Land Use	Area (Ha) 1999	In %	Area (Ha) 2008	In %	Area (Ha) 2013	In %	Area (Ha) 2016	In %	% Change (1999–2008)	% Change (2008–2013)	% Change (2013–2016)
Creek	1982.2	14.7	1843.2	13.7	1642.1	12.2	1538.2	11.4	−1	−1.5	−0.8
Deep water	1850.2	13.7	1720.8	12.8	1680.2	12.3	1581.3	11.7	−0.9	−0.5	−0.6
Dense mangrove	1152.5	8.59	986	7.3	850.5	6.3	810.2	6	−1.29	−1	−0.3
Mudflat	5847.6	43.5	5602.8	41.7	5503.5	41	5150.5	38.3	−1.8	−0.7	−2.7
Sand dunes	821.2	6.12	752.3	5.60	708.2	5.2	687.5	5.12	0.52	−0.4	−0.08
Sparse Mangrove	716.2	5.3	907.5	6.76	950.2	7.08	988.4	7.36	1.53	0.32	0.28
Vegetation	1043.8	7.78	882.5	6.57	782.5	5.8	1080.5	8.05	−1.13	−0.77	2.25

References

Ajithkumar, T.T., Thangaradjou, T., Kannan, L., 2008. Spectral reflectance properties of mangrove species of the Muthupettai mangrove environment, Tamil Nadu. J. Environ. Biol. 29 (5), 785–788.

Alongi, D., 2009. The Energetics of Mangrove Forests. Springer, New York, USA, p. 216. ISBN-13: 9781402042713.

Alongi, D.M., Boto, K.G., Robertson, A.I., 1992. In: Alongi, D.M., Robertson, A.I. (Eds.), Tropical Mangrove Ecosystem. American Geophysical Union, Washington, DC, pp. 251–292.

Arunachalam, S., Maharani, K., Chidambaram, S., Prasanna, M.V., Manivel, M., Thivya, C., 2011. A study on the land uses pattern change along the coastal region of Nagapattinam, Tamil Nadu. Int. J. Geomat. Geosci. 1, 700–720.

Beland, M., Goïta, K., Bonn, F., Pham, T.T.H., 2007. Assessment of land-cover changes related to shrimp aquaculture using remote sensing data: a case study in the Giao Thuy District, Vietnam. pp. 1491–1510.

Caratinal, C., Blasco, F., Thanikaimoni, G., 1973. Relation between the pollen spectra and the vegetation of South Indian Mangrove. Pollen Spores 15, 281–292.

Carter, H.N., Schmidt, S.W., Hiros, A.C., 2015. An international assessment of mangrove management. Diversity 7, 74–104.

Champion, H.G., Seth, S.K., 1968. A Revised Survey of the Forest Types of India. Government of India, Publication Division, New Delhi.

Clark, J.R., 1992. Integrated management of the coastal zone. technical paper no. 327, 13, Food and Agriculture Organization (FAO), Rome, Italy, p. 167.

Dan, T.T., Chen, C.F., Chiang, S.H., Ogwa, S., 2016. In: Mapping and Change Analysis by Using LANDSAT Imagery in Mangrove Forest.ISPRS Annals of the Photogrammetry, Remote Sensing and Spatial Information Sciences, pp. 3–8.

Dasgupta, R., Shaw, R., 2013. Changing perspective of mangrove management in India-an analytical overview. Ocean Mange. 80, 107–108.

Cornwell, R.B. (1937). Working plan for the forests of Sunderbans Division, 1933, 51 Vols. 1-2.

Gamble JS & Fischer CEC, 1957, Flora of Presidency of Madras, (Adlard & Sons Ltd., London), 1915-1930; Reprinted Edn, 3 Vols, (Botanical Survey of India, Calcutta).

Giri, C., Muhlhausen, J., 2008. Mangrove forest distributions and dynamics in Madagascar (1975-2005). Sensors 8, 2104–2117.

Gogate, M.G., (1984). East Coast Mangroves: An overview In: Proc. National Seminar on Resources, Development, and Environment of the Eastern Ghats, Andhra University Press. Waltair, A.P., India, Pp. 209–214.

Khairuddin, B., Yulianda, F., Kusmana, C., Yonvitner, 2015. In: Degradation mangrove by using Landsat 5 TM and Landsat 8 OLI image in Mempawah Regency, West Kalimantan Province year 1989–2014.The 2nd International Symposium on LAPAN-IPB Satellite for Food Security and Environmental Monitoring 2015, LISAT-FSEM.

Krishna Murthy, K., et al., 1981. A floristic study of halophytes to the Pichavaram mangroves. Bull. Bot. Surv. India 23, 114–120.

Ponnambalam, K., Chokkalingam, L., Subramaniam, V., Ponniah, J.M., 2012. Mangrove distribution and morphology changes in the Mullipallam Creek, Southeastern Coast of India. Int. J. Conserv. Sci. 3, 51–60.

Liu, K., Li, X., Shi, X., Wang, S., Hampshire, N., 2008. Monitoring Mangrove Forest Changes Using Remote Sensing And Gis Data With Decision-Tree Learning. Wetlands 28 (2), 336–346.

Lugo, A.E., Snedaker, S.C., 1974. The Ecology of Mangroves. J. Annu. Rev. Ecol. Evol. Syst. 5, 39–64.

Poonthip, S., Wanxiao, S., Tonny, J.O., 2008. Assessing the impact of the 2004 tsunami on mangroves using remote sensing and GIS techniques. Int. J. Remote Sens. 29 (12), 3553–3576.

Ricklefs, R.E., Schuter, D., 1993. Species Diversity in Ecological Systems. Chicago Press Ltd, London.

Santhi, D.R., Nagalakshmi, R., Nagamani, K., Manoharan, N., 2008. Multi-temporal studies on coastal wetlands of Vedaranyam coast using remote sensing and GIS. Int. J. Des. Manuf. Technol. 2 (1), 2008.

Satapathy, D.R., Krupadam, R.J., Pawan, K.L., Wate, S.R., 2007. The application of satellite data for the quantification of mangrove loss and coastal management in the Godavari estuary, east coast of India. Environ. Monit. Assess. 134 (2007), 453–469.

Selvam, V., 2006. Reflections of Recent Tsunami on Mangrove Ecosystems. In: Ramasamy, S.M., et al., (Ed.), Geomatics in Tsunami, p. 227, New Delhi: Department of Science and Technology; Tiruchirappalli, India: Centre for Remote Sensing, Bharathidasan University.

Son, N., Chen, C., Chang, N., Chen, C., Chang, L., Thanh, B., 2015. Mangrove Mapping and Change Detection in Ca Mau Peninsula, Vietnam, Using Landsat Data and Object-Based Image Analysis. IEEE J. Select. Top. Appl. Earth Observ. Remote Sens. 8 (2), 503–510.

Son, N.T., Thanh, B.X., Da, C.T., 2016. Monitoring mangrove forest changes from multi-temporal landsat data in Can Gio Biosphere Reserve, Vietnam. Wetlands. 36(3). https://doi.org/10.1007/s13157-016-0767-2.

State of Forest Report, 1999. Forest Survey of India. Ministry of Environment & Forest, Dehradun, India, p. 113.

Thomas, N., Lucas, R., Bunting, P., Hardy, A., Rosenqvist, A., Simard, M., 2017. Distribution and drivers of global mangrove forest change, 1996–2010. https://doi.org/10.1371/journal.pone.0179302.

Tomlinson, P.B., 1986. The Botany of Mangroves. Cambridge University Press, Cambridge, UK.

Zhang, K., Thapa, B., Ross, M., Gann, D., 2016. Remote sensing of seasonal changes and disturbances in mangrove forest: a case study from South Florida. Ecosphere. 7(6).

Further Reading

Clark, J.R., 1996. Coastal Zone Management Handbook. CRC Press LLC, Boca Raton, FL, USA, p. 694.

CHAPTER 21

Recent Morphobathymetrical Changes of the Vilaine Estuary (South Brittany, France): Discrimination of Natural and Anthropogenic Forcings and Assessment for Future Trends

Evelyne Goubert*, David Menier*, Camille Traini[†], Manoj Joseph Mathew*, Romain Le Gall*

*Géosciences Océan Laboratory UMR CNRS 6538, University of South Brittany, Vannes Cedex, France,
[†]University of Bayreuth, Institute of Geography, Bayreuth, Germany

1 Introduction

Since the transformation of the human lifestyle from nomadic to fixed locations, land-ocean transfers have evolved from a natural forcing toward a complex natural/anthropogenic forcing, from the Holocene to the Anthropocene (Crutzen and Stoermer, 2000; Meybeck and Vörösmarty, 2005). Coastal systems have therefore progressively evolved toward anthroposystems. Estuaries are systems quickly responding to various natural and anthropogenic forcings by sedimentological and morphobathymetrical variations (Ramkumar, 2003, 2004a, b, 2007; Ramkumar et al., 2001a, b; Blott et al., 2006; Harff et al., 2009; Harff and Zhang, 2016; Litle et al., 2017). Three kinds of forcings can be distinguished: (1) natural forcings unrelated to the consequences of human activities (geological settings, tidal cycles, global sea level, global climatic oscillations, etc.), (2) natural forcings whose characteristics currently depend on the consequences of global climate change induced by human activities (frequencies and intensities of winds/storms/waves/floods, sea level rise, etc.), and (3) anthropogenic forcings corresponding to the direct consequences of human activities. To better understand recent evolutions and future changes of estuarine environments, it is important to separate the control by natural forcings versus anthropogenic-induced forcings.

Numerous studies examined natural and anthropogenic forcings within large estuaries that are constantly modified by human activities and morphobathymetrical responses (Blott et al., 2006; Cuvilliez et al., 2009; Garel et al., 2014; Wu et al., 2016; Luan et al., 2017; Liu et al., 2017). Few other studies examined the impacts of cyclic natural forcings on the hydrodynamic and morphodynamic processes: tide and river flow (Prandle, 2004; Billy et al., 2012); storms and floods (Boudet et al., 2017); storms and sea level rise (Ding et al., 2013); and North Atlantic Oscillation at different time scales (Sorrel et al., 2010; Munoz Sobrino et al., 2014). In the Special Issue from the Estuarine and Coastal Sciences Association 55, Litle et al. (2017) showed the difficulties in distinguishing between human induced changes and natural variations (tides, river flows, climate oscillations, sea level rise, etc.).

For about 30 years, many studies have focused on understanding the Holocene evolution of coastal and estuarine environments of the French Atlantic coast (Allen and Posamentier, 1993, 1994; Proust et al., 2001; Menier, 2004; Chaumillon and Weber, 2006; Proust et al., 2010; Menier et al., 2014). Among the sites studied, the Vilaine Bay (South Brittany, France) is a heavily modified estuary due to the presence of a dam at 8 km from the confluence. The Vilaine Estuary is located at the widest part of the South Armorican continental shelf (north of the Bay of Biscay). Numerous works carried out on the Vilaine Bay showed two major periods of infilling (Sorrel et al., 2010; Traini et al., 2013; Traini et al., 2015): (1) a natural infilling with a tide-dominated estuary between a period of 10,000 to 5000 years BP, and a wave dominated estuary from 5000 to 3000 years BP, and (2) an anthropogenic forcing since the construction of the dam in the late 1960s. The first period is characterized by the natural forcings of Holocene sea level rise, the storminess, the North Atlantic Oscillations, and the geomorphological control (Traini et al., 2013; Sorrel et al., 2010). For the recent period (from 1960s), two steps can also be identified: increased filling from 1965 to 1995, and stabilization during the past 20 years. The forcings mentioned are the hydrodynamic forcings modified by the presence of the dam (Traini et al., 2015), but the cyclicities of hydrodynamic forcings (tide, wind, storms) are not considered.

The purpose of this chapter is to identify if the natural cyclicities can control the morphobathymetrical response of this anthropogenic estuary from the 1960s. Thus, the morphological variability of the Vilaine Estuary is analyzed by using data from bathymetric surveys conducted between 1960 and 2013. This has allowed drawing a few conclusions on the morphological evolution according to the influences of natural phenomena (tidal cycles, global sea level, storminess, global climatic oscillations), and human-induced forcings (regulation of the river discharge and reduction of the estuarine space by the dam).

2 Study Area

2.1 Regional Geological, Geomorphic, Oceanographic, and Climatic Settings

The Vilaine Estuary is a sheltered environment located along the French Atlantic coast, southeast of the Armorican Massif on the Armorican passive continental margin (Fig. 1). The estuary opens onto the Vilaine Bay that originated from the flooding of the last glacial incised valley during the Holocene transgression (Traini et al., 2013). In the bay, the water depths are not >25 m, and in the estuary, they do not exceed 10 m.

The Vilaine Bay is protected from the strong waves behind the Quiberon Peninsula and the Island Houat and Hoedic. The Vilaine Estuary is subjected to the actions of the swell, the agitation by the winds, the tidal currents, and the river discharges from the dam. It is a meso- to macrotidal estuary with semidiurnal tides (12 h 25 mn cycle), with an amplitude of 2.5 m in neap tide to >5 m in spring tide.

This coastal zone experiences a temperate oceanic climate with average temperatures of 12 °C, average rainfall of 800 mm per year, and dominant winds (Fig. 2) from the west in offshore areas (Belle-Ile). The coastline is oriented locally to the west/southwest sector near Vilaine Estuary. Over the shelf, the general currents are characterized by the dominant winds. The swells are associated with westerly winds/storms. The mean significant wave height is of 1–2 m for a mean period of 2–5 s (Tessier, 2006). Regional winds generating wave agitation blow dominantly from the west and northwest. The offshore swells are damped and/or the directions are modified by the presence of the Belle-Île Island and Quiberon Peninsula.

The influence of the North Atlantic Oscillation is not clear. The Vilaine Estuary in South Brittany is located in the transitional zone between the wetter and the drier areas, in case of NAO+ or NAO−. As the catchment area is in the north, the heavy rainfall and high flows appear to be due to the positive oscillation (Sorrel et al., 2010).

2.2 Geomorphological and Sedimentological Settings

The Vilaine River has a catchment area of 10,000 km^2 that drains two-thirds of Brittany, including the basins of the Rennes. The main feature of the bay-estuary system is the establishment from 1965 to 1970 of the Arzal Dam at 8 km from the mouth, reducing the oscillating volume and the horizontal space available for estuarine sedimentation.

The morphology of this estuary is characterized from east to west by (Fig. 1C: (a) the inner sector, situated between 8 km and 4 km from the mouth, comprising a narrow meandering channel with a depth of more than −5 m (under the LLTL, lowest low tide level), up to Tréhiguier, (b) The intermediate part from upstream of the Strado Mudflat to Halguen and Pen

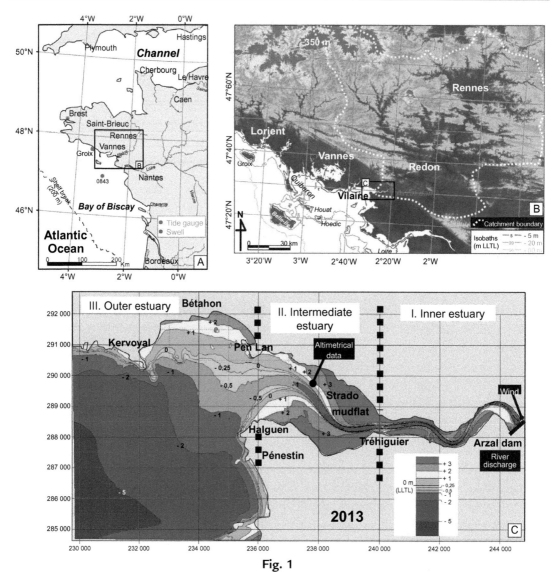

Fig. 1

Location of the Vilaine Estuary. (A) General map; (B) Regional map with the catchment area; (C) Morphobathymetrical map (2013) and location of measurements (altimetrical data, wind and river discharge).

Lan peaks. This zone is characterized by the Strado Mudflat at +3 m and by a channel whose depth decreases from −6 to −1 m and width increases, (c) The outer estuary corresponding to a very flat submarine delta, from −1 to −0.5 m whose distal part going to −7 m is reshaped by the swell. The general morphology of this estuary indicates that it is an intermediate domain between a partially wave-dominated estuary and tide-dominated estuary (Dalrymphe et al., 1992; Reison 1992; Dalrymple and Choi, 2007).

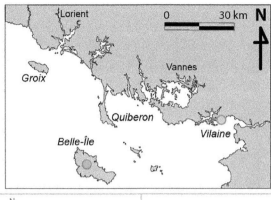

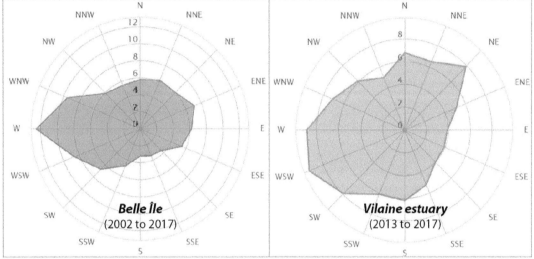

Fig. 2
Wind rose plot for Belle-Île and the Vilaine Estuary. *Data from Météo-France modified from Windfinder.com.*

The actual estuarine sedimentary facies are described in Fig. 3. The dominant sedimentary facies is composed of cohesive sediments (clay and silt). The sands are only dominant on the beaches. Because of the *meso* to macrotidal regime, of the surface of the estuary corresponds to intertidal mudflats that were occupied at the end of the 19th century by mussel farming in the intermediate and outer parts. After the construction of the dam, clam and cockle fishing developed on all mudflats.

2.3 The Main Features of the Actual Sedimentary Dynamic

According to several previous studies (Salomon and Lazure, 1988; Lazure and Salomon, 1991; Jouanneau et al., 1999; Tessier, 2006; Crave, 2009; Vested et al., 2013), the sedimentary particles that are currently deposited mainly come (a) from offshore, where the Loire is the

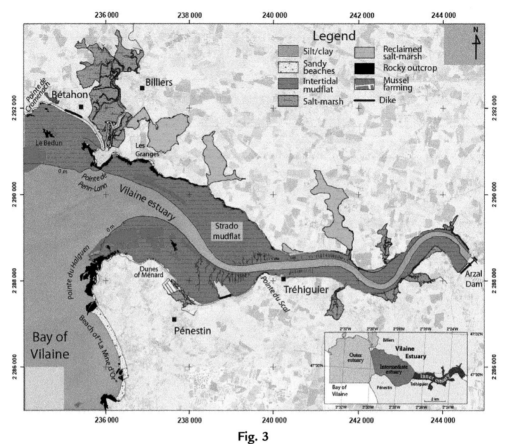

Fig. 3

Spatial distribution of sedimentary environments in the Vilaine Estuary. *Modified from Traini, C., Proust, J.N., Menier, D., Mathew, M.J., 2015. Distinguishing natural evolution and human impact on estuarine morpho-sedimentary development: a case study from the Vilaine estuary, France. Estuar. Coast. Shelf Sci. 163, 143–155.*

main contributor (Barbaroux and Gallene, 1973; Tessier, 2006), (b) or they are resuspended by the swell within the Vilaine Bay, (c) or they are resuspended by the agitation due to small windwaves inside the Vilaine Estuary. The particulate inputs of the Vilaine catchment are very low, and mainly organic (soil organic matter, diatoms) (Crave, 2009). The diversified organic matter inputs induce an important production of biodeposits and result in a greater cohesion of the fine fractions, and by the development of early diagenesis in anoxic and reducing conditions (Goubert, 1997). In the case of the altimetry of the intertidal mudflats in the inner and intermediate parts (Fig. 1C), erosion due to waves ($Hs > 20\,cm$) generated by westerly winds are observed (Goubert et al., 2010). River discharge and the tides only allow the amplification of the phenomenon (erosion or deposition). According to these hydrodynamic and sediment parameters, the Vilaine Estuary is a macrotidal sheltered coastal environment with moderate mixed energy influenced by wind/wave, tides, and fluvial inputs.

3 Data and Methods

3.1 Methodological Approach

From bathymetric data, digital elevation models were used to calculate the volumes deposited or eroded in different areas of the estuary between each bathymetric survey. For each period between two surveys, different hydrological and climatic parameters were analyzed: winds, river discharges, swell, sea-level, tidal amplitudes.

3.2 Bathymetric Data

Thirteen bathymetric surveys were carried out in the Vilaine Estuary from 1960 to 2013 with a single-beam echosounder (33 kHz/200–201 kHz). The reference level is the lowest low tide level (LLTL). Data are provided by The French Regional Facilities Office, French Naval Hydrographic and Oceanographic Service (SHOM), and Institution d'Aménagement de la Vilaine (IAV). The survey of 1960 is considered as the reference before the construction of the Arzal Dam from 1965 to 1970. Digital elevation model (DEM) and volume calculations were performed with MIKE©DHI.

3.3 Meteorological and Hydrologic Data

Winds (speed and orientation every 10 min) and river discharges (m^3/s hourly mean) measured at the dam (Fig. 1C) are collected and used in the study. Historical storms have been extracted from the Météo France website: http://tempetes.meteo.fr/. River discharges between 500 and 1000 m^3/s correspond to a decadal flood, and above 1000 m^3/s it is a centennial flood.

The swell data were collected from ANEMOC (Digital Atlas of Oceanic and Coastal Ocean States) of the 0843 point (Fig. 1A), which correspond to hindcast simulations with the TOMAWAC model EDF-LNHE from the various measurements carried out by the CETMEF. Only these calculated values are available for the period from 1979 to 2002. The data were processed in order to characterize the wave climate in terms of significant height (Hs) by quantifying the percentage of Hs > 4 m.

The sea level data were delivered by the Global Sea Level Observing System (GLOSS). It is measured by the tide gauge (Fig. 1A) of Brest from 1846 to 2014, by that of Groix from 1976 to 2014, and by satellites from 1993 (http://www.sonel.org/). The North Atlantic Oscillation Index, NAO DJFM index (December, January, February, March) was provided by James Hurrell from 1865 to 2014 (https://climatedataguide.ucar.edu/). The French Naval Hydrographic and Oceanographic Service (SHOM) provided the tidal coefficients (amplitudes): minimum of 20 for the very low neap tides and maximum of 120 for the very strong spring tides; the average is 75.

4 Results

4.1 Morphobathymetrical Evolution and Sedimentary Balance from 1960 to 2013

The main characteristics of morphobathymetric evolution from 1960 (Fig. 1) to 2013 (Fig. 4) are a strong meandering of the inner part of the estuary, the reduction of the width of the channel, and the progradation of the mudflats toward the center of the channel.

From the two bathymetric DEM, the map of the distribution of sediment accumulations was made, and sediment volumes deposited and eroded from the Arzal Dam to the Kervoyal Point (Fig. 5) were computed. Fig. 6 shows the spatio-temporal evolution of the infilling of the Vilaine Estuary. Since 1960, about 30 million m^3 (Mm^3) of sediment has been deposited on all four zones from the Arzal Dam to the Kervoyal Point. After a strong filling up until 1992, the Vilaine Estuary seemed to have registered a period of stability until 2005, with some fluctuations. But between 2005 and 2007, a generalized deposition of 3.7 Mm^3 was observed on all four zones. Between 2007 and 2013, the 6-year balance is stable overall, with a positive balance (1.6 Mm^3) on the two internal zones, but a very negative balance for the others external zones with an erosive balance of 5 Mm^3 is observed.

4.2 The Relationship Between the Spatio-Temporal Sedimentary Balance and the Climatic and Hydrodynamic Forcings

A schematic representation in Fig. 7 allows a temporal comparison between the morphobathymetric evolution of the Vilaine and the temporal chronicles of potential forcing.

The Cyclicities of the Hydrodynamic Forcings

The cyclicity of the NAO index does not seem to correlate with changes in the estuary. Two multiannualtidal cycles are present: the cycle at 18, 6 years (Saros cycle) is detected from the number of tides whose coefficients are higher than the average coefficient of 75, and the cycle of 4.5 years from the number of the great spring tides (coefficient > 106). On the other hand, the other parameters do not show cyclicities over the past 50 years.

Until 1992, the filling of the Vilaine was the direct consequence of the construction of the Arzal Dam. The factor that controlled sedimentary dynamics is the reduction in the horizontal space available for deposits in the estuarine zone. The construction of the Arzal Dam has reduced this area from 50 to 8 km. The rate of sedimentation in the estuarine zone has therefore increased, inducing the filling of the Vilaine Estuary. But, this period also corresponded to a phase with: (a) very low river discharges; (b) the weak tides in the Saros cycle from 1982 to 1990; (c) only two storms.

After 1992, the morphobathymetric changes may be correlated with the fluctuations of hydrodynamicforcings. From 1992 to 2005, different energetic events followed one another:

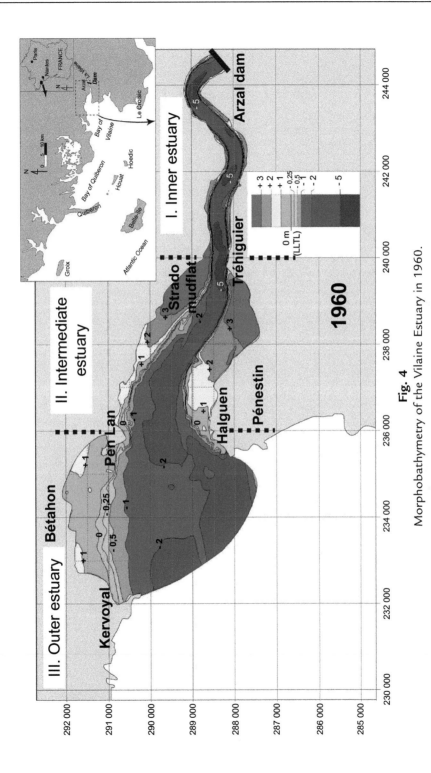

Fig. 4
Morphobathymetry of the Vilaine Estuary in 1960.

Chapter 21

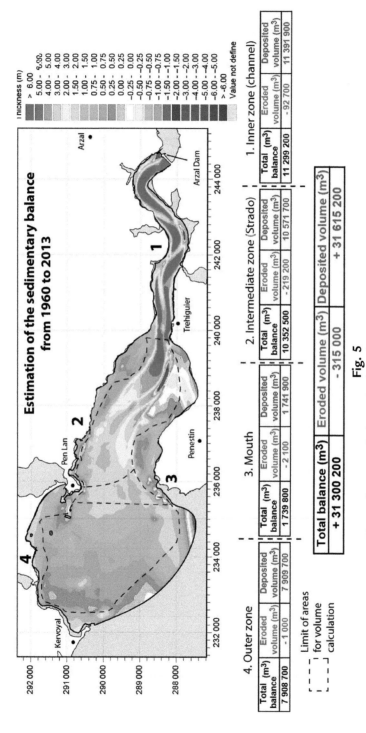

Fig. 5
The sedimentary balance in the Vilaine Estuary 1960 and 2013.

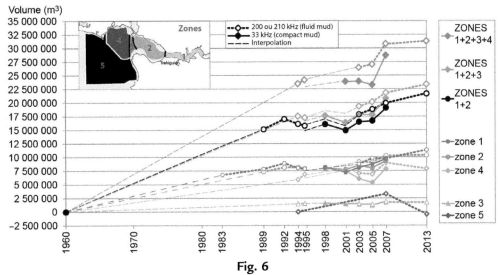

Fig. 6
The evolution of the infilling of the Vilaine Estuary from 1960 to 2013.

(a) high river discharges and floods (winter 1994 and winter 1995 with the centennial flood), (b) numerous high tides and corresponding strong swells/tides, (c) storms at the end of 1999 (Lothar and Martin), and late 2000/early 2001 floods.

The main factors controlling the morphobathymetric evolution of the Vilaine Estuary are storms, numerous high tides, and high flows. The different conjunctions of these energy events led to major phenomena of erosion. Conversely, the periods following these events and the less energetic periods show a return to a situation of equilibrium by different deposits on the mudflats, the inner channel, and the Strado intertidal mudflat, and on the outer estuary. It seems that the tidal conditions (number of high tides), either the intensity of the erosion, or the duration of return to a situation of equilibrium, or the location of the deposits follow the phases of erosion. Thus, globally, from 1992 to 2005, the very energetic hydrodynamic factors (flood, storms, tides) led to stability with phases of erosion followed by rapid redistribution. The sedimentary balance over this period is slightly negative.

From 2005 to 2007, low energy conditions led to the filling of the estuary. The factor that dominateed the sedimentary dynamics was mainly the tidal context with many tides with a coefficient lower than 75, as well as the low flows of 2006. These conditions of calm led to the resumption of the generalized filling accentuated by the slight creation of available space (erosion from 1992 to 2005). From 2007 to 2013, the particular context of cumulative very low river discharges, the succession of five major storms, including Xynthia, associated with very many spring tides led to a very strong erosion of the outer estuary and the infilling of the inner estuary, which was very sheltered during this period.

Chapter 21

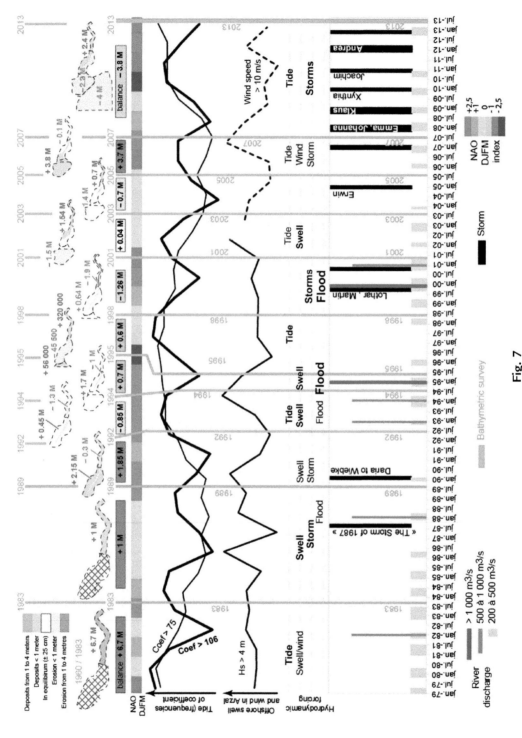

Fig. 7

Summary of deposits/erosion by zone and comparison with hydrodynamic forcings.

Thus, from 1992, the Vilaine Estuary seems to be currently functioning on the destabilization/resilience mode. Combinations of hydrodynamic events led to the juxtaposition of eroding zones and accretionary zones. In this case, the resilience corresponds to the redistribution of the sedimentary stocks according to the hydrodynamic parameters. Destabilization can also come from calm episodes resulting in a major accretion, especially on the intertidal mudflat (Strado) and in the inner channel. In this case, the resilience corresponded to the reincision of the navigation channel according to the forcing, either the river discharges, or the tide. This reincision therefore causes a lateral displacement of the channel at the meanders in the inner part.

5 Discussion

After the construction of the Arzal Dam, the destabilization of the Vilaine Estuary has been so significant that it is difficult to estimate the impact of natural forcings. This situation seems to be observed in all cases of significant construction in coastal areas (Garel et al., 2014; Lesourd et al., 2016; Liu et al., 2017). Nevertheless, this infilling may have been accelerated by the weak hydrodynamic conditions during this period. The duration of resilience to restore a new state of equilibrium was about 22 years.

After this major phase of infilling, no anthropogenic construction was done on the Vilaine Estuary. Thus, the estuary has reached a state of dynamic equilibrium according to the definition of Deng et al. (2016) for coastal environment. The causes of sedimentary dynamics could be related to cyclic variations of hydrodynamic events (Fig. 7): floods, storms, swell, tide. Different cyclicities are well known. The North Atlantic Oscillation (NAO) has a strong influence on Northern European storminess (positive NAO, Orme et al., 2016) and precipitation in southern Europe (negative NAO, Montaldo and Sarigu, 2017). But the location of the South Brittany (France) between northern and southern Europe cannot allow association of hydrodynamic events with positive or negative NAO phases for the past fifty years. In the case of the Vilaine Estuary, it seems that storms are associated to a rapid change of the NAO Index (Fig. 7). In the case of tides, the Saros cycle (18.6 years) is known to contribute significantly to coastal changes (Gratiot et al., 2008; Leroux, 2013). The tidal cycle at 4.5 years could explain the periodic changes in the Vilaine Estuary (mudflat versus channel and inner part versus outer part, Fig. 7).

6 Conclusions

The comparison of eroded or deposited sedimentary volumes in the Vilaine Estuary with different hydrodynamic forcings allowed us to study two phases of infilling in the estuary following the construction of a dam. The first phase of about 22 years corresponded to the rapid infilling of the estuary after the construction of the dam. The second phase has been a state of dynamic equilibrium due to various hydrodynamic forcings.

The current morphobathymetrical dynamic is therefore dependent on four main factors: (a) predictable, long-term tidal coefficients independent of human activities, (b) nonpredictable, but cyclic at the long timescale, rapid changes of the North Atlantic Oscillations, (c) two factors that are less predictable in the long term, but indirectly dependent on human activities: wind/swell (storm without precipitation) and river discharges (precipitation). Despite the current unpredictability of storms and precipitation, it is possible to characterize future periods in terms of tidal coefficients to consider different scenarios from the occurrences of storms and precipitation.

Since the construction of the Arzal dam in 1970, the average sea level measured by tide gauges and satellites showed an increase of 12 cm. With the rise of the sea level, the upper parts of the mudflats will be more and more often submerged. If this immersion phase corresponds to a phase of agitation (storm, wind), the vases stored at these mudflats will be remobilized and will return to the stock available for morphosedimentary dynamics. These sediments will fill the accommodation space created by the rise in sea level.

Acknowledgments

We thank the Institution d'Aménagement de la Vilaine (IAV), the French Naval Hydrographic and Oceanographic Service (SHOM), and the Danish Hydraulic Institut (DHI) for their different contributions to this synthesis.

References

Allen, G.P., Posamentier, H.W., 1993. Sequence stratigraphy and facies model of an incised valley fill: The Gironde estuary, France. J. Sediment. Petrol. 63 (3), 378–391.

Allen, G.P., Posamentier, H.W., 1994. Transgressive facies and sequence architecture in mixed tide and wave-dominated incised valleys: example from the Gironde estuary, France. In: Dalrymple, R., Zaitlin, B., Boyd, R. (Eds.), Incised Valley Systems: Origin and Sedimentary Sequences. Vol. 51, pp. 225–240. SEPM Special Publication.

Barbaroux, L., Gallene, B., 1973. Répartition des minérauxargileuxdans les sédimentsrécents de la Loire et du plateau continental. C. R. Acad. Sci. Paris 277D, 1609–1612.

Billy, J., Chaumillon, E., Fénièes, H., Poirier, C., 2012. Tidal and fluvial controls on the morphological evolution of a lobate estuarine tidal Bar: The Plassac tidal bar in the Gironde estuary (France). Geomorphology 169–170, 86–97.

Blott, S.J., Pye, K., van der Wal, D., Neal, A., 2006. Long-term morphological change and its causes in the Mersey estuary, NW England. Geomorphology 81, 185–206.

Boudet, L., Sabatier, F., Radakovitch, O., 2017. Modelling of sediment transport pattern in the mouth of the Rhone delta: role of storm and flood events. Estuar. Coast. Shelf Sci. 198, 568–582.

Chaumillon, E., Weber, N., 2006. Spatial variability of modern incised valleys on the French Atlantic coast: comparison between the Charente and the lay-Sèvre incised valleys. Incised valleys in time and space. SEPM Spec. Publ. 85, 57–85.

Crave, A. (coord), 2009. Inondationsdans le bassin de la Vilaine: paramètreshydrogéomorphologiques et flux terrigènes. Rapport pour l'Institutiond'Aménagementde la Vilaine, 245 p.

Crutzen, P.J., Stoermer, E.F., 2000. The "Anthropocene". Global Change, NewsLetter. The International Geosphere–Biosphere Programme (IGBP): A Study of Global Change 41, 17–18.

Cuvilliez, A., Deloffre, J., Lafite, R., Bessineton, C., 2009. Morphological responses of an estuarine intertidal mudflat to constructions since 1978 to 2005: The Seine estuary (France). Geomorphology 104, 165–174.

Dalrymphe, R.W., Zaitlin, B.A., Boyd, R., 1992. Estuarine facies models: conceptual basis and stratigraphic implications. J. Sediment. Petrol. 62 (6), 102–111.

Dalrymple, R.W., Choi, K., 2007. Morphologic and facies trends through the fluvial–marine transition in tide-dominated depositional systems: a schematic framework for environmental and sequence-stratigraphic interpretation. Earth Sci. Rev. 81 (3–4), 135–174.

Deng, J., Harff, J., Li, Y., Zhao, Y., Zhang, H., 2016. Morphodynamics at the coastal zone in the Laizhou Bay, Bohai Sea. J. Coast. Res. Spec. Issue 74, 59–69.

Ding, Y., Kuiry, S.N., Elgohry, M., Jia, Y., Altinakar, M.S., Yeh, K.C., 2013. Impact assessment of sea-level rise and hazardous storms on coasts and estuaries using integrated processes model. Ocean Eng. 71, 74–95.

Garel, E., Sousa, C., Ferreira, O., Morales, J.A., 2014. Decadal morphological response of an ebb-tidal delta and down-drift beach to artificial breaching and inlet stabilisation. Geomorphology 216, 13–25.

Goubert, E., 1997. Les Elphidium excavatum (Terquem), foraminifèresbenthiques, vivant enBaie de Vilaine d'octobre 1992 à septembre1996: morphologie, dynamique de population et relations avec l'environnement. Réflexions sur l'approcheméthodologique, la lignéeévolutive et sur l'utilisationenpaléoécologie. DoctoratUniversité Nantes, 300 p.

Goubert, E., Frénod, E., Peeters, P., Thuillier, P., Vested, H.J., Bernard, N., 2010. The use of altimetric data (Altus) in the characterization of hydrodynamic climates controlling hydrosedimentary processes of intertidal mudflat: The Vilaine estuary case (Brittany, France). Paralia 3, 6.17–6.31.

Gratiot, N., Anthony, E.J., Gardel, A., Gaucherel, C., Proisy, C., Wells, J.T., 2008. Significant contribution of the18,6 year tidal cycle to regional coastal changes. Nat. Geosci. 1, 169–172.

Harff, J., Zhang, H. (Eds.), 2016. Environmental processes and the natural and anthropogenic forcing in the Bohai Sea, eastern Asia. J. Coast. Res. Spec. Issue 74, 1–227.

Harff, J., Graf, G., Bobertz, B. (Eds.), 2009. Dynamics of natural and anthropogenic sedimentation (DYNAS). J. Mar. Syst. 75, 315–452.

Jouanneau, J.M., Weber, O., Cremer, B., Castaing, P., 1999. Fine-grained sediment budget on the continental margin of the Bay of Biscay. Deep-Sea Res. II 46, 2205–2220.

Lazure, P., Salomon, J.C., 1991. Coupled 2D and 3D modelling of coastal hydrodynamics. Oceanol. Acta 14 (2), 173–180.

Leroux, J., 2013. Chenauxtidaux et dynamique des prés-salésen régimeméga-tidal: approche multi-temporelle du siècle à l'événement de marée. Doctorat de l'Université de Rennes 1, 279 p.

Lesourd, S., Lesueur, P., Fisson, C., Dauvin, J.C., 2016. Sediment evolution in the mouth of the Seine estuary (France): a long-term monitoring during the last 150 years. C. R. Géosci. 348, 442–450.

Litle, S., Spencer, K.L., Schuttelaars, H.M., Millward, G.E. (Eds.), 2017. ECSA 55 unbounded boundaries and shifting baselines: estuaries and coastal seas in a rapidly changing world. Estuar. Coast. Shelf Sci. 198-B, 311–656.

Liu, Y., Xia, X., Chen, S., Jia, J., Cai, T., 2017. Morphological evolution of Jinshan trough in Hangzhou Bay (China) from 1960 to 2011. Estuar. Coast. Shelf Sci. 198, 367–377.

Luan, H.L., Ding, P.X., Wang, Z.B., Ge, J.Z., 2017. Process-based morphodynamic modeling of the Yangtze estuary at a decadal timescale: Controls on estuarine evolution and future trends. Geomorphology 290, 347–364.

Menier, D., 2004. Morphologie et remplissage des vallées fossils sud-armoricaines: apport de la stratigraphiesismique. Doctorat de l'Université de Bretagne Sud, Mémoires Géosciences Rennes 110, 202 p.

Menier, D., Augris, C., Briend, C., 2014. Les réseaux fluviatilesanciens du plateau continental de Bretagne Sud. Editions Quae. ISBN: 978-2-84433-174-8, 107 p.

Meybeck, M., Vörösmarty, C., 2005. Fluvial filtering of land-to-ocean fluxes: from natural Holocene variations to Anthropocene. C. R. Geosci. 337, 107–123.

Montaldo, N., Sarigu, A., 2017. Potential links between the North Atlantic oscillation and decreasing. J. Hydrol. 553, 419–437.

Munoz Sobrino, C., Garcia-Moreiras, I., Castro, Y., Martinez Carreno, N., de Blas, E., Fernandez Rodriguez, C., Judd, A., Garcia-Gill, S., 2014. Climate and anthropogenic factors influencing an estuarine ecosystem from NW Iberia: New high resolution multiproxy analyses from San Simón Bay (Ría de Vigo). Quat. Sci. Rev. 93, 11–33.

Orme, L.C., Reinhardt, L., Jones, R.T., Charman, D.J., 2016. Aeolian sediment reconstructions from the Scottish outer Hebrides: Late Holocene storminess and the role of the North Atlantic oscillation. Quat. Sci. Rev. 132, 15–25.

Prandle, D., 2004. How tides and river flows determine estuarine bathymetries. Prog. Oceanogr. 61, 1–26.

Proust, J.N., Menier, D., Guillocheau, F., Guennoc, P., Bonnet, S., Le Corre, C., Rouby, D., 2001. Les valléesfossiles de la Vilaine: nature et évolution du prismesédimentairecôtier du Pléistocènearmoricain. Bull. Soc. Géol. France 6, 737–749.

Proust, J.N., Renault, M., Guennoc, P., Thinon, I., 2010. Sedimentary architecture of the Loire River drowned valleys of the French Atlantic shelf. Bull. Soc. Géol. France 181 (2), 129–149.

Ramkumar, M., 2003. Geochemical and sedimentary processes of mangroves of Godavari delta: implications on estuarine and coastal biological ecosystem. Indian J. Geochem. 18, 95–112.

Ramkumar, M., 2004a. Dynamics of moderately well mixed tropical estuarine system, Krishna estuary, India: Part III nature of property-salinity relationships and quantity of exchange. Indian J. Geochem. 19, 219–244.

Ramkumar, M., 2004b. Dynamics of moderately well mixed tropical estuarine system, Krishna estuary, India: Part IV zones of active exchange of physico-chemical properties. Indian J. Geochem. 19, 245–269.

Ramkumar, M., 2007. Spatio-temporal variations of sediment texture and their influence on organic carbon distribution in the Krishna estuary. Indian J. Geochem. 22, 143–154.

Ramkumar, M., Pattabhi Ramayya, M., Rajani Kumari, V., 2001a. Dynamics of moderately well mixed tropical estuarine system, Krishna estuary, India: Part II Flushing time scales and estuarine mixing. Indian J. Geochem. 16, 75–92.

Ramkumar, M., Rajani Kumari, V., Pattabhi Ramayya, M., Gandhi, M.S., Bhagavan, K.V.S., Swamy, A.S.R., 2001b. Dynamics of moderately well mixed tropical estuarine system, Krishna estuary, India: Part I Spatio-temporal variations of physico-chemical properties. Indian J. Geochem. 16, 61–74.

Reison, G.E., 1992. Trangressive barrier island and estuarine systems. In: Facies models, response to sea level change. Love Printing Service Ltd, Ontario, Canadá, pp. 179–194.

Salomon, J.C., Lazure, P., 1988. Étude par modèlesmathématiques de quelques aspects de la circulation marine entre Quiberon et Noirmoutier. Rapport techanique, IFREMER/DERO-88.26-EL.

Sorrel, P., Tessier, B., Demory, F., Baltzer, A., Bouaouina, F., Proust, J.N., Menier, D., Traini, C., 2010. Sedimentary archives of the French Atlantic coast (inner bay of Vilaine, South Brittany): Depositional history and late Holocene climatic and environmental signals. Cont. Shelf Res. 30, 1250–1266.

Tessier, C., 2006. Caractérisation et dynamique des turbiditésen zone côtière: l'exemple de la région marine Bretagne Sud. Doctorat de l'Université Bordeaux I, 428 p.

Traini, C., Menier, D., Proust, J.N., Sorrel, P., 2013. Transgressive systems tract of a ria-type estuary: the late Holocene Vilaine River drowned valley (France). Mar. Geol. 337, 140–155.

Traini, C., Proust, J.N., Menier, D., Mathew, M.J., 2015. Distinguishing natural evolution and human impact on estuarine morpho-sedimentary development: a case study from the Vilaine estuary, France. Estuar. Coast. Shelf Sci. 163, 143–155.

Vested, H.J., Tessier, C., Christensen, B.B., Goubert, E., 2013. Numerical modelling of morphodynamics, Vilaine estuary. Ocean Dyn. 63, 423–446.

Wu, S., Cheng, H., Jun Xu, Y., Li, J., Zheng, S., 2016. Decadal changes in bathymetry of the Yangtze River estuary: humanimpacts and potential saltwater intrusion. Estuar. Coast. Shelf Sci. 182, 158–169.

CHAPTER 22

Impact of Seaweed Farming on Socio-Economic Development of a Fishing Community in Palk Bay, Southeast Coast of India

S. Rameshkumar, R. Rajaram
Department of Marine Science, Bharathidasan University, Tiruchirappalli, India

1 Introduction

Seaweeds or marine macroalgae are commercially important renewable marine living resources. They occur in the intertidal, shallow, and deep waters of the sea, up to 150 m in depth and also in estuaries. There are four classes of seaweeds, Chlorophyta (green algae), Phaeophyta (brown algae), Rhodophyta (red algae), and Cyanophyta (blue-green algae). Seaweeds are primary producers and provide shelter, nursery grounds, and food sources for marine organisms. Seaweeds are not only of high ecological importance, but are also of great economic importance. Dried thalli are used as human and animal food, and also as fertilizer. Extracted seaweed substances are used as stabilizers and stiffeners in the food, cosmetics, and pharmaceutical industries, and biotechnology (Jeeva and Kiruba, 2009; Wiencke and Bischof, 2012). Seaweeds grow abundantly along the Indian coastal waters, especially in Tamil Nadu, Gujarat, Lakshadweep, and Andaman and Nicobar Islands. There are also rich seaweed beds around Mumbai, Ratnagiri, Goa, Karwar, Varkala, Vizhinjam, Pulicat, Gulf of Mannar, and Chilka Lake. In India, 650 species of marine algae, with a maximum of 320 species of Rhodophyta, followed by 165 species of Chlorophyta and 150 species of Phaeophyta, have been recorded, among which, about 60 species are commercially important. About 302 species have been recorded on the Tamil Nadu coast. A total of 147 species of algae comprising 42 species of green algae, 31 species of brown algae, 69 species of red algae, and 5 species of blue green algae are distributed in the Gulf of Mannar. *Kappaphycus alvarezii* is economically important red tropical seaweed that has a higher demand for its cell wall polysaccharides (Bixler, 1996). Commercial cultivation of *K. alvarezii* originated in the Philippines in 1960 (Doty and Alvarez, 1975). Experimental farming had been carried out in several countries, including China, Venezuela, Japan, Fiji, the United States (Hawaii), the Maldives, Cuba, and India (Saito, 1971). In India, cultivation of this seaweed was initially started at Mandapam on the southeast coast of India (Eswaran et al., 2002). Commercial cultivation of *Kappaphycus* sp. has been successfully established along the Gulf of Mannar and Palk Bay coast of Tamil Nadu, India (Eswaran et al., 2002). This farming has become a viable alternative source of income for small-scale fishermen (Smith and Pestano-Smith, 1980; Smith, 1987).

Seaweed farming is a successful income earning occupation, and an economically viable alternative or supplemental livelihood option for coastal communities, mainly for the fisherwomen of the Ramanathapuram district of Tamil Nadu. The Self Help Groups model of seaweed farming emerged to be a successful livelihood option to the fishers who formed to take up seaweed farming. Currently, more than 50 such groups are successfully practicing seaweed cultivation. Subsequently, seaweed farming spread to neighboring districts such as Thanjavur, Pudukottai, and Tuticorin (Narayanakumar and Krishnan, 2013). Thus, it is hypothesized that seaweed farming has helped to improve their socio-economic conditions, as well as their standard of living of fishers' households.

K. alvarezii farming has also been considered as an example of community-based coastal management practices (Msuya et al., 2014; Poeloengasih et al., 2014). Because it is labor intensive, the method delivers both direct and indirect employment opportunities, and further improves community organization and cooperation. The production cycles are fast, spanning only 45 days; thus, economic returns on farming are high, as compared with other aquaculture and land-based agriculture. The engagement of women in tying seedlings, preparation of planting material, assistance in harvesting, drying, cleaning, and sorting biomass enables them to achieve economic independence.

The Palk Strait of the Tami Nadu coast has been found to be suitable for seaweed cultivation. There are long coastal stretches available with a potential number of farming activities that would create an alternative income/livelihood for the coastal poor. This chapter attempts to assess the impact of seaweed farming on their socio-economic status in Palk bay, on the southeast coast of India.

2 Commercial Importance of Seaweed

Seaweed is a multipurpose product that can be used for direct human consumption or processed into food additives, pet food, feeds, fertilizers, biofuel, cosmetics, and medicines (McHugh, 2003; Bixler and Porse, 2011). The species *Kappaphycus* sp., and *Eucheuma* sp., are primary raw materials for carrageenan. The species *Gracilaria* sp. is a primary raw material for agar-agar and carrageenan. These are thickening and jellying agents primarily used as food additives. Approximately 98% of seaweed production is obtained from five cultivated species: *Saccharina* sp., *Undaria* sp., *Porphyra* sp., *Kappaphycus* sp., and *Gracilaria* sp.

(Suo and Wang, 1992; Pereira and Yarish, 2008). *Kappaphycus* sp. is one of the most produced raw materials for the food and food polymer industries. Worldwide, the macroalgal production increases 5.7% every year. Approximately 18 million tons of macroalgae were produced from global capture and aquaculture in the year 2011 (FAO, 2014). About 96% of this came from aquaculture, mainly from Asian countries dominating seaweed culture production (FAO, 2014).

3 Significance of Seaweed Farming

There are a number of direct and indirect uses and indirect benefits of seaweed farming that include, but are not limited to: providing occupation for the coastal people, as it is an eco-friendly activity; providing a remedy for the nonavailability of the required quantity of seaweed for various uses; providing a continuous supply of raw material for seaweed-based industries; providing seaweeds of uniform quality for use in industry; reduction of the pressure on depletion of natural reserves and their ecosystem services; and above all, seaweed farming can be a tool for the reduction of coastal pollution, and may contribute as a major CO_2 sink.

4 Seaweed Farming in Tamil Nadu

The experimental field cultivation of seaweeds has been attempted only for a few economic seaweeds, such as, *Gracilaria edulis*, *Gelidella acerosa*, *Hypnea valentiae*, and *K. alvarezii*. However, *K. alvarezii* has been successfully cultivated in the Tamil Nadu coastal waters. Seaweed cultivation was first started at Mandapam on the southeast coast of India during 1995–97, and the commercial cultivation of *K. alvarezii* started in 2003 along the Tamil Nadu coast. At present, *K. alvarezii* production is carried out in three coastal districts of Tamil Nadu, namely Ramanathapuram, Pudukottai, and Thanjavur. The experimental farming and field cultivation of *K. alvarezii* were conducted in several Indian coastal areas for additional income generation to the coastal fishing communities. The agar-yielding seaweeds have been harvested since 1966 from the Gulf of Mannar Islands along the coastline from Rameswaram to Tuticorin, and the Sethubavachatram area in the Palk Bay region. The red seaweeds such as *Gelidiella acerosa* and *Gracilaria* sp. are collected throughout the year, and algin-yielding brown algae, such as *Sargassum* sp. and *Turbinaria* sp., are collected seasonally from August to January on the southeast coast of India.

5 Culture Methods

Seaweed cultivation was conducted by four different methods, including the bottom-culture method, long-line rope method, the single-rope floating raft technique (SRFT), and the floating bamboo raft method.

5.1 Long-Line Rope Method

This is a modified traditional off-bottom farming method, and it has few similarities to the raft method. Seedlings of about 100–150 g fresh weight are attached to a rope by raffia or a braider between anchors. The main rope is 6–8 mm in diameter, and about 20 m in length. The longline ropes are kept afloat in water with the help of floats tied at regular intervals, and anchoring is done on both ends. Timbers of casuarina, eucalyptus, bamboo, and so forth are used for anchoring. Poles are erected at 3 m intervals in a square fashion, and 12 mm polypropylene rope is tied in any two parallel directions (depending upon tidal and current directions). The seeded ropes (3 mm) are then tied at 10 cm intervals (Fig. 1A). Care needs to be taken to ensure that the seaweed always remains submerged (0.5 m below the surface), and receives sufficient sunlight. Generally, seeding in this case is done in the water to avoid seedling loss, which may occur if ropes are seeded on the shore and dragged into the sea. This farming technique is recommended for regions that experience moderate wave action, and particularly in areas with a low density of grazers. It is popularly adopted in the South Ramanathapuram, Pudukkottai, and Tuticorin districts of Tamil Nadu.

Fig. 1
(A) Long-line rope method; (B) net bag method; (C and D) floating bamboo raft culture method.

5.2 Net Bag Method

The net bag method was mainly followed for *Gracilaria dura*. The cultivation method was adopted according to the method (Veeragurunathan et al., 2016) described in the previous section, with certain changes. In this method, a 75-cm-long bag prepared from commercial fishnet is seeded with 200–300-g seedlings (Fig. 1B). The entire length of the net bag is covered with agro net (1.0 ± 0.2-mm mesh size and 90-μm thickness) in order to minimize grazing. The bag is then tied onto an 8-mm polypropylene rope, and the rope is tied on both sides to the bamboo poles erected vertically. On average, a biomass of 900 g fresh wt. can be harvested 45 days from each of the seedlings.

5.3 Floating Bamboo Raft Method

A bamboo raft (8–10 cm in diameter and 3×3 m) is used as the main frame for cultivation (Fig. 1C). The cultivation method was adopted from the previous standard method, with certain modifications (Eswaran et al., 2002). The angular portions are diagonally fixed with the help of supporting bamboo for keeping it intact. The clusters of rafts are tied with an anchor to secure them and maintain their buoyancy. Bottom netting is provided to avoid drifting of material, as well as to minimize grazing. The seeding is carried out on shore, and seeded rafts are transplanted into the open sea, subsequently. Harvesting is generally carried out after 45 days. This method is recommended where the water movement is gentle. It has been popularly adopted in the Tamil Nadu coastal regions.

5.4 Bottom-Culture Method

A square net ($2 \text{ m} \times 2 \text{ m}$ size) made of polypropylene rope (3-mm diameter) is seeded with an initial seedling weight of 1.5 kg fresh wt. net^{-1} (Fig. 1D). The seeded nets are tied at all four corners to vertical bamboo poles erected in the sea at 1.5-depth. Stone sinkers are used to anchor the seeded nets at 25 cm above the bottom. A biomass of 7 kg fresh wt. net^{-1} can be obtained at the end of 45-day growth.

6 Discussion

Seaweed cultivation provides a viable and sustainable alternative livelihood opportunity for small fishers. The commercial exploitation of seaweed in India was started in 1966. At present, seaweed is exploited in many localities of Tamil Nadu. The fishermen communities from various localities of the Palk regions benefit from employment in seaweed collection, in addition to their normal fishery activities. In addition, whenever the conditions are unfavorable for fishing, they are involved in the collection of seaweeds, such as *Gelidiella acerosa* and *G. edulis*. Approximately 2000 people are employed during the peak season (August-January) in the Palk Bay region alone. The potential seaweed production in Tamil Nadu state is 250,000 tons, which is the second highest in India (Munoz et al., 2004). Seaweed farming helps bring women into the mainstream of economic activity, sometimes for the first time, and empowers them to achieve a better lifestyle, enabling well-being of the whole family. A large number of primary workers are women (Fig. 2A–D). A thousand families are earning their livelihood through seaweed cultivation.

6.1 Role of Government and Nongovernment Sectors for the Development of Seaweed Farming in Tamil Nadu

As a result of seaweed farming in coastal management, and for providing employment to people, a lot of Self Help Groups, Village Youth Groups, NGOs, and various government and nongovernment agencies are promoting seaweed cultivation all over India, especially in Tamil Nadu. They encourage seaweed cultivation as an alternate livelihood option for the coastal poor. Seaweed farming in the southern districts of Tamil Nadu is being vigorously promoted by

Fig. 2
Women involved in seaweed farming on the Tamil Nadu coasts. (A) Single family members involved in seaweed farming. (B) Single village women's involved in seaweed farming. (C) Single women involved in seaweed farming. (D) Self Help Groups (SHGs) involved in seaweed farming. *From Radhika Rajasree, S.R., Gayathri, S., 2014. Women enterprising in seaweed farming with special references fisherwomen widows in Kanyakumari District Tamil Nadu India. J. Coast Dev. 17: 383.*

the Aquaculture Foundation of India (AFI). AquAgri Processing Private Limited (APPL) provides the necessary training, extends technical know-how, and guarantees buy-back for the all of the produce at a preagreed price. APPL has also introduced a new saving scheme (APPL-GIP-Growers' Investment Program), and incentive scheme to improve the residents' status. In *K. alvarezii* cultivation, the self-help group model promoted by the District Rural Development Agency (DRDA), Department of Biotechnology (DBT) and the Tamil Nadu State Fisheries Department with the assistance of Non-Governmental Organizations (NGOs) is found to be more effective. The Central Salt Marine Chemical Research Institute and Central Marine Fisheries Research Institute have developed culture techniques for some of the commercially important seaweed species in India, and the technologies are being transferred to the local people as well.

6.2 Employment Opportunities and Self-Help Groups (SHGs) Model in Seaweed Cultivation on the Tamil Nadu Coast

Seaweed mariculture has now become a potential employment-generating and income-earning activity that is practiced by more than a thousand members of SHGs, with the support of private investments, industries, financial institutions such as NABARD (through scheduled commercial banks), the National Fisheries Development Board (NFDB), and NGOs. The manpower required for fully harvesting the standing crop is considered the main opportunity for employment in this sector. However, the present employment in this sector is less than 10,000, of which more than 50% are women. *K. alvarezii* is a versatile species that grows fast; biomass doubles every 15 days. *K. alvarezii* cultivation takes place in several areas of the Tamil Nadu coast. About 600 families are cultivating this seaweed in the Ramanathapuram, Tuticorin, Tirunelveli, Kanayakumari, Pudukottai, and Thanjavur districts, and each family is earning around Rs 10,000 a month by spending Rs 63,000 for making 90 rafts. Due to this income potential, in many cases the entire family is becoming actively involved in cultivation. Studies conducted by seaweed farmers showed that about 2000 fishermen were involved in seaweed collection in six villages of the Ramanathapuram District. The employment potential for harvesting the standing crop of the country as a whole is worked out as 52,174 persons. Due to the potential of seaweed farming for the development of livelihoods, the CSMCRI has developed standard culture methods for a few commercial seaweeds, such as *G. edulis* and *G. debilis* (Fig. 3A–H) and Gracilaria dura (Fig. 4A).

The annual turnover of *K. alvarezii* seaweed farming alone can be safely estimated to be Rs 2.0 billion (Krishnan and Narayanakumar, 2010). From the years 2003 to 2009, *K. alvarezii* production has shown a steady increase from 147 tons to the maximum of 865 tons in the year 2009. At present, around 1000 families are dependent on *K. alvarezii* farming for their livelihood on the Tamil Nadu coast. In Ramanathapuram district, a total of 180 families from Sambai village and 70 families from Mangadu village directly depend on *K. alvarezii* farming for their livelihood. In this district there are around 8000 seaweed culture rafts floating in the

Fig. 3

Gracilaria edulis farming on the Mandapam coast. (A) Green strain of *G. debilis*. (B) Green strain of *G. edulis*. (C) Red strain of *G. debilis*. (D) Red strain of *G. edulis*. (E) Full growth of *G. debilis* green strain. (F) Full growth of *G. edulis* green strain. (G) Full growth of *G. debilis* red strain. (H) Full growth of *G. edulis* red strain. *From Veeragurunathan, V., Prasad, K., Singh, N., Malarvizhi, J., Mandal, S.K., Mantri, V.A., 2016. Growth and biochemical characterization of green and red strains of the tropical agarophytes Gracilaria debilis and Gracilaria edulis (Gracilariaceae, Rhodophyta). J. Appl. Phycol. 28, 3479–3489.*

Palk Bay region (Periyasamy et al., 2013). Palk Strait of the Ramnad District, Pudukkottai, and the Thanjavur and Tuticorin Districts were surveyed extensively and found suitable for seaweed cultivation. There are 500 km coastal stretches available with the potential for 5000 rafts, which would create an alternative income/livelihood to 2000 fishers or 100 SHG's (20 members each) of the coastal poor with turnover of Rs 100–500 million per annum (Periyasamy

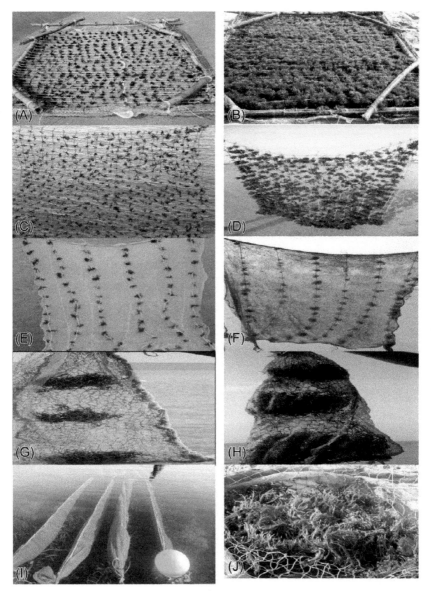

Fig. 4
Different culture methods of *Gracilaria dura* cultivation. (A) and (B) Raft culture method. (C) and (D) Net culture method. (E) and (F) Hanging rope method. (G) and (H) Net Bag method. (I) and (J) Net pouch method. *From Veeragurunathan, V., Eswaran, K.K., Saminathan, K.R., Mantri, V.A., Malarvizhi, J., Ajay, G., Jha, B., 2015. Feasibility of Gracilaria dura cultivation in the open sea on the Southeastern Coast of India. Aquaculture 438(1), 68–74.*

Fig. 5
Wild Sargassum species were collected at Seenippatharga in the Gulf of Mannar region. (A) Wild *Sargassum* sp. collected. (B) *Sargassum* sp. was deposited on shore. (C) *Sargassum* sp. were dried. (D) *Sargassum* sp. was ready to export.

et al., 2013). Corporate institutional and financial support led to the expansion of seaweed farming through the Self Help Groups (SHG) model (mostly women), starting on a small scale in the Ramanathapuram District of Tamil Nadu in 2000, which now gradually has spread to all coastal districts of the Palk Bay regions (Krishnan and Narayanakumar, 2010).

In addition to the cultivation, the *Sargasssum* sp. are collected on the Monoli and Musal Islands in the Gulf of Mannar regions (Fig. 5). The Sargassum species is commonly abundant during monsoon and premonsoon seasons. In the Gulf of Mannar regions, coastal village people are involved in wild harvest for commercial exports to seaweed industries for different parts of the country. A cost and revenue analysis of *K. alvarezii* made by Mantri et al. (2016) is presented in Table 1.

6.3 Environment Impact Assessment (EIA) Study for Invasive Seaweed Farming

The introduction of commercial invasive cultivars is gathering attention and is being considered as one of the potential challenges to algal cultivation as an economic development opportunity for coastal fishing communities. The ecological interactions of this species were first reported

Table 1 Expenditure and income of seaweed raft culture methods

Specifications	Details of Expenditure
Investments	
Number of cultivation periods	45
Total number of units	525
Cost of unit (Rs)	1000
Total Investments (Rs)	525,000
Financial Source	
Subsidy under different government schemes (Rs)	449,955
Biomass production/unit (kg)	150
Biomass production in a commercial farm/day (150 × 15) (kg)	2250
Biomass production/cycle (kg fr. wt)	101,250
Seed material used/cycle at 60 kg raft^{-1}	40,500
Biomass available for sale (kg fr. wt)	60,750
Biomass available for sale (kg dry wt)	6075
Income	
Cost of dry weed/kg (Rs)	35.30
Income from farm/cycle (Rs)	214,628
Income/beneficiary/cycle (Rs)	14,309
Income per month at Rs 318 day^{-1} for raft	9540

US$1 = Rs 66.70, 1€ = Rs 75.89 (as on June 8, 2016).

from the Hawaiian Islands (Woo et al., 1989), and subsequently Rodgers and Cox (1999) and Smith (2003) presented some of the first quantified evidence of significant negative impacts of nonindigenous marine algae in tropical waters. Conklin and Smith (2005) conducted a study to quantify *Kappaphycus* spp. Abundance, both spatially and temporally, and to investigate control options, including manual removal and the use of biocontrol agents in Kaneohe Bay, Hawaii, in the United States. Barrios (2005) reported that *K. alvarezii* vegetative tissues were borne as a crust of small cells radially arranged with a large-cell central medulla with rhizoidal filaments. Pereira and Verlecar (2005) reported *K. alvarezii* on the verge of becoming invasive in southern India, and its spread in the Gulf of Mannar. Chandrasekaran et al. (2008) reported the bioinvasion of *K. alvarezii* onto branching corals (*Acropora* sp.) in Kurusadai Island, the Gulf of Mannar Marine Biosphere Reserve, in south India. Rameshkumar and Rajaram (2016) reported the impacts of *K. alvarezii* on the plankton communities in the two localities of the Tamil Nadu coast. Rameshkumar and Rajaram (2017) reported the impacts of cultivation of *K. alvarezii* on the macro and meio benthos diversity in the Tuticorin coast, Gulf of Mannar, on the southeast coast of India. The potentially valuable invasive seaweed such as *K. alvarezii* in new ecosystems can create ecological problems. Hence, an EIA study should be conducted for the coastal ecological management and enhancement of the livelihood of coastal fishing communities.

7 Conclusions

Seaweed farming is a promising opportunity for uplifting the livelihood of the fishermen communities that inhabit coastal zones. This low cost occupation involves intensive manual labor and can sustain the entire family, including women, thereby not only helping the families' earning potential, but also supporting the community's holistic economic status. While harvesting natural varieties may lead to overexploitation, as with any other natural resource, artificial farming has only a few negative impacts on the natural environment. However, comprehensive studies on the impact of artificial farming need to be conducted. Seaweed farming also has the potential to be part of a coastal zone management strategy, and can be implemented elsewhere, including developing countries.

References

Barrios, J.E., 2005. Spread of exotic algae *Kappaphycus alvarezii* (Gigartinales: Rhodophyta) in the northeast region of Venezuela. Bol. Inst. Oceanogr. Venez. 44, 29–34.

Bixler, H.J., 1996. Recent developments in manufacturing and marketing carrageenan. Hydrobiologia 326/327, 35–57, 1996.

Bixler, H.J., Porse, H., 2011. A decade of change in the seaweed hydrocolloids industry. J. Appl. Phycol. 23, 321–335.

Chandrasekaran, S., Nagendran, N.A., Pandiaraja, D., Krishnankutty, N., Kamalakannan, B., 2008. Bioinvasion of *Kappaphycus alvarezii* on corals in the Gulf of Mannar, India. Curr. Sci. 94, 1167–1172.

Conklin, E.J., Smith, J.E., 2005. Abundance and spread of the invasive red algae, *Kappaphycus* spp., in Kaneohe Bay, Hawaii and an experimental assessment of management options. Biol. Invasions 7, 1029–1039.

Doty, M.S., Alvarez, V.B., 1975. Status, problems, advances and economics of Eucheuma farms. J. Mar. Technol. Soc. 9, 30–35.

Eswaran, K., Ghosh, P.K., Mairh, O.P., 2002. Experimental field cultivation of *Kappaphycus alvarezii* (Doty) Doty. ex. P. Silva at Mandapam region. Seaweed Res.Util. 24, 67–72.

FAO, 2014. Fisheries and Aquaculture Information and Statistics Service. 16 March 2014.

Jeeva, S., Kiruba, S., 2009. Bioremediating and biomediating potential of seaweeds. In: Abstracts of the National Seminar on Marine Resources: Sustainable Utilization and Conservation. Organized by Department of Plant Biology and Biotechnology St. Mary's College, Thoothukudi, p. 38.

Krishnan, M., Narayanakumar, R., 2010. Socio-Economic Dimensions of Seaweed Farming in India. Consultancy Report, FAO of UN (Personal Services Agreement), p. 103.

Mantri, V.A., Eswaran, K., Shanmugam, M., Ganesan, M., Veeragurunathan, V., Thiruppathi, S., Reddy, C.R.K., Seth, A., 2016. An appraisal on commercial farming of Kappaphycus alvarezii in India: Success in diversification of livelihood and prospects. J. Appl. Phycol.. https://doi.org/10.1007/s10811-016-0948-7.

McHugh, D.J., 2003. A guide to the seaweed industry. FAO Fisheries Technical paper, No 441, 105 p.

Msuya, F.E., Buriyo, A., Omar, I., Pascal, B., Narrain, K., Ravina, J.J.M., Mrabu, E., Wakibia, J.G., 2014. Cultivation and utilisation of red seaweeds in the Western Indian Ocean (WIO) Region. J. Appl. Phycol. 26, 699–705.

Munoz, J., Freile-Pelegrin, Y., Robledo, D., 2004. Mariculture of *Kappaphycus alvarezii* (Rhodophyta, Solieriaceae) color strains in tropical waters of Yucatan Mexico. Aquaculture 239, 161–177.

Narayanakumar, R., Krishnan, M., 2013. Socio-economic assessment of seaweed farmers in Tamil Nadu—a case study in Ramanathapuram District. Indian J. Fish. 60 (4), 51–57.

Pereira, R., Yarish, C., 2008. Mass production of marine macroalgae. In: Encyclopedia of Ecology, Elsevier, Oxford, pp. 2236–2247.

Pereira, N., Verlecar, X.N., 2005. Is Gulf of Mannar heading for marine bioinvasion? Curr. Sci. 89, 1309–1310.
Periyasamy, C., Anantharaman, P., Balasubramanian, T., 2013. Sustainable utilization of biological resources-seaweed farming a good option. Int. J. Appl. Biores. 18, 17–22.
Poeloengasih, C.D., Bardant, T.B., Rosyida, V.T., Maryana, R., Wahono, S.K., 2014. Coastal community empowerment in processing *Kappaphycus alvarezii*: a case study in Ceningan Island, Bali, Indonesia. J. Appl. Phycol. 26, 1539–1546.
Rameshkumar, S., Rajaram, R., 2016. In: Impacts of Cultivation of Seaweeds on Plankton and Benthos Populations on Open Sea Environment. Proceedings, 22nd International Seaweed Symposium. Denmark, Copenhagen, pp. 119–120.
Rameshkumar, S., Rajaram, R., 2017. Experimental cultivation of invasive seaweed *Kappaphycus alvarezii* (Doty) Doty with assessment of macro and meiobenthos diversity from Tuticorin coast, southeast coast of India. Reg. Stud. Mar. Sci. 9, 117–125.
Rodgers, S.K., Cox, E.F., 1999. Rate of spread of introduced rhodophytes *Kappaphycus alvarezii, Kappaphycus striatum,* and *Gracilaria salicornia* and their current distributions in Kaneohe Bay, Oahu, Hawaii. Pac. Sci. 53, 232–241.
Saito, Y., 1971. Seaweed aquaculture in the Northern Pacific. 1976. In: Pillai, T.V.R., Dill, W.A. (Eds.), Advances in Aquaculture.FAO Technical conference on Aquaculture, Kyota, Japan, pp. 7–16.
Suo, R., Wang, Q., 1992. Laminaria culture in China. INFO Fish. Int. 1, 40–42.
Smith, I.R., Pestano-Smith, R., 1980. Seaweed farming as alternative income for small-fishermen: a case study. *Proc. Indo-Pacific Fish Commun.* 19, 715–729.
Smith, I.R., 1987. The economics of small-scale seaweed production in the South China Sea region. FAO Fisheries Circular No., 806, p. 26.
Smith, J.E., 2003. Invasive macroalgae on tropical reefs: impacts, interactions, mechanisms and management. J. Phycol. 39. 53 pp.
Veeragurunathan, V., Prasad, K., Singh, N., Malarvizhi, J., Mandal, S.K., Mantri, V.A., 2016. Growth and biochemical characterization of green and red strains of the tropical agarophytes *Gracilaria debilis* and *Gracilaria edulis* (Gracilariaceae, Rhodophyta). J. Appl. Phycol. 28, 3479–3489.
Wiencke, C., Bischof, K., 2012. Seaweed Biology. Springer-Verlag, Berlin Heidelberg.
Woo, M., Smith, C., Smith, W., 1989. Ecological Interactions and Impacts of Invasive *Kappaphycus striatum* in Kaneohe Bay.Proceedings, First National Conference, Tropical Reef Marine Bioinvasions, p. 186.

Further Reading

FAO, 2013. Fisheries and Aquaculture Information and Statistics Service. 18 December 2013.
Radhika Rajasree, S.R., Gayathri, S., 2014. Women enterprising in seaweed farming with special references fisherwomen widows in Kanyakumari District Tamil Nadu India. J. Coast Dev. 17, 383.
Veeragurunathan, V., Eswaran, K., Malarvizhi, J., Gobalakrishnan, M., 2015. Cultivation of *Gracilaria dura* in the open sea along the southeast coast of India. J. Appl. Phycol. 27, 2353–2365.

CHAPTER 23

Habitat Risk Assessment Along Coastal Tamil Nadu, India—An Integrated Methodology for Mitigating Coastal Hazards

A. BalaSundareshwaran*, S. Abdul Rahaman*, K. Balasubramani[†], K. Kumaraswamy*, Mu. Ramkumar[‡]

Department of Geography, Bharathidasan University, Tiruchirappalli, India, [†]Department of Geography, Central University of Tamil Nadu, Tiruvarur, India, [‡]Department of Geology, Periyar University, Salem, India

1 Introduction

The coastal zone plays a significant role in the development of a region, yet poses vulnerability due to its very location where lithoatmospheric, bioatmospheric, and hydroatmospheric agents meet and interact with each other at varying spatial and temporal scales. The vulnerability is also enhanced due to the fact that the coastal zone is a region where two-thirds of the world's human population lives, and where physical, cultural, social, and economic zones converge. The population densities along the coastal regions are about three times higher than the global average (Small and Nicholls, 2003). It has been estimated that about 23% of the world's population lives both within 100 km distance from and within <100 m above sea level. Coastal regions are dynamic in nature, and are controlled by nearshore processes, beach morphology, and human activities (Chandrasekar et al., 2013). Coastal environments around the world tend to change very rapidly due to various erosional and depositional activities that take place at the boundary between land and sea (McCarthy et al., 2001). Rapid urbanization and economic development are the major driving factors of increased coastal risks (Wong et al., 2014). According to the Intergovernmental Panel on Climate Change, humans have increasingly started utilizing the coastal regions (Nicholls et al., 2007). Areas with high probabilities of urban expansion will be mainly located in coastal regions by 2030 (Seto et al., 2012; Li et al.,

2016). These traits make the coastal zone more vulnerable and risk-prone, which calls for concerted efforts to assess the type and intensity of risks involved and to be prepared with mitigation and rescue plans.

However, various coastal regions show unique and varied susceptibility to natural and other hazards (Kobayashi et al., 1993; Ferrario et al., 2014; Marchant, 2017) that make establishment of a universal method of coastal protection ineffective. Thus, it is necessary to identify susceptibility classes to natural coastal habitats based on their physical, social, economic, and ecological traits, in order to minimize the exposure to coastal hazards (Arkema et al., 2017; Guannel et al., 2015). Key vulnerabilities considered for assessment of coastal vulnerability include social, economic, and political dimensions, in addition to natural and environmental dimensions (Becken et al., 2014). Balica et al. (2012) focused on developing a Coastal City Flood Vulnerability Index based on exposure, susceptibility, and resilience to coastal flooding. The analytical hierarchical process for coastal vulnerability assessment based on both physical-geological parameters and socio-economic factors helps to derive realistic CVI. The socio-economic factors are those that either create obstacles, or empower an individual's or place's potential to respond or recover from a catastrophe; and the infrastructural factors are those that either intensify or reduce the impact of such a catastrophe (Borden et al., 2007). Arkema et al. (2015) claimed human activities would affect the flow of benefits in coastal management plans. Human society affects environmental change, but is also vulnerable to these changes (Armas and Gavris, 2013, Li et al., 2016). In addition, poor adaptation capacity, poor socio-economic and environmental conditions, and poor housing conditions are factors (Mallick 2014; Younus, 2017). In coastal areas, constant severe natural hazards influence people to formulate their own strategies to overcome hardship in their livelihood processes. Over time, indigenous communities have continued to rely heavily on their own inherited knowledge in observing the environment and dealing with natural disasters (Garai, 2017). The earthquake and tsunami of 2011 in Tohoku exposed that, even a developed county such as Japan, which is perceived to be resilient against natural disasters, was still very much vulnerable. Even though it takes many years to develop coastal cities and towns, a catastrophic natural disaster can destroy everything from human life to properties in an instant (Esteban et al., 2015). For mitigation and strengthening capacity, satellite remote sensing and spatial analysis are reliable possible solutions. Climate change scenarios and their implications on coastal vulnerability and hazards calls for the need to predict, assess, and model potential future impacts (Azaz, 2010; Roy and Blaschke, 2013; Yin et al., 2013; Thompson and Frazier, 2014; Hoque et al., 2017). Coastal vulnerability and ranking of the coastal regions in terms of their exposure to different climatic extreme events are the most debated issues, after the predictions of climate change theories. Accordingly, the coastal regions of the world have been categorized, and relative vulnerability indices have been developed (UNEP, 2005; IPCC, 2002; Gormitz et al., 1994; Turner et al., 1993).

Surrounded on three sides by major seas and encompassing about an 8000 km-long coastline, India is home to major coastal habitations and their attendant dense populations, which are inherently exposed to coastal natural hazards. About 80%–90% of global tropical cyclone deaths have been experienced along the coastal zones of the Bay of Bengal (Paul, 2009; Younus, 2017). Several previous studies have attempted vulnerability indexing of these areas due to cyclone and storm surge risks (Jayanthi, 1998; Kumar, 2003; Patwardhan et al., 2003; Sharma and Patwardhan, 2007). This fact can be exemplified by a statistic that, from the year 1737 onward alone, there were 23 major surge events in the Bay of Bengal, accounting for >10,000 human lives lost during each event (Murthy et al., 2006). Patwardhan et al. (2003) addressed vulnerability at the district level and computed the differential vulnerability indices of the coastal districts of India. They defined vulnerability in terms of three different components; that is, hazard, exposure, and adaptive capacity of the exposed area. Kumar (2003) defined vulnerability due to cyclones as a function of cyclone impact on the region, and resistance and resilience of the region to the impact, and computed the composite vulnerability index for Indian coastal districts by using the demographic, physical, economic, and social factors to construct the indices. Realizing the risks involved, there were previous attempts to assess the coastal vulnerability, including the state of Tamil Nadu. For example, Saxena et al. (2013) developed a coastal vulnerability index (CVI) for the Cuddalore District of Tamil Nadu. The study considered nine broad dimensions of vulnerability, namely, geographic, demographic, institutional, natural, social, safety infrastructure, physical, livelihood, and economic, which were weighted and derived results through the Analytic Hierarchy Process (AHP). Chandrasekar et al. (2013) discussed coastal vulnerability related to sea level changes and resultant erosion in southern Tamil Nadu. Geomorphology has been considered as an important parameter in understanding the dynamism of the coast (Rao et al., 2009; Kumar et al., 2010a,b; Mujabar and Chandrasekar, 2011; Rani et al., 2015). Murali et al. (2013) evaluated the Puducherry coast using the vulnerability assessment, and studied the tsunami of December 2004 and the Thane cyclone of 2011.

Recognizing the inherent vulnerabilities of coastal regions of India, and the need to have a disaster preparedness plan, including establishment of rescue and remedial measures for infrastructure, depending on physical, social, and economic parameters; an exclusive study for the coastal regions of Tamil Nadu State was attempted, and the results are discussed in this chapter (Table 1).

2 Study Area

The 1076 km-long coastline of Tamil Nadu state constitutes about 15% of the 8000 km coastline of India (Fig. 1). The study extends from Cape Comorin, the southernmost point of the Indian mainland at the Arabian Sea to Chennai at the Bay of Bengal. The state of Tamil Nadu is

Table 1 List of prominent disasters in the past at coastal Tamil Nadu, from 1991 to 2017

Type	Name	Year	Place	Grade	Units
Cyclone	BOB 09	14 November 1991	Tamil Nadu coast	CS	45 knots
Cyclone	BOB 07	13 November 1992	Tamil Nadu coast	SCS	55 knots
Cyclone	BOB 01	04 December 1993	Tamil Nadu coast	ESCS	90 knots
Cyclone	BOB 09	03 December 1996	Tamil Nadu coast	VSCS	65 knots
Cyclone	BOB 05	28 November 2000	Tamil Nadu coast	ESCS	102 knots
Tsunami		26 December 2004	Tamil Nadu coast		2.2 to 5 m run-up
Cyclone	Fanoos	08 December 2005	Tamil Nadu coast	CS	45 knots
Flood		04 December 2005	Chennai, Cuddalore		up to 7 m deep
Cyclone	Nisha	26 November 2008	Tamil Nadu coast	CS	45 knots
Cyclone	Jal	07 November 2010	Tamil Nadu coast	SCS	60 knots
Cyclone	Thane	29 December 2011	Tamil Nadu coast	VSCS	75 knots
Cyclone	Nilam	31 October 2012	Tamil Nadu coast	CS	45 knots
Cyclone	Madi	10 December 2013	Tamil Nadu coast	VSCS	65 knots
Flood		02 December 2015	Chennai, Cuddalore, Thoothukudi		up to 8 m deep
Cyclone	Roanu	21 May 2016	Tamil Nadu coast	CS	45 knots
Cyclone	Kyant	26 October 2016	Tamil Nadu coast	CS	40 knots
Cyclone	Nada	29 November 2016	Tamil Nadu coast	CS	40 knots
Cyclone	Vardah	11 December 2016	Tamil Nadu coast	VSCS	70 knots
Ennore Oil spill	Oil Spill	28 January 2017	Ennore Creek, Chennai		Accumulation over 3 km stretch
Cyclone	Ockhi	29 November 2017	Kanyakumari	CS	45 knots

CS, Cyclonic storm, **SCS**, Severe Cyclonic Strom, **VSCS**, Very Severe Cyclonic Strom, **ESCS**, Extreme Severe Cyclonic Strom.
Source: Collateral Data from Indian Metrological Data—best track, and UNSO—JTWC.

bounded by Karnataka and Andhra Pradesh on the north, the Bay of Bengal on the east, Kerala on the west, and the Indian Ocean on the south. Rivers flowing from the west to east in Tamil Nadu, namely, Palar, Ponnaiyar, Vellar, Cauvey, Vaigai, Vaippar, and Thamrabarani debauch at the coast into the Bay of Bengal. The state is comprised of 32 districts, of which 13 districts (viz. Chennai, Kancheepuram, Thiruvallur, Cuddalore, Villupuram, Thanjavur, Pudukottai, Thiruvarur, Nagapattinam, Ramanathapuram, Tuticorin, Tirunelveli, and Kanyakumari) fall along the coastal region, which is comprised of 51 coastal taluks. About one-fifth of the entire roadway network of the Tamil Nadu state is located in the coastal districts. The coastal taluks are comprised of a total population of 20,394,863 persons (Census, 2011).

3 Data and Methodology

To assess the coastal vulnerability, several data types (elevation, slope, geomorphic units, shoreline proximity, storm-affected areas, population density, household density, road proximity, literacy rate, built-up near to shoreline, land use/land cover, normalized

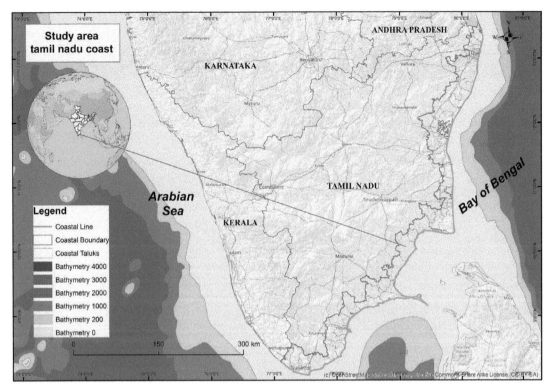

Fig. 1
Study area.

difference vegetative index, and rainfall) have been compiled from various repositories and sources (Table 2). A systems approach of the methodology is depicted in the flow chart (Fig. 2).

3.1 Analytical Hierarchy Process (AHP)

A multicriteria evaluation was adopted to assign weights and ranks through an Analytical Hierarchy Process (AHP). AHP involves building a hierarchy of decision elements (criteria) and then making comparisons between possible pairs in a matrix to give a weight for each element, and a consistency ratio. Developed by Saaty (1977, 1980), the factor weight of each criterion is determined by a pair-wise comparison matrix (Saaty, 1990, 1994; Saaty and Vargas, 2001; Rahaman and Aruchamy, 2017). Using this method, each criterion's layers were broken into smaller classes, and then these classes were compared based on their importance. For comparison, each class was rated against every other class by assigning a relative value between 1 and 9. This value and its description are shown in Table 3.

Table 2 Data obtained and their sources

No.	Data	Data Details	Source	Purpose
1.	Toposheets	Open Series Map 1:50,000 scale	Survey of India (SOI) http://www.surveyofindia.gov.in http://www.soinakshe.uk.gov.in	Basemap
		Physical Parameters		
2.	DEM	ASTER 30 m	USGS https://earthexplorer.usgs.gov	Relief (Elevation), Slope and Shoreline proximity
3.	Geomorphology	2005–2006 1:50,000 scale	NRSC Bhuvan Web Map Service (WMS) URL http://bhuvan5.nrsc.gov.in/bhuvan/wms	Geomorphic Units
4	Cyclone Best Tracks	1878–1977	UNSO–JTWC http://www.usno.navy.mil/NOOC/nmfc-ph/RSS/jtwc/best_tracks/ioindex.php	Storm affected area
		Socio-Economic Parameters		
6.	Population	2011	Census of India—2011 http://censusindia.gov.in	Population Density, Households Density and Literacy rate
7.	Settlements	2016	ESRI Basemap	Built-up near to shoreline
8.	Road Network	2016	ESRI Basemap and Open Street maps	Road Proximity
		Environmental and Climatic Parameters		
9.	Rainfall	2012–2015	Indian Meteorological Department (IMD) http://hydro.imd.gov.in/hydrometweb/(S(wak5jr455vd2xn55xuovqfej))/landing.aspx#	Annual Rainfall
10.	Land use/Land cover	2014/56 m	NRSCIRS-Resourcesat-2 LISS-III AWFIS	Landuse/Land cover
11	NDVI	2014	NRSCIRS-Resourcesat-2 LISS-III AWFIS	NDVI

In order to establish a pair-wise comparison matrix (A), factors of each level and their weights are shown as $A_1, A_2, ..., A_n$ and $w_1, w_2, ..., w_n$. The relative importance of a_i and a_j is shown as a_{ij}. The pair-wise comparison matrix of factors $A_1, A_2, ..., A_n$ as $A = [a_{ij}]$ is expressed as

$$A = \{a_{ij}\} n*n = \begin{pmatrix} 1 & a_{12} & ... & a_{1n} \\ \vdots & & \ddots & \vdots \\ a_{n1} & a_{n2} & ... & 1 \end{pmatrix} = \begin{pmatrix} 1 & \frac{w_1}{w_2} & .... & \frac{w_1}{w_n} \\ \frac{w_2}{w_1} & 1 & ... & \frac{w_2}{w_n} \\ \frac{w_n}{w_1} & \frac{w_n}{w_2} & & 1 \end{pmatrix} \quad (1)$$

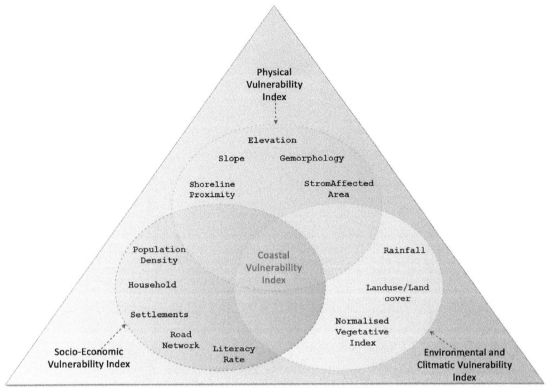

Fig. 2
Venn's diagram for coastal vulnerability index.

Table 3 Fundamental scales for pairwise comparisons

Scales	Degree of Preferences	Descriptions
1	Equally	Two activities contribute equally to the objective
3	Moderately	Experience and judgment slightly to moderately favor one activity over another
5	Strongly	Experience and judgment strongly or essentially favor one activity over another
7	Very strongly	An activity is strongly favored over another and its dominance is showed in practice
9	Extremely	The evidence of favoring one activity over another is of the highest degree possible of an affirmation
2, 4, 6, 8	Intermediate values	Used to represent compromises between the preferences in weights 1, 3, 5, 7 and 9

Source: Saaty, T.L., 1980. The Analytical Hierarchy Process. McGraw Hill, New York.

In this matrix, the element, $a_{ij} = 1/a_{ji}$ and thus, when $i = j$, $a_{ij} = 1$. A matrix is normalized using Eq. (2)

$$a'_{ij} = \frac{a_{ij}}{\sum_{i=1}^{n} a_{ij}} \qquad (2)$$

$$ij = 1, 2, 3, \ldots n.$$

and finally, weights of factors are computed using Eq. (3) as:

$$w_i = \left(\frac{1}{n}\right) \sum_{i=1}^{n} a'_{ij} \qquad (3)$$

In matrix-based pair-wise comparison, if the factor on the horizontal axis is more important than the factor on the vertical axis, the value varies between 1 and 9. Conversely, if the factor is less important, the value varies between the reciprocals 1/2 and 1/9 (Table 3). In AHP, for checking the consistency of the matrix, the consistency ratio is used, which depends on the number of parameters. The consistency ratio (CR) is obtained by comparing the consistency index (CI) with an average random consistency index (RI). The consistency ratio is defined using Eq. (4), and the CI is defined by using Eq. (5) as

$$CR = \frac{CI}{RI} \qquad (4)$$

$$CI = \frac{\lambda_{max} - n}{n - 1} \qquad (5)$$

Where, CI = consistency index, λ_{max} is the major or principal Eigen value of the matrix, and it is computed from the matrix, and n is the order of the matrix. The average random consistency index (RI, Table 4) is derived from a sample of randomly generated reciprocal matrices using the scales 1/9, 1/8,8 and 9.

The final result consists of the derived factor weights and class rating, and is calculated using the Consistency Ratio (CR). In AHP, the consistency used to build a matrix is checked by a consistency ratio, which depends on the number of criteria. For a 10 × 10 matrix, the CR must be <0.1 to accept the computed weights. The models with a CR >0.1 were automatically rejected, while a CR <0.1 was often acceptable.

Table 4 Random consistency index (RI)

N (number of factors)	1	2	3	4	5	6	7	8	9	10
RI	0.00	0.00	0.52	0.89	1.11	1.25	1.35	1.40	1.45	1.49

Source: Saaty, T.L., 1980. The Analytical Hierarchy Process. McGraw Hill, New York.

3.2 Fuzzy Linear Membership

The fuzzy linear transformation function applies a linear function between the minimum and maximum values. Anything below the minimum will be assigned as 0 (definitely not a member), and anything above the maximum as 1 (definitely a member). And the entire range of possibilities between 0 and 1 are assigned to some level of possible membership (the larger the number, the greater the possibility).

3.3 Coastal Vulnerability Index (CVI)

The coastal vulnerability index is drawn from three major indicators by assigning weights and ranks through an AHP-based pairwise comparison matrix (Table 5), such as physical indicators from the Physical Vulnerability Index (PVI), socio-economic indicators from the Socio-economic Vulnerability Index (SVI), and environmental and climatic factors from the Environmental and Climatic Vulnerability Index (ECVI), with an integrating process, which is mathematically described as:

$$\mathbf{PVI} = El_{ij} \times Sl_{ij} \times Sp_{ij} \times Gu_{ij} \times Sa_{ij} \qquad (6)$$

$$\mathbf{SVI} = Pd_{ij} \times Hh_{ij} \times Lr_{ij} \times St_{ij} \times Rn_{ij} \qquad (7)$$

$$\mathbf{ECVI} = Rf_{ij} \times LULC_{ij} \times NDVI_{ij} \qquad (8)$$

The resultant values from Eqs. (6)–(8), are substituted in Eq. (9) to derive the Coastal Vulnerability Index.

$$\mathbf{CVI} = \mathbf{PVI} + \mathbf{SVI} + \mathbf{ECVI} \qquad (9)$$

Where, El = elevation (relief), Sl = slope, Sp = shoreline proximity, Gu = geomorphic units, Sa = storm affected area, Pd = population density, H = household density, Lr = literacy rate, St = built-up near shoreline, Rd. = road proximity, Rf = rainfall annual, LULC = land use/land cover, and NDVI = normalized difference vegetation index and (i)-normalized weight for the criteria, (j)-rank for the subEcriteria. Where CVI = coastal vulnerability index, PVI = physical vulnerability index, SVI = socio-economic vulnerability indicators, and ECVI = environmental and climatic indicators.

Table 5 Pairwise comparison matrix of main criteria

	Physical	Socio-Economic	Environmental and Climate	Eigenvector
Physical	1	2	3	0.525
Socio-Economic	1/2	1	3	0.334
Environmental and Climate	1/3	1/3	1	0.142
$\lambda_{max} = 3.0538$	CI = 0.026	CR = 0.046		1

4 Results

4.1 Physical Indicators

4.1.1 Coastal relief (El)

'Relief' represents the average elevation of a region above the mean sea level. The ASTER GDEM was used for deriving the relief (elevation) of coastal Tamil Nadu. The study area was classified into six classes, 1000m to below 5, as very high to very low elevation range. The spatial distribution of these classes are given in Fig. 3. The highest elevation, with 100–1000m, is present over Thovala, and its surrounding taluks have been given the least weight (Table 6). The regions with elevations below 5m are present only in the outlet of the river, excluding the protruding part of Rameswaram, which has been given the highest weight.

4.1.2 Coastal slope (Sl)

The slope is the steepness of a region with respect to the plain, measured in percentage. The slope map of coastal Tamil Nadu is classified into five classes, with below 1% to >10% slope. The spatial distribution of slope classes is presented in Fig. 3. The slope percentage depicts a

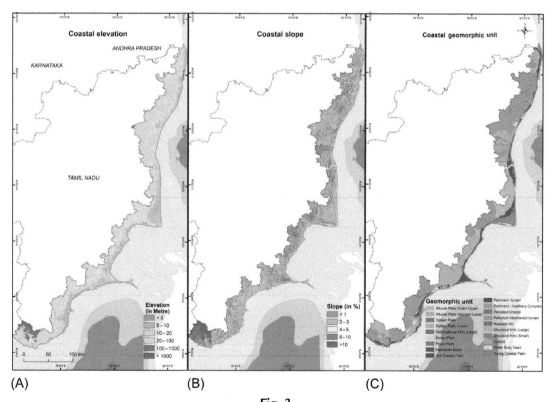

Fig. 3
(A) Elevation, (B) slope, (C) geomorphic unit.

Table 6 Normalized weigh for criteria and sub criteria

Criteria—Level I	Criteria—Level II	Normalized Weight	Criteria—Level III	Area (Ha)	Area%
Physical Indicators $W1 = 0.525$	Elevation (m) $W11 = 0.413$	0.216	< 5	86.85	0.36
			5–10	3468.61	14.30
			10–20	10,020.64	41.30
			20–100	9914.22	40.86
			100–1000	673.44	2.78
			>1000	98.59	0.41
	Slope (%) $W12 = 0.292$	0.153	<1	5709.18	23.53
			1–3	5518.18	22.74
			3–5	4196.56	17.30
			5–10	5640.66	23.25
			> 10	3197.78	13.18
	Geomorphic Unit $W13 = 0.057$	0.030	Alluvial Plain	7139.41	29.43
			Coastal Plain	4489.83	18.51
			Deltaic Plain	94.87	0.39
			Denudational Hills	64.05	0.26
			Eolian Plain	440.77	1.82
			Flood Plain	527.58	2.17
			Habitation Mask	0.82	0.00
			Pediplain	10,173.57	41.93
			Structural Hills	766.87	3.16
			Upland	465.90	1.92
			Water Body Mask	98.67	0.41
	Shoreline Proximity (km) $W14 = 0.133$	0.070	<1	1037.90	4.28
			1–2	945.58	3.90
			2–5	2637.69	10.87
			5–10	4203.17	17.32
			>10	15,438.02	63.63
	Strom Affected Area $W15 = 0.106$	0.056	Flood Affected Area	823.60	3.39
			Less Affected Area	12,547.75	51.72
			Medium Affected Area	6433.93	26.52
			Sever Affected Area	4457.09	18.37
			Flood Affected Area	823.60	3.39
Socio-Economic Indicators $W2 = 0.334$	Population Density (No. of Persons/km^2) $W21 = 0.406$	0.213	<500	12,545.79	51.71
			500–1000	9654.15	39.79
			1000–1500	1065.22	4.39
			1500–2000	278.43	1.15
			>2000	718.89	2.96
		0.139	<100	9673.34	39.87

(Continued)

Table 6 Normalized weigh for criteria and sub criteria—Cont'd

Criteria—Level I	Criteria—Level II	Normalized Weight	Criteria—Level III	Area (Ha)	Area%
	Household Density (No. of houses/km^2) $W22 = 0.265$		100–500	13,591.83	56.02
			500–1500	716.69	2.95
			1500–3000	60.12	0.25
			>300	0220.50	0.91
	Road Proximity (km) $W23 = 0.128$	0.067	<2	3279.48	13.52
			2–4	2197.43	9.06
			4–6	1925.30	7.94
			6–8	1779.97	7.34
			8–10	1718.86	7.08
			>10	13,361.31	55.07
	Literacy Rate (%) $W24 = 0.057$	0.030	0–1	14,506.33	59.79
			1–5	9587.52	39.52
			>5	168.62	0.69
	Built-up near to Shoreline (km) $W25 = 0.144$	0.076	<1	99.27	0.41
			1–2	78.05	0.32
			2–5	195.52	0.81
			5–10	315.17	1.30
			>10	893.39	3.68
			Non-settlement area	22,681.07	93.48
Environment and Climate Indicators $W3 = 0.142$	Landuse $W31 = 0.334$	0.111	Waterbody	894.76	3.69
			Built-up	2474.59	10.20
			Agriculture	3095.08	12.76
			Dense Scrub	3526.82	14.54
			Forest	5130.60	21.15
			Current Fallow	6072.11	25.03
			Costal/Inland Wetland	2860.21	11.79
			Cloud	208.18	0.86
	NDVI $W32 = 0.142$	0.70	0	5.27	0.02
			0–0.2	13,664.27	56.32
			0.2–0.4	9886.92	40.75
			0.4–0.6	705.90	2.91
	Rainfall (mm) $W33 = 0.525$	0.344	<750	5724.50	23.59
			750–1000	4174.03	17.20
			1000–1250	9373.92	38.64
			>1250	4989.91	20.57

Source: By author.

clear picture that there is not much steepness present; rather, there is a vast plain in the study area. The slope below 1% commonly represents depressions, tanks, rivers, and salt pans. The regions with a slope ranging from 2% to 5% are found distributed all over the study area. The least weight was assigned to the highest slopes.

4.1.3 Geomorphic unit (Gu)

Geomorphology is classified based on the National Remote Sensing Centre (NRSC) guidelines. The young and old coastal plains cover the entire tract of coastline from the northern tip of coastal Tamil Nadu to the southernmost tip, except very few spots where alluvial plains—older/upper, alluvial plains—younger/lower, eolian and deltaic plains cover the coastal tracts. The flood plain remains a prominent geomorphic unit along the river stretches of the north and south. However, the northern part of the study area has a wider floodplain, and southern rivers have very narrow floodplains. The coastal plains and floodplains were given the highest weight. The least weight was assigned to the structural and denudational hills, which are only predominant in the interior parts of the study area.

4.1.4 Shoreline proximity (Sp)

The shoreline proximity map shows the buffer of coastline toward the landmass. The shoreline proximity is delineated in a linear manner, with the premise that closer to the coast, the vulnerability is higher, and vice versa. Accordingly, the highest and lowest weights were assigned to the highest and lowest, respectively. The proximity to the shoreline is zoned into five classes from below 1 to above 10 km. Their spatial distribution and aerial extent are presented in Fig. 4 and Table 6, respectively.

4.1.5 Storm affected areas (Sa)

The study area was affected by cyclones about 38 times during the past three decades. The coastal track map shows the best track of cyclonic storms from 1991 to 2017 (Fig. 6A). The coastal taluks have been classified under four categories, namely, severe, medium, less, and flood-affected areas, and the spatial distribution of these categories is in Fig. 4B. The rank and weight assigned based on severity are presented in Table 6.

4.2 Socio-Economic Indicators

4.2.1 Population density (Pd)

The population density according to the 2011 census shows a huge difference in range. There are five classes of population density found in the coastal taluks of Tamil Nadu. The very high density of population (>2000 persons per sq. km) is noted in Chennai and its suburban Mathavaram, Ambattur, Alandur, Tambaram, and Shollinganallur taluks, and given the highest weight (Fig. 5A and Table 6). The medium density (1001–1500) is sparsely spread over three taluks; that is, Cuddalore, Thoothukudi, and Vilavancode. The low density (501–1000) was found in Ponneri, Sirperumbdur, Chengalpattu, Tirukalkundaram in the north part, Villupuram to Pattukottai in the central part, Ramanathapuram, Rameswaram, Tiruchendur, and Kalkulam in the south. The least weight was assigned to very low-density areas.

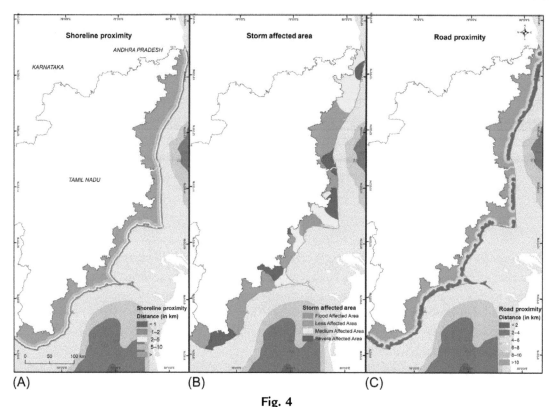

Fig. 4
(A) Shoreline proximity, (B) storm affected area, (C) road proximity.

4.2.2 Household density (Hh)

Very high household density (above 3000) is present in Chennai and Alandur, followed by high household density (1501–3000) at Tambaram. The medium household density (501–1500) is found at Mathavaram, Ambattur, and Sholinganallur in the north, and Agastheeswaram taluk in the southern tip of the study area. The low household density (101–500) corroborates with the lowest population density regions. The weights and ranks were given based on the influence toward the vulnerability (Table 6). The spatial distribution is presented in Fig. 5B.

4.2.3 Literacy rate (Lr)

The literacy rate is categorized into three classes: high (4.8–24.1), medium (1.7–4.7), and low (0–1.6). The high literacy along coastal Tamil Nadu is only present in Chennai. Medium literacy is present in the Ponneri, Alandur, Tambaram, Mathavaram, Ambattur, Sholinganallur, Sirperumbdur, Chengalpattu, Tindivanam, Villupuram, Panruti, Cuddalore, and Chidambaram taluks in the north; Pattukottai taluk in the central part; and Ramanathapuram, Thoothukudi, Agastheeswaram, Kalkulam, and Vilvancode taluks in the southern part of the study area.

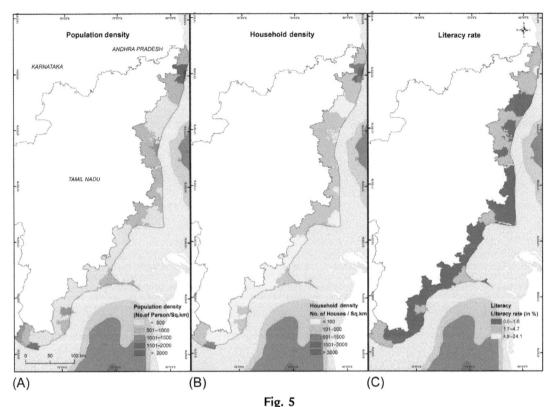

Fig. 5
(A) Population density, (B) household density, (C) literacy rate.

A low level of literacy is highly concentrated over the coastal regions, especially in southern coastal Tamil Nadu.

4.2.4 Built-up near shoreline (St)

The built-up categories in the study area are classified into built-up urban, built-up rural, and built-up industries/mining. The total counts of each cluster of build-up were collated for distance (1, 2, 5, 10 and 20 km) in km from the coastline, and each of the counts for the distance division were accounted for in individual classes. The vulnerability increases toward the coastline. The built-up areas within 1 km from the coastline were considered very highly vulnerable, while 20 km was considered the least vulnerable, and accordingly was given the appropriate weights (Fig. 6B and C; Table 6).

4.2.5 Road proximity (Rd)

Individual classes were established for road proximity <2, 2–4, 4–6, 6–8, 8–10 and >10 km for vulnerability. The spatial distribution of these classes are presented in Fig. 4B. Based on the road proximity, the coastal region can be distinctively trifurcated into three vulnerable zones,

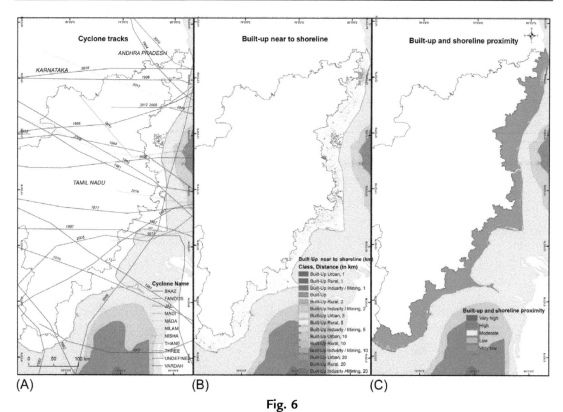

Fig. 6
(A) Cyclone tracks, (B) built-up near to shoreline, (C) built-up and shoreline proximity.

the very high, high to moderate, and low. Along the coastline, a buffer of 2 km is considered a very high vulnerability zone, followed by the high vulnerability of class 8–10, and above 10 km, which fall into the low vulnerability zone. The highest weight was given to the least proximal road, near the coastline; while the highest road proximity was given the least weight (Table 6), which is away from the coastline.

4.3 Environmental and Climatic Indicators

4.3.1 Land use/land cover (LULC)

The land is categorized into the following land use/land cover classifications: agriculture, built-up, current fallow, coastal/inland wetland, dense scrub, forest, waterbodies, and others/cloud (Fig. 7A). The agriculture and current fallow, forest, and dense scrub bifurcates the study area into northern and southern parts. The agriculture and current fallow are spread sparsely over the northern half of the study area from Ponneri to the Pattukottai taluks, and are concentrated densely in the southern half of the study area, except for the Tiruchendur and Kanniyakumari taluks. The forests are found in the northern half, and sparsely on other areas.

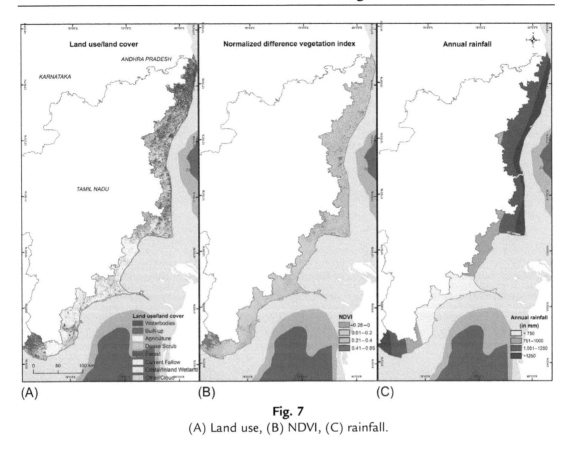

Fig. 7
(A) Land use, (B) NDVI, (C) rainfall.

The built-up area is highly concentrated at the metropolitan and suburban regions of Chennai, and shows a decreasing trend toward the south. The waterbodies are found everywhere, with a few exceptions at Chennai, Vedaranyam, and Thuthukudi taluks, where their sprawl and density are high. The highest weights were assigned to the built-up areas, waterbodies, and the coastal wetland areas; while forest and current fallow land were given the least weight (Table 6).

4.3.2 Normalized difference vegetation index (NDVI)

The Normalized Difference Vegetation Index shows the ranges from −0.28 to 0.65 (Fig. 7B). The areas with the steeper slope have a high vegetation cover. The highest weights were assigned to the nonvegetative areas, while the vegetative areas were given the least weight.

4.3.3 Annual rainfall (Rf)

The average annual rainfall data for the period of 5 years (2012–2016) was plotted using the spline interpolation technique. The rainfall distribution was divided into five classes: <750 mm to >1250 mm. The spatial distribution and areal extent of the rainfall have been given in Fig. 7C and Table 6, respectively. There are two distinct zones of very high (1250 mm) to high rainfall (1000–1250 mm) that were given the highest weight, and moderate (750–1000 mm) to low (below 750 mm) rainfall zones, which were given the least weight.

4.4 Coastal Vulnerability Index (CVI)

After a detailed study on the individual factors, all the indicators were categorized under the physical, socio-economic, and environmental and climatic indicators. They were then analyzed, weighted, and ranked according to their relevance/relative contribution to vulnerability. The resultant indices PVI, SVI, and CEVI were then combined using Eq. (9) to derive the CVI.

4.4.1 Physical vulnerability index (PVI)

In the physical indicator, each subcriterion was assigned with weight and rank based on AHP and fuzzy linear membership (Table 7), [*PVI = (0.216 * Elevation) + (0.153 * Slope) + (0.30 * Geomorphology) + (0.70 * Shoreline Proximity) + (0.56 * Storm Affected Area)*], which was categorized into four vulnerable classes, namely, very high vulnerable (VHV), high vulnerable (HV), moderate vulnerable (MV), and low vulnerable (LV). The results indicate that the HVH covers about 10.7% of the study area (Table 10), and the HV covers 16.7% of the coastline and interior land in the study area. The second-largest area of 34.9% is covered by the MV lying adjacent to the HV, followed by LV in the interior land portions with 37.7% area. The spatial distribution of PVI is shown in Fig. 8A and its description is presented in Table 10.

4.4.2 Socio-economic vulnerability index (SVI)

In the socio-economic indicator, each subcriterion was assigned with weights and ranks based on AHP and fuzzy linear membership (Table 8), which was categorized into four vulnerable classes, namely, very high vulnerable (VHV), high vulnerable (HV), moderate vulnerable (MV), and low vulnerable (LV). *SVI = (0.406 * Population Density) + (0.265 * Household*

Table 7 Pair wise comparison matrix of subcriteria with respect to physical indicator

	Elevation	Slope	Geomorphology	Shoreline Proximity	Strom Affected Area	Eigenvector
Elevation	1	3	4	4	3	0.413
Slope	1/3	1	5	3	5	0.292
Geomorphology	1/4	1/5	1	1/3	1/3	0.057
Shoreline Proximity	1/4	1/3	3	1	2	0.133
Strom Affected Area	1/3	1/5	3	1/2	1	0.106
$\lambda_{max} = 5.4147$	CI = 0.103	CR = 0.093				1

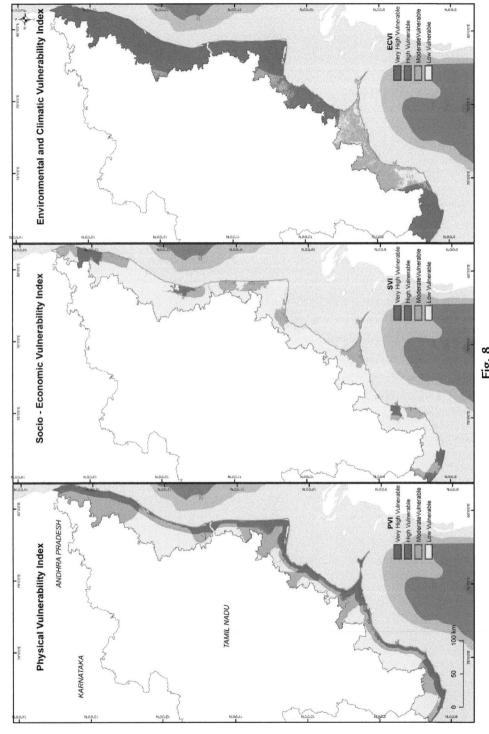

Fig. 8

(A) Physical vulnerability index, (B) socio-economic vulnerability index, (C) environmental and climatic vulnerability index.

Table 8 Pairwise comparison matrix of subcriteria with respect to socio-economic indicators

	Population Density	House Hold Density	Road Proximity	Literacy Rate	Settlement Near to Coast	Eigenvector
Population Density	1	2	4	5	3	0.406
House Hold Density	1/2	1	3	5	2	0.265
Road Proximity	1/4	1/3	1	5	1/2	0.128
Literacy Rate	1/5	1/5	1/5	1	1/2	0.057
Settlement Near to Coast	1/3	1/2	2	2	1	0.144
$\lambda_{max} = 5.2948$	CI = 0.073	CR = 0.066				1

Density) + (0.128 * Road Proximity) + (0.057 * Literacy Rate) + (0.144 * Built-up near to Coastline). The coastline and the rest of study area appear to be clearly out of danger in terms of the socio-economic factor, except for the Chennai and Agastheeswaram taluks. The HV and MV cover the river outlets and major deltas with 3.9% and 18.3% area, respectively. However, with a larger area under the LV of 74.2% area, the SVI necessitates the need for socio-economic planning. The LV owes to the higher literacy rate of 59.79%. The nonbuilt-up area from the shoreline, with 93.48% area, diminishes the socio-economic value due to the Coastal Regulation Zone (CRZ), 2011 planning in the coastal taluks. The spatial distribution of PVI is shown in Fig. 8B, and its description is presented in Table 10.

4.4.3 Environmental and climatic vulnerability index

In the environmental and climatic indicator, each of the criterion was assigned weights based on AHP and fuzzy linear membership (Table 9), and categorized into four vulnerable classes, namely, very high vulnerable (VHV), high vulnerable (HV), moderate vulnerable (MV), and low vulnerable (LV). *ECVI = (0.334* Land use/Land cover) + (0.142 * NDVI) + (0.525 * Rainfall).* The ECVI map sounds alarm bells for a major part of the study area. The spatial distribution of PVI is shown in Fig. 8C, and its description is presented in Table 10.

Table 9 Pairwise comparison matrix of subcriteria with respect to environmental and climatic indicators

	Land Use	NDVI	Rainfall	Eigenvector
Land use	1	3	1/2	0.334
NDVI	1/3	1	1/3	0.142
Rainfall	2	3	1	0.525
$\lambda_{max} = 3.0538$	CI = 0.026	CR = 0.046		1

Table 10 Coastal Vulnerability Area coverage in different indicators

Vulnerability Indicators Class	Physical Vulnerability Indicator (PVI)		Socio-Economic Vulnerability Indicator (SVI)		Environmental and Climatic Vulnerability Indicator (ECVI)	
	Area (in km²)	Area (in %)	Area (in km²)	Area (in %)	Area (in km²)	Area (in %)
Very High Vulnerability	2592.09	10.7	866.4471	3.6	5035.057	20.8
High Vulnerability	4058.76	16.7	943.1487	3.9	12,240.21	50.4
Medium Vulnerability	8466.34	34.9	4439.606	18.3	5636.779	23.2
Low Vulnerability	9145.17	37.7	18,013.16	74.2	1350.315	5.6

Source: By author.

5 Discussion

The study has evaluated multiple param and judged their relative contribution to coastal risk/protection. For example, the relief of a region is a deciding factor for various hazards. Usually, in the coastal areas, the lower the elevation, the more susceptible the region is to coastal hazards. The elevation trend illuminates the fact that 5- to 10-m elevated regions fall into the very high vulnerability class. The lower the value of the slope indicates more vulnerability to the study region (Pramanik et al., 2015). The slope provides information pertaining to the sea and tidal indicators, in accordance with possible prediction inundation during a cyclone, storm surge, tsunami, and so forth. The slope is also the major factor in deciding the infrastructure and build-up present along the coastal region. The geomorphic unit depicts the Earth's surface physiographic expression, the origin of the landform, and content of the material prevailing over the region. Most of the geomorphic agencies originate within the Earth's atmosphere, and are directed by the force of gravity (Thornbury and Schumm, 1969). Singarasubramanian et al. (2011) studied the geomorphic changes along the east coast of Tamil Nadu during and after the tsunami and observed that most of the coastal regions become steep in some places and shallow in other places due to the differential erosion and accretion of sediments by tsunami waves. A geomorphic classification makes it possible to separate active forms from inactive forms.

The shoreline proximity marks a risk boundary for the coastline. Cyclones originate over the equatorial regions of the earth, accompanied by heavy downpours and wind speeds >30 km/h. The coastal region of Tamil Nadu receives plenty of rainfall, under the wet spell of the north-eastern monsoon season, and the accompanying cyclonic rainfall along the coast. The rainfall appears to be a mixed blessing. Rainfall over a region is important for knowing the intensity of rainfall affecting the coastal slope and geomorphology of the study area, paving the way for an increased pace of coastal hazards. In general, the entire coastal taluks have been affected by cyclonic storms, except for the taluks from Vilathikulam to Vilavancode. The extreme southern parts of the taluks are free from cyclonic storms. As cyclones bring in gusty winds, which influence sea levels to swell and cause coastal inundation, storm surges, and flooding in the inland rivers; precipitating impacts that are multifold (Ramkumar, 2009).

The population of a region signifies the need for development and protection. The population helps in understanding the habitation growing around the populated areas. The coastal taluks are classified based on the density of population to identify the most vulnerable zone. It appears that density of population is associated with the delta plains of each river along the coastal taluks preceded by the urban metropolitan centers. The household is the dwelling in which one or more people live. The household depicts the number of individual buildings in the particular administrative unit, regardless of the population number. Thus, understanding the household density gives us the ratio of population density to the density of individual buildings shared by them. The road is part of the socio-economic infrastructure, and also provides economic benefits and supports development. There exists a close relationship between socio-economic progress and movement of people and goods. Literacy is the important key denoting social and economic progress. Widespread low literacy denotes low awareness, and more vulnerability to coastal hazards.

The coastal settlement represents the linear, as well as the long-time settled population, who are primarily settled along the coastal boundary. The coastal settlement originates based on the need of the particular coastal habitation. The coastal habitation is categorized based on communities dependent on fishing as a primary activity, a commercial area for governmental tourism and private hotels with beaches, and last, industrial units such as marinas, aqua farms, atomic power stations, and other areas with economic prospects. Thus, from this discussion, the results show the need for an integrated methodology for understanding the vulnerability and risk of coastal habitats (Table 11).

The CVI as a combined result from the PVI, SVI, and ECVI displays vulnerability based on 13 parameters. This is categorized into five vulnerable classes, namely, very high vulnerable (VHV), high vulnerable (HV), medium vulnerable (MV), low vulnerable (LV), and very low vulnerability (VLV) (Fig. 9). The VHV zones are spread over four places, Ponneri to Tirukalkundram, Cuddalore, Nagapattinam, and Agatheeswaram in the southern tip of Tamil Nadu. From Table 12 it follows that the VHV area occupies 5.5% of the study area, because of the cumulative contribution from low elevations, lower slopes, a high density of population and households, high rainfall, and severe cyclonic storm zones. The HV covers 25.2% of the study area, while about 28.6% of the area falls under the VL and VLV, and covers 25.8% of the area. Based on the CVI the index, the values were further reclassified into three Coastal Vulnerable Zones (CVZ), CVZ—I, II, and III, which cover 20.8, 53.4, and 25.8% of the study area, respectively.

Table 11 Risk factor and its relationship with hazard vulnerability

Coastal Vulnerability Indicators				Functional Relationship With Risk
Level I	Level II	Abbr.	Risk Factors	
Physical Indicators	Elevation	El	*Exposure*	Lower El = HR
	Slope	Sl	*Exposure + Susceptibility*	Lower Sl = HR
	Shoreline proximity	Sp	*Exposure + Susceptibility*	Lesser Sp = HR
	Geomorphic units	Gu	*Exposure + Susceptibility*	Types of Gu = HR
	Strom Affected Area	Sa	*Susceptibility*	High Occurrences of Sa = HR
Socio-Economic Indicators	Population density	Pd	*Exposure + Susceptibility*	Higher Pd = HR
	Household	Hh	*Exposure + Resilience*	Higher Hh = HR
	Built-up from the shoreline	St	*Exposure + Resilience*	Larger St = HR
	Road proximity	Rn	*Exposure + Susceptibility + Resilience*	Closer Rn = HR
	Literacy rate	Lr	*Resilience*	Lower Lr = HR
Environmental and Climatic Indicators	Rainfall	Rf	*Exposure + Susceptibility*	Higher Rf = HR
	Landuse/Land cover	LULC	*Exposure + Susceptibility + Resilience*	High—Built-up and Agriculture = HR
	Normalized Difference Vegetative Index	NDVI	*Exposure*	Low NDVI = HR

HR = High risk.
Source: By author.

6 Conclusions

This study was attempted based on the relative contributions of natural environmental, physical, and socio-economic parameters related to coastal safety, especially to the ecosystems, infrastructure, and lives of fauna and flora that inhabit the coastal zone. The relative contributions were judged based on the subjective and quantitative assessments of each of the parameters, which in turn were categorized, and their metrics were collated to produce a few measurable indices. The spatial patterns of these measured/attributed values were then analyzed to draw inferences on relative vulnerability classes. Results of the study on the assessment of coastal vulnerability utilizing the geospatial technology and multicriteria evaluation approach has demarcated the Chennai, Cuddalore, Nagapattinam, Thoothukudi, and Agatheeswaram areas to be highly vulnerable, and in need of urgent measures to thwart negative impacts.

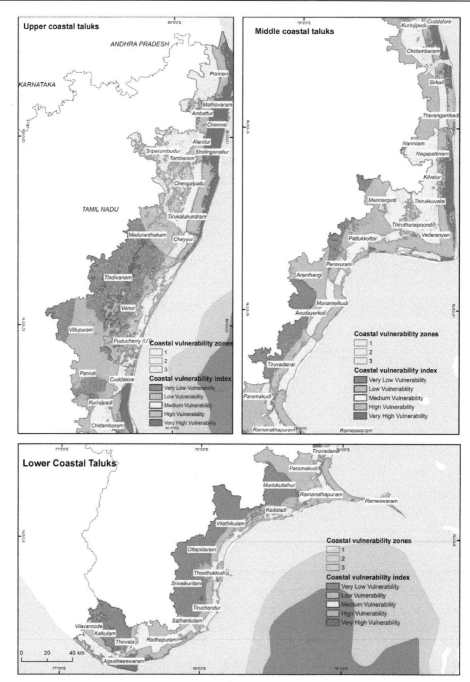

Fig. 9
Coastal vulnerability index and coastal vulnerability zones.

Table 12 Coastal Vulnerability Index Area and its Zone

Vulnerability Class	Area (in km²)	Area (in %)	Coastal Vulnerability Zone
Very High Vulnerability	1322.87	5.5	Zone–I
High Vulnerability	3703.38	15.3	
Medium Vulnerability	6116.40	25.2	Zone–II
Low Vulnerability	6853.43	28.2	
Very Low Vulnerability	6266.27	25.8	Zone–III

Source: By author.

Web Sources

NRSC, Bhuvan Portal. http://bhuvan.nrsc.gov.in/gis/thematic/index.php. Accessed 12.08.2017.
Open Street Maps. https://www.openstreetmap.org/. Accessed 23.08.2017.
Wikipedia, List of Tropical Cyclones that affected India. https://en.wikipedia.org/wiki/List_of_tropical_cyclones_that_affected_India#Tamil_Nadu. Accessed 19.11.2017.
Wikipedia, North Indian Ocean Tropical Cyclone. https://en.wikipedia.org/wiki/North_Indian_Ocean_tropical_cyclone. Accessed 04.12.2017.

Acknowledgments

AB and SA acknowledge the Department of Geography, Bharathidasan University, Tiruchirappalli for the facilities provided, the valuable comments provided by Dr. R. Jegankumar and Dr. P. Masilamani, and Dr. Edwin Jebakumar and Mr. S. Nitheshnirmal for their support of the research. The authors thank the reviewers for their critical comments and valid suggestions that helped improve the manuscript.

References

Arkema, K.K., Verutes, G.M., Wood, S.A., Clarke-Samuels, C., Rosado, S., Canto, M., Rosenthal, A., Ruckelshaus, M., Guannel, G., Toft, J., Faries, J., Silver, J.M., Griffin, R., Guerry, A.D., 2015. Embedding ecosystem services in coastal planning leads to better outcomes for people and nature. PNAS 112 (24), 7390–7395. https://doi.org/10.1073/pnas.1406483112.
Arkema, K.K., Griffin, R., Maldonado, S., Silver, J., Suckale, J., Guerry, A.D., 2017. Linking social, ecological, and physical science to advance natural and nature-based protection for coastal communities. Ann. N.Y. Acad. Sci. 1–22.
Armas, I., Gavris, A., 2013. Social vulnerability assessment using spatial multicriteria analysis (SEVI model) and the social vulnerability index (SoVI model) e a case study for Bucharest, Romania. Nat. Hazards Earth Syst. Sci. 13, 1481–1499. https://doi.org/10.5194/nhess-13-1481-2013.
Azaz, L., 2010. Capabilities of using remote sensing and GIS for tropical cyclones forecasting, monitoring, and damage assessment. Springer Netherlands. 22. https://doi.org/10.1007/978-90-481-3109-9.
Becken, S., Mahon, R., Rennie, H.G., Shakeela, A., 2014. The tourism disaster vulnerability framework: An application to tourism in Small Island destinations. Nat. Hazards 71 (1), 955–972. https://doi.org/10.1007/s11069-013-0946-x.
Balica, S.F., Wright, N.G., Meulen, F.V.D., 2012. A flood vulnerability index for coastal cities and its use in assessing climate change impacts. Nat. Hazards 64, 73–105. https://doi.org/10.1007/s11069-012-0234-1.
Borden, K.A., Schmidtlein, M.C., Emrich, C.T., Piegorsch, W.W., Cutter, S.L., 2007. Vulnerability of US cities to environmental hazards. J. Homel. Secur. Emerg. Manag. 4, 1–21. https://doi.org/10.2202/1547-7355.1279.

Census, 2011. Available at: http://censusindia.gov.in. Accessed on August 12, 2017.

Chandrasekar, N., Vivek, J., Saravanan, S., 2013. Coastal vulnerability and shoreline changes for southern tip of India-remote sensing and GIS approach. J. Earth Sci. Clim. Change 4, 144. https://doi.org/10.4172/2157-7617.1000144.

Coastal Regulation Zone Notification. 2011. Ministry of Environment and Forests, Gazette of India, Extraordinary, Part-II, Section 3, Sub-section (ii) of dated the 6th January.

Esteban, M., Takagi, H., Shibayama, T., 2015. Handbook of Coastal Disaster Mitigation for Engineers and Planners. Elsevier, Oxford.

Ferrario, F., Beck, M.W., Storlazzi, C.D., Micheli, F., Shepard, C.C., Airoldi, L., 2014. The effectiveness of coral reefs for coastal Hazard risk reduction and adaptation. Nat. Commun. 5, 3794.

Garai, J., 2017. Qualitative analysis of coping strategies of cyclone disaster in coastal area of Bangladesh. Nat. Hazards 85, 425–435. https://doi.org/10.1007/s11069-016-2574-8.

Gormitz, V.M., Daniels, R.C., White, T.W., Birdwell, K.R., 1994. The development of a coastal risk assessment database: vulnerability to sea level rise in the US south east. J. Coast. Res. 12, 327–338.

Guannel, G., Ruggiero, P., Faries, J., Arkema, K., Pinsky, M., Gelfenbaum, G., Guerry, A., Kim, C.K., 2015. Integrated modeling framework to quantify the coastal protection services supplied by vegetation. J. Geophys. Res. Oceans 120 (1), 324–345.

Hoque, M.A.A., Phinn, S., Roelfsema, C., Childs, I., 2017. Tropical cyclone disaster management using remote sensing and spatial analysis: a review. Int. J. Disaster Risk Reduct., 2–30. https://doi.org/10.1016/j.ijdrr.2017.02.008.

IPCC, 2002. Coastal Zone Management Sub-Group (CZMS): A Common Methodology for Assessing Vulnerability to Sea Level Rise. 2nd ed.. In: Global Climate Change and the Rising Challenge of the sea, IPCC-CZMS, Ministry of Transport, Public Works and Water Management, The Hague, The Netherlands.

Jayanthi, N., 1998. Cyclone Hazard, coastal vulnerability and disaster risk assessment along the Indian coasts. Vayu Mandal. 28 (1–4), 115–119.

Kobayashi, N., Raichle, A.W., Asano, T., 1993. Wave attenuation by vegetation. J. Waterw. Port Coast. Ocean Eng. 119 (1), 30–48.

Kumar, A., Narayana, A.C., Jayappa, K.S., 2010a. Geomorphology shoreline changes and morphology of spits along Southern Karnataka, West Coast of India: a remote sensing and statistics-based approach. J. Geomorphol. 120, 133–152. https://doi.org/10.1016/j.geomorph.02.023.

Kumar, K.K.S., 2003. Vulnerability and Adaptation of Agriculture and Coastal Resources in India to Climate Change, EERC Working Paper Series, NIP-4.

Kumar, T.S., Mahendra, R.S., Nayak, S., Radhakrishnan, K., Sahu, K.C., 2010b. Coastal vulnerability assessment for Orissa State, East Coast of India. J. Coast. Res. 26 (3), 523–534. https://doi.org/10.2112/09-1186.1.

Li, Y., Zhang, X., Zhao, X., Ma, S., Cao, H., Cao, J., 2016. Assessing spatial vulnerability from rapid urbanization to inform coastal urban regional planning. Ocean Coast. Manag. 123, 53–65.

Mallick, B., 2014. Cyclone shelters and their locational suitability: an empirical analysis from coastal Bangladesh. Disasters 38 (3), 654–671.

Marchant, M., 2017. Application of Coastal Vulnerability Index (CVI) on the Island of Oahu. Unpublished Thesis, Department of Global Environmental Science, University of Hawaii, Mānoa. http://www.soest.hawaii.eduoceanographyGESThesisMichelleMarchant.pdf. Accessed 21.04.2018.

McCarthy, J., Canziani, F., Leary, A., Dokken, J., White, S., 2001. Climate change: impacts, adaptation, and vulnerability. Intergovernmental Panel on Climate Change (IPCC) Report, 356.

Mujabar, P.S., Chandrasekar, N., 2011. Coastal Erosion Hazard and vulnerability assessment for southern coastal Tamil Nadu of India by using remote sensing and GIS. Nat. Hazards. https://doi.org/10.1007/s11069-011-9962-x.

Murali, R.M., Ankita, M., Amrita, S., Vethamony, P., 2013. Coastal Vulnerability Assessment of Puducherry Coast, India, Using the Analytical Hierarchical Process. 13 Copernicus Publications, pp. 3291–3311.

Murthy, K.S.R., Subrahmanyam, A.S., Murty, G.P.S., Sarma, K.V.L.N.S., Subrahmanyam, V., Rao, K.M., Rani, P.S., Anuradha, A., Adilakshmi, B., SriDevi, T., 2006. Factors guiding tsunami surge at the Nagapattinam-Cuddalore shelf, Tamil Nadu, East Coast of India. Curr. Sci. 90 (11), 1535–1538.

Nicholls, R.J., Hanson, S., Herweijer, C., Patmore, N., Hallegatte, S., Morlot, C., Jan, C.J., Muir Wood, R., 2007. Ranking of the world's cities most exposed to coastal flooding today and in the future: Executive summary. OECD (Environment).

Patwardhan, A., Narayan, K., Parthasarathy, D., Sharma, U., 2003. Impacts of Climate Change on Coastal Zone. In: Sukla, P.R., Sharma, S.K., Ravindranath, N.H., Garg, A., Bhattacharya, S. (Eds.), Climate Change and India: Vulnerability Assessment and Adaptation, Hyderabad. University Press, India, pp. 326–359.

Paul, B.K., 2009. Why relatively fewer people died? The Case of Bangladesh's Cyclone Sidr. Nat. Hazards 50, 289–304. https://doi.org/10.1007/s11069-008-9340-5.

Pramanik, M.K., Biswas, S.S., Mukherjee, T., Roy, A.K., Pal, R., 2015. Sea level rise and coastal vulnerability along the eastern coast of India through geo-spatial technologies. J. Geophys. Remote Sens. 4, 145. https://doi.org/10.4172/2169-0049.1000145.

Rahaman, S.A., Aruchamy, S., 2017. Geoinformatics based landslide vulnerable zonation mapping using analytical hierarchy process (AHP), a study of Kallar River sub watershed, Kallar watershed, Bhavani Basin, Tamil Nadu. Springer International Publishing Switzerland Model, Earth Syst. Environ. 3 (41), 1–13. https://doi.org/10.1007/s40808-017-0298-8.

Ramkumar, M., 2009. Types, Causes and Strategies for Mitigation of Geological Hazards. In: Mu, R. (Ed.), Geological Hazards: Causes, consequences & Methods of Containment. New India Publishing Agency, pp. 1–22.

Rani, N.N.V.S., Satyanarayana, A.N.V., Bhaskaran, P.K., 2015. Coastal vulnerability assessment studies over India: a review. Nat. Hazards 77 (1), 405–428. https://doi.org/10.1007/s11069-015-1597-x.

Rao, K.N., Subraelu, P., Rao, T.V., Malini, B.H., Ratheesh, R., Bhattacharya, S., Rajawat, A.S., 2009. Sea-level rise and coastal vulnerability: an assessment of Andhra Pradesh coast. J. Coast Conserv. 12, 195–207. https://doi.org/10.1007/s11852-009-0042-2.

Roy, D.C., Blaschke, T., 2013. Spatial vulnerability assessment of floods in the coastal regions of Bangladesh. Geomat. Nat. Haz. Risk, 1–24.

Saaty, T.L., 1977. A scaling method for priorities in Hierarchial structures. J. Math. Psychol. 15, 234–281.

Saaty, T.L., 1980. The Analytical Hierarchy Process. McGraw Hill, New York.

Saaty, T.L., 1990. The Analytic Hierarchy Process: Planning, Priority Setting, Resource Allocation, first ed. RWS Publications, Pittsburgh, p. 502.

Saaty, T.L., 1994. Fundamentals of Decision Making and Priority Theory with Analytic Hierarchy Process, first ed. RWS Publications, Pittsburgh, p. 527.

Saaty, T.L., Vargas, L.G., 2001. Models, Methods, Concepts, and Applications of the Analytic Hierarchy Process, 1st ed. Kluwer Academic, Boston, p. 333.

Saxena, S., Geethalakshmi, V., Lakshmanan, A., 2013. Development of habitation vulnerability assessment framework for coastal hazards: Cuddalore coast in Tamil Nadu, India—a case study. Weather Clim. Extremes 2, 48–57.

Seto, K.C., Gueneralp, B., Hutyra, L.R., 2012. Global forecasts of urban expansion to 2030 and direct impacts on biodiversity and carbon pools. Proc. Natl. Acad. Sci. 109 (40), 16083–16088. https://doi.org/10.1073/pnas.1211658109.

Sharma, U., Patwardhan, A., 2007. Methodology for identifying vulnerable hotspots to tropical cyclone hazards in India. Mitig. Adapt. Strateg. Glob. Chang. https://doi.org/10.1007/s11027-007-9123-4.

Singarasubramanian, S.R., Mukesh, M.V., Sujatha, K., 2011. Geomorphic and sedimentological changes in the East Coast of Tamil Nadu by Indian Ocean Tsunami-2004. Int. J. Environ. Sci. Dev. 2(3).

Small, C., Nicholls, R.J., 2003. A global analysis of human settlement in coastal zones. J. Coast. Res. 19(3).

Thompson, C.M., Frazier, T.G., 2014. Deterministic and probabilistic flood modeling for contemporary and future coastal and inland precipitation inundation. Appl. Geogr. 50, 1–14.

Thornbury, W.D., Schumm, S.A., 1969. Principles of geomorphology. J. Geol. 77 (6), 737–738. https://doi.org/10.1086/627477.

Turner, R.K., Doktor, P., Adger, N., 1993. Key Issues in the Economics of Sea level rise. CSERGE Working Paper. 93-04.

UNEP, 2005. Assessing Coastal Vulnerability: Developing a Global Index for Measuring Risk, UNEP/DEWA/RS.05-1, Nairobi, Kenya.

Wong, P.P., Losada, I.J., Gattuso, J.P., Hinkel, J., Khattabi, A., McInnes, K.L., Saito, Y., Sallenger, A., 2014. Coastal Systems and Low-Lying Areas. In: Field, C.B., Barros, V.R., Dokken, D.J., Mach, K.J., Mastrandrea, M.D., Bilir, T.E., Chatterjee, M., Ebi, K.L., Estrada, Y.O., Genova, R.C., Girma, B., Kissel, E.S., Levy, A.N., MacCracken, S., Mastrandrea, P.R., White, L.L. (Eds.), In: Climate Change In: Impacts, Adaptation, and Vulnerability. Part A: Global and Sectoral Aspects. Contribution of Working Group II to the Fifth Assessment Report of the Intergovernmental Panel on Climate Change. Cambridge University Press, Cambridge, pp. 361–409.

Yin, J., Yin, Z., Xu, S., 2013. Composite risk assessment of typhoon-induced disaster for China's coastal area. Nat. Hazards 69, 1423–1434. https://doi.org/10.1007/s11069-013-0755-2.

Younus, A.F.M., 2017. An assessment of vulnerability and adaptation to cyclones through impact assessment guidelines: a bottom-up case study from Bangladesh coast. Nat. Hazards 89, 1437–1459. https://doi.org/10.1007/s11069-017-3027-8.

Further Reading

Chandrasekar, N., Vivek, J., Soundaranayagam, J.P., Divya, C., 2011. Geospatial Analysis of Coastal Geomorphological Vulnerability along Southern Tamil Nadu Coast. Geospatial World Forum, Hyderabad.

Holgate, S.J., Matthews, A., Woodworth, L.P., Rickards, L.J., Tamisiea, M.E., Bradshaw, E., Foden, P.R., Gordon, K.M., Jevrejeva, S., Pugh, J., 2013. New data systems and products at the permanent Service for Mean Sea Level. J. Coast. Res. 29 (3), 493–504. https://doi.org/10.2112/JCOASTRES-D-12-00175.1.

National Disaster Management Plan, 2016. NDMA, GOI, Report.

Index

Note: Page numbers followed by *f* indicate figures and *t* indicate tables.

A

Analytical hierarchal process (AHP), 517, 519–522
  management criteria, 412
  protocol, 416*f*
Andhra Pradesh and Kerala coasts, coastal erosion, 162–163*f*, 164*t*
  geo-referenced images, 161
  global-scale activities, 166–168
Anthropogenic pressures, 131–132
Aquaculture Foundation of India (AFI), 506–507
ASTER GDEM, 524

B

Bathymetric data
  Bay of Saint Brieuc, 253*f*
  Vilaine Estuary, 491
Bay of Saint Brieuc
  coastal
    conservatory, 255
  physical and natural subsystem
    climate, 260–262
    geography and geomorphology, 255–257
    geology, 257–259
    hydrodynamics, 259–260
    sedimentology, 262–263
  protected natural areas, 254
  socio-economic subsystem, 263–275
Beach
  foreshore slope, 212–215
  morphodynamic state model, 211–212
  profile evolution, 192–196
  ridges, 209
Beach sediments, Miri coastal region
  environmental indices calculation, 289–290
  environmental significance
    biological effect of total trace metals, 315
    contamination factor, 313
    costal sediments *vs*. source rock/sediment, 316
    geoaccumulation index, 313–314
    paleontology environmental impact bio-monitoring, 316–318
    pollution load index, 314–315
  grain size, 290–293
  local and regional scale coastal zone management strategies, 318–319
  major oxides, 293
  provenance, 300–302
  rare earth elements, 294
  regional geology, 283–287
  sorting and recycling, 298–300
  statistical analysis
    correlation coefficient, 303–306
    factor analysis, 306–312
  trace elements, 293
  weathering and mobility of elements, 296–297
Beraya Beach
  geology, 70–75
  isopod *Sphaeroma triste* perforations, 92–93
  Sloping Shore Platform, 69
  subaerial gulley in soft sandstone, 88–89*f*
  subaerial gulleys and shallow runnels, 89–90
Berlin rules on water resources, 412–413
Blue Tears, 65–66
Breaker wave type
  beach morphodynamic state, 208, 211–212
  foreshore slope, 212–215
  particle settling velocity, 215
  wave climate, 212
Bungai Beach
  isopod *Sphaeroma triste* perforations, 92–93
  near-horizontal platforms, 91

C

Catastrophism, 20
Cauvery River. *See* River Cauvery
Change detection
  land use/cover, 476
  Muthupet mangrove forest, 477–480
Chromium (Cr), fractionation in sediments, 351–352
Cliff retreat, in South Brittany (France), 132, 133*f*

Index

Cliff retreat, in South Brittany (France) *(Continued)*
   accelerated clifftop retreat due to tree vegetation, 148–150, 150*f*
   anthropogenic factors, 139
   clifftop position determination, 135–136
   clifftop retreat mapping, 141–142
   end point rate method, 135–136
   exposed rocky cliffs, 138
   Geographical Information Systems (GIS), 131–132
   hierarchical clustering analysis (HCA), 141, 146, 149*f*
   high weathered cliffs, 138
   low weathered cliffs, 138
   microcliffs, 138
   multiple component analysis (MCA), 140–148, 148*f*
   in sheltered areas, 131–132
   spatial analysis, 139–140
   spatial database, 136–139, 137*f*
   tree and scrub vegetation cover, 142, 144*f*
   vegetation cover categories, 138
Climate change, sea-level rise, 403–404
Coastal aquifer management
   aquifer hydraulic conductivity, 239
   depth to water table, 239–240
   distance from shore/river, 241–242
   GALDIT method
     seawater intrusion, 234
     vulnerability index, 235
   saltwater intrusion status, 243–244
   vulnerability classes, 246*t*
Coastal City Flood Vulnerability Index, 516
Coastal defense systems, 114, 126–128
Coastal erosion, 113, 155
   Andhra Pradesh and Kerala coasts, 161, 162–163*f*, 164*t*, 166–168
   Asian Development Bank, 114
   Krishna Delta, 161–162, 164*f*, 168–170, 168*t*
   Terengganu coast
     relative sea level rise, 121, 122*f*
     seasonal surface current during monsoons, 121–122, 123*f*
     seasonal wave distribution, 122–125, 124*f*
   Visakhapatnam coast, 162–165, 165*f*, 170–172, 171–172*f*
Coastal habitats, 8–9
Coastal hazard, 516, 535–536, 537*t*
Coastal lagoons, 7
Coastal pollution, 383. *See also* Heavy metal pollution
Coastal processes, spatial and temporal scales, 3, 4*f*
Coastal vegetation systems, 5
Coastal vulnerability index (CVI)
   Chennai coast, 100
   and coastal vulnerability zones, 538*f*
   Cuddalore District of Tamil Nadu, 517
   environmental and climatic vulnerability index, 534
   habitat risk assessment, 523
   physical vulnerability index, 532
   socio-economic vulnerability index, 532–534
   Venn's diagram, 521*f*
Coastal vulnerable zones (CVZ), 536, 538*f*
Coastal water, 331
Coastal zone, 435–437, 440, 447
Coastline variation analysis, 189–190
Cuddalore Coast, shoreline changes, 101*f*
   automatic delineation method, 102
   band ratio approach, 103
   DSAS statistical parameters, 102, 104
   end point rate (EPR), 103–105, 109*f*
   erosion/accretion, 105
   K-means clustering method, 103
   MNDWI technique, 100, 102–103
   net shoreline movement (NSM), 103–105, 108*f*
   shoreline change detection, 104–105, 110*f*
   shoreline delineation images, 104–106*f*, 105
CVI. *See* Coastal vulnerability index (CVI)

D

Dam Arzal, 487, 491–492
Deep-sea hiatus, 42
Digital Shoreline Analysis System (DSAS), 118
Domed–collapse controlled caves, 82–83

E

Ecosystem, human health, 331
Environmental and climatic vulnerability index (ECVI), 534
Environmental management systems, 5–6
Environmental significance, Miri coastal region
   biological effect of total trace metals, 315
   contamination factor, 313
   costal sediments *vs.* source rock/sediment, 316
   geoaccumulation index, 313–314
   paleontology and environmental impact through bio-monitoring, 316–318
   pollution load index, 314–315
Environment impact assessment (EIA) study, 114, 510–511
Erosion
   control (*see* Coastal erosion)
   vs. hydrodynamic forcings, 496*f*
Estuary
   tide-dominated, 486
   Vilaine (*see* Vilaine Estuary)
Exxon Eustatic Curve, 32–34, 34*f*

Index

F
Fairbridge eustatic curve, 24, 24f
Fluvial dynamics, 433–434, 440, 451
Foreshore slope, 212–215
France
    Bay of Saint Brieuc, 252 (*see also* Bay of Saint Brieuc)
    ICZM, 252

G
GALDIT method
    Cuddalore, 247f
    index model, 234
    seawater intrusion, 234
    vulnerability index, 235
Gâvres-Penthièvre beach-dune system
    beach changes over short time scale, 186–189
    beach profile evolution, 192–196
    coastal dynamics, revealed by profiling, 196–198
    coastline variation analysis, 189–190
    geomorphological setting, 184f
    sediment cell boundaries, 190–192
    sediment transport, 192
        control of beach morphology, 199–202
    temporal coastline variations, 196
    waves and currents, 185–186
Geoid, 25–26
Geoidal eustasy, 43
Geospatial technology, 100
Glacial isostatic uplift, 15
Global warming, 3–5
Gopalpur Harbour, India, 114
Grain size, monthly and seasonal changes, 218–219t
Grand Solar Cycles (GSC), 26
Groundwater
    coastal aquifer management, 239
    contamination, 332
    inundation, 403–404
    isotope studies, $\delta^{18}O$ values, 373
    Pondicherry region, 362
    seawater intrusion, 233–234
Gulf of Morbihan cliffs, 132–134, 141f. *See also* Cliff retreat, in South Brittany (France)

H
Habitat risk assessment
    AHP, 517, 519–522
    CVI, 517, 523, 532–534 (*see also* Coastal vulnerability index (CVI))
    cyclones, 535–536
    environmental and climatic indicators
        annual rainfall data, 531–532
        land use/land cover, 530–531
        normalized difference vegetation index, 531
    fuzzy linear transformation, 523
    past disasters at coastal Tamil Nadu, from 1991 to 2017, 518t
    physical indicators
        coastal relief, 524, 525–526t
        coastal slope, 524–526
        geomorphic unit, 527
        shoreline proximity, 527
        storm affected areas, 527
    population densities, 515–516
    risk factor, hazard vulnerability, 537t
    socio-economic indicators
        built-up near shoreline, 529
        cyclone tracks, 530f
        household density, 528
        literacy rate, 528–529
        population density, 527
        road proximity, 529–530
    urbanization and economic development, 515–516
    vulnerability due to cyclones, 517
Heavy metal pollution
    chromium, 348–349
    cluster analysis, 393
    concentration in sediment samples, 387–389
    metal concentration in water samples, 386–387
    multivariate statistical analysis, 386
    occurrences of untreated effluents, 392
Hierarchical clustering analysis (HCA), 141, 146, 149f
Hydrogeochemistry
    Chadda's hydrogeochemical process evaluation, 371
    isotope studies, 371–374
    statistical analysis
        correlation, 375
        factor analysis, 375–376
        postmonsoon season, 377
        premonsoon season, 376–377
Hydrogeology, 65

I
Indian subcontinent
    ILR program
        Cenozoic-recent river basins, 443–444
        climate, 444–445
        coastal hydrodynamics, 446–448
        geology, 435–440
        geomorphology, 440–443
        tectono-geomorphic evolution, 443–444
    NE monsoon, 208
    SW monsoon, 208
Infilling, spatio-temporal evolution, 492, 495f
Integrated coastal zone management (ICZM), 114, 251–252
Integrated shoreline management plan (ISMP), 114
Intergovernmental Panel on Climate Change (IPCC), 406
Interlinking of rivers (ILR)
    deltaic and coastal regions, 456–460
    hazard mitigation, 460–463
    political impediments on nationalization and ILR, 450
    potential threat, 452–456
    river basins and geological impediments, 450–451

Index

Interlinking of rivers (ILR) *(Continued)*
  self-regulating river dynamics, 452
  spatio-temporal variations of precipitation, 449–450
  sustainable development, 463–465
International Law Association (ILA), 412–413
Ion exchange, 371, 377

K

Krishna Delta, coastal erosion, 168*t*
  erosion-accretion trend, 164*f*
  regional activities, 168–170
  river-borne sediments, 161

L

Land use planning, 6
Last Glaciation Maximum (LGM), 28
Legal subsystem, Bay of Saint Brieuc
  coastal conservatory, 255
  protected natural areas, 254

M

Malaysia
  Bernam River, 281–282
  Miri coast of Sarawak, 282 *(see also* Miri coastal region, beach sediments)
Mangroves, 7
  change detection, 477–480
  spectral indices
    normalized difference vegetation index, 476
    normalized difference water index, 476
Marine inundation, 404*f*
Metal fractionation, 348
Mike 21-SW, 118
Miri coastal region, beach sediments, 289–290
Mörner eustatic curve, 24–25
Morphobathymetrical evolution and sedimentary balance from 1960 to 2013, 492

N

National Coastal Erosion, 114
National Remote Sensing Centre (NRSC) guidelines, 527
Nearshore region, 212
Normalized Difference Water Index (NDWI), Muthupet Mangrove forest, 478*f*
Northwest Borneo Coastline (Sarawak, Malaysia)
  coastal erosion, 66
    arches, 85–86
    caves, 80–85
    geology, 70–75
    mass wasting erosion features, 75–80
    mesoscale erosion features, 87–92
    microscale erosion features, 92–93
    rock falls, 80
  episodic blue bioluminescence phenomenon, 65–66
  Koppen-Geiger climate classification category, 68

O

Ocean level variables, 18*t*
Oceanographic settings, Vilaine Estuary, 487
Ocean water, 331
Oxygen isotope record, equatorial Pacific core, 28*f*

P

Palaeomagnetic chronology, equatorial Pacific core, 28*f*
Peliau cliffs area, 84*f*
  caves, 80–85, 81*t*
  geology, 70–75
  infiltration role, 93–94
  interbedded lithology, 93
  joint controlled caves, 94
  joint enlargements, 87
  near-horizontal platforms, 91
  rock falls, 80
  rock formations, lithologics, and structural data, 71–72*t*
  tafoni, 88–89*f*, 91

Physical vulnerability index (PVI), 532
Pollution
  East Coast of Southern India
    antibiotic resistant strains, 335, 338–340
    epidemiological survey, 333–335
  heavy metal (*see* Heavy metal pollution)
Postglacial rise, in sea level, 15
Pre-quaternary sea level changes
  in Cretaceous era
    eustatic amplitude ranges, 46
    geophysical explanations, 46–47
    sequence boundaries, 45
    updated eustatic curve, 44, 44*f*
  Late Phanerozoic sea level curve, 36
  Mid-Cretaceous, 36
    gravitational drop motions, 38, 41*f*
    land/sea distribution and geoid relief, 37–38, 37*f*
    sea level graphs, 38, 39–40*f*
    transgressions and regressions, 41*f*
  Paris 1980 illusory consensus, 35
Prismatic—joint controlled caves, 83
Provenance, sediment, 300–302

Q

Quaternary geological formations, 115
Quaternary sea level changes
  eustasy, 24–26
  eustatic variables, 31, 32*f*
  glacial/interglacial cycles, 27–28
  global glacial isostasy problem
    global *vs.* regional loading model, 29, 30*f*
    linear viscosity model, 29
  satellite altimetry, 31
  postglacial period, 24–26
Quiberon Peninsula cliffs, 134, 135*f*, 143*f*. *See also* Cliff retreat, in South Brittany (France)

Index

R

Recycling
  beach sediments, 301f
  sorting, 298–300
Regional Ocean Modelling System (ROMS), 118–119
Resistant sandstone cliffs, 70
Risk-assessment code (RAC), 351–352
River Cauvery
  chromium fractionation, 352–355, 353f
    Cr-F1 fraction, 356–358
    F1 and F2 factions, 356
    mobility factor assessment, 355t
    non-residual fractions, 355
  heavy metal contaminants, 348–349
Rocky coasts, 65
Rotational eustasy, 26, 52–54

S

Safe yield, 411
Saltwater intrusion, 404f
Sanitation
  coastal community, 344
  food sources, 331–332
Satellite altimetry, 31, 52
Sea level changes, 15
  Earth's rotation, 16
  late 20th century achievements, 22
  observations in 18th century, 16–19
  observations in 19th century
    elevation theory, 20
    geoidal eustasy, 21
    glacial isostasy, 21–22
    ice ages and glacial eustasy, 21
    stratigraphy, 20
    tectono-eustasy, 20–21
  pre-quaternary, 32–48
  quaternary, 23–32
  rise
    dune and swale topography, 405
    groundwater extraction, 403
    low impact development (LID) strategies, 405
  tertiary, 48–51

Seasonal variation
  chemical concentration, 367t
  electrical conductivity, spatial distribution, 366–370, 369f
  Piper classification, 368–370, 370f
  postmonsoon season, 368–370
  premonsoon season, 368
Seawater intrusion (SWI), 233–234
Seaweed farming
  Chlorophyta, 501–502
  culture methods
    bottom-culture method, 505
    floating bamboo raft method, 505
    long-line rope method, 504
    net bag method, 505
  Cyanophyta, 501–502
  employment opportunities and self-help groups (SHGs), 507–510
  environment impact assessment (EIA) study, 510–511
  government and non-government sectors, 506–507
  Kappaphycus alvarezii, 501–502
  Palk Strait, 502
  Phaeophyta, 501–502
  Rhodophyta, 501–502
Sediment mean size measurement, 118
Sediment transport, 192
  and control of beach morphology, 199–202
  evolution of coastal landscapes, 181–182
Sequential extraction
  Cr fractionation, 351–352
  heavy metals, 348
  metal fractionation data, 352
Shepard eustatic curve, 24, 24f
Shoreline changes, Cuddalore Coast, 101f
  automatic delineation method, 102
  band ratio approach, 103
  DSAS statistical parameters, 102, 104
  end point rate (EPR), 103–105, 109f
  erosion/accretion, 105
  K-means clustering method, 103
  MNDWI technique, 102
  net shoreline movement (NSM), 103–105, 108f
Shoreline management plans, 2–3
South China Sea, 115–116, 125
South Brittany (France), cliff retreat in. *See* Cliff retreat, in South Brittany (France)
Storm, 495
  and sea level rise, 486
  swells, 487
Subaerial weathering processes, 131–132

T

Tabular-lithologically controlled caves, 82–83
Taman Selera beach
  geology, 70–75
  intertidal runnels, 88–89f, 92
Terengganu coast, Malaysia
  coastal defense structures, 126–128
  coastal erosion
    causes, 120–125
    control, 126–128
  coastal processes, 125–126
  grain size distributions, 119, 121t
  magnitude of coastal evolution, 119
  methodology, 118–119
  monsoon seasons, 117f, 121–125
  physical setting method, 119
  relative sea level rise, 121
  seasonal wave distribution, 122–125
  sediment mean size measurement, 118
  shoreline evolution, 118
  tidal elevation, 122f
Terengganu International Airport, 115–116, 116f
Tertiary sea level changes
  Messinian Salinity Crisis, 50
  Pliocene era sea level highstand, 50–51
  South American east coast, 48–49

Tide
- cyclic natural forcings, 486
- natural infilling, 486
- semidiurnal, 487
- spring, 495
- weak, 492

Trace metal
- analysis, 335
- distribution, 341f
- Tamilnadu coastal groundwater, 343f, 344t

Transitional coastal zones, economic value, 1

Tusan cliffs area, 94
- arches, 85–86, 86t
- bedding plane enlargements, 87
- caves, 80–85, 81t
- coastal cliff retreat components, 80t
- erosion pockets, 87–88, 90f
- geology, 70–75
- sandstone porosity, 70–74
- subaerial gulleys and shallow runnels, 89–90
- wave cut notches, 91

U

Uniformitarianism, 20

V

Vilaine Estuary
- bathymetric data, 491
- climatic settings, 487
- destabilization, 497
- geomorphic settings, 487
- meteorological and hydrologic data, 491
- morphobathymetrical evolution, 492
- sedimentary dynamic features, 489–490

Visakhapatnam coastal erosion, 171–172f
- sentinel MSI image, 165f

Vulnerability index, 233–235

W

Water
- bacteriological examination, 332
- coastal aquifer, 340–341

CPI Antony Rowe
Chippenham, UK
2018-11-29 15:47